Introducing
MECHANICS

BRIAN JEFFERSON
TONY BEADSWORTH

OXFORD
UNIVERSITY PRESS

Great Clarendon Street, Oxford OX2 6DP

Oxford University Press is a department of the University of Oxford.
It furthers the University's objective of excellence in research, scholarship,
and education by publishing worldwide in

Oxford   New York

Athens   Auckland   Bangkok   Bogotá   Buenos Aires   Calcutta
Cape Town   Chennai   Dar es Salaam   Delhi   Florence
Hong Kong   Istanbul   Karachi   Kuala Lumpur   Madrid
Melbourne   Mexico City   Mumbai   Nairobi   Paris   São Paulo
Singapore   Taipei   Tokyo   Toronto   Warsaw

and associated companies in   Berlin   Ibadan

© Brian Jefferson and Tony Beadsworth 2000

The moral rights of the author have been asserted

Database right Oxford University Press (maker)

First published 2000

British Library Cataloguing in Publication Data

Data available

ISBN 0 19 914710 8

Typeset and illustrated by Tech-Set Ltd, Gateshead, Tyne and Wear
Printed and bound in Great Britain by
Butler & Tanner Ltd, Frome and London

Contents

Preface   v

1 Modelling   1
Modelling reality   1
The modelling process   3
Another example of modelling   4
Conventional terms   6

2 Vectors   8
Notation   8
Properties of vectors   9
Components of a vector   15
Vectors in three dimensions   21
Scalar product   25

3 Kinematics   31
Terminology   31
Average speed   31
Average acceleration   33
Distance and displacement, speed and
    velocity   34
Uniform speed, velocity and acceleration   34
Displacement–time graphs   35
Velocity–time graphs   36
Motion with uniform acceleration   41
Free fall under gravity   45

4 Force   50
The force of gravity   50
Types of force   51
Drawing diagrams   54
Forces at a point: modelling by vectors   57

Examination Questions:
Chapters 2 to 4   66

5 Newton's Laws of Motion   75
Newton's first law   75
Newton's second law   76
Newton's third law   81
Connected particles   85
Systems with related accelerations   92

6 Calculus in Kinematics   97
Velocity at an instant   97
Acceleration as a derivative   104
Variable forces   106

7 Kinematics in Two and
Three Dimensions   111
Differentiating and integrating a vector
    with respect to a scalar   111
Motion in two and three dimensions   116
Motion with uniform acceleration   121
Projectiles   123
Exploring the model further   127
Revising the model   132

Examination Questions:
Chapters 5 to 7   135

8 Relative Motion   144
Notation   146
Solving relative velocity problems   147
Interception, collision and closest
    approach   156

9 Friction   164
Experiments   164
Static and dynamic friction   169
Situations involving acceleration   175
Angle of friction   179

10 Turning Effects of Forces   183
Experiment   183
Moment of a force   184
Parallel forces: resultants and couples   190
Non-parallel forces: resultants and
    couples   195
Equivalent force systems   205
Appendix   209

Examination Questions:
Chapters 8 to 10   212

11 Centre of Mass   221
Experiment   221
Centre of gravity, centre of mass
    and centroid   222
Calculus methods: laminae and solids
    of revolution   236
Calculus methods: arcs and shells   240
Sliding and toppling   245
Appendix   249

12	**Work, Energy and Power**	**250**
	Work	250
	Energy	257
	Conservation of mechanical energy	260
	Power	266

13	**Momentum and Impulse**	**275**
	Change of momentum	275
	Conservation of linear momentum	281
	Elastic impact	289
	Loss of kinetic energy	293
	Oblique impact	295

14	**Jointed Rods and Frameworks**	**301**
	Jointed rods	301
	Frameworks	307
	Method of sections	309
	Graphical methods	312

| | **Examination Questions:** | |
| | **Chapters 11 to 14** | **318** |

15	**Dimensional Analysis**	**330**
	Introduction	330
	Units and dimensions	331
	Dimensions in calculus	332
	Dimensionless quantities	333
	Dimensions of constants	334
	Dimensional consistency	334
	Finding a formula	335

16	**Circular Motion**	**340**
	Linear and angular speed	340
	Angular displacement, speed and acceleration	340

	Relation between linear and angular measures	341
	Modelling circular motion	342
	Problems involving non-horizontal forces	351
	Circular motion with non-uniform speed	359

17	**Elasticity**	**370**
	Elastic strings and springs	370
	Other ways of expressing the linear model	377
	Tension, work and energy	381

18	**Oscillatory Motion**	**392**
	Modelling vertical oscillations	392
	Simple harmonic motion	393
	Associated circular motion	403
	Simple pendulum	405
	Incomplete oscillations	409
	Damped oscillations	411

19	**Differential equations**	**420**
	Definitions and classification	420
	Forming differential equations	421
	First-order equations	423
	Second-order linear equations with constant coefficients	433

| | **Examination Questions:** | |
| | **Chapters 15 to 19** | **441** |

| | **Answers** | **456** |

| | **Index** | **472** |

Preface

Our intention with this book has been to produce a text covering the mechanics content of all the single-subject pure mathematics and mechanics specifications for A-level which will come into force in September 2000. We have not followed the syllabus of any one examination board, but have sought to develop the subject in such a way as to be accessible to all students.

We have tried to combine the best of the current approach, emphasising modelling and the 'real world' relevance of the subject, with some of the virtues of the more traditional texts. We have endeavoured to make modelling considerations the basis of the discussion of most topics,and, where appropriate, we have developed topics from the starting point of a practical problem or experiment. We have, however, not allowed this to compromise the need for a degree of mathematical rigour, and have included sufficient questions leading to solutions in algebraic form to satisfy those with a taste for such problems.

A special feature of the text is the reference to a number of spreadsheets, used to analyse the data from suggested experiments or to explore the implications of certain models. These can be downloaded from the Oxford University Press website (http://www.oup.co.uk/mechanics). While we do not claim any great sophistication for these, it is hoped they will be a useful resource in helping students gain a 'feel' for the subject.

The order in which topics have been covered is approximately that in which we choose to proceed in our own teaching. Naturally, this will not accord with everyone's approach, and the text contains a degree of cross-referencing to assist those wishing to 'dip in'.

The opening chapter of the book introduces the ideas of modelling and the modelling cycle, and emphasises the need to specify the assumptions made when developing a model and the importance of testing the predictions of that model against experimental data. In the next chapter, we develop the vector tools which underpin much of the subject. Chapter 3 explores the basic ideas of kinematics. This is followed by two chapters covering the concept of force and the all important Newton's laws. We then return to kinematics for a further three chapters, dealing with motion in two and three dimensions, the use of calculus, the concept of relative velocity.

Chapter 9 explores the problem of modelling friction, starting from a simple experimental approach. We then examine the concept of the moment of a force in Chapter 10, and consider the conditions necessary for equilibrium. Moments are then applied in the next chapter to finding centre of mass.

Chapters 12 and 13 are devoted to work, energy and power and to momentum respectively. The final six chapters deal with the 'harder' topics of frameworks, circular motion, elasticity and simple harmonic motion, together with a discussion of dimensional analysis and an introductory treatment of differential equations.

We anticipate that most students will use this book with the guidance of a teacher, but every effort has been made to make it readable and accessible to those using it for self-study or for revision. The exposition of topics proceeds by small steps and with a large number of worked examples to reinforce the ideas. The exercises are designed to give practice in the rote application of techniques, but also contain questions of a more testing nature. In addition, there are five sections containing a substantial selection of recent examination questions.

We are grateful to AEB, EDEXCEL, MEI, NEAB, NICCEA, OCR and WJEC for permission to use their questions. The answers provided for these questions are the sole responsibility of the authors.

We would like to express our thanks the Nigel Watts of King's School, Bruton, for the idea of 'modelling a skipper' used in the first chapter.

Thanks are also due to Rob Fielding and James Nicholson for checking the answers to the exercises and the examination questions. Finally, we owe an enormous debt of gratitude to John Day for his painstaking and detailed work in editing the book, and for his help and suggestions, which have contributed in no small measure to the final product.

Brian Jefferson
Tony Beadsworth
April 2000

1 Modelling

I cannot bring a world quite round, although I patch it as I can.
WALLACE STEVENS

A group of people on holiday with Explorer Tours proposes to drive directly across a stretch of desert from their present position A to a camp site at B. They consult their map of the region (scale 1 cm : 1 km), which clearly marks A and B, to decide how far they will need to drive.

They measure the straight line AB on their map with a ruler and find it to be 18.6 cm. They conclude that they will need to drive 18.6 km.

When they reach B, they check the distance they have travelled and find that it is 19.2 km.

Modelling reality

These people followed a process which is fundamental to the application of mathematics to real problems. They started with the real problem ...

'How far will we drive in going from A to B?'

set up a mathematical model ...

'The line AB on the map is a scale drawing of the journey.'

and from this model they obtained a solution to the problem. They then checked their solution against reality.

Simplifying assumptions

In setting up the model, the group made three simplifying assumptions.

- Using the map distance takes no account of hills and valleys. The model assumes that the journey is flat. That is, that any extra distance caused by hills is insignificant in relation to the length of the journey. The model would therefore tend to underestimate the actual distance driven.

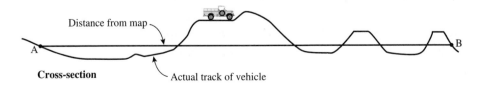

1

- The model assumes that the journey is in an exact straight line. In practice, it is likely that there are rocks and other obstacles which they need to circumvent. So, again, the model is likely to produce an underestimate.

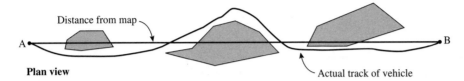

- The model assumes that the journey is so short that they can safely ignore any distortions caused by the fact that the line AB on the map is a flat projection of a journey taking place on the curved surface of the Earth. All map projections distort shapes and distances, the nature of the distortion depending on the particular method of projection used.

Comparison with reality: errors

Having solved their model, the group then did the journey, enabling them to compare their solution with the actual distance travelled. In making this comparison, they would need to be aware of sources of error, both in their prediction and in their measurement of reality.

- Their measurement of AB on the map is at best correct to one decimal place. This would place their predicted distance, D_P km, in the interval $18.55 \leqslant D_P < 18.65$. In addition, identifying their starting and finishing points on the map could only be an approximate affair, perhaps extending the error bounds to $18.5 \leqslant D_P < 18.7$.

- They found the actual distance using the odometer on their vehicle. This displayed 24 924.6 km at the start and 24 943.8 km at the end of their journey. These values are truncated to the nearest one decimal place below, which would put the start reading, S km, and the finish reading, F km, in the intervals $24\,924.6 \leqslant S < 24\,924.7$ and $24\,943.8 \leqslant F < 24\,943.9$ respectively.

- The minimum value of $(F - S)$ is therefore $24\,943.8 - 24\,924.7 = 19.1$ km and its maximum value is $24\,943.9 - 24\,924.6 = 19.3$ km. The actual distance, D_A km, would therefore have error bounds $19.1 \leqslant D_A < 19.3$. Even this assumes that the inevitable inaccuracy in the odometer mechanism was small enough to be insignificant over a short journey.

Was the model good enough?

Once the errors had been quantified as far as possible, the group would be able to decide whether their model was a sufficiently accurate representation of reality for their purpose. If not, they would need to re-examine the assumptions they made and modify the model. They might, for example, be able to obtain a larger-scale map and measure a route including detours around likely obstacles.

The modelling process

All applications of mathematics to real-world problems follow the same process as the one described above, which comprises the following eight steps:

1 **Specify the real problem** This should be a clear statement of the situation and should specify the results required in the solution.
2 **Make simplifying assumptions** All the factors which might affect the result should be considered and a decision made as to which should be taken into account in the model and which should be ignored. We may also make assumptions about the way in which certain variables are related. For example, we might decide to assume that air resistance is proportional to velocity.
3 **Set up the mathematical model** In the example given, this was a scale drawing, but it would more usually be a set of equations describing the behaviour of the simplified system.
4 **Solve the mathematical model** The equations should be solved to obtain the outcome which would result from the simplified system.
5 **Decide what really happens** This may involve setting up an experiment or obtaining data from published sources.
6 **Quantify the likely errors** There may be errors in the values used in the model and/or in the results obtained from the experiment. Error bounds should be established for all such values and the effects on the outcome should be quantified.
7 **Compare with reality** The results from the model should be compared with those obtained in reality to decide if the model provides a sufficiently accurate representation of the real situation. The errors mentioned in **6** need to be taken into account in this comparison.

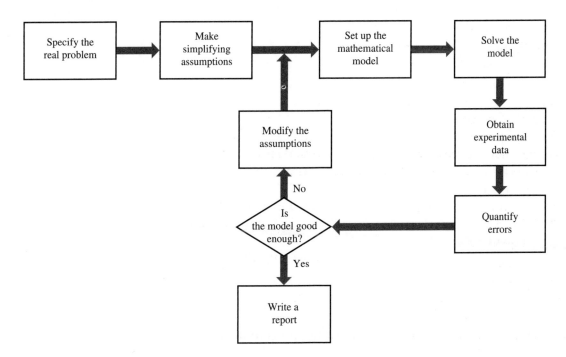

8 Modify the model If the model does not give an adequate representation of the real situation, it is necessary to re-examine the assumptions on which it was based. A new model should then be set up to allow for the effect of one or more of the factors which had previously been ignored. The whole process should then be repeated, perhaps several times, until a sufficiently accurate model is obtained.

This process is summarised in the flowchart on page 3.

In this book, we concentrate on problems involving forces and the motion of objects, but the process of mathematical modelling is common to all situations in which mathematics is applied to real-world problems.

Another example of modelling

The problem is to model the motion of a person skipping.

We first need to state the precise questions which we wish to answer, for example:

- What is the relationship between the speed of the rope and the height of the jump?
- Are there limitations on these quantities for a given person and rope?

Our next task is to list all the factors which we think might have a bearing on the problem. This list can be as long and the factors as fanciful as you wish. It is better to include something a bit daft than to fail to take account of an important factor. Here is a possible list – you can probably think of several more items.

> Length of rope
> Mass of rope
> Flexibility of rope
> Thickness of rope
> Whether the rope drags on the ground
> Gravity
> Height of person
> Mass of person
> Size of feet, length of arms and other physical proportions
> Movement of arms and therefore the locus of rope
> Speed of rope
> Height of jump. Do we measure this as the movement of the person's centre of gravity or as the gap between the feet and the ground (bending of legs)?
> Amount of time feet need to stay in contact with the ground in the jumping process
> Air resistance
> How 'bouncy' the ground is

Once we have our list, we must decide what assumptions to make.

For a first, simple model we might decide that a rope, which is curved and has mass all the way along, is too complex. It would be easier mathematically to replace it with a thin, rigid rod attached to two strings of negligible mass. In addition, it would be simpler if we supposed that the rope is being made to rotate at a constant speed in a circle around a fixed point in space, with the ground being a tangent to the circle.

The simplest way to model the person would be as a cuboid of uniformly dense material rising and falling without any change of shape. This would spend a fixed proportion of each cycle in contact with the ground and the rest moving vertically under gravity.

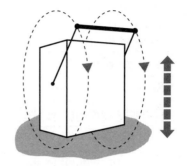

In this model, any resistance to the motion of the rope or the person would be ignored.

The important variables are the length, r, of the strings; the speed, v, of the rod around the circle; the height, h, of the jump; the time, t, from the start of the motion; and the proportion, p, of time spent in contact with the ground. On the assumption that the rope is at the bottom of the circle when the person is at the top of the jump, we could write equations connecting r, v, h, p and t. These equations would form the model and by manipulating them we could find solutions predicting the position of the rope and the person for any value of t.

There would be a lower limit on the rate of skipping because the value of v would have to be great enough to prevent the rope going slack at the top of the circle.

There would also be an upper limit because time would be needed for the person to get sufficiently high off the ground to allow the passage of the rope.

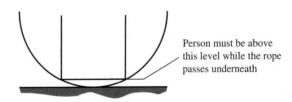

Person must be above this level while the rope passes underneath

Our task would now be to observe people skipping, first to decide on a reasonable value for p and then to test out the predictions of our model about the relation between the height of jump and the speed of the rope.

It is unlikely that the model would be very good, so we would need to reassess our assumptions. Observing skippers would help us to decide which assumptions to modify. We would continue to refine our model and test against observation until we regarded the predictions as sufficiently accurate.

Conventional terms

When stating problems in mathematics, we often use terms which imply that certain assumptions are being made. For example, you will see questions referring to a string as *light*. This would indicate that the mass of the string is sufficiently small for it to be ignored.

The common terms are given in the table below.

Term	Applies to	What is disregarded
Inextensible	Strings, rods	Stretching
Light	Strings, springs, rods	Mass
Particle	Object of negligible size	Rotational motion
Rigid	Rods	Bending
Small	Object of negligible size	Rotational motion
Smooth	Surfaces, pulleys	Friction

Exercise 1

1 Do you think it would be reasonable to disregard air resistance in the following situations?
   a) A marble dropped from an upstairs window.
   b) A table tennis ball dropped from an upstairs window.
   c) A marble dropped from an aircraft at 2000 metres altitude.
   d) A shot being putt.
   e) A rocket firework being set off.
   f) A child on a swing.
   g) A person walking.
   h) A person cycling.

2 Do you think it would be reasonable to disregard friction in the following situations?
   a) Skiing downhill.
   b) A child going down a slide.
   c) Raising an object on a rope passing over a tree branch.
   d) Raising an object on a rope passing over a pulley.
   e) A car being driven in a straight line.
   f) A car being driven round a curve.

3 In the sport of bungee jumping, participants jump from a platform with an elastic rope attached to their ankles. The other end of the rope is attached to the platform. Participants free fall until the elastic stretches. The tension built up in the elastic slows them down and eventually brings them to a temporary stop. Often the jump takes place over water and the participants have the choice of whether to come to a stop before they hit the water, whether to get their hair wet or whether to plunge into the water to a depth chosen by them. The problem is to work out the correct length of rope to satisfy their desires.

In modelling this problem the following list of factors was drawn up. Separate them into three lists:

A Those which can be totally ignored in forming a mathematical model.

B Those which cannot be ignored but which you think would be too difficult to include in an initial model.

C Those which should probably be included in an initial model.

In each case, while at this stage you do not have enough knowledge to answer this question with 100% confidence, try to justify your inclusion of each item in its list.

a) The weight of the person.
b) The height of the person.
c) The height of the platform.
d) The elasticity of the bungee rope.
e) The number of ropes used.
f) The accuracy with which the measurements can be made.
g) The weight of the bungee rope.
h) The weather conditions.
i) The depth of the water.
j) The style of jumping.
k) Air resistance.
l) The clothing worn.
m) The maximum stress the body can take.
n) The way the bungee rope deforms when it is stretched.
o) How fast the water is flowing.
p) How fast the person wants to be moving when he/she hits the water.
q) Whether there is a cross wind.
r) How the rope is tied to the ankles.
s) Any more you can think of.

4 For each of the following situations, make a list of the factors which you think might have a bearing on the outcome.

a) The amount of water falling on a person crossing an open space in the rain.
b) The motion of a boat crossing a river.
c) A tennis player serving.
d) A toy car free-wheeling from rest down a slope.
e) A child swinging on a rope tied to a tree branch.

2 Vectors

When modelling physical systems, we use a number of quantities, such as force, displacement, velocity, acceleration and momentum, which share a common property: namely, all of these quantities can be specified completely only by stating **both** their **magnitude** (size) and their **direction**. Such quantities are called **vectors**.

A **vector quantity** is one which has both **magnitude** and **direction**.

We also use other quantities, such as distance, speed, work and power, which are completely specified by their magnitude. Such quantities are called **scalars**.

A **scalar quantity** is one which has **only magnitude**.

Because of this shared vector property, the mathematical techniques used for combining and manipulating displacements work equally well when we wish to combine and manipulate forces or velocities. We therefore need to spend some time becoming familiar with the language and mathematics of vectors.

Notation

The simplest vector quantity to illustrate is a displacement, or translation, for a given distance in a given direction. This can be represented by a directed line segment.

The line segment shown in the diagram represents a translation from A to B. To show that it is a translation rather than just the distance AB, an arrow is put over the pair of letters to give $\overrightarrow{AB}$. This convention is the more widely used, particularly by the examination boards. (The other way to represent a directed line segment is to print its pair of letters in a bold face to give, in our example, **AB**.)

An alternative way of labelling vectors is to use a single letter in bold type, such as **a**. This would be handwritten as <u>a</u>.

Magnitude

The magnitude of the vector $\overrightarrow{AB}$ is shown as AB or $|\overrightarrow{AB}|$.

The magnitude of the vector **a** is shown as $|\mathbf{a}|$ or a.

Unit vector

A vector with a magnitude of 1 unit is called a **unit vector**. The unit vector in the direction of a vector **a** is usually labelled **â** (often referred to as 'a hat').

Properties of vectors

Equality of vectors

Vectors are equal if they have the same magnitude and direction. For example, in the parallelogram shown on the right, $\overrightarrow{AB} = \overrightarrow{DC}$ and $\overrightarrow{AD} = \overrightarrow{BC}$.

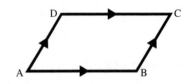

Addition of vectors

In the triangle on the right, we can see that if we combine the translations $\overrightarrow{AB}$ and $\overrightarrow{BC}$, the effect would be the same as the single translation $\overrightarrow{AC}$.

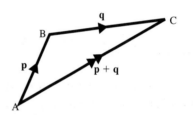

We say that $\overrightarrow{AC}$ is the **vector sum** of $\overrightarrow{AB}$ and $\overrightarrow{BC}$, and write

$$\overrightarrow{AB} + \overrightarrow{BC} = \overrightarrow{AC}$$

$\overrightarrow{AC}$ is called the **resultant** of $\overrightarrow{AB}$ and $\overrightarrow{BC}$.

(Note the use of a double arrowhead in the diagram to signify that the vector is a resultant.)

You should be clear that this does **not** mean that the lengths AB + BC = AC. Think of the + symbol as meaning 'followed by', so $\overrightarrow{AB} + \overrightarrow{BC} = \overrightarrow{AC}$ means

translation $\overrightarrow{AB}$ followed by translation $\overrightarrow{BC}$ is equivalent
to translation $\overrightarrow{AC}$

The justification for using the + symbol will be clear when we consider vectors in component form.

Zero vector

If we were to combine the displacements $\overrightarrow{AB}$ and $\overrightarrow{BA}$, the resultant would be a vector with zero magnitude (its direction would be undefined). We call this the **zero vector** and write it as **0** (handwritten $\underline{0}$).

Negative vectors

As $\overrightarrow{AB} + \overrightarrow{BA} = \mathbf{0}$, it is reasonable to write $\overrightarrow{BA} = -\overrightarrow{AB}$.

The translations $\overrightarrow{AB}$ and $\overrightarrow{BA}$ are the exact opposites of each other.

In general, the vector $-\mathbf{a}$ has the same magnitude as $\mathbf{a}$ but the opposite direction.

Multiplying by a scalar

If the translation $\mathbf{a}$ is applied twice, the effect is a translation twice as far in the same direction:

$$\mathbf{a} + \mathbf{a} = 2\mathbf{a}$$

$$|2\mathbf{a}| = 2\,|\mathbf{a}|$$

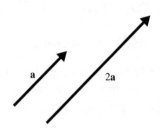

In general, $k\mathbf{a}$ is a vector parallel to $\mathbf{a}$ and with magnitude $k\,|\mathbf{a}|$:

$$|k\mathbf{a}| = k|\mathbf{a}|$$

Commutativity

Translation $\mathbf{p}$ followed by translation $\mathbf{q}$ has the same resultant as $\mathbf{q}$ followed by $\mathbf{p}$. In the diagram

$$\mathbf{p} + \mathbf{q} = \overrightarrow{AB} + \overrightarrow{BC} = \overrightarrow{AC}$$

$$\mathbf{q} + \mathbf{p} = \overrightarrow{AD} + \overrightarrow{DC} = \overrightarrow{AC}$$

$$\Rightarrow \quad \mathbf{p} + \mathbf{q} = \mathbf{q} + \mathbf{p}$$

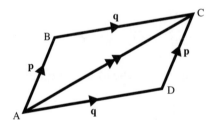

That is, vector addition is **commutative**.

Associativity

If we add several vectors, the order in which we bracket them does **not affect** the resultant. In the diagram:

$$(\mathbf{p} + \mathbf{q}) + \mathbf{r} = \overrightarrow{AC} + \overrightarrow{CD} = \overrightarrow{AD}$$

$$\mathbf{p} + (\mathbf{q} + \mathbf{r}) = \overrightarrow{AB} + \overrightarrow{BD} = \overrightarrow{AD}$$

$$\Rightarrow \quad (\mathbf{p} + \mathbf{q}) + \mathbf{r} = \mathbf{p} + (\mathbf{q} + \mathbf{r})$$

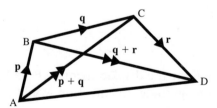

That is, vector addition is **associative**.

Subtraction

Subtracting the vector $\overrightarrow{BC}$ is equivalent to adding $-\overrightarrow{BC}$, or $\overrightarrow{CB}$. In the diagram, $\mathbf{p} = \overrightarrow{AB}$ and $\mathbf{q} = \overrightarrow{AD}$, so:

$$\mathbf{p} - \mathbf{q} = \overrightarrow{AB} - \overrightarrow{AD}$$

$$= \overrightarrow{AB} + \overrightarrow{DA}$$

$$= \overrightarrow{DA} + \overrightarrow{AB}$$

$$= \overrightarrow{DB}$$

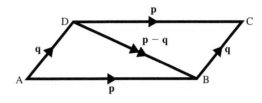

Example 1 In the diagram, ABEF and BCDE are squares. Vector $\overrightarrow{AB} = \mathbf{p}$ and vector $\overrightarrow{AE} = \mathbf{q}$. Find **a)** $\overrightarrow{AC}$ **b)** $\overrightarrow{AD}$ **c)** $\overrightarrow{AF}$ **d)** $\overrightarrow{EC}$

SOLUTION

a) $\overrightarrow{AC} = 2\overrightarrow{AB} \quad \Rightarrow \quad \overrightarrow{AC} = 2\mathbf{p}$

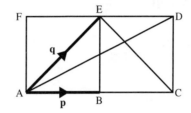

b) $\overrightarrow{AD} = \overrightarrow{AE} + \overrightarrow{ED} = \mathbf{q} + \overrightarrow{ED}$

But $\overrightarrow{ED} = \overrightarrow{AB} = \mathbf{p}$

$\Rightarrow \quad \overrightarrow{AD} = \mathbf{q} + \mathbf{p}$

Note Any route from A to D gives the required result. For example, we could have said

$$\overrightarrow{AD} = \overrightarrow{AB} + \overrightarrow{BD}$$

As $\overrightarrow{BD} = \overrightarrow{AE} = \mathbf{q}$, this gives

$$\overrightarrow{AD} = \mathbf{p} + \mathbf{q}$$

c) $\overrightarrow{AF} = \overrightarrow{AE} + \overrightarrow{EF} = \mathbf{q} + \overrightarrow{EF}$

But $\overrightarrow{EF} = \overrightarrow{BA} = -\mathbf{p}$

$\Rightarrow \quad \overrightarrow{AF} = \mathbf{q} - \mathbf{p}$

d) $\overrightarrow{EC} = \overrightarrow{EA} + \overrightarrow{AC} = -\mathbf{q} + 2\mathbf{p}$

Example 2 The diagram shows a cuboid ABCDEFGH, with $\overrightarrow{AB}$, $\overrightarrow{BC}$ and $\overrightarrow{CG}$ corresponding to the vectors $\mathbf{p}$, $\mathbf{q}$ and $\mathbf{r}$, as shown. M is the mid-point of GH. Find, in terms of $\mathbf{p}$, $\mathbf{q}$ and $\mathbf{r}$, the following vectors:

a) $\overrightarrow{AC}$ **b)** $\overrightarrow{DF}$ **c)** $\overrightarrow{BM}$

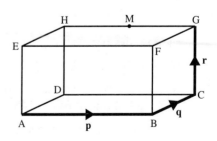

SOLUTION

a) $\overrightarrow{AC} = \overrightarrow{AB} + \overrightarrow{BC} = \mathbf{p} + \mathbf{q}$

b) $\overrightarrow{DF} = \overrightarrow{DC} + \overrightarrow{CB} + \overrightarrow{BF} = \mathbf{p} - \mathbf{q} + \mathbf{r}$

c) $\overrightarrow{BM} = \overrightarrow{BC} + \overrightarrow{CG} + \overrightarrow{GM} = \mathbf{q} + \mathbf{r} - \frac{1}{2}\mathbf{p}$

Example 3 ABCD is a parallelogram. E is the mid-point of AC. Vector $\overrightarrow{AB} = \mathbf{p}$ and vector $\overrightarrow{AD} = \mathbf{q}$. Find **a)** $\overrightarrow{BE}$ **b)** $\overrightarrow{BD}$

What can be deduced from the result?

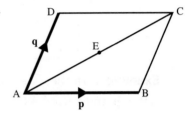

SOLUTION

a) First notice that

$$\overrightarrow{AC} = \overrightarrow{AB} + \overrightarrow{BC} = \mathbf{p} + \mathbf{q}$$

and

$$\overrightarrow{AE} = \frac{1}{2}\overrightarrow{AC} = \frac{1}{2}(\mathbf{p} + \mathbf{q})$$

Therefore,

$$\overrightarrow{BE} = \overrightarrow{BA} + \overrightarrow{AE}$$
$$= -\mathbf{p} + \frac{1}{2}(\mathbf{p} + \mathbf{q})$$
$$= \frac{1}{2}(\mathbf{q} - \mathbf{p})$$

b) $\overrightarrow{BD} = \overrightarrow{BC} + \overrightarrow{CD} = \mathbf{q} - \mathbf{p}$

We can see from this that $\overrightarrow{BE} = \frac{1}{2}\overrightarrow{BD}$. This means that BE is half the length of BD and BED is a straight line. That is, E is the mid-point of BD. This proves that the diagonals of a parallelogram bisect each other. (Many standard geometrical theorems can be proved by vector methods in this way.)

Example 4 An expedition in the Sahara travels 10 km on a bearing of 080° and then 8 km on a bearing of 045°. What is the expedition's final position in relation to its starting point?

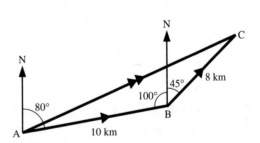

SOLUTION

The resultant of the two stages of the journey is the vector $\overrightarrow{AC}$ in the diagram.

In triangle ABC, we have AB = 10 km, BC = 8 km and $A\widehat{B}C = 145°$.

By the cosine rule:

$$AC^2 = AB^2 + BC^2 - 2 \times AB \times BC \times \cos A\widehat{B}C$$

$$= 10^2 + 8^2 - 2 \times 10 \times 8 \times \cos 145°$$

$$= 295.08$$

$$\Rightarrow \quad AC = 17.18 \, \text{km}$$

By the sine rule:

$$\frac{AC}{\sin A\widehat{B}C} = \frac{BC}{\sin B\widehat{A}C}$$

$$\Rightarrow \quad \frac{17.18}{\sin 145°} = \frac{8}{\sin B\widehat{A}C}$$

$$\Rightarrow \quad \sin B\widehat{A}C = \frac{8 \sin 145°}{17.18} = 0.2671$$

$$\Rightarrow \quad B\widehat{A}C = 15.5°$$

So, $\overrightarrow{AC}$ is a displacement of 17.18 km on a bearing of $080° - 015.5° = 064.5°$.

Example 5 A swimmer, who can swim at $0.8 \, \text{m s}^{-1}$ in still water, wishes to cross a river flowing at $0.5 \, \text{m s}^{-1}$.

a) If she aims straight across the river, what will be her actual velocity?
b) If she wishes to travel straight across, in what direction should she aim and what will be her actual speed?

SOLUTION

We need to make the simplifying assumption that the water flows at a uniform speed at all points on the crossing. We can then represent the velocities by the vector diagrams on the right.

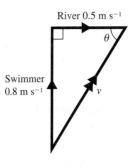

a) The swimmer's actual speed is v, so $v = \sqrt{0.8^2 + 0.5^2} = 0.943 \, \text{m s}^{-1}$
Her direction θ is given by

$$\tan \theta = \frac{0.8}{0.5} \quad \Rightarrow \quad \theta = 58°$$

So the swimmer travels at $0.943 \, \text{m s}^{-1}$ at an angle of $58°$ to the direction of the river.

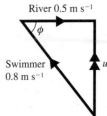

b) The direction of the swimmer's aim is given by

$$\cos \phi = \frac{0.5}{0.8} \quad \Rightarrow \quad \phi = 51.3°$$

Her actual speed is $u = \sqrt{0.8^2 - 0.5^2} = 0.625\,\text{m s}^{-1}$.

So, she should aim upstream at 51.3° to the bank. She will then travel straight across the river at $0.625\,\text{m s}^{-1}$.

Exercise 2A

1 ABCE is a rectangle. CDEF is a rhombus. G is the mid-point of AB. $\overrightarrow{AF} = \mathbf{p}$ and $\overrightarrow{EB} = \mathbf{q}$.

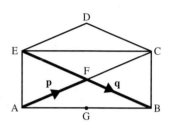

a) Find in terms of **p** and **q**:

  i) $\overrightarrow{AB}$ **ii)** $\overrightarrow{CB}$ **iii)** $\overrightarrow{DB}$

b) Show that $\overrightarrow{EB} + \overrightarrow{CA} = 2\overrightarrow{DF}$.

2 The diagram shows a regular hexagon ABCDEF with $\overrightarrow{AB} = \mathbf{p}$ and $\overrightarrow{BC} = \mathbf{q}$. Find in terms of **p** and **q**:

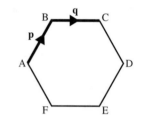

  a) $\overrightarrow{AD}$ **b)** $\overrightarrow{AC}$ **c)** $\overrightarrow{CE}$ **d)** $\overrightarrow{BE}$ **e)** $\overrightarrow{EA}$

3 The diagram shows a trapezium ABCD with AB parallel to DC and twice as long. E is the mid-point of BC. $\overrightarrow{AD} = \mathbf{p}$ and $\overrightarrow{DC} = \mathbf{q}$. Find in terms of **p** and **q**:

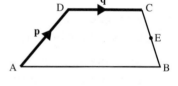

  a) $\overrightarrow{AB}$ **b)** $\overrightarrow{AC}$ **c)** $\overrightarrow{CD}$ **d)** $\overrightarrow{DB}$ **e)** $\overrightarrow{AE}$ **f)** $\overrightarrow{ED}$

4 The diagram shows a tetrahedron OABC with $\overrightarrow{OA} = \mathbf{a}$, $\overrightarrow{OB} = \mathbf{b}$ and $\overrightarrow{OC} = \mathbf{c}$. D is the mid-point of AB and E is on BC so that the ratio BE : EC = 2 : 1. Find in terms of **a**, **b** and **c**:

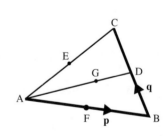

  a) $\overrightarrow{AC}$ **b)** $\overrightarrow{AB}$ **c)** $\overrightarrow{AD}$ **d)** $\overrightarrow{BC}$

  e) $\overrightarrow{BE}$ **f)** $\overrightarrow{OE}$ **g)** $\overrightarrow{DE}$

5 Use vector methods to show that the line joining the mid-points of two sides of a triangle is parallel to the third side and half its length.

6 The diagram shows triangle ABC with D, E and F the mid-points of BC, AC and AB respectively. G is the point on AD such that the ratio AG : GD = 2 : 1. Vector $\overrightarrow{AB} = \mathbf{p}$ and $\overrightarrow{BC} = \mathbf{q}$.

a) Find in terms of **p** and **q**:

  i) $\overrightarrow{DB}$ **ii)** $\overrightarrow{DA}$ **iii)** $\overrightarrow{BG}$ **iv)** $\overrightarrow{GE}$

  Explain what your results indicate about the points B, G and E.

b) Prove the equivalent result for points C, G and F.

7 ABCD is a quadrilateral. E and F are the mid-points of the diagonals AC and BD respectively.

  a) Show that $\overrightarrow{AB} + \overrightarrow{AD} = 2\overrightarrow{AF}$ and $\overrightarrow{CB} + \overrightarrow{CD} = 2\overrightarrow{CF}$.

  b) Hence show that $\overrightarrow{AB} + \overrightarrow{AD} + \overrightarrow{CB} + \overrightarrow{CD} = 4\overrightarrow{EF}$.

8 ABCD is a rectangle. Show that $\overrightarrow{AB} + 2\overrightarrow{BC} + 2\overrightarrow{CD} + 7\overrightarrow{DA} + 2\overrightarrow{BD} = 3\overrightarrow{CA}$.

9 In each of the following cases, find the magnitude and direction of the resultant of the two given vectors.

  a) A displacement of magnitude 3.5 km on a bearing 050° and a displacement of magnitude 5.4 km on a bearing of 128°.

  b) A displacement of magnitude 26 km on a bearing of 175° and a displacement of 18 km on a bearing of 294°.

  c) Velocities of 15 km h^{-1} due north and 23 km h^{-1} on a bearing of 253°.

  d) Forces of 355 N on a bearing of 320° and 270 N on a bearing of 025°.

10 A boat travels from point A on a bearing of 075° for 25 km and then travels a further 18 km to point B. If the distance AB is 14 km, find the direction in which the boat travelled on the second stage of the journey.

11 Two ships A and B set out from port O simultaneously. The first travels due north at 16 km h^{-1}, the second due east at 13 km h^{-1}.

  a) The vector $\overrightarrow{AB}$ represents the displacement of B from A. Express this in terms of $\overrightarrow{OA}$ and $\overrightarrow{OB}$.

  b) Find the magnitude and direction of $\overrightarrow{AB}$ after **i)** 1 hour, **ii)** 3 hours, **iii)** t hours.

  c) The ships' radios have a range of 120 km. For how long will the ships remain in contact?

  d) For how long would they remain in contact if B had travelled north-east?

12 A boat which can travel at 5 m s^{-1} in still water is crossing a river 200 m wide. The rate of flow of the river is 2 m s^{-1}, assumed uniform at every point in the river. Points A and B are directly opposite each other across the river.

  a) If the boat leaves A and steers towards B, at what speed will it travel and at what point will it reach the opposite bank?

  b) If the boat needs to travel towards B, in what direction should it be steered and at what speed will it travel?

  c) If the boat needs to land at point C, 150 m upstream of B, in what direction should it be steered and how long will it take to reach the bank?

Components of a vector

Vectors need to be specified in relation to a frame of reference. In the previous section, we used compass directions as our frame of reference, but more commonly we use *x*- and *y*-axes. (We will confine ourselves to two dimensions for the present.)

We have already defined a unit vector as a vector with a magnitude of 1 unit. We now define **i** and **j** as the unit vectors in the positive *x*-direction and positive *y*-direction respectively. All other vectors can then be formed by combining multiples of **i** and **j**.

For example, in the diagram, the vector $\overrightarrow{OP} = \overrightarrow{OA} + \overrightarrow{AP}$.
But $\overrightarrow{OA} = 3\mathbf{i}$ and $\overrightarrow{AP} = 2\mathbf{j}$, which give

$$\overrightarrow{OP} = 3\mathbf{i} + 2\mathbf{j}$$

The 3 and the 2 are called respectively the **x-component** and the **y-component** of $\overrightarrow{OP}$.

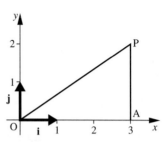

An alternative notation is $\overrightarrow{OP} = \begin{pmatrix} 3 \\ 2 \end{pmatrix}$, which is called a **column vector**.

When a vector is given as a magnitude and an angle, we can convert it into component form. This is called **resolving the vector into components**.

In the diagram, the vector $\overrightarrow{OP}$ has magnitude *r* and direction θ to the positive *x*-direction.

From the triangle, $x = r\cos\theta$ and $y = r\sin\theta$,

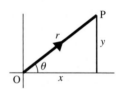

$$\Rightarrow \quad \overrightarrow{OP} = r\cos\theta\,\mathbf{i} + r\sin\theta\,\mathbf{j}$$

When we are given the vector in component form

$$\overrightarrow{OP} = x\mathbf{i} + y\mathbf{j}$$

its magnitude is given by

$$r = \sqrt{x^2 + y^2}$$

and its direction by θ, where $\tan\theta = \dfrac{y}{x}$.

Example 6 Express each of the vectors shown on the left in component form.

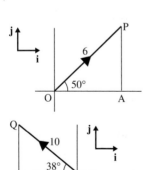

SOLUTION

$OA = 6\cos 50° = 3.857$
$AP = 6\sin 50° = 4.596$

$\Rightarrow \quad \overrightarrow{OP} = 3.857\mathbf{i} + 4.596\mathbf{j}$

$OB = 10\cos 38° = 7.88$
$BQ = 10\sin 38° = 6.157$

$\Rightarrow \quad \overrightarrow{OQ} = -7.88\mathbf{i} + 6.157\mathbf{j}$

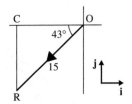

$$OC = 15\cos 43° = 10.97$$
$$CR = 15\sin 43° = 10.23$$

$$\Rightarrow \quad \overrightarrow{OR} = -10.97\mathbf{i} - 10.23\mathbf{j}$$

Example 7 Find the magnitude and direction of the following vectors:

a) p $= 2\mathbf{i} + 5\mathbf{j}$ **b) q** $= 3\mathbf{i} - 2\mathbf{j}$ **c) r** $= -\mathbf{i} - 2\mathbf{j}$

SOLUTION

a)

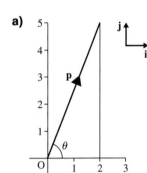

$$|\mathbf{p}| = \sqrt{2^2 + 5^2} = 5.385$$

$$\tan\theta = \frac{5}{2} \quad \Rightarrow \quad \theta = 68.2°$$

b)

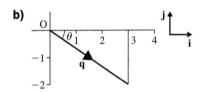

$$|\mathbf{q}| = \sqrt{3^2 + (-2)^2} = 3.606$$

$$\tan\theta = \frac{2}{3} \quad \Rightarrow \quad \theta = 33.7°$$

c)

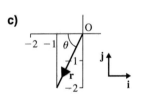

$$|\mathbf{r}| = \sqrt{(-1)^2 + (-2)^2} = 2.236$$

$$\tan\theta = 2 \quad \Rightarrow \quad \theta = 63.4°$$

There are two things to note.

- In Example 7, the direction is given as an angle indicated in the diagram. More formally, the direction would be given as a **rotation** θ from the positive x-direction, with $-180° < \theta \leqslant 180°$ and the anticlockwise sense taken as positive. The answers would then be **a)** 68.2°, **b)** −33.7°, **c)** −116.6°.

- Your calculator may have functions for converting between rectangular and polar coordinates. These may be used to convert between components (x, y) and magnitude and direction (r, θ). This can act as a check on your results, but cannot be a substitute for clearly shown working. It is, in any case, no help with problems couched in algebraic rather than numerical terms.

Combining vectors in component form

When we add vectors $\mathbf{p} = 3\mathbf{i} + \mathbf{j}$ and $\mathbf{q} = 2\mathbf{i} + 3\mathbf{j}$, we can see from the diagram that the resultant is

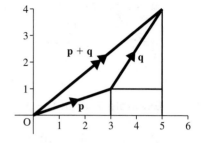

$$\mathbf{p} + \mathbf{q} = 5\mathbf{i} + 4\mathbf{j}$$

So,

$$(3\mathbf{i} + \mathbf{j}) + (2\mathbf{i} + 3\mathbf{j}) = 5\mathbf{i} + 4\mathbf{j}$$

That is, when adding vectors we add the corresponding components. This justifies the use of the $+$ symbol for combining vectors. It can be generalised to give the following results.

- $(a\mathbf{i} + b\mathbf{j}) + (c\mathbf{i} + d\mathbf{j}) = (a + c)\mathbf{i} + (b + d)\mathbf{j}$
- $(a\mathbf{i} + b\mathbf{j}) - (c\mathbf{i} + d\mathbf{j}) = (a - c)\mathbf{i} + (b - d)\mathbf{j}$
- $k(a\mathbf{i} + b\mathbf{j}) = ka\mathbf{i} + kb\mathbf{j}$ where k is a scalar

Example 8 Given $\mathbf{p} = 12\mathbf{i} + 5\mathbf{j}$ and $\mathbf{q} = 3\mathbf{i} - 4\mathbf{j}$, find:

a) i) $\mathbf{p} - \mathbf{q}$ **ii)** $2\mathbf{p} + 3\mathbf{q}$

b) a vector parallel to $\mathbf{p}$ and with magnitude 39

c) unit vector $\hat{\mathbf{q}}$

SOLUTION

a) i) $\mathbf{p} - \mathbf{q} = 9\mathbf{i} + 9\mathbf{j}$

    ii) $2\mathbf{p} + 3\mathbf{q} = (24\mathbf{i} + 10\mathbf{j}) + (9\mathbf{i} - 12\mathbf{j})$
$$= 33\mathbf{i} - 2\mathbf{j}$$

b) $|\mathbf{p}| = \sqrt{12^2 + 5^2} = 13$

    A vector $k\mathbf{p}$ is parallel to $\mathbf{p}$ and with k times the magnitude, so the required vector is

$$3\mathbf{p} = 36\mathbf{i} + 15\mathbf{j}$$

c) $|\mathbf{q}| = \sqrt{3^2 + 4^2} = 5$

    $\Rightarrow$ $\hat{\mathbf{q}} = \frac{1}{5}\mathbf{q} = 0.6\mathbf{i} - 0.8\mathbf{j}$

Unit vectors

We can generalise the process used in part **c** of Example 8 to find the unit vector in the direction of a given vector $\mathbf{a}$:

$$\hat{\mathbf{a}} = \frac{\mathbf{a}}{|\mathbf{a}|}$$

Resultant vectors

Example 9 Find the magnitude and direction of the resultant of the forces shown in the diagram. (N stands for newton, the unit of force, which we will meet on page 50.)

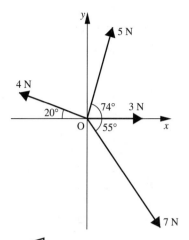

SOLUTION

To find the resultant, **R**, we express each force in component form and add.

$$\mathbf{R} = \begin{pmatrix} 3 \\ 0 \end{pmatrix} + \begin{pmatrix} 5\cos 74° \\ 5\sin 74° \end{pmatrix} + \begin{pmatrix} -4\cos 20° \\ 4\sin 20° \end{pmatrix} + \begin{pmatrix} 7\cos 55° \\ -7\sin 55° \end{pmatrix}$$

$$= \begin{pmatrix} 4.63 \\ 0.44 \end{pmatrix}$$

We now find the magnitude and direction of **R**.

From the triangle on the right, we have

$$R = \sqrt{4.63^2 + 0.44^2} = 4.66\,\text{N}$$

$$\tan\theta = \frac{0.44}{4.63} \quad \Rightarrow \quad \theta = 5.4°$$

Equality of vectors

Suppose we have $\mathbf{p} = a\mathbf{i} + b\mathbf{j}$ and $\mathbf{q} = c\mathbf{i} + d\mathbf{j}$, and that $\mathbf{p} = \mathbf{q}$.

$$\Rightarrow \quad \mathbf{p} - \mathbf{q} = \mathbf{0}$$

$$\Rightarrow \quad (a - c)\mathbf{i} + (b - d)\mathbf{j} = \mathbf{0}$$

$$\Rightarrow \quad a - c = 0 \quad \text{and} \quad b - d = 0$$

$$\Rightarrow \quad a = c \quad \text{and} \quad b = d$$

Exercise 2B

1 Given vectors $\mathbf{a} = 2\mathbf{i} - \mathbf{j}$, $\mathbf{b} = -2\mathbf{i} + 3\mathbf{j}$ and $\mathbf{c} = 4\mathbf{i} + \mathbf{j}$, calculate:

a) $\mathbf{a} + \mathbf{b}$ **b)** $\mathbf{a} - \mathbf{c}$ **c)** $2\mathbf{b} - \mathbf{a}$ **d)** $2\mathbf{a} + 3\mathbf{c}$ **e)** $|\mathbf{a}|$ **f)** $|\mathbf{b} + \mathbf{c}|$

2 Given vectors $\mathbf{p} = 3\mathbf{i} + u\mathbf{j}$, $\mathbf{q} = v\mathbf{i} - 4\mathbf{j}$ and $\mathbf{r} = 4\mathbf{i} - 6\mathbf{j}$, find:

a) the values of u and v if $\mathbf{p} - \mathbf{q} = \mathbf{r}$
b) the value of u if $\mathbf{p}$ and $\mathbf{r}$ are parallel.

3 Given $\mathbf{a} = -3\mathbf{i} + 4\mathbf{j}$, find:

a) a vector parallel to $\mathbf{a}$ and with magnitude 20
b) the unit vector $\hat{\mathbf{a}}$ in the direction of $\mathbf{a}$.

4 Express each of the following vectors in the form $x\mathbf{i} + y\mathbf{j}$.

a)

[diagram: vector of magnitude 7 at 34° above positive x-axis]

b)

[diagram: vector of magnitude 9.6 at 78° from positive x-axis]

c)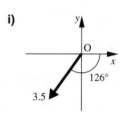

[diagram: vector of magnitude 12 at 57° from y-axis]

d)

[diagram: vector of magnitude 4 along positive y-axis]

e)

[diagram: vector of magnitude 8.3 at 64° from negative x-axis]

f)

[diagram: vector of magnitude 24 at 115° from positive x-axis]

g)

[diagram: vector of magnitude 6.2 at 55° below negative x-axis]

h)

[diagram: vector of magnitude 14 at 27° below positive x-axis]

i)

[diagram: vector of magnitude 3.5 at 126° measured, pointing down-left]

5 Find the magnitude, r, and the direction, θ, of the following vectors, where θ is the anticlockwise rotation from the positive x-direction and $-180° < \theta \leqslant 180°$.

a) $5\mathbf{i} + 2\mathbf{j}$ **b)** $7\mathbf{i} + 9\mathbf{j}$ **c)** $-5\mathbf{j}$ **d)** $-2\mathbf{i} + 3\mathbf{j}$

e) $3\mathbf{i} - 5\mathbf{j}$ **f)** $-6\mathbf{i} - 5\mathbf{j}$ **g)** $-2\mathbf{i}$

6 A ship sails from O to A, a distance of 20 km on a bearing of 072°, and then from A to B, a distance of 28 km on a bearing of 024°. Take east to be the x-direction and north the y-direction.

a) Express the displacement vectors $\overrightarrow{OA}$ and $\overrightarrow{AB}$ in component form.

b) Hence find the resultant displacement $\overrightarrow{OB}$ in component form.

c) Find the magnitude and direction of the resultant displacement.

7 Two ships, A and B, leave harbour at O. A travels 35 km on a bearing of 280° and B travels 50 km on a bearing of 030°. Take east to be the x-direction and north the y-direction

a) Express the displacement vectors $\overrightarrow{OA}$ and $\overrightarrow{OB}$ in component form.

b) Express the vector $\overrightarrow{AB}$ in terms of $\overrightarrow{OA}$ and $\overrightarrow{OB}$, and hence find $\overrightarrow{AB}$ in component form.

c) How far and on what bearing is ship B from ship A?

8 Starting simultaneously from the same spot, O, Alvin and Bernard set out across a field in the fog. Alvin travels with velocity $\mathbf{i} + 2\mathbf{j}\,\mathrm{m\,s^{-1}}$ and Bernard with velocity $3\mathbf{i} + \mathbf{j}\,\mathrm{m\,s^{-1}}$. To avoid losing each other, they hold opposite ends of a 90 m string.

a) At what speed is Alvin travelling?

b) Find the directions of travel of the two people and hence the angle between their paths.

c) Find the position vectors $\overrightarrow{OA}$ and $\overrightarrow{OB}$ at time t seconds, and hence the vector $\overrightarrow{AB}$ at that time.

d) Find the value of t for which the string becomes taut.

Vectors in three dimensions

The techniques just used for handling vectors in two dimensions can easily be extended to three dimensions. The only change is that we now need a frame of reference with three perpendicular axes: the x-, y- and z-axes. The unit vectors in these directions are **i**, **j** and **k** respectively.

If we draw the positive x- and y-axes as usual, there are two possible directions for the positive z-direction: namely, 'out of' the page or 'into' the page. The convention is the former, so that if the page is lying on a desk, the positive z-axis points straight upwards. Axes like these form a right-hand set, so called because if a right-hand screw (for example, a bottle top) positioned parallel to the z-axis is rotated in the conventional positive (anticlockwise) sense, it will travel in the positive z-direction.

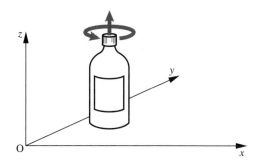

Drawing 3-D axes can be done in several ways. The two most common are shown below.

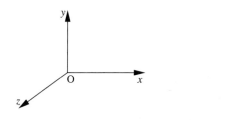
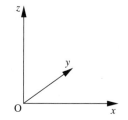

A vector with magnitude r will have components $r\cos\alpha$, $r\cos\beta$ and $r\cos\gamma$, where α, β and γ are the angles it makes with the positive x-, y- and z-directions respectively.

Therefore, with reference to the diagram, we have

$$OA = r\cos\alpha$$
$$OB = r\cos\beta$$
$$OC = r\cos\gamma$$

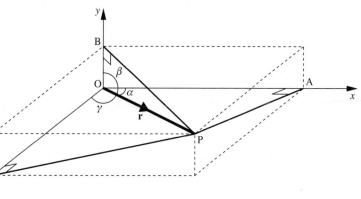

When we know a vector in component form, $\mathbf{r} = a\mathbf{i} + b\mathbf{j} + c\mathbf{k}$, we can find its magnitude as follows.

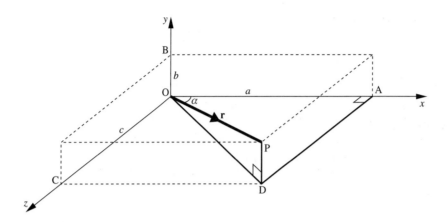

In triangle OAD, we have

$$OD^2 = OA^2 + AD^2$$
$$= a^2 + c^2$$

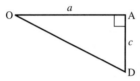

In triangle OPD, we have

$$OP^2 = OD^2 + PD^2$$
$$= a^2 + c^2 + b^2$$

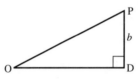

So, for $\mathbf{r} = a\mathbf{i} + b\mathbf{j} + c\mathbf{k}$, we have

$$|\mathbf{r}| = \sqrt{a^2 + b^2 + c^2}$$

If the vector makes angles of α, β and γ with the axes, then from triangle OAP we get

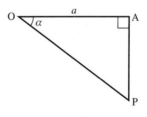

$$\cos \alpha = \frac{OA}{OP}$$
$$= \frac{a}{\sqrt{a^2 + b^2 + c^2}}$$

Similarly, we have

$$\cos \beta = \frac{b}{\sqrt{a^2 + b^2 + c^2}} \quad \text{and} \quad \cos \gamma = \frac{c}{\sqrt{a^2 + b^2 + c^2}}$$

It follows that

$$\cos^2\alpha + \cos^2\beta + \cos^2\gamma = 1$$

Example 10 Given $\mathbf{a} = 2\mathbf{i} - 2\mathbf{j} + \mathbf{k}$ and $\mathbf{b} = 3\mathbf{i} + 4\mathbf{j} + 6\mathbf{k}$, find:

a) $\mathbf{a} - \mathbf{b}$ **b)** $2\mathbf{a} + 3\mathbf{b}$ **c)** $|\mathbf{a}|$ **d)** $\hat{\mathbf{a}}$

SOLUTION

a) $\mathbf{a} - \mathbf{b} = -\mathbf{i} - 6\mathbf{j} - 5\mathbf{k}$

b) $2\mathbf{a} + 3\mathbf{b} = (4\mathbf{i} - 4\mathbf{j} + 2\mathbf{k}) + (9\mathbf{i} + 12\mathbf{j} + 18\mathbf{k})$

$\qquad\qquad = 13\mathbf{i} + 8\mathbf{j} + 20\mathbf{k}$

c) $|\mathbf{a}| = \sqrt{2^2 + (-2)^2 + 1^2}$

$\qquad = 3$

d) $\hat{\mathbf{a}} = \frac{1}{3}\mathbf{a}$

$\qquad = \frac{2}{3}\mathbf{i} - \frac{2}{3}\mathbf{j} + \frac{1}{3}\mathbf{k}$

Example 11 A helium balloon is released from O. It rises at a constant $2\,\mathrm{m\,s^{-1}}$ and is blown by a steady wind of $10\,\mathrm{m\,s^{-1}}$ coming from a bearing of $250°$. How far is it from O after 20 seconds?

SOLUTION

Taking east as the x-direction and north as the y-direction, the velocity of the wind is

$$\begin{pmatrix} 10\cos 20° \\ 10\sin 20° \\ 0 \end{pmatrix}$$

The still air velocity of the balloon is $\begin{pmatrix} 0 \\ 0 \\ 2 \end{pmatrix}$

Therefore, the resultant velocity is

$$\begin{pmatrix} 10\cos 20° \\ 10\sin 20° \\ 0 \end{pmatrix} + \begin{pmatrix} 0 \\ 0 \\ 2 \end{pmatrix} = \begin{pmatrix} 10\cos 20° \\ 10\sin 20° \\ 2 \end{pmatrix} = \begin{pmatrix} 9.397 \\ 3.420 \\ 2 \end{pmatrix}$$

After 20 seconds, the displacement vector is

$$20 \times \begin{pmatrix} 9.397 \\ 3.420 \\ 2 \end{pmatrix} = \begin{pmatrix} 187.94 \\ 68.40 \\ 40 \end{pmatrix}$$

The distance is the magnitude of this displacement vector, which is

$$\sqrt{187.94^2 + 68.40^2 + 40^2} = 203.96\,\mathrm{m}$$

Example 12 A vector makes angles of 66° and 53° with the positive x- and y-directions respectively. Find the angle that it makes with the z-direction.

SOLUTION

We know that

$$\cos^2\alpha + \cos^2\beta + \cos^2\gamma = 1$$
$$\Rightarrow \quad \cos^2 66° + \cos^2 53° + \cos^2\gamma = 1$$
$$\Rightarrow \quad \cos^2\gamma = 1 - \cos^2 66° - \cos^2 53°$$
$$= 0.472$$
$$\Rightarrow \quad \cos\gamma = \pm 0.687$$
$$\Rightarrow \quad \gamma = 46.6° \quad \text{or} \quad 133.4°$$

Exercise 2C

1 Given vectors $\mathbf{a} = 3\mathbf{i} - \mathbf{j} + 2\mathbf{k}$, $\mathbf{b} = 6\mathbf{i} - 3\mathbf{j} - 2\mathbf{k}$ and $\mathbf{c} = \mathbf{i} + \mathbf{j} - 3\mathbf{k}$, find:

a) i) $\mathbf{a} - \mathbf{b}$ **ii)** $2\mathbf{a} + 5\mathbf{c}$ **iii)** $|\mathbf{b}|$ **iv)** $|\mathbf{c} - \mathbf{a}|$ **v)** $\hat{\mathbf{b}}$

     vi) the angles between $\mathbf{b}$ and the positive x-, y- and z-directions

b) a vector parallel to $\mathbf{b}$ and with magnitude 28

c) the values of p, q and r when

$$p\mathbf{a} + q\mathbf{c} = 3\mathbf{i} - 5\mathbf{j} + r\mathbf{k}$$

2 The diagram shows a vector $\mathbf{r}$ with magnitude 8, making angles of 27°, 85° and 64° with the positive x-, y- and z-directions respectively. Express $\mathbf{r}$ in component form $a\mathbf{i} + b\mathbf{j} + c\mathbf{k}$.

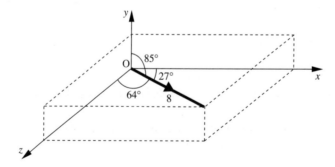

3 A vector $\mathbf{p}$ has magnitude 12 and makes angles of 68° and 75° with the positive y- and z-directions respectively. Find the two possible values for the angle which $\mathbf{p}$ makes with the positive x-direction. Hence find the two possible vectors $\mathbf{p}$ in component form.

4 A vector $a\mathbf{i} + b\mathbf{j} + c\mathbf{k}$ has magnitude r and makes angles α, β and γ with the positive x-, y- and z-directions respectively. Complete the following table. In some cases, there are two possible solutions.

	r	α	β	γ	a	b	c
a)	7	38°	69°	60°			
b)	6	115°	49°	128.7°			
c)	9	54°	72°				
d)					4	6	12
e)					−3	4	5
f)	4				2	−1	
g)		33°	59°		5		
h)		75°	42°				−8

5 Eamonn Chute is a wildlife photographer. He spots a butterfly which is flying so that t seconds after he sees it, its displacement (in metres) from his camera is

$$(1 + 0.2t)\mathbf{i} + 0.05t^2\mathbf{j} + 0.5\sqrt{t}\,\mathbf{k}$$

After 6 seconds, he presses the shutter and gets a perfect picture.

a) At what angles to the three axes did he point his camera?
b) At what distance did he set the focus?

Scalar product

Although vectors cannot be multiplied together in the usual sense, there are two ways in which vectors **a** and **b** can be combined which are reminiscent of multiplication and which serve useful functions. These are the **scalar product** (or **dot product**) and the **vector product** (or **cross product**). The vector product is outside the scope of this book, but we do need to know how the scalar product is defined.

First, we need to be clear what we mean by the **angle between two vectors.**

It is the angle formed at a point where the two vectors are **both directed away from** the point.

In the diagrams, θ is the angle between **a** and **b**, and ϕ is the angle between **c** and **d**.

For two vectors **a** and **b** the scalar product (dot product) is defined as

$$\mathbf{a.b} = |\mathbf{a}|\,|\mathbf{b}|\cos\theta$$

where θ is the angle between **a** and **b**.

Notice that **a.b** is a scalar quantity – hence the name scalar product.

The scalar product can be interpreted as the magnitude of the component of **a** in the direction of **b** (namely, $|\mathbf{a}|\cos\theta$) multiplied by the magnitude of **b**.

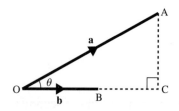

$$OC = |\mathbf{a}|\cos\theta$$
$$OB = |\mathbf{b}|$$
$$\Rightarrow \quad \mathbf{a.b} = OC \times OB$$

You will meet this interpretation when we deal with the work done by a force (page 254).

We can also express the scalar product in terms of the components of the two vectors. This can be deduced as follows.

Consider triangle OAB, with angle θ between the vectors **a** and **b**, as shown.

Let $\mathbf{a} = a_1\mathbf{i} + a_2\mathbf{j}$ and $\mathbf{b} = b_1\mathbf{i} + b_2\mathbf{j}$, which give

$$\mathbf{b} - \mathbf{a} = (b_1 - a_1)\mathbf{i} + (b_2 - a_2)\mathbf{j}$$

By the cosine rule:

$$OA^2 + OB^2 - 2 \times OA \times OB \times \cos\theta = AB^2$$
$$\Rightarrow \quad |\mathbf{a}|^2 + |\mathbf{b}|^2 - 2|\mathbf{a}|\,|\mathbf{b}|\cos\theta = |\mathbf{b} - \mathbf{a}|^2$$
$$\Rightarrow \quad (a_1^2 + a_2^2) + (b_1^2 + b_2^2) - 2|\mathbf{a}|\,|\mathbf{b}|\cos\theta = (b_1 - a_1)^2 + (b_2 - a_2)^2$$
$$\Rightarrow \quad |\mathbf{a}|\,|\mathbf{b}|\cos\theta = a_1 b_1 + a_2 b_2$$
$$\Rightarrow \quad \mathbf{a.b} = a_1 b_1 + a_2 b_2$$

If we had been using three dimensions, a similar process would give

$$\mathbf{a.b} = a_1 b_1 + a_2 b_2 + a_3 b_3$$

Finding the angle between vectors

From the definition of scalar product, $\mathbf{a.b} = |\mathbf{a}|\,|\mathbf{b}|\cos\theta$, we can obtain the result

$$\cos\theta = \frac{\mathbf{a.b}}{|\mathbf{a}|\,|\mathbf{b}|}$$

Example 13 Find the angle θ between $\mathbf{a} = 2\mathbf{i} + 4\mathbf{j} - \mathbf{k}$ and $\mathbf{b} = \mathbf{i} + 2\mathbf{j} + 3\mathbf{k}$.

SOLUTION

Using the component version of the scalar product gives

$$\mathbf{a.b} = 2 \times 1 + 4 \times 2 + (-1) \times 3 = 7$$

Also, we have

$$|\mathbf{a}| = \sqrt{2^2 + 4^2 + (-1)^2} = \sqrt{21}$$
$$|\mathbf{b}| = \sqrt{1^2 + 2^2 + 3^2} = \sqrt{14}$$
$$\Rightarrow \quad \cos\theta = \frac{7}{\sqrt{21} \times \sqrt{14}} = 0.4082 \quad \Rightarrow \quad \theta = 65.9°$$

Example 14 Triangle ABC has $A(3, 2, 5)$, $B(1, 4, -2)$ and $C(-2, 0, 3)$. Find the angle ABC.

SOLUTION

The required angle is that between vectors $\overrightarrow{BA}$ and $\overrightarrow{BC}$, namely θ.

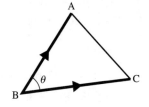

From the coordinates of A, B and C, we get

$$\overrightarrow{BA} = 2\mathbf{i} - 2\mathbf{j} + 7\mathbf{k} \quad \text{and} \quad \overrightarrow{BC} = -3\mathbf{i} - 4\mathbf{j} + 5\mathbf{k}$$
$$\Rightarrow \quad \overrightarrow{BA}.\overrightarrow{BC} = 2 \times (-3) + (-2) \times (-4) + 7 \times 5 = 37$$

Also, we have

$$BA = \sqrt{2^2 + (-2)^2 + 7^2} = \sqrt{57}$$
$$BC = \sqrt{(-3)^2 + (-4)^2 + 5^2} = \sqrt{50}$$
$$\Rightarrow \quad \cos\theta = \frac{37}{\sqrt{57} \times \sqrt{50}} = 0.6931 \quad \Rightarrow \quad \theta = 46.1°$$

Properties of the scalar product

- $\mathbf{a.a} = |\mathbf{a}|\,|\mathbf{a}| \cos 0° = a^2$ ($\mathbf{a.a}$ is sometimes referred to as $\mathbf{a}^2$.)

  In particular, for unit vectors: $\hat{\mathbf{a}}.\hat{\mathbf{a}} = 1$, $\mathbf{i.i} = 1$, $\mathbf{j.j} = 1$, $\mathbf{k.k} = 1$

- When $\mathbf{a}$ and $\mathbf{b}$ are perpendicular, then $\mathbf{a.b} = |\mathbf{a}|\,|\mathbf{b}| \cos 90° = 0$

  In particular, $\mathbf{i.j} = 0$, $\mathbf{j.k} = 0$, $\mathbf{k.j} = 0$

Example 15 Given points $A(2, -3, 4)$, $B(4, 1, 0)$, $C(8, 9, -8)$ and $D(0, -2, -5)$, show that ABC is a straight line with BD perpendicular to it.

SOLUTION

From the coordinates of A, B and C, we get

$$\overrightarrow{AB} = 2\mathbf{i} + 4\mathbf{j} - 4\mathbf{k} \quad \text{and} \quad \overrightarrow{AC} = 6\mathbf{i} + 12\mathbf{j} - 12\mathbf{k}$$

$$\Rightarrow \quad \overrightarrow{AC} = 3\overrightarrow{AB}$$

This means that $\overrightarrow{AB}$ and $\overrightarrow{AC}$ have the same direction and, as both start from A, ABC must be a straight line.

Also, we have

$$\overrightarrow{BD} = -4\mathbf{i} - 3\mathbf{j} - 5\mathbf{k}$$

Therefore,

$$\overrightarrow{AC}.\overrightarrow{BD} = 6 \times (-4) + 12 \times (-3) + (-12) \times (-5) = 0$$

$$\Rightarrow \quad \overrightarrow{AC} \text{ is perpendicular to } \overrightarrow{BD}$$

Example 16 Use vector methods to show that the diagonals of a rhombus intersect at right angles.

SOLUTION

Let ABCD be a rhombus, with $\overrightarrow{AB} = \overrightarrow{DC} = \mathbf{a}$ and $\overrightarrow{AD} = \overrightarrow{BC} = \mathbf{b}$, as shown. We therefore have

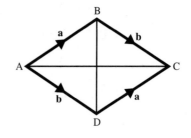

$$\overrightarrow{AC} = \mathbf{a} + \mathbf{b} \quad \text{and} \quad \overrightarrow{DB} = \mathbf{a} - \mathbf{b}$$

$$\Rightarrow \quad \overrightarrow{AC}.\overrightarrow{DB} = (\mathbf{a} + \mathbf{b}).(\mathbf{a} - \mathbf{b})$$

$$= \mathbf{a}.\mathbf{a} + \mathbf{b}.\mathbf{a} - \mathbf{a}.\mathbf{b} - \mathbf{b}.\mathbf{b}$$

$$= a^2 - b^2$$

But as the figure is a rhombus, $a = b$. Therefore, we have

$$\overrightarrow{AC}.\overrightarrow{DB} = 0 \quad \Rightarrow \quad \text{AC is perpendicular to DB}$$

Note Two assumptions were made in Example 16: namely, that the scalar product is **commutative** (that is, $\mathbf{a}.\mathbf{b} = \mathbf{b}.\mathbf{a}$) and it is **distributive over addition** (that is, we can multiply out brackets).

Both of these properties can be shown from the component form of the scalar product.

$$\mathbf{a}.\mathbf{b} = a_1b_1 + a_2b_2 + a_3b_3$$

$$= b_1a_1 + b_2a_2 + b_3a_3 \quad \text{(as multiplication of ordinary}$$
$$\text{numbers \textbf{is} commutative)}\}$$

$$\Rightarrow \quad \mathbf{a}.\mathbf{b} = \mathbf{b}.\mathbf{a}$$

$$\mathbf{a.(b+c)} = a_1(b_1 + c_1) + a_2(b_2 + c_2) + a_3(b_3 + c_3)$$
$$= a_1b_1 + a_1c_1 + a_2b_2 + a_2c_2 + a_3b_3 + a_3c_3$$
(as multiplication **is** distributive over addition)
$$= a_1b_1 + a_2b_2 + a_3b_3 + a_1c_1 + a_2c_2 + a_3c_3$$
$$\Rightarrow \quad \mathbf{a.(b+c) = a.b + a.c}$$

Exercise 2D

1 Find the scalar product of the following pairs of vectors.

a) $3\mathbf{i} - \mathbf{j}$ and $4\mathbf{i} + 2\mathbf{j}$

b) $\mathbf{i} + 3\mathbf{j}$ and $6\mathbf{i} - 2\mathbf{j}$

c) $2\mathbf{j}$ and $\mathbf{i} + \mathbf{j}$

d) $\mathbf{i} + 2\mathbf{j} + 3\mathbf{k}$ and $2\mathbf{i} - \mathbf{j} + \mathbf{k}$

e) $4\mathbf{i} + 5\mathbf{j} - 2\mathbf{k}$ and $3\mathbf{i} - 2\mathbf{j} + \mathbf{k}$

f) $-4\mathbf{i} + \mathbf{k}$ and $3\mathbf{i} + \mathbf{j}$

What do you deduce about the pairs of vectors in parts **b** and **e**?

2 Find the scalar product of each pair of vectors with magnitudes and angle as shown below.

a)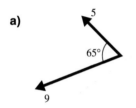

b)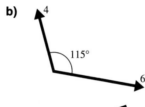

c)

d)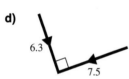

e)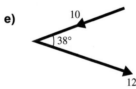

f)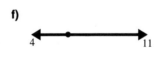

3 Find the angles between the following pairs of vectors.

a) $\begin{pmatrix} 2 \\ 3 \end{pmatrix}$ and $\begin{pmatrix} -4 \\ 1 \end{pmatrix}$

b) $\begin{pmatrix} 1 \\ -2 \\ 2 \end{pmatrix}$ and $\begin{pmatrix} -3 \\ 4 \\ 5 \end{pmatrix}$

c) $2\mathbf{i} + \mathbf{j}$ and $\mathbf{i} - 3\mathbf{j}$

d) $5\mathbf{i} + 3\mathbf{j} + \mathbf{k}$ and $\mathbf{i} - \mathbf{j} + 2\mathbf{k}$

e) $4\mathbf{i} - \mathbf{j} + 5\mathbf{k}$ and $-2\mathbf{i} + \mathbf{k}$

f) $3\mathbf{j} + \mathbf{k}$ and $\mathbf{i} - 2\mathbf{j}$

g) $2\mathbf{i} + 3\mathbf{j} + 6\mathbf{k}$ and $3\mathbf{i} - \mathbf{j} - \mathbf{k}$

h) $a\mathbf{i} + b\mathbf{j}$ and $c\mathbf{k}$

4 Triangle PQR has P(2, 5), Q(4, 8) and R(−1, 6). Use the scalar product to find the angle PQR.

5 Triangle ABC has A(−1, 3, −3), B(2, 4, 6) and C(3, 0, −5). Use the scalar product to find the angle ACB.

6 Two altitudes of a triangle ABC meet at H, as shown. Relative to some origin O the position vectors of A, B, C and H are **a**, **b**, **c** and **h**. Show that

$$(\mathbf{h} - \mathbf{a}).(\mathbf{b} - \mathbf{c}) = (\mathbf{h} - \mathbf{b}).(\mathbf{c} - \mathbf{a}) = 0$$

Deduce that $(\mathbf{h} - \mathbf{c}).(\mathbf{b} - \mathbf{a}) = 0$ and hence that the three altitudes of a triangle are concurrent.

7 Two objects are moving so that their velocities at time t seconds are given by

$$\mathbf{v}_1 = t\mathbf{i} + 2\mathbf{j} - t^{-1}\mathbf{k} \quad \text{and} \quad \mathbf{v}_2 = (2 - t)\mathbf{i} + (t + 1)\mathbf{j} + 3t\mathbf{k}$$

a) Find the angle between their directions of motion when $t = 2$.
b) Find the times at which their directions of motion are perpendicular to each other.

3 Kinematics

The spirit of the time shall teach me speed.
WILLIAM SHAKESPEARE

Kinematics is the branch of mechanics which deals with motion. So, in this chapter, we are concerned with the relationships between the **position**, the **velocity** and the **acceleration** of bodies, and how these change with **time**. We are not concerned with the causes of these changes of motion.

Terminology

Apart from time, the characteristics of motion are vector quantities, and direction is as important as magnitude. However, we sometimes just refer to their magnitude, and the terminology we use indicates this difference.

Displacement   This is a vector quantity which states the position of an object relative to some chosen origin.

Distance   This is a scalar quantity which states how far the object has travelled. This is **not** the same as its position. For example, if you throw a ball up in the air and then catch it, the distance it travels might be 10 m (5 m up and 5 m down) but at the end the ball's displacement from its starting point is 0 m.

Velocity   This is a vector quantity which states how fast the object is moving, and in which direction.

Speed   This is a scalar quantity which just states how fast the object is moving. It is the magnitude of the velocity vector.

Acceleration   This term is used both for the vector quantity *rate of change of velocity* and for the scalar quantity *rate of change of speed*.

Average speed

In order to formulate a basic model for motion, we consider, for the present, only scalar quantities. All motion is assumed to take place along a straight line. We also model the objects as particles, which enables us to ignore their size.

The **average speed** of an object is defined as

$$\text{Average speed} = \frac{\text{Distance travelled}}{\text{Time taken}}$$

Units

The SI unit of speed is **metres per second**. The notation for this is $\mathrm{m\,s^{-1}}$ (occasionally m/s).

Speed is sometimes given in other units, such as kilometres per hour ($\mathrm{km\,h^{-1}}$).

Example 1 A train travels from Penzance to London, stopping at Plymouth, Exeter and Bristol en route. The distances (in km) and the times of arrival at the stations are shown in the table. Find the average speed for each section of the journey and the average speed for the whole journey. Ignore the time spent stopped in the stations.

Station	Distance travelled (km)	Time
Penzance	0	0600
Plymouth	121	0706
Exeter	195	0736
Bristol	321	0839
London	521	0954

SOLUTION

Penzance–Plymouth: Distance = 121 km, time = 66 min = 1.1 h. Therefore, we have

$$\text{Average speed} = \frac{121}{1.1} = 110\,\mathrm{km\,h^{-1}}$$

Plymouth–Exeter: Distance = 74 km, time = 30 min = 0.5 h. Therefore, we have

$$\text{Average speed} = \frac{74}{0.5} = 148\,\mathrm{km\,h^{-1}}$$

Exeter–Bristol: Distance = 126 km, time = 63 min = 1.05 h. Therefore, we have

$$\text{Average speed} = \frac{126}{1.05} = 120\,\mathrm{km\,h^{-1}}$$

Bristol–London: Distance = 200 km, time = 75 min = 1.25 h. Therefore, we have

$$\text{Average speed} = \frac{200}{1.25} = 160\,\mathrm{km\,h^{-1}}$$

Plymouth–London: Distance = 521 km, time = 234 min = 3.9 h. Therefore, we have

$$\text{Average speed} = \frac{521}{3.9} = 133.6\,\mathrm{km\,h^{-1}}$$

Note The average speed of a journey of several stages cannot be found by calculating the mean of the speeds for the individual stages. It must be found from the **total distance** travelled and the **total time** taken.

Average acceleration

Following the ideas above, we define the **average acceleration** of an object by

$$\text{Average acceleration} = \frac{\text{Change in speed}}{\text{Time taken to change}}$$

Units

The SI unit of acceleration is **metres per second per second** or metres per second^2. The notation for this is m s^{-2} (occasionally m/s^2).

Example 2 A car increases its speed from $10\,\text{m s}^{-1}$ to $25\,\text{m s}^{-1}$, taking 5 seconds to do so. Find its average acceleration.

SOLUTION

$$\text{Average acceleration} = \frac{\text{Change in speed}}{\text{Time taken to change}}$$

$$= \frac{25 - 10}{5} = 3\,\text{m s}^{-2}$$

Exercise 3A

1 In a charity walk, a group walked for 2 hours, covering a distance of 6 km. They then stopped for lunch, which took another hour, and afterwards walked the final leg of 12 km in 3 hours. Find their average speed over the whole journey.

2 A canoe race takes place on a river. Competitors have to paddle downstream for 18 km and then return upstream to their starting point. A competitor takes 2 hours to complete the downstream leg and returns at 8 km per hour.

a) Find the average speed for the downstream leg.
b) Find the time taken for the upstream leg.
c) Find the average speed for the whole race.

3 In the 1988 Olympic Games in Seoul, the winning times for the women's track events were as follows:

100 m	10.54 s	800 m	1 min 56.10 s
200 m	21.34 s	1500 m	3 min 53.96 s
400 m	48.65 s	3000 m	8 min 26.53 s

Find the average speed for each of these events in m s^{-1}.

4 A person drove a distance of 18 km from her home to the motorway before joining it for the rest of her journey. Her average speed for the first part of the journey was 54 km h^{-1} and her overall average speed was 80 km h^{-1}. If her journey took a total of 72 minutes, find her average speed for the second part of the journey.

5 In a motor rally stage, part of the journey is through populated areas and speeds are restricted. The rest is through a forest track where there are no restrictions on speed. One competitor took 48 minutes to complete the stage at an average speed of 100 km h^{-1}. The slow section of the stage took 18 minutes and the average speed for the forest section was 125 km h^{-1}. Find the average speed for the slow section of the stage.

Distance and displacement, speed and velocity

In the questions in Exercise 3A, you were not required to consider the direction in which motion was taking place. The canoeist in Question **2**, for example, followed an 'out and back' course, whilst the driver in Question **4** followed a route which essentially kept going forwards. In both cases, you worked out the distance travelled, 36 km in the case of the canoeist and 96 km in the case of the car driver.

The **displacement** of an object is its position relative to some origin (usually its starting point). For motion in a straight line, a displacement can be in one of two directions, which are defined respectively as **positive** and **negative**. Hence, in the case of the canoeist, since the course returned to the starting point, the final displacement was zero. In the case of the car driver, the final displacement would have a magnitude of 96 km if the road from her starting point were straight. This is unlikely, and without more information we could not tell what her final displacement is. However, we often make the modelling assumption that journeys take place in a straight line.

We must also distinguish between positive and negative velocities. If in the example of the canoeist, we took the outward leg of the race to be in the positive direction, the velocity would be positive for this leg and negative for the return leg. Of course, the speed, being a scalar quantity, would be positive for both legs.

Uniform speed, velocity and acceleration

When an object moves a given distance in a given time, its speed may not be constant. In calculating its average speed, we are finding the constant speed necessary to achieve the same distance in the same time. In reality, objects rarely travel at a constant speed for any length of time, but the idea is a useful modelling assumption. We use the terms **uniform speed** and **uniform velocity** to indicate that we have made the assumption that the speed or velocity do not change over the period of time in which we are interested.

In a similar way, we can make the modelling assumption that a body has **uniform acceleration**.

Displacement–time graphs

When an object travels with uniform velocity, its displacement changes by the same amount in each equal time period. Drawing a graph of this will give a straight line.

Example 3 A boy walks at $4\,\mathrm{km\,h^{-1}}$ for 3 hours. Draw his displacement–time graph.

SOLUTION

We make a table of the displacement at intervals of 1 hour and plot a graph of the results.

Time (h)	0	1	2	3
Displacement	0	4	8	12

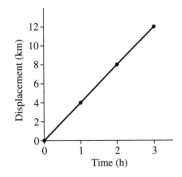

As expected, the graph is a straight line.

It is interesting to look at the gradient of the line.

Using points (0, 0) and (3, 12), we have

$$\text{Gradient} = \frac{12 - 0}{3 - 0} = 4$$

In calculating the gradient, we are dividing 12 km by 3 h, so the unit of the result is $\mathrm{km\,h^{-1}}$.

The gradient is therefore $4\,\mathrm{km\,h^{-1}}$, which is the boy's velocity.

For any straight-line displacement–time graph,

$$\text{Gradient} = \frac{\text{Change of displacement}}{\text{Time taken to change}} = \text{Velocity}$$

When the graph is not a straight line, the gradient of the curve at any point gives the velocity of the object at that instant.

The gradient of the displacement–time graph is the velocity of the object.

Velocity–time graphs

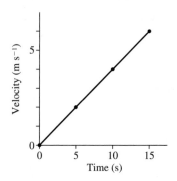

This graph shows the velocity of an object (in $m\,s^{-1}$) plotted against time (in s).

$$\text{Gradient} = \frac{6-0}{15-0} = 0.4$$

Here, we are dividing $6\,m\,s^{-1}$ by 15 s, so the unit for this gradient is $m\,s^{-2}$, which represents acceleration.

Hence, the acceleration of this object is $0.4\,m\,s^{-2}$. As the gradient is constant, the object is moving with uniform acceleration.

For any straight-line velocity–time graph,

$$\text{Gradient} = \frac{\text{Change of velocity}}{\text{Time taken to change}} = \text{Acceleration}$$

When the graph is not a straight line, the gradient of the curve at any point gives the acceleration of the object at that instant.

[The gradient of the velocity–time graph is the acceleration of the object.

Another important property of the velocity–time graph relates to the area between the graph and the time axis.

If a cyclist travels at a uniform velocity of $8\,km\,h^{-1}$ for 3 h, the distance travelled is 24 km. The velocity–time graph is shown on the right.

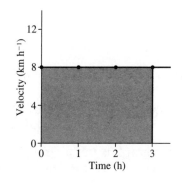

The area of the shaded rectangle is given by

$$\text{Area} = 3 \times 8 = 24$$

The unit for this area is $h \times km\,h^{-1} = km$.

This means that the area, 24 km, represents the displacement of the cyclist from his starting point.

It can be shown that this result also holds true for non-uniform velocity.

[The area between the velocity–time graph and the time axis corresponds to the displacement of the object during its motion.

Example 4 The graph shows the displacement (in km) of a cyclist from a town A, plotted against time (in hours).

a) What assumptions have been made in drawing the graph?

b) What happened in the different stages of the journey?

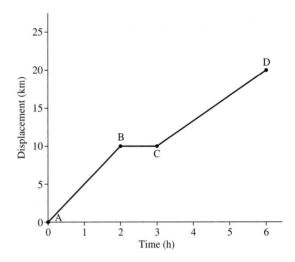

SOLUTION

a) Since this is a displacement–time graph, the gradient at each point represents the velocity of the cyclist at that time. The sections of the graph are straight and so we have assumed that the velocity of the cyclist was uniform during each stage. (This is, of course, most unlikely in practice.)

As displacement is a vector quantity but appears on the graph merely as 'distance from A', we have assumed that the journey took place along a straight line.

We have also assumed that as each stage started and finished, the change of velocity was instantaneous.

b) Examining each section of the graph in turn, we find:

AB   The journey from A to B took 2 hours and the displacement was 10 km. The velocity of the cyclist, corresponding to the gradient of the graph, was

$$\text{Velocity} = \frac{10}{2} = 5 \, \text{km h}^{-1}$$

BC   The gradient of this part of the graph is zero. The cyclist had zero velocity and so must have had an hour's rest.

CD   The gradient of this section of the graph, and therefore the cyclist's velocity, was

$$\text{Velocity} = \frac{20 - 10}{6 - 3} = 3.33 \, \text{km h}^{-1}$$

Example 5 A car accelerates at a uniform rate of $3 \, \text{m s}^{-2}$ for 4 seconds. It then travels at constant velocity for 5 seconds before slowing down to stop at a constant rate of $2 \, \text{m s}^{-2}$. Draw a velocity–time graph for the motion and from it find the total distance travelled in the journey.

SOLUTION

The graph is shown on the right.

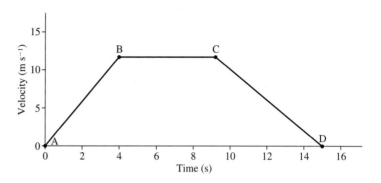

The stages of the journey are:

AB   A straight-line graph corresponding to uniform acceleration. The velocity after 4 seconds is $3 \times 4 = 12\,\mathrm{m\,s^{-1}}$, and so the coordinates of B are (4, 12).

BC   The car travels at a uniform velocity and so the gradient (the acceleration) is zero.

CD   The car slows at a constant rate of $2\,\mathrm{m\,s^{-2}}$. The section CD on the graph is, therefore, a straight line with a gradient of -2. So, the coordinates of D are (15, 0).

The displacement of the car for the whole journey is given by the area of the trapezium ABCD. So, we have

$$\text{Displacement} = \tfrac{1}{2}(5 + 15) \times 12 = 120\,\mathrm{m}$$

As the car is moving forwards throughout the journey (the velocity is positive throughout), this also represents the distance that the car travels.

Example 6 A bus starts on its route from town A to town B, 20 km away. When it reaches B, it stops for 15 minutes before returning to A. If the bus's speed is assumed to be a constant $40\,\mathrm{km\,h^{-1}}$, draw the displacement–time and velocity–time graphs for the journey and interpret the graphs.

SOLUTION

We model the journey assuming the road AB is a straight line, with A as the origin and the direction from A to B as positive.

Displacement–time graph   The journey from A to B is represented by the section PQ of the graph. The velocity is constant and so the graph is a straight line with gradient $40\,\mathrm{km\,h^{-1}}$.

The section QR represents the 15-minute rest stop at B.

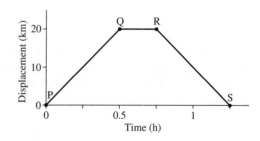

The return journey is represented by the section RS. This has a gradient of $-40\,\text{km}\,\text{h}^{-1}$.

Velocity–time graph The section PQ represents a journey at a uniform velocity of $40\,\text{km}\,\text{h}^{-1}$.

The section QR represents the rest stop when the velocity is zero.

The section RS represents the return journey, when the velocity is $-40\,\text{km}\,\text{h}^{-1}$.

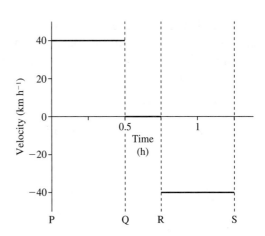

Interpretation The section PQ of the displacement–time graph has a positive gradient of 40, corresponding to the velocity of $40\,\text{km}\,\text{h}^{-1}$.

The section RS of the displacement–time graph has a negative gradient of -40, corresponding to the velocity of $-40\,\text{km}\,\text{h}^{-1}$ as the bus moves back towards the origin.

The area of the velocity–time graph above the time axis (from P to Q) represents the displacement $\overrightarrow{AB}$. So, we have

$$\text{Area} = 40 \times 0.5 = 20\,\text{km}$$

The area of the velocity–time graph below the time axis (from R to S) represents the displacement $\overrightarrow{BA}$. So, we have

$$\text{Area} = -40 \times 0.5 = -20\,\text{km}$$

The total (resultant) **displacement** from A is $20 + (-20) = 0\,\text{km}$

The total **distance travelled** is $|20| + |-20| = 40\,\text{km}$

Exercise 3B

1 A woman, walking a dog along a straight path, stops and releases the dog at a point A. She then continues to walk forward at a constant speed of $1.4\,\text{m}\,\text{s}^{-1}$. The dog runs $100\,\text{m}$ forward in 10 seconds, stops and sniffs for 10 seconds, then runs forward a further $50\,\text{m}$ in 20 seconds. It then spots another dog $100\,\text{m}$ the other side of A and runs back to join it at $5\,\text{m}\,\text{s}^{-1}$.

a) Draw a displacement–time graph for the dog and the woman.
b) From your graph, estimate where and when the dog passes the woman.
c) Find the average speed and the average velocity of the dog for the whole time period described.

2 A cyclist starts from town A to town B, 60 km away. In the first hour, he travels 20 km. He then rests for 15 minutes before completing the second stage to 20 km at the same speed. He then has a second 15-minute rest period before completing his final stage at the same speed as the previous two.

a) Draw a displacement–time graph representing the journey.

A second cyclist starts from B to cycle to A. She leaves B at the same time as the first cyclist leaves A and travels non-stop at $16 \, km \, h^{-1}$.

b) On the same axes, draw a displacement–time graph for the second cyclist, measuring the displacement of this cyclist **from A**.

c) At what time and where do the two cyclists pass each other?

3 A boat starts from rest at a point A and accelerates in a straight line at a constant rate of $0.5 \, m \, s^{-2}$ for 12 seconds. The propeller is then put into reverse, decelerating the boat to rest uniformly in 15 seconds. The propeller stays in reverse, making the boat accelerate back towards A. It reaches a speed of $6 \, m \, s^{-1}$ in 10 seconds and then continues at that speed.

a) Draw a velocity–time graph for the boat.

b) What was the furthest distance from A that the boat travelled?

c) What was the total time between the boat's leaving A and its returning to A?

4 The displacement–time graph shows the progress of a villager doing her weekly shopping on foot.

a) Describe the motion during each stage of the journey.

b) Draw the corresponding velocity–time graph.

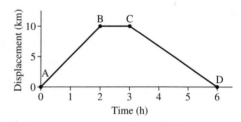

5 Chulchit drives to work along a straight road of length 8 km. His car accelerates and decelerates at $2.5 \, m \, s^{-2}$ and his preferred cruising speed is $90 \, km \, h^{-1}$.

a) Assuming that he has a clear run with no hold-ups, sketch a velocity–time graph of his journey and hence calculate his journey time.

Chulchit hears on the radio that there is to be a $36 \, km \, h^{-1}$ speed limit for 2 km because of road works somewhere along his route, but there is no information as to where it will be.

b) Investigate the effect of the positioning of the road works on Chulchit's journey time and find his maximum and minimum best-journey times while the speed limit is in effect.

Motion with uniform acceleration

When designing mathematical models of motion, we often assume that the acceleration of a body is uniform (constant). This simplifies the model and allows us to derive a useful set of equations.

Suppose that a body starts with velocity u when $t = 0$, and accelerates at a uniform rate a for a time t, when its velocity has become v. Call the displacement achieved during this motion s. Usually, t is in seconds, s in metres, u and v in $\mathrm{m\,s^{-1}}$, and a in $\mathrm{m\,s^{-2}}$, but the equations we derive will be true for any consistent set of units.

As the body has uniform acceleration, the velocity–time graph is a straight line (constant gradient), as shown.

The acceleration is the gradient of the graph, given by

$$\text{Gradient} = \frac{v - u}{t - 0}$$

So, we have

$$a = \frac{v - u}{t}$$

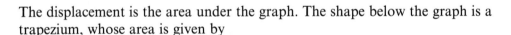

Rearranging this equation, we get

$$v = u + at \qquad [1]$$

The displacement is the area under the graph. The shape below the graph is a trapezium, whose area is given by

$$\text{Area} = \tfrac{1}{2}(u + v)t$$

So, we have

$$s = \tfrac{1}{2}(u + v)t \qquad [2]$$

From our graph, we have derived two important formulae:

$$v = u + at \qquad [1]$$

$$s = \tfrac{1}{2}(u + v)t \qquad [2]$$

Notice that each formula involves four of the five variables s, u, v, a and t. Formula [1] has no s, whilst formula [2] has no a.

We can combine these formulae to produce three further formulae covering the other possible sets of four variables. You should now do this for yourself in the exercise below.

Exercise 3C

1 Use formula [1] to eliminate v in formula [2] to give formula [3].

2 Use formula [1] to eliminate u in formula [2] to give formula [4].

3 Use formula [1] to eliminate t in formula [2] to give formula [5].

These five formulae can be used to solve problems provided that we can make the modelling assumption that acceleration is uniform.

Note When acceleration is **not uniform**, these formulae are **not valid** and should **not be used**.

When attempting to solve a problem, you should list those variables whose values you know, together with the variable you are trying to find. This will guide you to the correct formula to use. In practice, some of the formulae are more frequently used than others. Rearranged in order of usefulness they are:

$$v = u + at$$

$$s = ut + \tfrac{1}{2}at^2$$

$$v^2 = u^2 + 2as$$

$$s = \tfrac{1}{2}(u + v)t$$

$$s = vt - \tfrac{1}{2}at^2$$

Example 7 A car, travelling at $15\,\mathrm{m\,s^{-1}}$, accelerates at $3\,\mathrm{m\,s^{-2}}$ for 5 seconds. What then is its speed?

SOLUTION

We know $u = 15\,\mathrm{m\,s^{-1}}$, $a = 3\,\mathrm{m\,s^{-2}}$ and $t = 5\,\mathrm{s}$. We need to find v.

The formula containing these variables is $v = u + at$.

Substituting the known values, we get

$$v = 15 + 3 \times 5 = 30\,\mathrm{m\,s^{-1}}$$

Example 8 A car, travelling at $10\,\mathrm{m\,s^{-1}}$, accelerated at $4\,\mathrm{m\,s^{-2}}$ until it was travelling at $18\,\mathrm{m\,s^{-1}}$.

a) How far did it travel during its acceleration?
b) How long did it take to effect this change?

SOLUTION

a) We know $u = 10\,\mathrm{m\,s^{-1}}$, $v = 18\,\mathrm{m\,s^{-1}}$ and $a = 4\,\mathrm{m\,s^{-2}}$. We need to find s.

The formula containing these variables is $v^2 = u^2 + 2as$.

Substituting the known values, we get

$$18^2 = 10^2 + 2 \times 4 \times s$$
$$\Rightarrow \quad s = 28$$

So, the distance travelled during acceleration is 28 m.

b) With u, v and a as in part **a**, we need to find t.

The formula containing these variables is $v = u + at$.

Substituting the known values, we get

$$18 = 10 + 4 \times t$$
$$\Rightarrow \quad t = 2$$

So, the change took 2 seconds to effect.

Example 9 A particle passes a point O with a velocity of $10\,\mathrm{m\,s^{-1}}$ and an acceleration of $-4\,\mathrm{m\,s^{-2}}$.

a) Find the velocity of the particle after 1 s, 2 s, 3 s, 4 s, 5 s.
b) Find the displacement of the particle at the same times.
c) Draw a velocity–time graph and a displacement–time graph for the motion during the first 5 seconds.

SOLUTION

a) Using $v = u + at$, where $u = 10\,\mathrm{m\,s^{-1}}$ and $a = -4\,\mathrm{m\,s^{-2}}$, we obtain

$$v = 10 + (-4) \times t$$
$$\Rightarrow \quad v = 10 - 4t$$

Substituting the values of t into the above, we have

t	1	2	3	4	5
v	6	2	-2	-6	-10

b) Using $s = ut + \frac{1}{2}at^2$, we get

$$s = 10t + \frac{1}{2} \times (-4) \times t^2$$
$$\Rightarrow \quad s = 10t - 2t^2$$

Substituting the values of t into the above, we have

t	1	2	3	4	5
s	8	12	12	8	0

c) The velocity–time and displacement–time graphs, plotted from the values in the two tables, are given below.

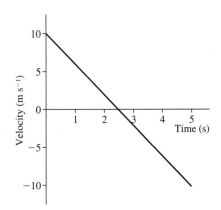
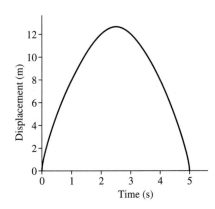

Note The terms **deceleration** and **retardation** are sometimes used to describe a **negative** acceleration. So, for example, the negative acceleration of $-4\,\mathrm{m\,s^{-2}}$ in Example 9 would be described as a deceleration or retardation of $4\,\mathrm{m\,s^{-2}}$, omitting the minus sign.

Exercise 3D

1 A train leaves a station and accelerates uniformly at a rate of $3\,\mathrm{m\,s^{-2}}$ for 30 seconds.

a) How far does it travel during this period?

b) How fast is it travelling at the end of the period?

2 A stone is dropped from the top of a tower. It takes 5 seconds to reach the ground, by which time it is moving at $50\,\mathrm{m\,s^{-1}}$.

a) What is its acceleration?

b) How high is the tower?

3 A body starts from rest with uniform acceleration and in 10 seconds has moved a distance of 150 m.

a) What is its acceleration?

b) How fast is it moving at the end of this period?

4 A train leaves a station from rest with a constant acceleration of $0.2\,\mathrm{m\,s^{-2}}$. It reaches its maximum speed after 2 minutes and maintains this speed for a further 4 minutes, when it slows down with an acceleration of $-1.5\,\mathrm{m\,s^{-2}}$. How far apart are the two stations?

5 A train accelerates uniformly from rest for 1 minute when its velocity is $30\,\text{km}\,\text{h}^{-1}$. It maintains this speed until it is 500 m from the next station when it slows down to a stop. Find the accelerations during the first and last phases of the journey.

6 A car crosses a speed hump with a velocity of $4\,\text{m}\,\text{s}^{-1}$. It then accelerates at a rate of $2.5\,\text{m}\,\text{s}^{-2}$ to a speed of $9\,\text{m}\,\text{s}^{-1}$ when the driver applies the brakes, causing an acceleration of $-3\,\text{m}\,\text{s}^{-2}$, reducing the speed of the car to $4\,\text{m}\,\text{s}^{-1}$ to cross the next hump.

a) How far apart are the humps?
b) How long does the car take to travel from one hump to the next?

7 A lift ascends from rest with an acceleration of $0.5\,\text{m}\,\text{s}^{-2}$ before slowing with an acceleration of $-0.75\,\text{m}\,\text{s}^{-2}$ for the next stop. If the total journey time is 10 seconds, what is the distance between the two stops?

8 Two particles, P and Q, are moving along the same line in the same direction. P is 10 m behind Q. P starts from rest and has an acceleration of $2\,\text{m}\,\text{s}^{-2}$. Q has a uniform velocity of $3\,\text{m}\,\text{s}^{-1}$.

a) How long does P take to catch Q?
b) How far has P travelled in doing so?

Free fall under gravity

A body falling through the air is subject to gravity and to air resistance. Air resistance varies, depending on the shape, size and speed of the falling body, and, if we choose to include it in our model, the solution of the problem is complex. However, for a small body falling at a relatively slow speed, the effect of air resistance is quite small, and so we often make the following modelling assumptions:

- The body is a point mass.
- Air resistance can be ignored.
- The motion of the body is along a vertical line.
- The acceleration due to gravity is constant.

With these assumptions, the only force acting on the body is its weight, and its acceleration is that due to gravity. This is denoted by g and is usually taken as equal to $9.8\,\text{m}\,\text{s}^{-2}$. (Some examination boards use $g = 10\,\text{m}\,\text{s}^{-2}$ in some or all of their questions, and usually tell you this either in the candidates' instructions or in the questions. In the absence of such information, you are advised to use $g = 9.8\,\text{m}\,\text{s}^{-2}$.)

With the above modelling assumptions, the formulae for motion with uniform acceleration derived on page 42 are valid and can be used to solve all free-fall problems.

Example 10 A stone is dropped vertically from the top of a tower, 20 m high. How long does it take to reach the ground and what is its velocity as it hits the ground? (Take $g = 10\,\mathrm{m\,s^{-2}}$.)

SOLUTION

Taking the origin as the top of the tower, and downwards as the positive direction, we have $u = 0\,\mathrm{m\,s^{-1}}$, $a = g = 10\,\mathrm{m\,s^{-2}}$ and $s = 20\,\mathrm{m}$.

To find the time, t, to reach the ground, we use $s = ut + \frac{1}{2}at^2$, which gives

$$20 = \frac{1}{2} \times 10 \times t^2$$
$$\Rightarrow \quad t^2 = 4$$
$$\Rightarrow \quad t = 2 \quad \text{or} \quad -2$$

In the context of the question, the negative value is inappropriate, and so $t = 2\,\mathrm{s}$.

To find the velocity, v, with which the stone hits the ground, we use $v^2 = u^2 + 2as$, which gives

$$v^2 = 2 \times 10 \times 20 = 400$$
$$\Rightarrow \quad v = 20 \quad \text{or} \quad -20$$

In the context of the question, the negative value is inappropriate, and so $v = 20\,\mathrm{m\,s^{-1}}$.

Note Having found the value of t in the first part of Example 10, we could have used $v = u + at$ to find the value of v in the second part.

Example 11 A ball was thrown vertically upwards from ground level with a velocity of $28\,\mathrm{m\,s^{-1}}$.

a) What was its maximum height above the ground?
b) How long did it take to return to the ground?

SOLUTION

Taking the point from which the ball was thrown as the origin, and upwards as the positive direction, we have $u = 28\,\mathrm{m\,s^{-1}}$ and $a = -g = -9.8\,\mathrm{m\,s^{-2}}$.

a) At the top of the ball's flight $v = 0\,\mathrm{m\,s^{-1}}$ and $s = h\,\mathrm{m}$, the maximum height.

To find h, we use $v^2 = u^2 + 2as$, which gives

$$0 = 28^2 + 2 \times (-9.8) \times h$$
$$\Rightarrow \quad h = 40$$

So, the maximum height reached is 40 m.

b) When the ball returns to ground level, $s = 0$ m.

To find t, the time taken to return to ground level, we use $s = ut + \frac{1}{2}at^2$, which gives

$$0 = 28t + \frac{1}{2} \times (-9.8) \times t^2$$

$$\Rightarrow \quad 0 = t(28 - 4.9t)$$

$$\Rightarrow \quad t = 0 \quad \text{or} \quad 5.71$$

In the context of the question, the solution $t = 0$ represents the time at which the ball was thrown, and the solution $t = 5.71$ represents the time at which the ball returned to ground level. So, the ball took 5.71 s to return to ground level.

Example 12 A ball was thrown vertically upwards from ground level with a velocity of $20\,\text{m s}^{-1}$. A boy was leaning out of a second-floor window, 8 m above the ground, trying to catch it. When the ball was on its way up, he missed it but managed to catch it on its way down.

a) What was the time of flight of the ball?
b) How fast was it travelling when it was caught?

Take $g = 10\,\text{m s}^{-2}$.

SOLUTION

Assuming upwards is positive, and taking the origin to be at ground level, we have $u = 20\,\text{m s}^{-1}$ and $a = -g = -10\,\text{m s}^{-2}$.

a) When the ball was caught, $s = 8$ m.

To find the time of flight, we use $s = ut + \frac{1}{2}at^2$, which gives

$$8 = 20t + \frac{1}{2} \times (-10) \times t^2$$

$$\Rightarrow \quad 0 = 5t^2 - 20t + 8$$

$$\Rightarrow \quad t = 0.451 \quad \text{or} \quad 3.549 \quad \text{(using the quadratic equation formula)}$$

In the context of the question, the solution $t = 0.451$ represents the time that the ball took to reach the window on the way up, and the solution $t = 3.549$ represents the time that the ball took to reach the window on the way down.

So, the time of flight of the ball is 3.549 s.

b) To find the velocity of the ball as it was caught, we use $v = u + at$, which gives

$$v = 20 + (-10) \times 3.549 = -15.49$$

. In the context of the question, the negative velocity indicates that the
. ball was travelling downwards, so it was travelling at a speed of
. $15.49 \, \mathrm{m \, s^{-1}}$ when it was caught.

Note We could have used $v^2 = u^2 + 2as$ to calculate the speed. The solutions
to this equation would have been $\pm 15.49 \, \mathrm{m \, s^{-1}}$. The positive solution would
represent the velocity on the way up and the negative solution would represent
the velocity on the way down.

Exercise 3E

1 A stone is dropped from the top of a cliff 50 m high.

    a) How long does it take to reach the beach below?
    b) What is its velocity when it hits the beach?

2 A ball is thrown vertically upwards with a speed of $15 \, \mathrm{m \, s^{-1}}$.

    a) What will be the greatest height reached by the ball.
    b) How long does it take to reach maximum height?
    c) How long does it take to reach ground level again?

3 A ball was thrown vertically upwards. It just touched a cable 20 m above the ground.

    a) What was the initial speed of the ball?
    b) How long did the ball take to reach ground level again?
    c) What was the velocity of the ball when it hit the ground?

4 A stone was thrown vertically upwards with a speed of $5 \, \mathrm{m \, s^{-1}}$ from the top of a cliff, 60 m
high, so that it fell to the beach below.

    a) What was the greatest height reached by the stone?
    b) What was the velocity of the stone when it hit the beach?
    c) How long did it take for the stone to hit the beach?

5 A stone was thrown vertically upwards with a speed of $10 \, \mathrm{m \, s^{-1}}$. One second later, another
stone was thrown vertically upwards from the same point and with the same speed.

    a) How high were the stones when they met?
    b) How long after the first stone was thrown did the stones meet?

6 A boy dropped a stone from the top of a multistorey car park. At the same time, his friend
threw a second stone vertically upwards from the ground below, with a speed of $30 \, \mathrm{m \, s^{-1}}$. The
two stones met 1.5 s later. How high was the top of the car park?

7 A body falls from rest from the top of a tower. During the last second of its motion it falls $\frac{7}{16}$
of the whole distance. Show that the time taken for the descent is independent of the value of
g, and find the height of the tower in terms of g.

8 A stone falls past a window, 2.5 m high, in 0.5 s. Taking $g = 10 \, \text{m s}^{-2}$, find the height from which the stone fell.

9 An object is thrown vertically downwards with speed V. During the sixth second of its motion, it travels a distance h. Find V in terms of h and g.

10 An object is projected vertically upwards with a velocity of $u \, \text{m s}^{-1}$, and after t s a second object is projected upwards from the same point and with the same velocity. Find, in terms of u, t and g, the time which elapses between the second object's projection and the collision between the objects.

4 Force

Don't fight forces; use them.
RICHARD BUCKMINSTER FULLER

Everyone has an intuitive idea of what is meant by force. When you cycle, you are aware of the force of air resistance. When you lift an object, you are aware of the force of gravity. When you water-ski, you are subject to the tension force of the towline. When you kneel down for any length of time, you suffer from the reaction force of the floor pressing on your knees.

Force is a vector quantity, because its effect is dependent on its magnitude and its direction. Additionally, the effect of a force varies depending on the point at which it is applied – its **line of action**. For example, attaching a crane hook to the end of a girder or to its middle gives different results on lifting. In the first case, the force would have a rotational effect. In this chapter, we model objects as masses concentrated at a single point, so rotational effects are not involved. Other situations are discussed on pages 183–211.

As far as the motion of an object is concerned, the important thing is the total effect – the **resultant** – of the forces acting on it. In this chapter, we consider only those situations where the resultant force is zero and so does not affect the motion of the object.

The SI unit of force is the **newton** (N). This is defined as the force needed to accelerate an object of mass 1 kg with an acceleration of $1\,\mathrm{m\,s^{-2}}$. (This is explored more fully on page 76.)

The force of gravity

It is important that you are clear about the difference between **mass** and **weight**.

The mass of an object, measured in kilograms, depends only on the amount of matter forming the object. (A kilogram was originally defined as the mass of 1 litre of pure water.) The mass of an object is the **same** wherever it is placed in the universe.

When an object is placed near another object, they are each subject to a force arising from their gravitational attraction for each other. The size of this force depends on their masses and the distance between their centres of mass. You will be aware that an object at or near the surface of the Earth is subject to the force of gravity acting downwards (the Earth is acted upon by an equal force acting upwards). This force is the weight of the object, measured in newtons. The weight of an object would be **different** if we placed the object on the Moon, although its mass would be unchanged.

In Newton's model of the universe, the force in newtons between two objects of masses m_1 and m_2 (in kilograms) separated by a distance d metres is given by

$$F = \frac{Gm_1m_2}{d^2}$$

G is called the **gravitational constant**, which is $6.67 \times 10^{-11}\,\mathrm{N\,m^2\,kg^{-2}}$.

The mass of the Earth is $5.98 \times 10^{24}\,\mathrm{kg}$. For a small object of mass m kg near the surface of the Earth, the distance d between its centre and the Earth's centre is the radius of the Earth, which is approximately $6.37 \times 10^6\,\mathrm{m}$. This gives the force acting on the object as

$$F = \frac{(6.67 \times 10^{-11}) \times (5.98 \times 10^{24})\,m}{(6.37 \times 10^6)^2}$$

$$= 9.8m \text{ newtons}$$

This means that an object of mass m kg near the surface of the Earth is subject to a constant downward force – its weight – of $9.8\,m$ N. As we will see on page 80, the 9.8 is the **acceleration due to gravity** and is represented by g:

$$g \approx 9.8\,\mathrm{m\,s^{-2}}$$

For most purposes, we take g to be constant everywhere on the Earth's surface. This assumes that the Earth can be thought of as a sphere which is uniformly dense or is at least formed of concentric shells of uniform density.

Types of force

In addition to weight, there are other forces which may act on objects. The four considered here are **tension**, **thrust**, **normal reaction** and **friction**.

Tension forces

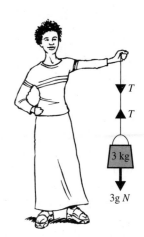

When a string is pulled, it will exert a tension force opposing the pull. If we can assume that the string is **light** (that is, its mass is negligible compared with the rest of the system), the tension force will be the **same** throughout the string.

The woman in the diagram is holding an object of mass 3 kg suspended on a light string.

The tension, T, exerts an upward force on the object and an equal downward force on the woman's hand.

When the object is stationary, the tension must balance the weight, so

$$T = 3g$$

Similarly, the woman's hand is stationary, so she must be exerting an upward force of $3g$ N to balance the downward pull of the tension in the string.

The tension in a light string passing over a pulley can still be taken as constant throughout its length provided the pulley is **smooth**. That is, any friction forces are so small as to be negligible.

Example 1 Two boxes, A and B, each of mass 40 kg, are connected by a light rope. A second rope is attached to B and passes over a smooth pulley. The other end of this rope is then fixed to the ground at C. What forces act on **a)** box A, **b)** box B and **c)** the ground at C?

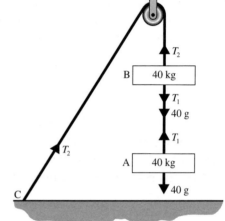

SOLUTION

a) The forces on box A are its weight, $40g$ N, acting downwards, and the tension T_1, acting upwards. The box is stationary, so

$$T_1 = 40g \text{ N}$$

b) The forces on box B are its weight, $40g$ N, the tension T_1 acting downwards, and the tension T_2 acting upwards. The box is stationary, so

$$T_2 = T_1 + 40g$$
$$= 80g \text{ N}$$

c) The pulley is smooth, so the tension is the same throughout the rope joining box B to C. This means that the rope is pulling at the ground with a force of $80g$ N.

Forces in rods: thrust

In the example where the woman was supporting the 3 kg mass, the string could have been replaced by a light rod without altering the forces involved. Rods can, therefore, also be under tension. However, unlike strings, it makes sense to push the end of a rod. The rod will exert an opposing push, called a **thrust**. Like tension, thrust is the **same** throughout a light rod.

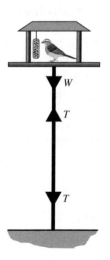

The diagram shows a bird table supported by a light, vertical rod.

The rod exerts an upward thrust force, T, on the table, which is equal to the downward force supplied by the weight, W, of the table and the bird.

At the bottom of the rod, there is an equal downward thrust force exerted on the ground by the rod. This will be countered by an equal force exerted by the ground on the rod.

Contact forces

There are two types of force which occur as a result of contact between objects: **normal reaction** and **friction**.

Normal reaction

Consider a cup resting on a table. As it is in equilibrium, the downward force of its weight, W, must be balanced by an upward force, R, exerted by the table.

Similarly, if a trap door of weight W is open and resting on a support, as shown, the support must provide a force R to help keep the trap door in equilibrium. This force is at right angles to the trap door. (There will also be a force at the hinge, but this is not a contact force as the door is fastened to the hinge.)

In each case, the reaction force is at right angles to (normal to) the plane of contact, and so such a force is called a normal reaction.

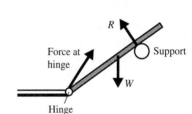

Friction

When you try to drag an object along the ground and start by pulling gently on the rope, the object will not move. The force you are exerting is being balanced by the force of friction. When you gradually increase the pulling force, P, the friction force, F, increases to match it until it reaches its maximum. If you pull harder than this, the object will move, although there will still be a friction force resisting its motion.

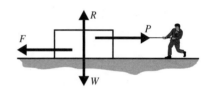

Friction **always** acts in a direction opposite to that in which the object is moving or tending to move.

The maximum friction depends on the nature of the surfaces in contact and the normal reaction between these surfaces. (Modelling friction is covered on pages 164–81.) In some cases, the friction force is small enough to be ignored. This is said to be a **smooth** contact.

Resistive forces also occur when an object moves through the air or through a liquid. In some situations, these forces are small enough for us to ignore them in our model. In other situations, we must take them into account. For example, in modelling the motion of a ball-bearing falling a short distance in air, we could probably safely ignore air resistance, but for a sheet of paper falling through air or a ball-bearing falling in a tank of oil, we would have to take account of the resistance. Like contact friction, the resistance of air or a liquid always acts to oppose the motion of the object.

Drawing diagrams

When solving problems in mechanics, it is vital that you draw clear, careful diagrams of a good size, and mark in all the forces involved in your model, together with relevant lengths and angles. This will help you to analyse the problem and to explain clearly the steps in your solution.

Example 2 A large box of mass m kg is being towed up a rough 30° slope using a rope at 20° to the slope. Draw a diagram to show the forces acting on the box.

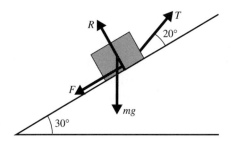

SOLUTION

The forces acting on the box are its weight, mg, the tension, T, in the rope, the normal reaction, R, and the friction, F, which acts down the slope to oppose the motion.

Example 3 The same box is now allowed to slide down the slope controlled by the rope. Draw the forces in this situation.

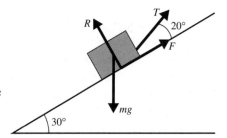

SOLUTION

The only difference in this situation is that the box is moving down the slope and so the friction force acts up the slope to oppose it.

Example 4 A car of mass M kg is towing a trailer of mass m kg on a light rigid towbar. There are resistances to motion F_C and F_T on the car and trailer respectively. Draw diagrams to show the forces acting on the car and on the trailer

a) when the car exerts a driving force P N
b) when the car exerts a braking force B N.

SOLUTION

a)

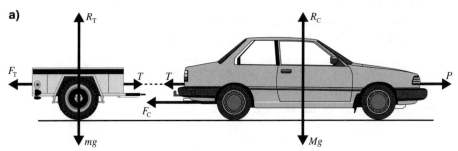

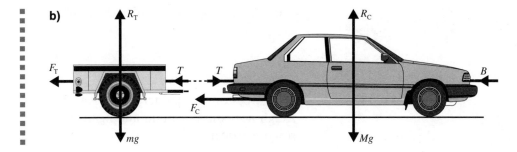

Notes

- In case **a**, the towbar is in tension. In case **b**, it is under thrust.

- We usually show the tension/thrust forces only when we consider the car and trailer separately. But when we consider the car and trailer as a single system, the tension/thrust forces are **internal forces**, which are not shown.

- Strictly, the upward reaction on the car consists of four reactions, one at each wheel. But we usually show them as a single combined reaction unless we are examining the forces on the wheels.

- Strictly, the forward driving force, P, is actually the friction between the driven wheels and the ground, which acts in a forward direction to prevent the wheels from spinning. However, for most purposes, we show this as a generalised forward driving force.

Exercise 4A

1 Copy each of the following diagrams and mark in the forces indicated.

The forces acting on this brick which is sliding down a rough inclined plank.

The forces on this shelf, supported symmetrically on two brackets.

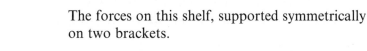

The forces on the shelf, supported by one bracket and an inclined wire.

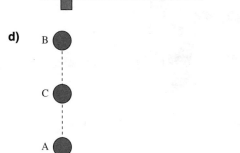

The forces on this ball, which has been thrown vertically, at A (on the way up), at B (top of its flight) and at C (on the way down).

e)

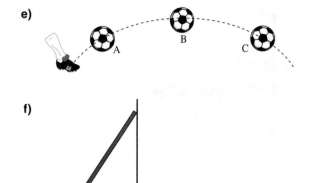

The forces on this football at A, B and C.

f)

The forces on this ladder which is resting on rough horizontal ground and against a rough vertical wall.

2 A block A, of mass M_A, rests on a rough horizontal table. It is connected to a second block B, of mass M_B, by a light string. The string passes over a smooth pulley at the edge of the table and B hangs suspended.

a) Draw a diagram to show the forces acting on each of the blocks.

b) Assuming that the system remains at rest, state the tension in the string and the friction force acting on the block A.

3 A man of mass m stands in a lift of mass M which is supported by a cable. Draw separate diagrams to show the forces acting **a)** on the lift and **b)** on the man.

4 A uniform ladder of mass m and length $4a$ rests with one end on rough horizontal ground. The ladder leans against a rough garden wall so that one quarter of its length protrudes above the wall. Draw a diagram to show the forces acting on the ladder.

5 A glass rod rests in a smooth hemispherical bowl so that part of the rod extends beyond the rim of the bowl. Draw a diagram to show the forces acting on the rod.

6 A small object of mass m is suspended by a light string, the other end of which is tied to a fixed support A. The object moves in a horizontal circle below A (this arrangement is known as a conical pendulum). Draw a diagram to show the forces acting on the object.

7 In this diagram, the pulleys are smooth, the strings light and inextensible.

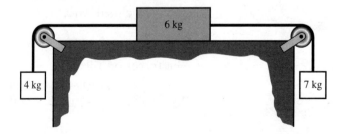

Draw a diagram to show the forces on the 6 kg mass, and calculate the friction force acting on it if the system is at rest.

8 The diagram shows a light rod AB
hinged to a vertical wall at A, and a
light string BC attached to the rod at B
and the wall at C. A mass of 2 kg is
suspended from B. Draw a diagram to
show the forces acting at the point B.

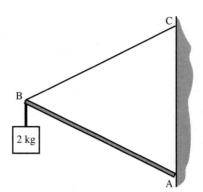

Forces at a point: modelling by vectors

Force is a vector quantity, having both magnitude and direction. So, vector
techniques can be used to study systems of forces acting at a single point.

Example 5 The diagram shows
four horizontal dog leads OA, OB,
OC and OD, each tied to the same
post at O. The dogs are pulling on
the leads with forces as shown.
Find their combined effect.

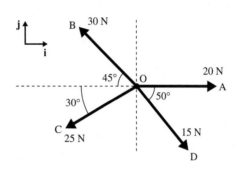

SOLUTION

The four forces are equivalent to a single force, the **resultant**. The
magnitude and direction of the resultant can be found from a scale
drawing or by calculation.

Scale drawing

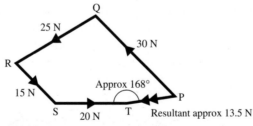

Scale: 1mm ≡ 1 N

The forces are represented by the displacements $\overrightarrow{PQ}$, $\overrightarrow{QR}$, $\overrightarrow{RS}$ and $\overrightarrow{ST}$ in
the drawing. The resultant force is then represented by the displacement
$\overrightarrow{PT}$. You should try drawing it to scale. By measurement you should find
that the resultant force has a magnitude of approximately 13.5 N and acts
at approximately −168° to the direction OA in the first diagram.

Note A small scale is used here because of the limitation on space. In practice, you should use the **largest convenient scale** to maximise the accuracy of your results.

Calculation Taking the vectors **i** and **j** to be in the directions shown in the first diagram, we can express each of the forces in component form and add them directly.

If **R** is the resultant force, we have

$$\mathbf{R} = (20\mathbf{i}) + (-30\cos 45°\,\mathbf{i} + 30\sin 45°\,\mathbf{j}) + (-25\cos 30°\,\mathbf{i} - 25\sin 30°\,\mathbf{j})$$

$$+ (15\cos 50°\,\mathbf{i} - 15\sin 50°\,\mathbf{j})$$

$$= -13.22\mathbf{i} - 2.78\mathbf{j}$$

The magnitude of **R** is then

$$|\mathbf{R}| = \sqrt{13.22^2 + 2.78^2} = 13.51\,\text{N}$$

and its direction is given by θ, where

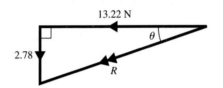

$$\tan\theta = \frac{2.78}{13.22} \quad \Rightarrow \quad \theta = 11.86°$$

Forces in equilibrium

When a system of forces acting at a single point has a resultant of zero, the forces are said to be in **equilibrium**. The system of forces in Example 5 would be in equilibrium if we added a fifth force with magnitude 13.51 N acting at 11.86° to OA. This additional force is sometimes referred to as the **equilibrant** for the system.

When a system of forces is in equilibrium, the polygon produced when making a scale drawing is closed, because the total effect of the displacements would be to return to the starting point. This is referred to as the **polygon of forces**. Although it is rare to use scale drawing to solve a problem, knowing that the forces can be represented by a closed polygon can be useful, especially in cases involving three forces acting at a point, when we have a **triangle of forces**.

Consider an object of mass 12 kg suspended by two light, inextensible strings AB and BC. The task is to calculate the tensions in the two strings.

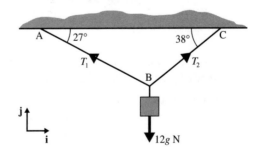

Method 1: triangle of forces

The three forces are in equilibrium and so they can be represented by the sides of a triangle, as shown.

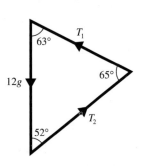

Using the sine rule, we get

$$\frac{T_1}{\sin 52°} = \frac{T_2}{\sin 63°} = \frac{12g}{\sin 65°}$$

which gives

$$T_1 = \frac{12g \sin 52°}{\sin 65°} = 102\,\text{N}$$

$$T_2 = \frac{12g \sin 63°}{\sin 65°} = 116\,\text{N} \quad \text{(to the nearest 1 N)}$$

Method 2: resolving forces

We express the forces in component form, taking the **i**- and **j**-directions horizontally and vertically, as shown in the diagram at the foot of page 58. The resultant force is zero, so we have

$$(-T_1 \cos 27°\,\mathbf{i} + T_1 \sin 27°\,\mathbf{j}) + (T_2 \cos 38°\,\mathbf{i} + T_2 \sin 38°\,\mathbf{j}) + (-12g\,\mathbf{j}) = \mathbf{0}$$

$$\Rightarrow \quad (-T_1 \cos 27° + T_2 \cos 38°)\mathbf{i} + (T_1 \sin 27° + T_2 \sin 38° - 12g)\mathbf{j} = \mathbf{0}$$

$$\Rightarrow \quad -T_1 \cos 27° + T_2 \cos 38° = 0 \quad \text{and} \quad T_1 \sin 27° + T_2 \sin 38° - 12g = 0$$

We now write the above as simultaneous equations:

$$-0.891\,T_1 + 0.788\,T_2 = 0 \qquad [1]$$

$$0.454\,T_1 + 0.616\,T_2 = 117.6 \qquad [2]$$

Solving [1] and [2] gives $T_1 = 102\,\text{N}$ and $T_2 = 116\,\text{N}$ (to the nearest 1 N).

In this method, we often omit the vector equation and just write down the two component equations directly. We need to indicate where the equations come from, so the solution would look like this:

Resolving horizontally gives $\quad -T_1 \cos 27° + T_2 \cos 38° = 0$

Resolving vertically gives $\quad T_1 \sin 27° + T_2 \sin 38° - 12g = 0$

followed by solving the simultaneous equations.

Lami's theorem

In Method 1, it is possible to write down equations equivalent to those obtained from the sine rule without drawing the triangle.

Using the fact that the $\sin\theta = \sin(180° - \theta)$,

$$\frac{T_1}{\sin 52°} = \frac{T_2}{\sin 63°} = \frac{12g}{\sin 65°}$$

is equivalent to

$$\frac{T_1}{\sin 128°} = \frac{T_2}{\sin 117°} = \frac{12g}{\sin 115°}$$

and these angles are those opposite the appropriate forces, as shown in the diagram.

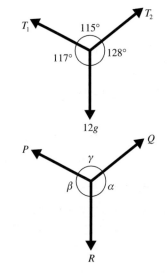

This rule can be stated in general:

For any set of three concurrent forces P, Q, R in equilibrium as shown

$$\frac{P}{\sin \alpha} = \frac{Q}{\sin \beta} = \frac{R}{\sin \gamma}$$

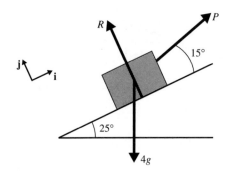

This is known as **Lami's theorem**.

Example 6 A block of mass 4 kg lies at rest on a smooth plane which is inclined at $25°$ to the horizontal. It is kept in place by a light string which is angled at $15°$ to the plane. Find the tension, P, in the string and the normal reaction, R, of the plane on the block.

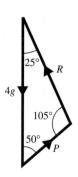

SOLUTION

There are three ways to solve this problem: constructing a triangle of forces, applying Lami's theorem or resolving the forces.

Triangle of forces The forces are in equilibrium and can therefore be represented by the sides of a triangle, as shown.

Using the sine rule, we obtain

$$\frac{P}{\sin 25°} = \frac{R}{\sin 50°} = \frac{4g}{\sin 105°}$$

which gives

$$P = \frac{4g \sin 25°}{\sin 105°} = 17.15\,\text{N}$$

$$R = \frac{4g \sin 50°}{\sin 105°} = 31.09\,\text{N}$$

Lami's theorem The angles between the forces are as shown. Then, according to Lami's theorem, we have

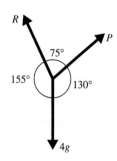

$$\frac{P}{\sin 155°} = \frac{R}{\sin 130°} = \frac{4g}{\sin 75°}$$

which gives

$$P = \frac{4g\sin 115°}{\sin 75°} = 17.15\,\text{N}$$

$$R = \frac{4g\sin 130°}{\sin 75°} = 31.09\,\text{N}$$

Resolving the forces Take the **i**- and **j**-directions parallel to and perpendicular to the plane, as shown in the diagram on page 60.

Resolving in the **i**-direction gives $\quad P\cos 15° - 4g\sin 25° = 0$ $\qquad$ [1]

Resolving in the **j**-direction gives $\quad P\sin 15° + R - 4g\cos 25° = 0$ $\qquad$ [2]

From [1], we obtain

$$P = \frac{4g\sin 25°}{\cos 15°} = 17.15\,\text{N}$$

Substituting $P = 17.15\,\text{N}$ in [2], we have

$$17.15\sin 15° + R - 4g\cos 25° = 0$$
$$\Rightarrow \quad R = 31.09\,\text{N}$$

Example 7 The diagram shows three forces in equilibrium. Find the value of P and the size of angle θ.

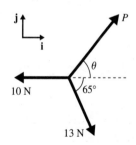

SOLUTION

The solution of this problem, using either a triangle of forces or Lami's theorem, involves the use of trigonometric formulae which you may not yet have met.

Using resolution of forces makes use of the formulae

$$\sin^2\theta + \cos^2\theta = 1 \quad \text{and} \quad \frac{\sin\theta}{\cos\theta} = \tan\theta$$

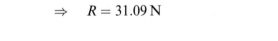

Take the **i**- and **j**-directions as shown in the diagram.

Resolving in the **i**-direction, we get

$$P\cos\theta + 13\cos 65° - 10 = 0 \quad \Rightarrow \quad P\cos\theta = 10 - 13\cos 65° \qquad [1]$$

Resolving in the **j**-direction, we get

$$P\sin\theta - 13\sin 65° = 0 \quad \Rightarrow \quad P\sin\theta = 13\sin 65° \qquad [2]$$

Squaring [1] and [2] and adding the results, we have

$$P^2(\cos^2\theta + \sin^2\theta) = (10 - 13\cos 65°)^2 + (13\sin 65°)^2$$

$$\Rightarrow \quad P^2 = 159.12$$

$$\Rightarrow \quad P = 12.6\,\text{N}$$

And dividing [2] by [1], we get

$$\frac{\sin\theta}{\cos\theta} = \frac{13\sin 65°}{10 - 13\cos 65°}$$

$$\Rightarrow \quad \tan\theta = 2.615$$

$$\Rightarrow \quad \theta = 69.1°$$

More than three forces

Although any set of concurrent forces in equilibrium can be represented by the sides of a closed polygon, when four or more forces are involved the only practical method of solution is to resolve the forces.

Example 8 An object A of weight W is suspended by light strings AB and AC attached to points B and C. BC is horizontal and of length 5 m. AB and AC are 4 m and 3 m respectively. A horizontal force, P, is applied to the object so that the tension in AB is twice that in AC. Find, in terms of W, the value of P and the tension in AC.

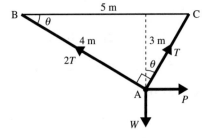

SOLUTION

As ABC is a 3-4-5 triangle, $B\widehat{A}C$ is 90°. It follows that $\cos\theta = \frac{4}{5}$ and $\sin\theta = \frac{3}{5}$.

For the forces acting on the object:

Resolving vertically gives

$$T\cos\theta + 2T\sin\theta - W = 0$$

$$\Rightarrow \quad \tfrac{4}{5}T + \tfrac{6}{5}T = W$$

$$\Rightarrow \quad T = \tfrac{1}{2}W$$

Resolving horizontally gives

$$P + T\sin\theta - 2T\cos\theta = 0$$

$$\Rightarrow \quad P = \tfrac{8}{5}T - \tfrac{3}{5}T = T$$

$$\Rightarrow \quad P = \tfrac{1}{2}W$$

Example 9 A smooth ring of mass 3 kg is threaded on a light string 64 cm long. The ends of the string are attached to points A and B on the same level, where AB is 48 cm. A force, P, is applied to the ring so that it rests vertically below B. Find the value of P and the tension in the string.

SOLUTION

Since the ring is smooth, the tension is the same throughout the length of the string.

Let BC be x cm, so that AC is $(64 - x)$ cm.

By Pythagoras' theorem:

$$x^2 + 48^2 = (64 - x)^2$$

$$\Rightarrow \quad x = 14$$

So, AC $= 50$ cm and BC $= 14$ cm, which give

$$\sin \theta = \tfrac{7}{25} \quad \text{and} \quad \cos \theta = \tfrac{24}{25}$$

For the forces acting on the ring:

Resolving vertically gives $\qquad T + T\sin\theta - 3g = 0$

$$\Rightarrow \quad \frac{32}{25}T = 3g$$

$$\Rightarrow \quad T = 22.97 \, \text{N}$$

Resolving horizontally gives $\qquad P - T\cos\theta = 0$

$$\Rightarrow \quad P = \frac{24}{25}T = 22.05 \, \text{N}$$

Exercise 4B

1 For each of the following systems of forces, find the resultant and state the force which would have to be added to the system to maintain equilibrium.

a)

b)

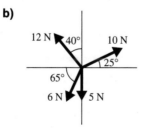

c)

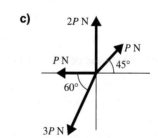

2 Given that each of the following systems of forces is in equilibrium, find the unknown forces *P* and *Q*.

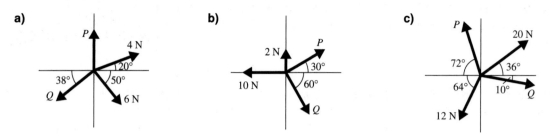

a)

b)

c)

3 ABCD is a rectangle with AB and AD of lengths 2*a* and *a* respectively. Forces of 2 N, 3 N and 3√5 N act along AB, AD and AC respectively, in the directions indicated by the order of the letters. Find the force which would need to be applied at A to maintain equilibrium.

4 Each of the following systems of forces is in equilibrium. For each one, draw a triangle of forces and from it calculate the unknown forces and angles.

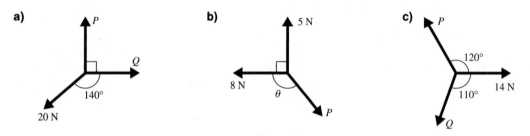

a)

b)

c)

5 Rework the problems in Question **4** using resolution of forces.

6 For each of the following systems of forces in equilibrium, use Lami's theorem to calculate the unknown forces.

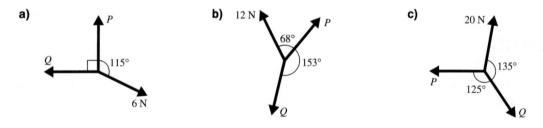

a)

b)

c)

7 A particle of mass 4 kg is suspended from a point A on a vertical wall by means of a light, inextensible string of length 130 cm.

a) A horizontal force, *P*, is applied to the particle so that it is held in equilibrium a distance 50 cm from the wall. Find the value of *P* and the tension in the string.

b) By drawing a triangle of forces, or otherwise, find the magnitude and direction of the minimum force which would hold the particle in this position, and the tension in the string which would result.

8 The diagram shows a cylinder of mass 8 kg lying at rest on two smooth planes inclined at angles of 40° and 50° to the horizontal. Calculate the reaction forces exerted by the planes on the cylinder.

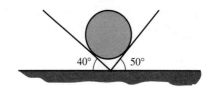

9 Two hooks A and B are fixed to a ceiling, where AB is 2.5 m.

a) A small object of mass 3 kg is suspended from A and B by two light inextensible strings so that it is 2 m from A and 1.3 m from B. Calculate the tensions in the strings.

b) The two strings are now replaced by a single string of length 3.3 m which is threaded through a smooth ring attached to the object. A horizontal force, P, is applied to the object so that it rests in the same position in relation to A and B. Calculate the value of P and the tension in the string.

10 A string is threaded through a smooth ring of weight W and is tied to two points A and B on the same level. The ring is pulled by a horizontal force, P, so that the two parts of the string are inclined at angles of 60° and 30° respectively to the vertical.

a) Draw the two possible configurations which fit the given facts.

b) Show that in each case the tension in the string is $W(\sqrt{3} - 1)$.

c) Find the two possible values of P.

11 A light string of length a is attached to two points A and B on the same level and a distance b apart, where $b < a$. A smooth ring of weight W is threaded on the string and is pulled by a horizontal force, P, so that it rests in equilibrium vertically below B. Show that the tension in the string is

$$\frac{W(a^2 + b^2)}{2a^2}$$

and find the force P.

12 A particle of weight W is attached by a light inextensible string of length a to a point A on a vertical wall. The particle is supported in equilibrium by a light rigid strut of length b attached to a point B on the wall at a distance a vertically below A. Show that the tension in the string is W and find the thrust in the rod.

Examination questions

Chapters 2 to 4

Chapter 2

1 The vectors **a** and **b** are given by:

$$\mathbf{a} = \mathbf{i} - \mathbf{j} + \mathbf{k} \quad \text{and} \quad \mathbf{b} = \mathbf{j} + 3\mathbf{k}$$

  i) Find the vector **c** (in terms of **i**, **j** and **k**) such that

$$\mathbf{a} - 2\mathbf{b} = \mathbf{b} + \mathbf{c}$$

  ii) Find the unit vector in the direction of $4\mathbf{a} - 3\mathbf{b}$.   (NICCEA)

2 The vectors **a** and **b** are given by:

$$\mathbf{a} = 4\mathbf{i} + p(\mathbf{j} - \mathbf{k}) \quad \text{and} \quad \mathbf{b} = q\mathbf{i} + r\mathbf{j} - q\mathbf{k}$$

  i) Find the values of p, q and r for which $\mathbf{a} - 2\mathbf{b} = 6\mathbf{i} - 2\mathbf{j} - 6\mathbf{k}$.
  ii) Taking $p = 4$, find the unit vector in the direction of **a**.   (NICCEA)

3 The vectors **a** and **b** are defined as follows:

$$\mathbf{a} = \mathbf{i} + \mathbf{j} + 4\mathbf{k} \qquad \mathbf{b} = 3(\mathbf{i} - \mathbf{j})$$

  i) Find the magnitudes of **a**, **b** and **c**, where **c** is the resultant of **a** and **b**.
  ii) Deduce that **a**, **b** and **c** can represent the sides of a right-angled triangle.
  iii) Find the unit vector in the direction of $3\mathbf{a} + \mathbf{b}$.   (NICCEA)

4 a) Two vectors, **a** and **b**, are given by:

$$\mathbf{a} = (\lambda + 2)\mathbf{i} + 2\mathbf{j} + (\lambda - 4)\,\mathbf{k} \quad \text{and} \quad \mathbf{b} = 7\mathbf{i} + \mathbf{j} + 2\mathbf{k}$$

  Find the values of λ for which the magnitudes of **a** and **b** are equal.
  b) i) Find a unit vector in the direction of $3\mathbf{i} - 12\mathbf{j} + 4\mathbf{k}$.
     ii) Hence, find the force, **F**, of magnitude 104 N, in this direction.   (NICCEA)

5 Two forces $\mathbf{F}_1 = (2\mathbf{i} + 3\mathbf{j})$ N and $\mathbf{F}_2 = (\lambda\mathbf{i} + \mu\mathbf{j})$ N, where λ and μ are scalars, act on a particle. The resultant of the two forces is **R**, where **R** is parallel to the vector $\mathbf{i} + 2\mathbf{j}$.

  a) Find, to the nearest degree, the acute angle between the line of action of **R** and the vector **i**.
  b) Show that $2\lambda - \mu + 1 = 0$.

Given that the direction of $\mathbf{F}_2$ is parallel to **j**,

  c) find, to three significant figures, the magnitude of **R**.   (EDEXCEL)

6 Three forces $\mathbf{F}_1$, $\mathbf{F}_2$ and $\mathbf{F}_3$ act on a particle and $\mathbf{F}_1 = (-3\mathbf{i} + 7\mathbf{j})$ newtons, $\mathbf{F}_2 = (\mathbf{i} - \mathbf{j})$ newtons and $\mathbf{F}_3 = (p\mathbf{i} + q\mathbf{j})$ newtons.

a) Given that this particle is in equilibrium, determine the value of p and the value of q.

The resultant of the forces $\mathbf{F}_1$ and $\mathbf{F}_2$ is $\mathbf{R}$.

b) Calculate, in N, the magnitude of $\mathbf{R}$.
c) Calculate, to the nearest degree, the angle between the line of action of $\mathbf{R}$ and the vector $\mathbf{j}$.

(EDEXCEL)

Chapter 3

7 Two humps are to be installed on a road to prevent traffic reaching speeds of greater than $12\,\mathrm{m\,s}^{-1}$ between the humps. Assume that:

i) the speed of cars when they cross the humps is effectively zero
ii) after crossing a hump they accelerate at $3\,\mathrm{m\,s}^{-2}$ until they reach a speed of $12\,\mathrm{m\,s}^{-1}$
iii) as soon as they reach a speed of $12\,\mathrm{m\,s}^{-1}$ they decelerate at $6\,\mathrm{m\,s}^{-2}$ until they stop.

a) A simple model ignores the lengths of the cars. Use this to find the distance between the humps.
b) One factor that has not been taken into account is the length of the cars. Revise your answer to part **a** to take this into account, giving your answer to the nearest metre. You must state clearly any assumptions that you make.   (AEB 96)

8 A racing car emerging from a bend reaches a straight stretch of road. The start of the straight stretch is the point O and there are two marker points, A and B, further down the road. The distance OA = 64 m and the distance OB = 250 m. The car passes O at time 0 s and, moving with constant acceleration, passes A and B at times 2 s and 5 s respectively. Find

a) the acceleration of the car
b) the speed of the car at B.   (EDEXCEL)

9 Two athletes, Sam and Tom, are in a race. Sam runs at a constant speed of $8.8\,\mathrm{m\,s}^{-1}$. When Sam is 180 m from the finishing tape, Tom is 10 m behind him. At this moment, Tom, who was running at $8.5\,\mathrm{m\,s}^{-1}$, begins to accelerate at a constant rate of $0.2\,\mathrm{m\,s}^{-2}$. When his speed reaches $9.3\,\mathrm{m\,s}^{-1}$, he ceases to accelerate and continues to run with this speed.

a) i) Find the time taken for Tom to accelerate from $8.5\,\mathrm{m\,s}^{-1}$ to $9.3\,\mathrm{m\,s}^{-1}$.
   ii) Find the distance Tom runs during this time.
b) Determine
   i) which athlete wins the race
   ii) how far ahead of the other athlete the winning athlete is when he passes the finishing tape.
c) In this question the athletes have been modelled as particles. Comment on this assumption with reference to your answers to part **b**.   (NEAB)

10 Two stunt drivers drive their cars along a straight horizontal road. The first car is travelling at $30\,\mathrm{m\,s}^{-1}$ and is followed by the second car, 22 m directly behind, travelling at the same speed.

At time $t = 0$ seconds, the driver of the first car applies the brakes and the car decelerates at $4\,\mathrm{m\,s^{-2}}$. Two seconds later, the second car brakes and decelerates at $5\,\mathrm{m\,s^{-2}}$.

i) Find the time when the cars collide.
ii) Find the speeds of the cars at the moment of impact.   (NICCEA)

11 Two sprinters compete in a 100 m race, crossing the finishing line together after 12 seconds. The two models, **A** and **B**, as described below, are models for the motions of the two sprinters.

Model A   The sprinter accelerates from rest at a constant rate for 4 seconds and then travels at a constant speed for the rest of the race.

Model B   The sprinter accelerates from rest at a constant rate until reaching a speed of $9\,\mathrm{m\,s^{-1}}$ and then travels at this speed for the rest of the race.

a) For **model A**, find the maximum speed and the initial acceleration of the sprinter.
b) For **model B**, find the time taken to reach the maximum speed and the initial acceleration of the sprinter.
c) Sketch a distance–time graph for each of the two sprinters on the same set of axes. Describe how the distance between the two sprinters varies through the race.   (AEB 98)

12 A train passes a signal box B with speed $8\,\mathrm{m\,s^{-1}}$, and t seconds later its speed is $v\,\mathrm{m\,s^{-1}}$ and its displacement from B is x metres. The (t, v) graph shown on the right, consisting of three line segments, models the train's journey from the time it passes B until it comes to rest at the next station S.

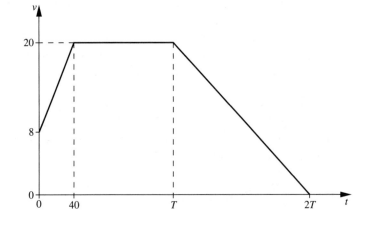

i) Find the acceleration of the train when $t = 10$.
ii) Given that the average speed for the journey from B to S is $13.4\,\mathrm{m\,s^{-1}}$, find T.
iii) Sketch the (t, x) graph for the first 40 s after the train passes B.   (OCR)

13 A car and a van are at rest on a straight, horizontal road with the van 25 m in front of the car. At time $t = 0$ seconds, the van moves off with an acceleration of $1.5\,\mathrm{m\,s^{-2}}$. At time $t = 5$, the car moves off in the same direction with an acceleration of $2\,\mathrm{m\,s^{-2}}$.

i) Sketch on the same diagram the velocity–time graphs which model the motion of the two vehicles.
ii) After how many seconds does the car catch up with the van?   (NICCEA)

14 A motorcyclist drives at a steady speed of $30\,\mathrm{m\,s^{-1}}$ along a straight road in a built-up area, thus breaking the speed limit. A policeman sits in a police car on the road and notes the motorcycle's speed as it passes. The policeman takes 5 s to get ready and then sets off in pursuit of the motorcycle, which maintains its constant speed. The police car accelerates from rest with uniform acceleration $3\,\mathrm{m\,s^{-2}}$ until it reaches a maximum speed of $45\,\mathrm{m\,s^{-1}}$, after which it continues at this steady speed.

a) Draw on the same diagram speed–time graphs to illustrate the movements of the motorcycle and the police car.

T seconds after the motorcycle passes the police car, the police car has reached its maximum speed, but has not yet overtaken the motorcycle.

b) Find an expression for the distance travelled by the police car in terms of T.
c) Hence, or otherwise, find the time that elapses from the moment the motorcyclist passes the police car to the moment when the police car draws level with the motorcycle. (EDEXCEL)

15 A lift travels from rest from the ground floor and comes to rest again at a car park 15 metres above the ground floor. The motion of the lift takes place in three stages. In the first stage the lift moves with a constant acceleration; it then moves with a constant velocity of $4\,\mathrm{m\,s^{-1}}$; and finally it moves with a constant retardation until it comes to rest. The times for the three stages of the motion are $1\frac{1}{2}$, t and 1 seconds, respectively.

a) Sketch a velocity–time graph to show the motion of the lift.
b) Hence, or otherwise, calculate the time for which the lift is in motion.
c) Calculate the average velocity of the lift during the motion. (NEAB)

16 A ball is dropped from the top of a high building and is observed as it passes the various floors of the building. The distance between consecutive floors of the building is d metres and the velocity of the ball as it passes the fifth floor is $u\,\mathrm{m\,s^{-1}}$.

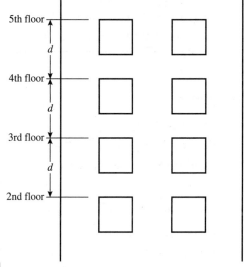

a) The ball is observed to take $\frac{1}{2}$ second to fall between the fifth and fourth floors. Taking the value of g to be $10\,\mathrm{m\,s^{-2}}$, show that $4d = 2u + 5$.
b) i) The ball takes a further $\frac{3}{10}$ second to fall between the fourth and third floors. Find another equation in d and u.
   ii) Hence show that $d = 3$ and $u = 3\frac{1}{2}$.
c) Find, giving your answers correct to two significant figures,
   i) the speed of the ball as it passes the second floor
   ii) the time for the ball to fall between the third and second floors. (NEAB)

17 A girl wishes to estimate the depth d m of a mine shaft. She drops a stone down the shaft and finds that there is an interval of 6 seconds between the instant she dropped the stone and the instant she heard the stone hit the bottom of the shaft.

a) She decides to make a first estimate by assuming that the stone took 6 seconds to drop from the top to the bottom of the shaft. Calculate her first estimate for d.
b) She then used her first estimate for d, together with the fact that the speed of sound is $332\,\mathrm{m\,s^{-1}}$ to estimate the actual time taken by the stone to drop. She then used this time to find a second estimate for d. Calculate her second estimate for d.
c) Obtain, but do not attempt to solve, the equation satisfied by d when the time taken by sound to travel from the bottom to the top of the shaft is taken into account. (WJEC)

18

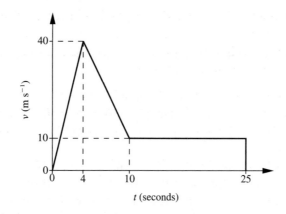

The diagram shows an approximate (t, v) graph for the motion of a parachutist falling vertically; $v\,\mathrm{m\,s^{-1}}$ is the parachutist's downwards velocity at time t seconds after he jumps out of the plane. Use the information in the diagram

i) to give a brief description of the parachutist's motion throughout the descent
ii) to calculate the height from which the jump was made.

The mass of the parachutist is 90 kg. Calculate the upwards force acting on the parachutist, due to the parachute, when $t = 7$.

State two ways in which you would expect an accurate (t, v) graph for the parachutist's motion to differ from the approximate graph shown in the diagram.   (OCR)

Chapter 4

19 A smooth plane is inclined at an angle $10°$ to the horizontal. A particle P of mass 2 kg is held in equilibrium on the plane by a horizontal force of magnitude F newtons, as shown in the figure on the right.

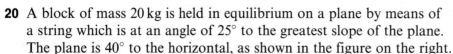

Find, to three significant figures,

a) the normal reaction exerted by the plane on P
b) the value of F.   (EDEXCEL)

20 A block of mass 20 kg is held in equilibrium on a plane by means of a string which is at an angle of $25°$ to the greatest slope of the plane. The plane is $40°$ to the horizontal, as shown in the figure on the right.

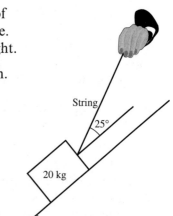

The situation is first modelled by assuming that the plane is smooth.

i) Draw a diagram showing the forces acting on the block.
ii) Show that the tension in the string is about 139 N and find the normal reaction of the plane on the block.

An experiment shows that when the tension in the string is increased to 172 N, the block is still in equilibrium. The model is now refined to take account of friction.

iii) Draw a diagram showing the forces acting on the block and calculate the frictional force.
iv) Without further calculations, state with a reason whether the normal reaction of the plane on the block is the same in parts **ii** and **iii**.   (MEI)

21 A camping lamp P, of mass 1.2 kg, is supported by two light wires fixed inside a tent. The camping lamp hangs in equilibrium with the wires inclined to the horizontal at angles of 20° and 25°, as shown in the diagram. Find the tension in each wire.   (OCR)

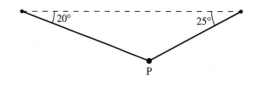

22 The diagram on the right shows a crate C of weight 2000 N suspended in equilibrium by two cables AC and BC attached to two fixed points A and B on the same horizontal level. The cables AC and BC are inclined at 30° and 45° to the horizontal respectively. Modelling the crate as a particle and the cables as light inextensible strings, find the tension in the cable BC.

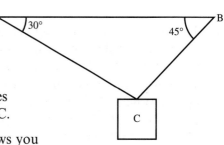

State what modelling assumptions the adjective 'light' allows you to make about the tensions in the cables.   (WJEC)

23

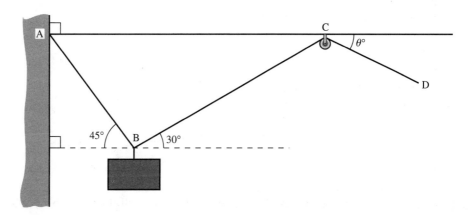

Two light, inextensible strings are attached to a small case of mass 12 kg at B. One string is fixed to a point A. The other string passes over a small smooth pulley at C and is held at D. The points A and C are at the same height and AB and BC are at 45° and 30°, respectively, to the horizontal. The string section CD is at $\theta°$ to the horizontal, as shown in the figure above. The system is in equilibrium.

i) Draw a diagram showing all of the forces acting on the mass at B.
ii) By considering a triangle representing the forces acting at B, or by considering the equations for the equilibrium of the mass at B in two directions, calculate the tensions in the string sections AB and BC.
iii) The position of D is moved so that θ increases but B remains in the same position. What effect does this have on the tension in the string section CD?

The end D of the string section CD is now pulled so that the mass at B rises. It is then held in equilibrium.

iv) Describe what effect this has on the tensions in the string sections AB and BC. [Note that you are not required to do any further calculations.] Explain briefly why the system cannot be in equilibrium with B on the same level as A and C.   (MEI)

24 Two horizontal wires are attached to a point P of a vertical post, and forces of magnitude 8 N and 10 N are applied along these wires as shown in the diagram. The forces are inclined at 30° to each other. Find the magnitude of the resultant of these forces, and find the angle which the resultant makes with the 8 N force.   (OCR)

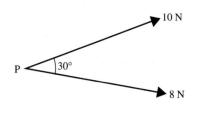

25 Two forces, acting in a vertical plane, have a horizontal resultant of magnitude R newtons. One of the forces has magnitude 6 N and acts at an angle of θ above the horizontal. The other force has magnitude 4 N and acts at an angle of 30° below the horizontal, as shown in the diagram on the right.

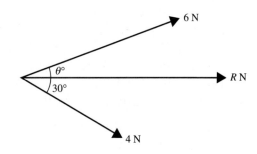

i) By resolving vertically, or otherwise, find the value of θ.

ii) Calculate the value of R.   (OCR)

26

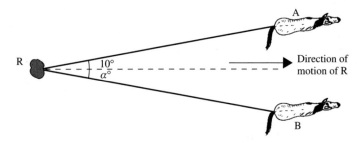

Two cart-horses, A and B, pull a small heavy rock R in a straight line over rough horizontal ground by means of two ropes attached to the rock. The horses are separated from each other so that each rope makes a small angle with the direction of motion of the rock, as shown in the figure above.

Throughout the motion, the rope attached to A has a tension of 800 N and makes an angle of 10° with the direction of motion of R, and the rope attached to B has a tension of 500 N and makes an angle of α° with the direction of motion of R. The rock is modelled as a particle, the ropes are assumed to be horizontal, and air resistance is assumed to be negligible. Using the model,

a) find, to the nearest whole number, the value of α.

Given that the horses drag the rock very slowly along the ground at a constant speed,

b) find, to three significant figures, the resistance to motion experienced by the rock.

c) Suggest one reason why it is reasonable to ignore air resistance in the situation described.

d) Suggest one refinement of the model, in relation to the ropes, which should be incorporated to make the model a more accurate reflection of the situation.   (EDEXCEL)

27 A particle is in equilibrium under the action of three coplanar forces. Two of the forces are represented in the figure on the right.

 i) Find the magnitude of the third force.
 ii) Find the angle between the third force and the force of magnitude 450 N. (NICCEA)

28 Coplanar forces of magnitudes 1 N, 2 N, 3 N act on a particle, as shown in the diagram on the right; the angle between the directions of each pair of the forces is 120°. Show by calculation that the magnitude of the resultant of the three forces is $\sqrt{3}$ N, and find the angle between the direction of the resultant and the direction of the 1 N force. (OCR)

29 The diagram on the right shows four coplanar forces of magnitudes F N, G N, 20 N and 100 N acting at a point O in the directions shown. Given that the forces are in equilibrium, find F and G. (WJEC)

30 Four horizontal wires are attached to the top of a telegraph pole. The tensions in the wires are as shown on the right.

 i) Find the magnitude of the resultant of these tensions.
 ii) Find the angle which this resultant makes with OA. (NICCEA)

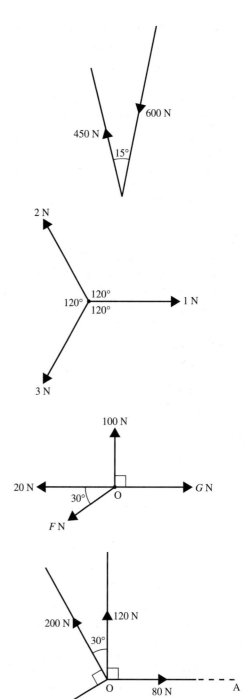

31 An object, of mass 40 kg, is supported in equilibrium by four cables. The forces, in newtons, exerted by three of the cables, $\mathbf{F}_1$, $\mathbf{F}_2$ and $\mathbf{F}_3$, are given in terms of the unit vectors $\mathbf{i}$, $\mathbf{j}$ and $\mathbf{k}$ as $\mathbf{F}_1 = 80\mathbf{i} + 20\mathbf{j} + 100\mathbf{k}$, $\mathbf{F}_2 = 60\mathbf{i} - 40\mathbf{j} + 80\mathbf{k}$ and $\mathbf{F}_3 = -50\mathbf{i} - 100\mathbf{j} + 80\mathbf{k}$. The unit vectors $\mathbf{i}$ and $\mathbf{j}$ are perpendicular and horizontal and the unit vector $\mathbf{k}$ is vertically upwards.

a) Find $\mathbf{F}_4$, the force exerted by the fourth cable, in terms of $\mathbf{i}$, $\mathbf{j}$ and $\mathbf{k}$. Also find its magnitude to the nearest newton.

b) Find the angle between $\mathbf{F}_1$ and $\mathbf{F}_4$.   (AEB 98)

32

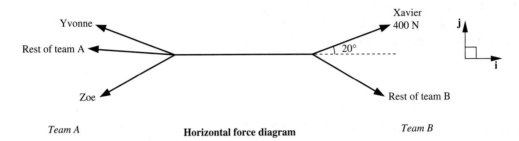

Team A   **Horizontal force diagram**   *Team B*

In a tug-of-war competition, two teams A and B are pulling a rope and are in equilibrium. The situation is modelled by assuming that the rope has negligible mass and all the forces acting on it are in the same horizontal plane. The horizontal unit vectors $\mathbf{i}$ and $\mathbf{j}$ are in the direction of the rope and perpendicular to it as shown in the figure above. When the answers to the following questions are vectors they should be given in terms of $\mathbf{i}$ and $\mathbf{j}$.

The tension in the section of the rope between the teams is 3000 N.

i) Write down the force being exerted on the rope by the whole team of B.

Xavier in team B is pulling with a force of 400 N at an angle of $20°$ to the $\mathbf{i}$ direction, as shown in the figure.

ii) Calculate the combined force on the rope which the rest of team B is exerting.

Yvonne in team A is pulling with a force $(-240\mathbf{i} + 100\mathbf{j})$ N.

iii) Calculate the magnitude and direction of the force with which Yvonne is pulling.

Zoe is also in team A. She and Yvonne are pulling on the same point of the rope, as shown in the figure, with a combined force of 750 N in the direction $-24\mathbf{i} - 7\mathbf{j}$.

iv) Calculate the force with which Zoe is pulling.

v) State whether the following assumptions are required for the model used, giving **brief** reasons for your answers.

    a) The rope has negligible mass.
    b) The rope is inextensible.   (MEI)

33 The rhombus ABCD has $B\hat{A}D = 50°$. Forces of magnitude 5 N, 3 N, 6 N and 2 N act along the edges AB, BC, CD and AD respectively of this rhombus, as shown in the figure on the right.

i) Find the magnitude of the resultant of these forces.

ii) With the aid of a diagram, find the angle which the resultant force makes with the edge AD of the rhombus.   (NICCEA)

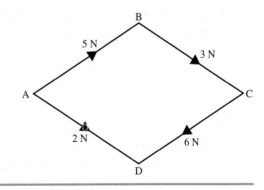

5 Newton's laws of motion

Nature, and Nature's laws lay hid in night:
God said, 'Let Newton be!' and all was light.
ALEXANDER POPE

In Chapter 3 we considered the motion of objects, and in Chapter 4 the ways in which forces combine. The most basic consideration in mechanics is the connection between these. That is, what forces are involved in the motion of a body and what effect will a given set of forces have on that motion.

Until the 17th century, the accepted model for the motion of objects had not changed significantly since the time of Aristotle some 2000 years earlier. Essentially, Aristotle believed that for an object to be in motion there must at all times be a force causing that motion. In other words, that force is linked to velocity. For example, when a ball is thrown, Aristotelian mechanics would say that there is a force pushing the ball along throughout its flight. (This is a surprisingly common notion to this day – even in *Star Trek* they seem to keep the engines running all the time, which is quite unnecessary once the required speed is reached!)

Newton's contribution was to formalise the idea developing in his time that force was linked not to velocity but to change of velocity: that is, to acceleration. He then calculated how objects should behave according to this model and showed that the results were a good match with observed reality. All the ideas developed in the present book are based on the Newtonian model. Although they are called Newton's *laws*, you should be aware that they are just a mathematical model, albeit a rather good one, of the observable world.

We should mention that an alternative model – the theory of relativity – was developed by Einstein in the early 1900s. The predictions from the two models only differ significantly at the atomic and astronomical levels, where Einstein's model is superior. So, for most purposes, Newton's model continues to be the basis for analysing the motion of systems.

Newton's first law

Every object remains at rest or moves with constant velocity unless an external force is applied.

This is really talking about an imbalance of forces. If you start to pull a stationary trolley and someone else pulls in the opposite direction with an equal force, the trolley will stay at rest. If you are pulling the trolley along and the force you are exerting is exactly the same as the friction forces on the trolley, it will move at a constant velocity. It will only speed up, slow down or change direction if all the forces acting on it combine to give a resultant force.

The next thing needed is to establish the relationship between the magnitude of the force and the acceleration produced. This problem is addressed by Newton's second law. Newton couched this in terms of momentum, but for our purposes we can state it as follows.

Newton's second law

When an object undergoes acceleration, the force needed to produce it is in the direction of the acceleration, and is proportional both to the acceleration and to the mass of the object.

This accords well with common sense in that

- for a given object a larger acceleration will require a larger force and
- the more massive the object, the greater the force needed to achieve a given acceleration.

Symbolically, Newton's second law is expressed as

$$\mathbf{F} \propto m\mathbf{a} \quad \text{or} \quad \mathbf{F} = km\mathbf{a} \quad \text{where } k \text{ is a constant}$$

Notice that this is a relation between vectors because the force and the acceleration have the same direction.

The SI unit of force is the newton, which is defined as the force needed to accelerate a 1 kg mass at $1\,\mathrm{m\,s^{-2}}$. With this definition, the value of k in the above equation is 1, and Newton's second law becomes

$$\mathbf{F} = m\mathbf{a}$$

This equation is called the **equation of motion** of the body.

Note   It cannot be too strongly emphasised that here **F** is the **resultant** of the forces acting on the body.

Example 1 The engine of a car of mass 900 kg produces a driving force of 2000 N. There are resistive forces of 650 N. Find the acceleration of the car on a level road.

SOLUTION

Acceleration takes place in a horizontal direction, so the only forces to consider are those acting horizontally. Resolving horizontally, we get

$$\text{Resultant force} = 2000 - 650 = 1350\,\mathrm{N}$$

The equation of motion is $F = ma$, giving

$$1350 = 900a$$

$$\Rightarrow \quad a = 1.5\,\mathrm{m\,s^{-2}} \quad \text{horizontally}$$

Note Strictly, the equation of motion in Example 1 is a vector equation and would formally be written as

$$1350\mathbf{i} = 900\mathbf{a}$$

$$\Rightarrow \quad \mathbf{a} = 1.5\mathbf{i}\,\mathrm{m\,s^{-2}}$$

However, when motion is in a straight line, it is common practice to write the equation in scalar form, as shown.

Example 2 A horizontal force of 50 N is applied to a sledge of mass 20 kg resting on level snow. The sledge accelerates at $2.2\,\mathrm{m\,s^{-2}}$. Find the friction force acting on the sledge.

SOLUTION

Let the friction force be F.

Resolving horizontally gives the resultant force as $50 - F$ N. So, applying Newton's second law gives

$$50 - F = 20 \times 2.2$$

$$\Rightarrow \quad F = 6\,\mathrm{N}$$

Note that there are other forces acting: namely, the weight of the sledge and the normal reaction of the surface. But as there is no vertical acceleration, these forces have a resultant vertical force of zero and so have been ignored.

Example 3 An object of mass 10 kg is acted on by forces $3\mathbf{i} + 6\mathbf{j}$, $2\mathbf{i} - 3\mathbf{j}$ and $\mathbf{i} + 2\mathbf{j}$ relative to some coordinate system. Find the acceleration of the object.

SOLUTION

The resultant force acting on the object is

$$(3\mathbf{i} + 6\mathbf{j}) + (2\mathbf{i} - 3\mathbf{j}) + (\mathbf{i} + 2\mathbf{j}) = 6\mathbf{i} + 5\mathbf{j}$$

Let the acceleration of the object be $\mathbf{a}$. Then, by Newton's second law, we have

$$6\mathbf{i} + 5\mathbf{j} = 10\mathbf{a}$$

$$\Rightarrow \quad \mathbf{a} = 0.6\mathbf{i} + 0.5\mathbf{j}$$

CHAPTER 5 NEWTON'S LAWS OF MOTION

which is the acceleration in vector component form. We can now, if required, find the magnitude and direction of the acceleration:

$$|\mathbf{a}| = \sqrt{0.6^2 + 0.5^2} = 0.781\,\mathrm{m\,s^{-2}}$$

If θ is the angle with the **i**-direction, then we have

$$\tan\theta = \frac{0.5}{0.6} \quad \Rightarrow \quad \theta = 39.8°$$

Example 4 A cyclist exerts a driving force of 120 N while travelling at a constant $4\,\mathrm{m\,s^{-1}}$. The combined mass of cyclist and machine is 80 kg.

a) Find the resistance force acting.
b) If the cyclist increases the driving force to 140 N, find the distance travelled in the next 3 seconds, stating any assumptions made.

SOLUTION

a) For a constant velocity, the resultant forward force is zero. Therefore, we have

$$\text{Resistance force} = 120\,\mathrm{N}$$

b) Assuming that the resistance force remains constant, and applying Newton's second law, we obtain

$$140 - 120 = 80a \quad \text{where } a \text{ is the acceleration}$$

$$\Rightarrow \quad a = 0.25\,\mathrm{m\,s^{-2}}$$

Using $s = ut + \frac{1}{2}at^2$, where s is the distance travelled, $u = 4\,\mathrm{m\,s^{-1}}$, $t = 3\,\mathrm{s}$ and $a = 0.25\,\mathrm{m\,s^{-2}}$, we obtain

$$s = 4 \times 3 + \frac{1}{2} \times 0.25 \times 9 = 13.125\,\mathrm{m}$$

Exercise 5A

1 A body of mass 40 kg is acted upon by a resultant force of 90 N. Find the acceleration of the body.

2 Find the force needed to accelerate a body of mass 25 kg at $2.1\,\mathrm{m\,s^{-2}}$.

3 A body is acted upon by a resultant force of 24 N and undergoes acceleration of $3.6\,\mathrm{m\,s^{-2}}$. What is the mass of the body?

4 The following table shows information about a vehicle moving on a level road. Find the missing quantities.

	Driving force (N)	Resistance force (N)	Mass (kg)	Acceleration (m s^{-2})
a)	1200	800	500	
b)	2000	600		3.5
c)	900		650	0.8
d)		250	800	1.3
e)	500	800	750	

5 A car of mass 700 kg is acted upon by a driving force of 2200 N and a constant resistance of 800 N. The car starts from rest and travels along a horizontal road. After 6 seconds, the driver depresses the clutch and the car coasts to rest.

a) What was the greatest speed achieved by the car?
b) How far did the car travel altogether?

6 Find, in vector component form, the acceleration of a body of mass 4 kg acted upon by forces $5\mathbf{i} + \mathbf{j}$, $2\mathbf{i} + 7\mathbf{j}$ and $-4\mathbf{i} - 3\mathbf{j}$.

7 A body of mass 2 kg is acted on by forces $2\mathbf{i} + 4\mathbf{j}$, $3\mathbf{i} - 5\mathbf{j}$ and an unknown force $\mathbf{P}$. Find the force $\mathbf{P}$ when the acceleration of the body is $2\mathbf{i} - \mathbf{j}$.

8 Find the magnitude of the resultant force needed to give an object of mass 5 kg an acceleration of $(2\mathbf{i} - 3\mathbf{j})\,\text{m s}^{-2}$.

9 A horse is towing a truck along rails. The horse is attached to the truck by means of a rope of negligible mass which is horizontal and makes an angle of 20° with the direction of the rails. The truck has a mass of 1200 kg and its motion is opposed by a resistance force of 300 N. Find the tension in the rope if the acceleration of the truck is 0.3 m s^{-2}.

10 Find the magnitude and direction of the acceleration of each of the objects illustrated.

a)

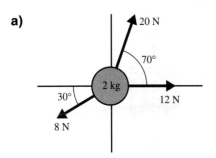

b)

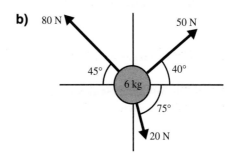

11 Rory, Aurora and Raoul are three lions fighting over a piece of meat of mass 12 kg. Each lion exerts a horizontal pull. Rory pulls with a force of 800 N. Aurora, who is 120° to Rory's right, exerts a force of 400 N. Raoul is 140° to Rory's left. The meat accelerates in Rory's direction.

a) Find the force which Raoul is exerting

b) Find the magnitude of the acceleration.

12 A boat of mass 3 tonnes is steered due east with its engines exerting a driving force of 4000 N. A wind blowing from the south exerts a force of 1200 N. There is a resistance of 2000 N opposing motion. Find the magnitude and direction of the boat's acceleration.

Weight

We can now see the reason for the relation between mass and weight introduced on page 51. An object allowed to fall freely (ideally in a vacuum) near the Earth's surface is observed to accelerate at about 9.8 m s^{-2}. The value varies slightly depending on where on the Earth the experiment is conducted. It is called the **acceleration due to gravity** and is denoted by g.

Because the object is accelerating, there must be a downward force, W, acting on it. If the mass of the object is m kg, by Newton's second law we have

$$W = mg$$

The force W is called the **weight** of the object, It must be stressed that the mass of an object does not vary but its weight depends on the gravitational acceleration it experiences.

For example, an object of mass 10 kg has a weight of $10 \times 9.8 = 98$ N near the Earth's surface. If the object were taken to the Moon, its mass would still be 10 kg but, as gravitational acceleration on the Moon is about 1.6 m s^{-2}, its weight would be $10 \times 1.6 = 16$ N.

Example 5 A crane lifts a 120 kg object on the end of its cable of negligible mass. At first, the object accelerates at 2 m s^{-2}. It then travels at a uniform speed and finally it slows to rest with an acceleration of -1.2 m s^{-2}. Find the tension in the cable at each stage of its motion.

SOLUTION

The weight of the object is $120 \times 9.8 = 1176$ N

Stage 1 Resolving upwards (taking upwards as the positive direction) and using Newton's second law, we get

$$T - 1176 = 120 \times 2 \quad \Rightarrow \quad T = 1416\,\text{N}$$

Stage 2 There is no acceleration and thus no resultant force. The tension and the weight must be equal. Therefore, $T = 1176\,\text{N}$.

Stage 3 Resolving upwards and using Newton's second law, we get

$$T - 1176 = 120 \times -1.2 \quad \Rightarrow \quad T = 1032\,\text{N}$$

Newton's third law

For every action there is an equal and opposite reaction.

This formally states an idea that you met on page 53: namely, that if an object A exerts a force on a second object B (either by direct contact or at a distance by magnetic attraction, gravitation etc), then B will exert a force of the same magnitude and opposite direction on A.

The effect of this is that, if both A and B are part of the system under consideration, the force of A on B and the force of B on A cancel out. They are forces **internal to the system** and do not affect the acceleration of the system. They only become important when we examine the acceleration of object A (or B) alone.

Example 6 A man of mass 90 kg is standing in a lift of mass 300 kg which is accelerating upwards at $0.6\,\text{m s}^{-2}$. Find the tension in the lift cable and the reaction between the man and the floor of the lift.

SOLUTION

When finding the tension in the cable, the forces between the man and the lift are internal and need not be considered. The system is just the mass of 390 kg being raised by the cable.

Resolving upwards (taking upwards as the positive direction) and using $F = ma$, we obtain

$$T - 390\,g = 390 \times 0.6$$

$$\Rightarrow \quad T = 4056\,\text{N}$$

When finding the reaction between the floor and the man, we regard the system as being the man of mass 90 kg acted on by a reaction force R.

Resolving upwards and using $F = ma$, we obtain

$$R - 90\,g = 90 \times 0.6$$

$$\Rightarrow \quad R = 936\,\text{N}$$

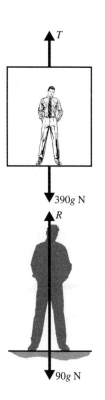

T

$390g$ N

R

$90g$ N

Example 7 An engine of mass 10 tonnes is pulling a truck of mass
3 tonnes. The resistance forces acting on the engine and the truck are
4000 N and 1500 N respectively. The driving force of the engine is
14 000 N. Find the acceleration of the system and the tension in the
coupling between the engine and the truck.

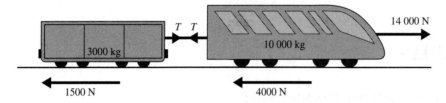

SOLUTION

When finding the acceleration, we take the engine and truck as a complete
system. So, we need consider only the driving force and the resistances,
because the tension in the coupling is an internal force. We can also
ignore the vertical forces.

Resolving horizontally and using $F = ma$, we get

$$14\,000 - 4000 - 1500 = (10\,000 + 3000)a$$

$$\Rightarrow \quad a = 0.654\,\mathrm{m\,s}^{-2}$$

To find the tension in the coupling, we consider just the
forces acting on the truck. So, the tension becomes an
external force, as shown.

Resolving horizontally and using $F = ma$, we get

$$T - 1500 = 3000 \times 0.654$$

$$\Rightarrow \quad T = 3461.5\,\mathrm{N}$$

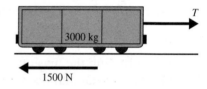

Note that we could have just as easily considered the forces on the engine.
This would have given

$$14\,000 - 4000 - T = 10\,000 \times 0.654$$

$$\Rightarrow \quad T = 3461.5\,\mathrm{N}$$

Example 8 An object of mass 8 kg is being towed by a light
string up a slope inclined at 20° to the horizontal. The string
is inclined at 30° to the slope. There is a frictional resistance
of 40 N. The object is accelerating up the slope at $0.8\,\mathrm{m\,s}^{-2}$.

a) Find the tension in the string.
b) Find the normal reaction exerted by the slope on the object.

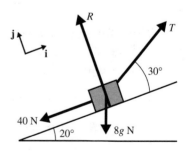

SOLUTION

a) Resolving in the **i**-direction and using $F = ma$, we get

$$T \cos 30° - 40 - 8g \sin 20° = 8 \times 0.8$$

$$\Rightarrow \quad T = 84.5 \, \text{N}$$

b) Resolving in the **j**-direction there is no acceleration. Therefore, the resultant is zero:

$$R + T \sin 30° - 8g \overset{cos}{\sin} 20° = 0$$

Substituting the value for T in the above equation gives $R = 31.4 \, \text{N}$.

Exercise 5B

1 Each of the following involves an object of mass of 20 kg moving vertically on the end of a cable. It is assumed that the only forces acting are the weight of the object and the tension in the cable.

a) Find the acceleration of the object when the tension is
   i) 250 N   ii) 150 N
b) Find the tension in the cable when the object is
   i) moving upwards at a constant speed of $5 \, \text{m s}^{-1}$
   ii) moving downwards with a constant speed of $4 \, \text{m s}^{-1}$
   iii) accelerating upwards at $2 \, \text{m s}^{-2}$
   iv) moving upwards and slowing uniformly from $6 \, \text{m s}^{-1}$ to $2 \, \text{m s}^{-1}$ in 6 seconds
   v) moving downwards and slowing uniformly from $6 \, \text{m s}^{-1}$ to rest in 8 metres.

2 An object of mass 40 kg is suspended by a light string from the ceiling of a lift of mass 200 kg.

a) The lift accelerates upwards at $1.2 \, \text{m s}^{-2}$. Find the tension in the lift cable and the tension in the string.
b) The string breaks if it suffers a tension of more than 700 N. Find the greatest possible tension in the lift cable if the string remains intact.

3 An object of mass 50 kg is placed on the floor of a lift. Find the reaction between the object and the floor when the lift is

a) accelerating upwards at $1.2 \, \text{m s}^{-2}$
b) moving upwards at a constant $3.5 \, \text{m s}^{-1}$
c) moving upwards but slowing uniformly from $5 \, \text{m s}^{-1}$ to $2 \, \text{m s}^{-1}$ in 4 seconds
d) accelerating downwards at $2 \, \text{m s}^{-2}$.

4 Bathroom scales actually measure the reaction force between the scales and the person standing on them, but the dial is calibrated to show the mass of the person assuming that the scales are placed in a horizontal position on the surface of the Earth. This means that if the reaction is R, the dial shows the value $R \div 9.8$.

What reading will the dial show if a person of mass 80 kg stands on the scales

a) on a level surface on the Moon where the acceleration due to gravity is $1.6 \, \mathrm{m \, s^{-2}}$
b) on the horizontal floor of a lift accelerating upwards at $1.5 \, \mathrm{m \, s^{-2}}$
c) on the horizontal floor of a lift accelerating downwards at $0.8 \, \mathrm{m \, s^{-2}}$
d) on the horizontal floor of a lift travelling upwards at a constant $3 \, \mathrm{m \, s^{-1}}$
e) on a surface sloping at $25°$ to the horizontal.

5 An object of mass 20 kg hangs from a spring balance in a lift. Its apparent mass is 24 kg. What is the acceleration of the lift?

6 Two objects of mass 3 kg and 4 kg are connected by a light inextensible string and both can be raised and lowered on the end of a second string, as shown. Find the tensions in the two strings when the system is

a) at rest
b) moving upwards at a constant speed of $2 \, \mathrm{m \, s^{-1}}$
c) moving upwards with acceleration $3 \, \mathrm{m \, s^{-2}}$.

7 An object of mass 5 kg is suspended by means of two identical light strings from a rod of mass 3 kg, with the strings making angles of $30°$ with the horizontal. The rod is suspended by another light string, as shown.

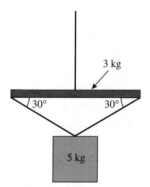

a) Find the tensions in the strings if the system is accelerating upwards at $1.5 \, \mathrm{m \, s^{-2}}$
b) The same type of string is used throughout, with a breaking strain of 120 N. What is the maximum possible upward acceleration of the system and which string will break if this is exceeded?

8 An object of mass 12 kg is pulled up a smooth slope, inclined at $45°$ to the horizontal, by a string parallel to the slope.

a) If the tension in the string is 120 N, find the acceleration of the object.
b) If the tension is then reduced so that the object has an acceleration down the slope of $2 \, \mathrm{m \, s^{-2}}$, find the new tension.

9 An object of mass 3 kg is placed in a tank of oil and allowed to sink. The resistance force acting on the object is kv newtons, where v is its velocity in $\mathrm{m \, s^{-1}}$ and k is a constant. It is observed that when $v = 3$ the acceleration is $2 \, \mathrm{m \, s^{-2}}$.

a) Find the value of k.
b) Find the acceleration when $v = 2$.
c) Find the maximum speed achieved by the object.

10 An object of mass 200 kg is being raised by means of a cable whose breaking strain is 2240 N. Find the shortest time in which the object can be raised a distance of 39.2 m, starting and finishing at rest. [You may find a velocity–time graph helpful.]

11 A car of mass 800 kg is towing a caravan of mass 300 kg along a horizontal road. The resistance forces (assumed constant) on the car and the caravan are 700 N and 1200 N respectively.

a) The car exerts a driving force of 3000 N. Find the acceleration of the system and the tension in the coupling.

b) Find the force in the coupling when the system is travelling at a constant speed of 50 km h^{-1}

c) Find the force in the coupling when the car exerts a braking force of 2000 N.

12 Two identical blocks A and B, each of mass m, are connected together by a light string, S_1, and are placed on a smooth plane inclined at 30° to the horizontal, as shown. A second string, S_2, is attached to block A and is used to tow the blocks up the slope, with S_2 inclined at 30° to the slope.

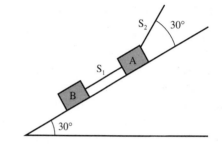

a) Show that if block A is to remain in contact with the slope, the tension in S_2 cannot exceed $mg\sqrt{3}$ and hence find the maximum possible acceleration of the system.

b) Show that the ratio between the tensions in S_1 and S_2 is $\sqrt{3} : 4$, independent of the acceleration of the system.

13 A vehicle of mass $3m$ is towing a trailer of mass m up an inclined plane by means of a towrope. The resistance force on the vehicle is R and on the trailer is $2R$. A braking force P is applied to the vehicle, and both it and the trailer slow down with the towrope still taut. Find the tension in the towrope in terms of P and R, and hence show that for the towrope to remain under tension, the braking force must be less than $5R$.

Connected particles

This term is usually reserved for situations where objects are connected by strings passing over pulleys or other supports.

First, let us examine the simplest situation: two objects connected by a string which passes over a single pulley. Below is a list of the factors which may have an effect on the motion of the system.

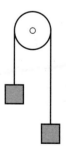

- **Mass of the objects**   Clearly this is a crucial factor in determining what happens. In particular, it is the comparison between the masses which matters. Two masses which are very different are more likely to swamp some of the other factors listed below than are two similar masses.

- **Size of the objects**   This will only have a significant influence if they are large enough for air resistance to have a noticeable effect.

- **Mass of the string**   A string with significant mass has different tensions at different points along its length. In addition, the amount of mass on either side of the pulley changes as the string moves. The importance of these effects depends on how massive the string is compared with the objects.

- **'Stretchiness' of the string**   If the string changes length when it is put under tension, the effect will be that the two objects will not necessarily be going at the same speed or have the same acceleration.

- **Friction at the pulley**   If there were no friction, the tensions in the string on either side of the pulley would be equal, but friction in the pulley would cause them to be different.

- **Mass and radius of the pulley**   These can be considered together because what really matters is the **moment of inertia**. This is a concept you will not encounter unless you are doing further mechanics, but essentially it determines how much turning force is needed to accelerate the pulley. It depends on the size of the pulley and the distribution of mass within it. A pulley with a significant moment of inertia would 'use up' some of the available force to get it turning. We could, of course, replace the pulley with a fixed peg, but then friction might become more significant.

- **How the system is set in motion**   We assume that the system is released with the strings hanging vertically. If this is not so, there may be some pendulum-like movement which could affect the motion.

To set up a model to allow for all the above factors would be very complex. The usual thing is to work with the simplest model and only introduce other factors if the model fails adequately to match what happens experimentally.

The simple model makes the following assumptions.

- The objects are **particles**. That is, they are small enough to be treated as mass concentrated at a single point.

- The string is **light**. That is, its mass is so small compared with that of the objects that we can regard it as having no mass.

- The string is **inextensible**. Its length alters so little under tension that we can treat the length as effectively constant.

- The pulley is **smooth**. That is, the frictional resistance in the pulley is so small compared with other forces that we can treat it as zero friction.

- The pulley is **light**. That is, the force needed to accelerate the pulley is negligible.

With these assumptions, we can state that the tension is the same throughout the string and that the motions of the objects are the same in terms of acceleration, speed and distance travelled, albeit in opposite directions.

Example 9 Particles of mass 3 kg and 5 kg are attached to the ends of a light, inextensible string passing over a smooth pulley. The system is released from rest. Find the acceleration of the system and the tension in the string.

SOLUTION

First, we write down the equation of motion for each of the masses separately.

For the 5 kg mass: $5g - T = 5a$ [1]

For the 3 kg mass: $T - 3g = 3a$ [2]

We then solve these simultaneous equations.

Adding [1] and [2], we get

$$2g = 8a$$

$$\Rightarrow \quad a = \tfrac{1}{4}g = 2.45 \, \text{m s}^{-2}$$

Substituting the value for a in [2], we obtain

$$T - 3g = \tfrac{3}{4}g$$

$$\Rightarrow \quad T = 3\tfrac{3}{4}g = 36.75 \, \text{N}$$

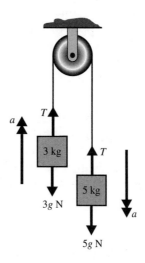

Note   In Example 9 we chose the positive direction separately for each object rather than set a universal positive direction. This is common practice.

Testing the model

If you have access to suitable equipment, you might try testing the model to see how good its predictions are. The main limitation will probably be your ability to time the motion of the system accurately enough. The masses used in Example 9 would not be suitable, because they are rather large and because the predicted acceleration is quite high. The system would move 2 m in about 1.3 seconds, which is too short a time to measure accurately.

You will need to experiment to find the most workable arrangement. For example, masses of 95 grams and 100 grams would be expected to move 2 m in about 4 seconds. Work out the predicted acceleration and travel times for various combinations of masses and then compare them with your experimental findings.

If the system moves more slowly than predicted (it should **not** move faster), you may need to adjust the model. The most likely additional factor to allow for is the friction in the pulley. The simplest model for this is to assume that it makes the tensions differ by a fixed amount F N. The equations of motion in the previous example would become

For the 5 kg mass: $5g - (T + F) = 5a$ [1]

For the 3 kg mass: $T - 3g = 3a$ [2]

Use your experimental results to find an estimated value for *F*. You can then use this refined model to predict the behaviour of the system with a new pair of masses and test this prediction against reality.

It is unlikely that you could go beyond this with a simple experiment, but in theory you could successively refine the model either by examining whether the friction is dependent on the speed of the system, or by including some of the other factors listed on page 85–6, until the agreement between prediction and practice is as close as desired.

Example 10 A block of mass 4 kg rests on a horizontal table. It is attached by means of a light, inextensible string to a particle of mass 9 kg. The string passes over a smooth pulley at the edge of the table, as shown. There is a frictional resistance of 20 N opposing the motion of the block. Find the acceleration of the system, the tension in the string and the resultant force acting on the pulley.

SOLUTION

First, write down the equations of motion for the two objects.

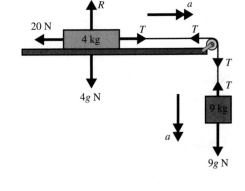

For the 9 kg mass: $9g - T = 9a$ [1]

For the 4 kg mass: $T - 20 = 4a$ [2]

Solving these equations gives

$$a = 5.25\,\mathrm{m\,s^{-2}} \quad \text{and} \quad T = 41.0\,\mathrm{N}$$

To find the force on the pulley, we need to realise that each part of the string exerts a tension force on the pulley, as shown in the diagram. Therefore, we have

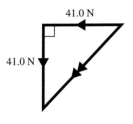

$$\text{Resultant force} = \sqrt{41.0^2 + 41.0^2} = 58.0\,\mathrm{N}$$

acting at 45° to the horizontal.

Example 11 A block of mass 3 kg is attached by light, inextensible strings to particles of mass 5 kg and 8 kg. The strings pass over smooth pulleys at either end of a smooth horizontal table, as shown. Find the acceleration of the system and the tensions in the two strings.

SOLUTION

This time the tensions in the two strings are different. So we have three equations of motion.

For the 8 kg mass: $\quad 8g - T_1 = 8a \qquad$ [1]

For the 3 kg mass: $\quad T_1 - T_2 = 3a \qquad$ [2]

For the 5 kg mass: $\quad T_2 - 5g = 5a \qquad$ [3]

We then solve these simultaneous equations.

Adding [1] and [2], we get

$$8g - T_2 = 11a \qquad [4]$$

Adding [3] and [4], we get

$$3g = 16a \quad \Rightarrow \quad a = \tfrac{3}{16}g = 1.84 \,\mathrm{m\,s^{-2}}$$

Substituting in [1], we have

$$8g - T_1 = 1\tfrac{1}{2}g \quad \Rightarrow \quad T_1 = 6\tfrac{1}{2}g = 63.7 \,\mathrm{N}$$

Substituting in [3], we have

$$T_2 - 5g = \tfrac{15}{16}g \quad \Rightarrow \quad T_2 = 5\tfrac{15}{16}g = 58.2 \,\mathrm{N}$$

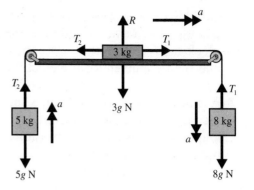

Exercise 5C

1 Two particles of mass 5 kg and 7 kg are connected by a light, inextensible string passing over a smooth pulley. Find

  a) the acceleration of the system
  b) the tension in the string
  c) the force on the pulley.

2 Two particles of mass 2 kg and 3 kg are connected by a light, inextensible string passing over a smooth pulley. The system is released from rest with the 3 kg particle a distance of 4 m above the ground. Find the acceleration of the system and the speed at which the 3 kg particle hits the ground.

3 Two particles of mass m and $2m$ are connected by a light, inextensible string passing over a smooth pulley. Find the acceleration of the system and the tension in the string.

4 A block of mass 3 kg rests on a smooth table. It is connected by a light, inextensible string passing over a smooth pulley at the edge of the table to a 2 kg particle hanging freely. Find the acceleration of the system and the tension in the string.

5 A block of mass 4 kg rests on a table. It is connected by a light, inextensible string passing over a smooth pulley at the edge of the table to a 5 kg particle hanging freely. There is a friction force of 20 N acting on the block. Find the acceleration of the system and the tension in the string.

6 A block of mass 2 kg rests on a smooth table. It is connected by a light, inextensible string passing over a smooth pulley at the edge of the table to a 3 kg particle hanging freely. The block starts from rest at a distance 1.5 m from the pulley. Find the acceleration of the system and the time taken for the block to reach the pulley.

7 A block of mass 4 kg rests on a smooth plane inclined at 20° to the horizontal. It is connected by a light, inextensible string passing over a smooth pulley at the top of the slope to a 3 kg particle hanging freely. Find the acceleration of the system and the tension in the string.

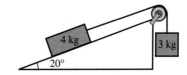

8 A block of mass m rests on a smooth plane inclined at 30° to the horizontal. It is connected by a light, inextensible string passing over a smooth pulley at the top of the slope to a second particle of mass m hanging freely. Show that the system accelerates at $\frac{1}{4}g\,\mathrm{m\,s^{-2}}$, and find the tension in the string.

9 Blocks of mass 3 kg and 2 kg are connected by a light, inextensible string and are placed on a smooth, horizontal table as shown, with the string taut. The 2 kg block is connected by a similar string passing over a smooth pulley at the edge of the table to a 4 kg particle hanging freely. Find the acceleration of the system and the tensions in the strings.

10 A block of mass 5 kg placed on a smooth, horizontal table is connected by light, inextensible strings passing over smooth pulleys at opposite ends of the table to particles of mass 4 kg and 7 kg hanging freely. Find the acceleration of the system and the tensions in the strings.

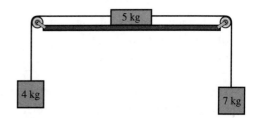

11 A block of mass $2m$ placed on a smooth, horizontal table is connected by light, inextensible strings passing over smooth pulleys at opposite ends of the table to particles of mass m and $3m$ hanging freely. Find the acceleration of the system and the tensions in the strings.

12 A smooth plank AB of length r is fixed with A on horizontal ground and B a distance h above the ground. A block of mass m is placed on the plank and is connected by a light, inextensible string passing over a pulley at B to a particle of mass m hanging freely. The system is set in motion with the block at A and the particle at B. Find the acceleration of the system and show that the time taken for the particle to reach the ground is

$$2\sqrt{\frac{rh}{g(r-h)}}$$

13 Particles A and B of mass 2 kg and 5 kg respectively are connected by a light, inextensible string passing over a smooth pulley. Initially the system is at rest with A on the ground and B at 3 m above the ground. The system is released. Find

a) the acceleration of the system
b) the speed with which the system is moving when B hits the ground
c) how much further A will rise before coming instantaneously to rest.

14 A block of mass 3 kg placed on a smooth, horizontal table is connected by light, inextensible strings passing over smooth pulleys at opposite ends of the table to particles of mass 1 kg and 4 kg hanging freely. The system starts from rest and moves 2 m, at which point the descending particle strikes the ground and stops. Find

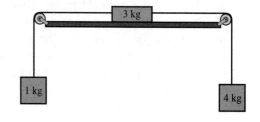

a) the speed at which the system is moving when this happens
b) the further distance which the rest of the system moves before coming instantaneously to rest.

15 A particle of mass m rests on a smooth plank AB inclined at $60°$ to the horizontal. It is connected by a light, inextensible string passing over a pulley at B to a particle of mass M hanging freely. A rests on horizontal ground and B is a distance h above the ground. The system is released from rest with the particles at A and B respectively. Show that the speed at which the descending particle hits the ground is

$$\sqrt{\frac{gh(2M - m\sqrt{3})}{M + m}}$$

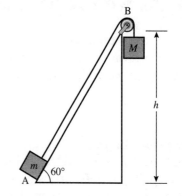

If the ascending particle just reaches B before coming instantaneously to rest, show that $M : m = 2 : \sqrt{3}$.

16 The diagram shows an object A of mass 5 kg connected by a light, inextensible string passing over a smooth pulley to a box B of mass 4 kg. There is an object C of mass 2 kg resting on the horizontal floor of the box. Find

a) the acceleration of the system
b) the reaction between C and the floor of the box.

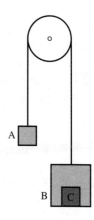

17 The diagram shows a version of what was known as Attwood's machine, which was used as a means of estimating g. Two objects, both of mass M, are connected by a light string passing over a smooth pulley. The system starts from rest with one of the masses at A as shown, and a small rider of mass m attached to it. The system moves a distance h, at which point the mass M passes through a ring B which removes the rider. The system continues to move at uniform speed and the mass is timed in its descent from B to C, a distance h. If the system takes a time t in moving from B to C, show that

$$g = \frac{h(2M+m)}{2mt^2}$$

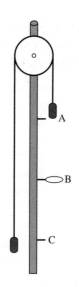

Systems with related accelerations

In the situations examined so far, the various parts of the system have had the same acceleration. When this is not the case, it is necessary to find the relationship between the various accelerations before the problem may be solved.

Example 12 In the diagram, the pulley A is free to move and the pulley B is fixed. A light, inextensible string passes over pulley B and carries pulley A on one end and a particle of mass 6 kg on the other. A second, similar string passes over pulley A and carries particles of mass 2 kg and 4 kg. The pulleys are light and smooth. Find the tensions in strings and the accelerations of the three masses.

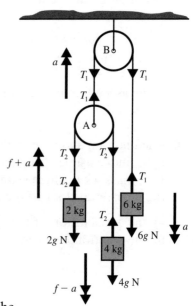

SOLUTION

Suppose the 6 kg mass is moving downwards with acceleration a. (If it is, in fact, moving upwards, the value of a will turn out to be negative.) Pulley A would then have acceleration a upwards.

If the whole system comprised pulley A and the particles it carries, those particles would have an acceleration, f say, relative to A. As A also has an acceleration, the total acceleration of the 2 kg mass would be $(f+a)$ upwards and of the 4 kg mass $(f-a)$ downwards, as shown.

There are four unknown quantities in this problem – a, f, T_1 and T_2 – so we need four equations. We find the equations of motion for the three particles and for pulley A. Notice that as pulley A is described as **light**, we treat it as having zero mass.

For the 6 kg mass:	$6g - T_1 = 6a$	[1]
For the 4 kg mass:	$4g - T_2 = 4(f - a)$	[2]
For the 2 kg mass:	$T_2 - 2g = 2(f + a)$	[3]
For pulley A:	$T_1 - 2T_2 = 0$	[4]

We now solve these equations.

$2 \times$ Eqn [1] $+ 3 \times$ Eqn [2]	$\Rightarrow$	$24g - 2T_1 - 3T_2 = 12f$	[5]
Eqn [1] $- 3 \times$ Eqn [3]	$\Rightarrow$	$12g - T_1 - 3T_2 = -6f$	[6]
Eqn [5] $+ 2 \times$ Eqn [6]	$\Rightarrow$	$48g - 4T_1 - 9T_2 = 0$	[7]

Substituting from [4] into [7], we get

$$48g - 17T_2 = 0 \quad \Rightarrow \quad T_2 = \frac{48}{17}g = 27.7 \, \text{N}$$

Substituting back into [4], we get

$$T_1 = \frac{96}{17}g = 55.3 \, \text{N}$$

Substituting back into [1], we get

$$a = \frac{1}{17}g = 0.576 \, \text{m s}^{-2}$$

Substituting back into [2], we get

$$f = \frac{6}{17}g = 3.46 \, \text{m s}^{-2}$$

Example 13 The diagram shows fixed pulleys A and C and another, B, which is free to move. The pulleys are light and smooth. A light, inextensible string passes round all three pulleys, as shown, and carries particles of mass 2kg and 4 kg. Pulley B carries a load of 5 kg. Find the tension in the string and the accelerations of the three masses.

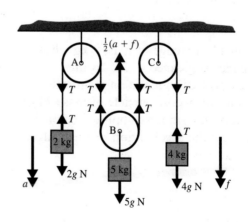

SOLUTION

Suppose the system were released from rest. In a time t, the 2 kg mass would fall a distance $\frac{1}{2}at^2$ and the 4 kg mass a distance $\frac{1}{2}ft^2$. The string around pulley B would therefore shorten by $\frac{1}{2}(a+f)t^2$. This would be shared evenly between the two sections of that string and so pulley B would rise by $\frac{1}{4}(a+f)t^2$. If the acceleration of pulley B is a_B, we have

$$\frac{1}{2}a_Bt^2 = \frac{1}{4}(a+f)t^2$$
$$\Rightarrow \quad a_B = \frac{1}{2}(a+f)$$

We can now write down the three equations of motion.

For the 2 kg mass: $2g - T = 2a$ [1]

For the 4 kg mass: $4g - T = 4f$ [2]

For the 5 kg mass: $2T - 5g = \dfrac{5}{2}(a+f)$ [3]

We solve these equation.

From [1]: $a = g - \dfrac{1}{2}T$ From [2]: $f = g - \dfrac{1}{4}T$

Substituting into [3], we get

$$2T - 5g = \frac{5}{2}\left(2g - \frac{3}{4}T\right) \quad \Rightarrow \quad T = \frac{80}{31}g = 25.3\,\text{N}$$

Substituting back, we obtain

$$a = \frac{9}{31}g = 2.85\,\text{m s}^{-2} \quad \text{and} \quad f = \frac{11}{31}g = 3.48\,\text{m s}^{-2}$$

Related accelerations also occur where objects are moving on surfaces which are themselves free to move.

Example 14 A wedge of mass 4 kg whose sloping face is inclined at 30° to the horizontal is free to move on a smooth horizontal surface. A particle of mass 2 kg is placed on the smooth sloping face of the wedge. Find the acceleration of the wedge.

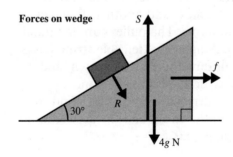

Forces on wedge

SOLUTION

Suppose the acceleration of the wedge is f and the acceleration of the particle is a **relative to the wedge**. R is the reaction between the particle and the wedge.

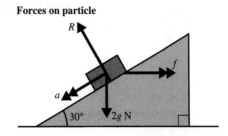

Forces on particle

The first diagram (page 94, bottom) shows the forces acting on the wedge. If we resolve horizontally and use $F = ma$, we get

$$R \sin 30° = 4f \qquad [1]$$

The second diagram (above right) shows the forces acting on the particle. Its acceleration has a component downwards of $a \sin 30°$ and to the left of $(a \cos 30° - f)$. Resolving in these directions, we get

$$2g - R \cos 30° = 2a \sin 30° \qquad [2]$$

and $$R \sin 30° = 2(a \cos 30° - f) \qquad [3]$$

From [1], we have $R = 8f$. Substituting into [2] and [3], we have

$$2g - 4f\sqrt{3} = a \qquad [4]$$

and $$6f = a\sqrt{3} \qquad [5]$$

We now solve these equations.

$$\text{Eqn } [5] - \sqrt{3} \times \text{Eqn } [4] \quad \Rightarrow \quad 18f - 2g\sqrt{3} = 0$$

$$\Rightarrow \quad f = \frac{g\sqrt{3}}{9} = 1.89 \, \text{m s}^{-2}$$

Exercise 5D

1 In each of the following diagrams the pulleys are light and smooth, the strings are light and inextensible and the surfaces are smooth. Find the accelerations and tensions.

a)

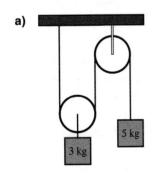

b)

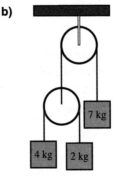

c)

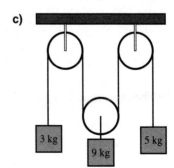

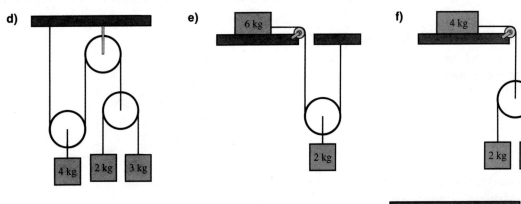

2 Particles of mass m_1 and m_2 are connected by a
light, inextensible string passing over three light,
smooth pulleys, two fixed and one free to move,
as shown. A particle of mass m_3 is suspended from
the movable pulley. The system is released from rest.
Show that if the movable pulley remains stationary

$$m_3 = \frac{4m_1 m_2}{(m_1 + m_2)}$$

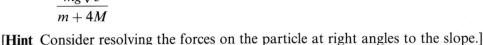

3 The diagram shows a particle of mass m resting
on the smooth inclined surface of a 30° wedge of
mass M, which in turn rests on a smooth horizontal
surface. The system is released from rest. Show that
the acceleration of the wedge is

$$\frac{mg\sqrt{3}}{m + 4M}$$

[**Hint** Consider resolving the forces on the particle at right angles to the slope.]

4 The diagram shows particles of mass 3 kg and 5 kg
connected by a light, inextensible string passing over a
smooth pulley at the vertex of a wedge of mass 10 kg.
The particles rest on the smooth sloping surfaces of the
wedge, which are inclined at 45° to the horizontal. The
wedge is free to move on a smooth horizontal plane.
Find the acceleration of the wedge and the reaction
between the wedge and the horizontal plane.

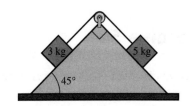

5 A smooth fixed plane is inclined at 30° to the horizontal. A wedge of mass M and angle 30° is
held on the surface so that its upper face is horizontal, and a particle of mass m rests on this
face. The system is released from rest. Show that the resultant acceleration of the particle is

$$\frac{(M + m)g}{4M + m}$$

6 Calculus in kinematics

Mark this, that there is change in all things.
TERENCE

Velocity at an instant

On page 35, we looked at displacement–time graphs. In particular, when the velocity of a body is constant, the graph is a straight line and the velocity is represented by the gradient of the graph. So, we have

$$\text{Velocity} = \frac{\text{Change in displacement}}{\text{Time taken to change}}$$

When the displacement–time graph is not linear, the gradient changes. But it is still true that the gradient at a point on the graph represents the velocity at that instant. In calculus, we find the gradients of graphs by differentiation and we can use this technique here. That is, we write

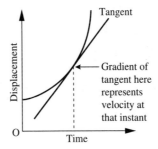

$$v = \frac{\mathrm{d}s}{\mathrm{d}t}$$

and, similarly, for acceleration

$$a = \frac{\mathrm{d}v}{\mathrm{d}t}$$

Any derivative formed by differentiating with respect to **time** creates a **rate of change**. Thus velocity is the rate of change of displacement and acceleration is the rate of change of velocity.

Just as we can differentiate these variables, we can also integrate them. On pages 36–9, we give examples in which the area under the velocity–time curve represents change in displacement. This area can be found by integration.

In general,

$$v = \int a \, \mathrm{d}t \quad \text{and} \quad s = \int v \, \mathrm{d}t$$

These results allow us to solve kinematics problems **whatever the acceleration** of the body, whereas the equations on page 42 are valid only when the acceleration is uniform.

Example 1 A particle, travelling along a straight wire from a point A, has a velocity of $6\,\mathrm{m\,s^{-1}}$ at $t = 0$ and an acceleration of $-6t\,\mathrm{m\,s^{-2}}$.

a) Find its velocity when $t = 1, 2, 3$ seconds.
b) Find its position when $t = 1, 2, 3$ seconds.
c) Find how far it has travelled during the first 3 seconds of its motion.

SOLUTION

a)
$$v = \int a\,\mathrm{d}t = \int -6t\,\mathrm{d}t$$

$$= -3t^2 + c \quad (c \text{ is the constant of integration})$$

When $t = 0$, $v = 6 \Rightarrow c = 6$, which gives

$$v = -3t^2 + 6$$

t	0	1	2	3
v	6	3	-6	-21

b)
$$s = \int v\,\mathrm{d}t = \int (-3t^2 + 6)\,\mathrm{d}t$$

$$= -t^3 + 6t + k \quad (k \text{ is the constant of integration})$$

When $t = 0$, $s = 0 \Rightarrow k = 0$, which gives

$$s = -t^3 + 6t$$

t	0	1	2	3
s	0	5	4	-9

c) The particle travels from A and passes through B when $t = 1$ (see diagram below). At that moment, it is still moving to the right at $3\,\mathrm{m\,s^{-1}}$. But during the second second, it stops momentarily, before returning through C and A and finally travelling to D. After this time, it continues to travel to the left in the diagram and at an increasing speed.

To find the total distance travelled, we need to find out where it comes to rest. At this point, $v = 0$. Therefore, we have

$$-3t^2 + 6 = 0$$

$$\Rightarrow \quad t^2 = 2 \quad \Rightarrow \quad t = 1.414 \quad \text{or} \quad -1.414$$

We take $t = 1.414$, because its negative value is inappropriate in the context of the problem.

The position of the particle is given by

$$s = -t^3 + 6t$$
$$= -1.414^3 + 6 \times 1.414 = 5.6569$$

The total distance travelled in the first 3 seconds is, therefore,

$$(5.6569 + 5.6569 + 9)\,\text{m} = 20.314\,\text{m}$$

An extended model

Example 2 A car is travelling between two sets of traffic lights. Starting from rest at the first set of lights, it accelerates up to a maximum speed before slowing down to a stop at the second set of traffic lights. The motion of the car is modelled by the following formula for s, the displacement of the car from its starting point:

$$s = \frac{1}{45}t^2(45 - t)$$

a) Find an expression for the velocity of the car and from this find the times at which the car is stationary.
b) Find an expression for the acceleration of the car and from this find the time when the acceleration is 0.
c) Find the distance between the traffic lights.
d) Find the maximum velocity of the vehicle.

SOLUTION

a)
$$v = \frac{\mathrm{d}s}{\mathrm{d}t} = \frac{\mathrm{d}}{\mathrm{d}t}\left(\frac{1}{45}t^2(45 - t)\right)$$

$$\Rightarrow \quad v = 2t - \frac{1}{15}t^2$$

When $v = 0$, $t = 0$ or $t = 30$.

b)
$$a = \frac{\mathrm{d}v}{\mathrm{d}t} = \frac{\mathrm{d}}{\mathrm{d}t}\left(2t - \frac{1}{15}t^2\right)$$

$$\Rightarrow \quad a = 2 - \frac{2}{15}t$$

When $a = 0$, $t = 15$.

c) The distance between the lights is given by

$$s(30) - s(0) = \frac{1}{45} \times 30^2(45 - 30) - 0 = 300$$

So, the distance between the lights is 300 m.

d) The maximum velocity occurs when the acceleration is zero and is given by

$$v(15) = 2 \times 15 - \frac{1}{15} \times 15^2 = 15$$

So, the maximum velocity is $15\,\mathrm{m\,s^{-1}}$.

Revising the model

We are not told what assumptions were made in formulating the model in Example 2. However, we notice that when the car reaches the second set of traffic lights, whilst it has a velocity of zero (it has stopped moving), it has an acceleration given by $a(30) = -2\,\mathrm{m\,s^{-2}}$. In this situation, the car will start to move backwards! In fact, the original formula indicates that when $t = 45$, $s = 0$ and the car will have returned to the first set of lights. We can limit the effects of this formula by stating that it is valid only for $0 \leqslant t \leqslant 30$.

Clearly, this is a fault in the way that the model was devised. Ideally, we want a situation where $a = 0$ when $t = 0$ and $t = 30$, and also at the point of maximum velocity between the lights.

The expression for acceleration can be modelled by a cubic (there are other alternatives) of the form

$$a = Kt(t - 15)(t - 30)$$

where K is a constant to be determined. This satisfies the condition that $a = 0$ when $t = 0, 15, 30$.

In order to find the velocity, we use

$$v = \int a \, dt = \int Kt(t - 15)(t - 30) \, dt$$

$$= \int K(t^3 - 45t^2 + 450t) \, dt$$

$$\Rightarrow \quad v = K(\tfrac{1}{4}t^4 - 15t^3 + 225t^2) + c \quad (c \text{ is the constant of integration})$$

The required initial conditions, $v = 0$ when $t = 0$, give $c = 0$. Hence, we have

$$v = K(\tfrac{1}{4}t^4 - 15t^3 + 225t^2)$$

$$\Rightarrow \quad v = \tfrac{1}{4}Kt^2(t - 30)^2$$

The vehicle comes to rest when $v = 0$. The solutions to this are $t = 0, 0, 30, 30$. Thus, there are only two occasions in the motion when the vehicle is stationary, both of which correspond to traffic lights. So, another of our conditions has been satisfied.

We are now in a position to investigate the displacement of the car.

$$s = \int v \, dt = \int K(\tfrac{1}{4}t^4 - 15t^3 + 225t^2) \, dt$$

$$\Rightarrow \quad s = \frac{K}{20}(t^5 - 75t^4 + 1500t^3) + c'$$

The initial conditions $s = 0$ when $t = 0$ give $c' = 0$. So, we have

$$s = \frac{K}{20}(t^5 - 75t^4 + 1500t^3)$$

The other boundary condition is when $t = 30$, $s = 300$. This gives $K = \frac{1}{675}$.

The vehicle reaches its maximum velocity when $a = 0$ at $t = 15$. This gives $v = 12\,656.25K\,\mathrm{m\,s^{-1}}$.

Using $K = \frac{1}{675}$, the maximum velocity is $18.75\,\mathrm{m\,s^{-1}}$. This is a reasonable value, but perhaps a little in excess of the speed limit in an area where most traffic lights are to be found.

Our final model has produced the following formulae:

$$a = \frac{1}{675}t(t - 15)(t - 30)$$

$$v = \frac{1}{2700}(t^4 - 60t^3 + 900t^2)$$

$$s = \frac{1}{13\,500}(t^5 - 75t^4 + 1500t^3)$$

The graphs of these functions, for $0 \leqslant t \leqslant 30$, are shown below.

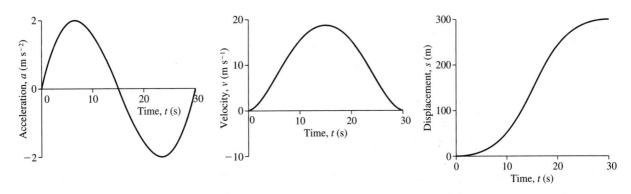

In the solution to this problem, we identified significant features of the motion by substituting appropriate values of s, v and a. The techniques are common to many problems and are generalised below.

- It is usual to take the start of the motion as the origin, so that $s = 0$ when $t = 0$. If we know s in terms of t, we can then find the times at which the object returns to its starting point by putting $s = 0$ and solving the resulting equation. We would, of course, expect $t = 0$ to be one of the roots of the equation.

On the rare occasion that the starting point is not the origin but at some displacement $s = k$ ($k \neq 0$), we would, of course, form the required equation by putting $s = k$. The equation would still have $t = 0$ as one of its roots.

- An object comes instantaneously to rest when its velocity is zero. This corresponds to a stationary point (a maximum, minimum or point of inflexion) on the displacement–time graph.

  A maximum or minimum point indicates that the direction of the object's motion reverses. For example, the motion of an object bouncing up and down on the end of a piece of elastic has two extreme positions (at the top and bottom of the oscillation), and by setting $v = 0$ we can find the times at which it reaches these positions.

  If we were to extend the displacement–time graph for our model of the motion of the car beyond $t = 30$, we would find that it has a point inflexion at $t = 30$. This means that the model has the car slowing to rest but immediately starting forward again.

- A moving object achieves its maximum (or minimum) velocity when the acceleration is zero. In our model for the motion of the car, this happened when $t = 15$.

  Exceptionally, zero acceleration could indicate a point of inflexion on the velocity–time graph.

Exercise 6A

1 A particle, moving in a straight line, starts from rest at O and has an acceleration (in $\mathrm{m\,s^{-2}}$) at time t given by $a = 30 - 6t$.

   a) Find its velocity and position at time t.
   b) Find its velocity and position after 5 seconds.
   c) Find the greatest positive displacement of the particle from O.
   d) Find how long the particle takes to return to O.

2 A particle, moving in a straight line, has a velocity given by $v = 6t - 3t^2 \, \mathrm{m\,s^{-1}}$.

   a) Find its change in position from time $t = 1$ to time $t = 3$.
   b) Find the distance it travels from time $t = 1$ to time $t = 3$.

3 The velocity of a particle, travelling along a straight line, is given by $v = 4t + 6$, where the positive direction is to the right. At time $t = 0$, it is 8 m to the left of point A.

   a) Find an expression for the position of the particle at time t.
   b) Find at what times the particle is at the point A.
   c) i) Find how long it takes the particle to reach the point A.
      ii) What is the significance of the second solution to part **b**?

4 A particle, P, moves along a straight wire. At time t seconds, its displacement, s metres from a fixed point, O, on the wire is given by

$$s = 5\sin\left(\frac{\pi t}{3}\right)$$

a) Find the position of the particle when $t = 1, 2, 3, 4, 5$ and 6 seconds.
b) Find the velocity of the particle at the same times.
c) Find the acceleration of the particle at the same times.
d) What will happen to the particle at times after $t = 6$?

5 A particle, P, is moving along a straight wire. At time t seconds, its displacement, s metres from a fixed point, O, on the wire is given by

$$s = t(t^2 - 16)$$

a) Find the time(s) when P is at the fixed point O.
b) Find the time(s) when P is not moving.
c) Find the displacement of P from O when P is stationary.
d) Find the acceleration of P when $t = 5$.

In parts **a** and **b**, interpret any apparently inappropriate roots of your equations.

6 A particle is moving along a line and has an acceleration given by $a = (2t - 5)\,\mathrm{m\,s^{-2}}$. When $t = 4$, the particle has a velocity of $2\,\mathrm{m\,s^{-1}}$ and has a displacement of $+8\,\mathrm{m}$.

a) Find an expression for its velocity at time t.
b) Find when the particle is at rest.
c) Find where the particle is when it is at rest.

7 A bird leaves its nest flying along a straight line to an adjacent tree, where it collects some food without landing. It then returns to its nest along the same line. Its position is modelled by the formula

$$s = 30t - t^2$$

where s is measured in metres and t is measured in seconds.

a) How long does the journey take?
b) How far away is the second tree?
c) Criticise the model.

8 A particle is moving along a straight line and its position, measured from the point O, is given by the formula

$$s = t^3 - 2t^2 - t + 2$$

where s is measured in metres and t is measured in seconds.

a) Find the times when the particle is at O.
b) Find the velocities and accelerations at the times when the particle is at O.

9 A ball is thrown straight up in the air. Its height at time t s, h m, measured from the ground, is given by

$$h = 4 + 8t - 5t^2$$

a) Find how long it takes the ball to reach the ground.
b) Find an expression for its velocity, v.
c) Find its velocity when it reaches the ground.
d) Find the maximum height reached by the ball.

e) Explain why the expression $\int_0^t v \, dt$ does not represent the distance travelled by the ball.

10 A safety device, designed to bring a moving body to a stop, moves so that the displacement, s, of the body from the datum point is modelled by

$$s = 6 + 6e^{-t}$$

a) How far from the datum point is the body when $t = 0, 1, 2, 3, 4, 5$ seconds?
b) What is the velocity of the body at the times given?
c) What is the acceleration of the body at the times given?
d) Are there any shortcomings with this model?

11 A particle moves so that its acceleration at time t is given by

$$a = -\frac{120}{t^2}$$

a) Find an expression for the velocity of the particle given that, at time $t = 1$, it has a velocity of $60 \, \text{m s}^{-1}$.
b) Find an expression for the displacement of the particle given that, at $t = 1$, it is at the origin.
c) Describe the motion of this particle as t increases.

Acceleration as a derivative

So far, the variables of motion – s (displacement) and v (velocity) – have been expressed as functions of time and we have used the relationships

$$v = \frac{ds}{dt} \quad \text{and} \quad s = \int v \, dt$$

$$a = \frac{dv}{dt} \quad \text{and} \quad v = \int a \, dt$$

From these we can derive two further expressions for the acceleration, the first of which is

$$a = \frac{d}{dt}(v) = \frac{d}{dt}\left(\frac{ds}{dt}\right) = \frac{d^2s}{dt^2}$$

That is, acceleration is the second derivative of displacement with respect to time. At this stage, this merely reinforces what we already know: namely, that integrating acceleration once gives velocity, and a second time gives displacement.

The second expression relates acceleration to velocity and displacement. Using the chain rule for differentiation, we obtain

$$a = \frac{dv}{dt} = \frac{dv}{ds}\frac{ds}{dt} = v\frac{dv}{ds}$$

This second form can be used to solve problems when acceleration is given as a function of v and/or s.

The original form of the expression for a, as dv/dt, can also be used when acceleration is given as a function of v, or as a function of both v and t in special circumstances.

Example 3 A particle moves in such a way that its acceleration is given by

$$a = \frac{1}{6v}$$

When $t = 0$, $s = 0$ and $v = 10$. Find a relationship between the velocity and **a)** time and **b)** displacement.

SOLUTION

a) Using $a = \dfrac{dv}{dt}$, we get

$$\frac{dv}{dt} = \frac{1}{6v}$$

We can solve this **differential equation** by separating its variable (see pages 423–5), giving

$$\int 6v\,dv = \int 1\,dt$$

$$\Rightarrow \quad 3v^2 = t + c$$

The initial conditions, $v = 10$ when $t = 0$, give

$$3(10)^2 = 0 + c \quad \Rightarrow \quad c = 300$$

Hence, we obtain

$$3v^2 = t + 300$$

b) Using $a = v\dfrac{dv}{ds}$, we get

$$v\frac{\mathrm{d}v}{\mathrm{d}s} = \frac{1}{6v}$$

Separating the variables and integrating, we have

$$\int 6v^2 \, \mathrm{d}v = \int 1 \, \mathrm{d}s$$

$$\Rightarrow \quad 2v^3 = s + k$$

The initial conditions, $v = 10$, $s = 0$ when $t = 0$, give

$$2(10)^3 = 0 + k \quad \Rightarrow \quad k = 2000$$

Hence, we obtain

$$2v^3 = s + 2000$$

Example 3 involved solving differential equations by the technique known as the **separation of variables**. This and other methods for solving differential equations are covered on pages 420–40.

Variable forces

Bodies undergo variable acceleration as a result of their being subjected to variable forces. These can arise in many ways. For example, the motion of a body in a resistive medium is opposed by a force which is usually a function of the velocity of the body. Again, the force acting on a body as a result of magnetic or gravitational attraction is inversely proportional to the square of the body's displacement (the **inverse square law**). Such situations are explored in Examples 4 and 5.

Example 4 A small body of mass m is $10\,000$ km above the surface of the Earth and is travelling towards it at a speed of $200\,\mathrm{m\,s^{-1}}$. Find the velocity of the body when it reaches a point 1000 km above the Earth's surface. (The mass of the Earth is 5.98×10^{24} kg, its radius is 6.37×10^6 m and the universal gravitational constant, G, is $6.67 \times 10^{-11}\,\mathrm{N\,m^2\,kg^{-2}}$.)

SOLUTION

We saw on page 51 that the gravitational force between objects of masses m_1 and m_2 separated by a distance d is given by

$$F = \frac{Gm_1m_2}{d^2}$$

Let the distance of the body from the centre of the Earth be s m. The force acting on it is, therefore

$$\frac{6.67 \times 10^{-11} \times 5.98 \times 10^{24}m}{s^2} = \frac{3.99 \times 10^{14}m}{s^2}$$

Expressing the acceleration of the body as $v\dfrac{dv}{ds}$ and applying Newton's second law, we have

$$mv\frac{dv}{ds} = -\frac{3.99 \times 10^{14}m}{s^2}$$

The acceleration is negative because s has its positive direction away from the Earth. Separating the variables and integrating, we get

$$\int v\,dv = -\int \frac{3.99 \times 10^{14}}{s^2}\,ds$$

$$\Rightarrow \quad \frac{v^2}{2} = \frac{3.99 \times 10^{14}}{s} + c$$

The body is initially $10\,000\,000 + 6\,370\,000 = 16\,370\,000$ m from the centre of the Earth and travelling at $200\,\text{m s}^{-1}$. Substituting these values, we have

$$\frac{200^2}{2} = \frac{3.99 \times 10^{14}}{1.637 \times 10^7} + c \quad \Rightarrow \quad c = -24\,353\,855 \quad \text{(to nearest integer)}$$

which gives

$$\frac{v^2}{2} = \frac{3.99 \times 10^{14}}{s} - 24\,353\,855$$

When the body is 1000 km from the surface,

$$s = 1\,000\,000 + 6\,370\,000 = 7\,370\,000$$

Hence, we have $\dfrac{v^2}{2} = \dfrac{3.99 \times 10^{14}}{7.37 \times 10^6} - 24\,353\,855 = 29\,784\,548$

$$\Rightarrow \quad v = 7716.8$$

So, the body is travelling towards the Earth at $7720\,\text{m s}^{-1}$ (to 3 sf).

Example 5 A particle of mass m moves in a straight line on a horizontal surface against a resistance of magnitude kv^2, where v is the speed of the particle and k is a constant. If the initial speed is V, show that the displacement, s, of the particle from its initial position at time t is given by

$$s = \frac{m}{k}\ln\left(1 + \frac{k}{m}Vt\right)$$

SOLUTION

Expressing the acceleration as $\dfrac{dv}{dt}$ and applying Newton's second law, we have

$$m\frac{dv}{dt} = -kv^2$$

Separating the variables and integrating, we obtain

$$\int \frac{1}{v^2}\,dv = -\frac{k}{m}\int dt$$

$$\Rightarrow \quad -\frac{1}{v} = -\frac{kt}{m} + c$$

When $t = 0$, $v = V$, giving $c = -\dfrac{1}{V}$. So, we have

$$-\frac{1}{v} = -\frac{kt}{m} - \frac{1}{V} \quad \Rightarrow \quad v = \frac{ds}{dt} = \frac{mV}{m + kVt}$$

Separating the variables and integrating, we obtain

$$\int ds = \int \frac{mV}{m + kVt}\,dt$$

$$\Rightarrow \quad s = \frac{m}{k}\ln(m + kVt) + C$$

When $t = 0$, $s = 0$, giving $C = -\dfrac{m}{k}\ln m$. So, we have

$$s = \frac{m}{k}\ln(m + kVt) - \frac{m}{k}\ln m$$

$$\Rightarrow \quad s = \frac{m}{k}\ln\left(1 + \frac{k}{m}Vt\right)$$

as required.

Exercise 6B

1 A particle, P, is moving with a constant acceleration, a, in a straight line. When s, the displacement of the particle from a fixed point, O, is zero, the particle has velocity u.

By writing the acceleration as $v\dfrac{dv}{ds}$, derive the formula $v^2 = u^2 + 2as$.

2 A particle, moving along a line, has an acceleration $-v\,\mathrm{m\,s^{-2}}$, where $v\,\mathrm{m\,s^{-1}}$ is the velocity at time t seconds. When the particle is $10\,\mathrm{m}$ from a fixed point, O, on the line, it is travelling at $40\,\mathrm{m\,s^{-1}}$.

   a) Find an expression for the velocity of the particle in terms of its displacement, $s\,\mathrm{m}$, from O.
   b) What happens to the distance of the particle from O as time increases?

3 A body moves along a straight line away from a fixed point, O. Its acceleration is modelled by the formula $a = -4s$, where s is the displacement of the body from O. If the body starts from O with a velocity of $3\,\mathrm{m\,s^{-1}}$, find an algebraic relationship between v, the velocity of the body, and s.

4 The acceleration of a particle, moving along a straight line, is given by $a = -2/v$, where $v\,\text{m s}^{-1}$ is the velocity of the particle at time t seconds after starting its motion. Initially, it has a velocity of $10\,\text{m s}^{-1}$ and is at a fixed point, O, on the line.

a) Find an expression for the velocity, v, of the particle in terms of the time, t.

b) Find an expression for the velocity, v, of the particle in terms of the displacement, $s\,\text{m}$, from O.

5 A particle moves along a straight line with an acceleration $-2v^2\,\text{m s}^{-2}$, where $v\,\text{m s}^{-1}$ is the velocity of the particle at time t seconds after starting its motion. Initially, the particle is at a fixed point O, on the line, travelling with a velocity of $2\,\text{m s}^{-1}$.

a) Find an expression for the velocity, v, in terms of the displacement, s.

b) Find an expression for the velocity, v, in terms of the time, t.

6 A body in space has an acceleration towards the Earth given by $-\dfrac{gR^2}{s^2}$, where g is the acceleration due to gravity at the surface of the Earth, R is the radius of the Earth and s is the distance of the body from the centre of the Earth.

a) By writing a as $v\dfrac{\mathrm{d}v}{\mathrm{d}s}$, integrate the equation and find the relationship between v and s.

b) Assuming that the body has zero velocity when its height **above the surface of the Earth** is H, show that the velocity of the body as it lands on Earth can be written as

$$v = \sqrt{\frac{2gHR}{R+H}}$$

c) If H is small compared with R, show that this reduces to same value as given by the equations for constant acceleration.

d) What is the equivalent result if H is very large compared with R?

7 The acceleration, a, of a body falling through the air, with air resistance acting on the body, is modelled by the equation

$$a = 10 - 2v$$

where v is the velocity of the body. The body is dropped from rest when $t = 0$ and $s = 0$.

a) i) By writing a as $\dfrac{\mathrm{d}v}{\mathrm{d}t}$, integrate the equation and show, provided $v < 5$, that

$$-\ln(5 - v) = 2t + c$$

where c is an arbitrary constant.

ii) Using the initial conditions, find the value of c and hence find an expression for v in terms of t.

iii) Integrate this equation and find an expression for s in terms of t satisfying the initial conditions.

b) i) Show that

$$\frac{v}{5 - v} \equiv \frac{5}{5 - v} - 1$$

ii) By writing a as $v\dfrac{\mathrm{d}v}{\mathrm{d}s}$ in the original equation, integrate the equation and show, provided $v < 5$, that

$$-5\ln(5 - v) - v = 2s + k$$

where k is an arbitrary constant.

iii) Using the initial conditions, find the value of k and hence show that

$$\ln\left(\frac{5}{5 - v}\right) = \frac{2s + v}{5}$$

8 A body moves along a straight line so that its acceleration towards a fixed point, E, is given by $a = -20\cos 2t$, where t is the time after the body has been released. Initially, the body is at rest 5 m from E.

a) Integrate the equation to find v in terms of t.

b) Integrate your equation from part **a** to find s in terms of t.

c) Show that the acceleration can be written as $a = -4s$.

d) By writing a as $v\dfrac{\mathrm{d}v}{\mathrm{d}s}$, integrate the equation in part **c** to obtain a relationship between v and s.

e) Show that your result from part **d** is consistent with your results from parts **a** and **b**.

9 A uniform, heavy rope of length 4 m lies on a straight line perpendicular to the edge of a smooth horizontal suface with 0.5 m of its length hanging vertically from the edge. The rope is released from rest and slides completely over the edge without encountering any obstacle. If the mass per unit length of the rope is m kg, and the length of the rope overhanging the edge at time t s is x m, write down the equation of motion of the rope, and hence find the speed with which it leaves the surface.

10 A particle of mass 2 kg is attached to one end of an elastic rope, the other end being fixed to a point on a smooth horizontal plane. The particle is held on the plane so that the rope is stretched 3 m beyond its usual length, and is then released from rest. Given that the force exerted by the rope when it is stretched by an amount x m is $400x$ N, find the speed of the particle at the moment when the rope goes slack.

11 A particle of mass m is attracted towards a point O by a force of magnitude $\dfrac{km}{x^2}$, where x is its displacement from O and k is a constant. If the particle is released from rest at a distance a from O, show that when it is halfway to O its speed is $\sqrt{\dfrac{2k}{a}}$.

12 A body falls from rest in a medium whose resistance is kv^2 N per unit mass. Show that after it has fallen a distance x, its speed is

$$\sqrt{\frac{g(1 - \mathrm{e}^{-2kx})}{k}}$$

7 Kinematics in two and three dimensions

Space may produce new worlds.
JOHN MILTON

On pages 97–108, we investigated the motion of bodies moving in one dimension. In this chapter, we extend these ideas to two- and three-dimensional motion and develop a model for motion through the air under the effect of gravity. Before we do this, we develop the necessary mathematical tools by looking at the effect of differentiating and integrating a vector with respect to a scalar.

Differentiating and integrating a vector with respect to a scalar

The variables of motion, **r**, **v** and **a**, are vectors and can therefore be written in terms of the unit vectors, **i**, **j** and **k**.

In one-dimensional motion, when we differentiate the displacement of a body with respect to time, we obtain an expression for its velocity. Similarly, when we differentiate the velocity of a body with respect to time, we obtain an expression for its acceleration. Extending this idea to vectors in general, we can write

$$\mathbf{v} = \frac{d}{dt}(\mathbf{r}) \quad \text{or} \quad \mathbf{v} = \frac{d\mathbf{r}}{dt}$$

and

$$\mathbf{a} = \frac{d}{dt}(\mathbf{v}) \quad \text{or} \quad \mathbf{a} = \frac{d\mathbf{v}}{dt}$$

The left-hand expressions emphasise what is being done. The right-hand expressions are what we usually write. The two forms are interchangeable.

However, these expressions do not tell us how to perform the differentiations or the related integrations.

If we write $\mathbf{r} = x\mathbf{i} + y\mathbf{j} + z\mathbf{k}$, then we have

$$\mathbf{v} = \frac{d\mathbf{r}}{dt} = \frac{d}{dt}(x\mathbf{i} + y\mathbf{j} + z\mathbf{k})$$

$$\Rightarrow \quad \mathbf{v} = \frac{d}{dt}(x\mathbf{i}) + \frac{d}{dt}(y\mathbf{j}) + \frac{d}{dt}(z\mathbf{k})$$

In this expression, $\mathbf{v}$, x, y and z are variables, but $\mathbf{i}$, $\mathbf{j}$ and $\mathbf{k}$ are constants. Therefore, we have

$$\mathbf{v} = \left[\frac{d}{dt}(x)\right]\mathbf{i} + \left[\frac{d}{dt}(y)\right]\mathbf{j} + \left[\frac{d}{dt}(z)\right]\mathbf{k}$$

$$\Rightarrow \quad \mathbf{v} = \frac{dx}{dt}\mathbf{i} + \frac{dy}{dt}\mathbf{j} + \frac{dz}{dt}\mathbf{k}$$

This last expression shows that, in order to differentiate the vector $x\mathbf{i} + y\mathbf{j} + z\mathbf{k}$, we simply differentiate each of the components in turn and treat the unit vectors like any other constant.

Example 1 Differentiate the vector $\mathbf{r} = (4t)\mathbf{i} + (3t - 5t^2)\mathbf{j}$ with respect to t.

SOLUTION

Writing $\mathbf{r} = x\mathbf{i} + y\mathbf{j}$, we have

$$x = 4t \quad \text{and} \quad y = 3t - 5t^2$$

from which we obtain

$$\frac{dx}{dt} = 4 \quad \text{and} \quad \frac{dy}{dt} = 3 - 10t$$

$$\Rightarrow \quad \frac{d\mathbf{r}}{dt} = 4\mathbf{i} + (3 - 10t)\mathbf{j}$$

Note The intermediate steps of differentiation are shown in Examples 1 and 2 as explanation, but would not in practice be written down.

Example 2 Differentiate the vector $\mathbf{v} = 4\cos t\,\mathbf{i} + 4\sin t\,\mathbf{j}$ with respect to t.

SOLUTION

Writing $\mathbf{v} = x\mathbf{i} + y\mathbf{j}$, we have

$$x = 4\cos t \quad \text{and} \quad y = 4\sin t$$

from which we obtain

$$\frac{dx}{dt} = -4\sin t \quad \text{and} \quad \frac{dy}{dt} = 4\cos t$$

$$\Rightarrow \quad \frac{d\mathbf{v}}{dt} = -4\sin t\,\mathbf{i} + 4\cos t\,\mathbf{j}$$

We can treat the integration of a vector in a similar way by writing

$$\mathbf{v} = \int \mathbf{a}\,dt \quad \text{and} \quad \mathbf{r} = \int \mathbf{v}\,dt$$

Putting $\mathbf{a} = x\mathbf{i} + y\mathbf{j} + z\mathbf{k}$, we have

$$\mathbf{v} = \int \mathbf{a}\,dt = \int (x\mathbf{i} + y\mathbf{j} + z\mathbf{k})\,dt$$

$$\Rightarrow \quad \mathbf{v} = \int \mathbf{a}\,dt = \int x\mathbf{i}\,dt + \int y\mathbf{j}\,dt + \int z\mathbf{k}\,dt$$

In this expression, $\mathbf{v}$, $\mathbf{a}$, x, y and z are variables but $\mathbf{i}$, $\mathbf{j}$ and $\mathbf{k}$ are constants. Therefore, we have

$$\mathbf{v} = \int \mathbf{a}\,dt = \left(\int x\,dt\right)\mathbf{i} + \left(\int y\,dt\right)\mathbf{j} + \left(\int z\,dt\right)\mathbf{k}$$

Thus, when we want to integrate a vector with respect to a scalar, we integrate each component and treat the unit vectors like any other constant.

Example 3 Integrate the vector $\dfrac{d\mathbf{r}}{dt} = 6t\mathbf{i} + (2t - 9t^2)\mathbf{j}$ with respect to t.

SOLUTION

We write

$$\frac{d\mathbf{r}}{dt} = \frac{dx}{dt}\mathbf{i} + \frac{dy}{dt}\mathbf{j}$$

which gives

$$\frac{dx}{dt} = 6t \quad \text{and} \quad \frac{dy}{dt} = 2t - 9t^2$$

Integrating these two expressions, we obtain

$$x = 3t^2 + c_1 \quad \text{and} \quad y = t^2 - 3t^3 + c_2$$

where c_1 and c_2 are constants of integration. Therefore, we have

$$\mathbf{r} = (3t^2 + c_1)\mathbf{i} + (t^2 - 3t^3 + c_2)\mathbf{j}$$

Notice that the final result in Example 3 can be written as

$$\mathbf{r} = 3t^2\mathbf{i} + (t^2 - 3t^3)\mathbf{j} + (c_1\mathbf{i} + c_2\mathbf{j})$$

or $\qquad \mathbf{r} = 3t^2\mathbf{i} + (t^2 - 3t^3)\mathbf{j} + \mathbf{c}$

In the second version, each of the components has been integrated individually, ignoring the need for an arbitrary constant, which has been added at the end. This constant of integration **must be a vector**. As usual, we find the value of this vector constant from the initial conditions of the particular problem.

Example 4 Integrate the vector $\dfrac{\mathrm{d}\mathbf{r}}{\mathrm{d}t} = 6t^2\mathbf{i} + (3t^2 - 8t)\mathbf{j}$ with respect to t.

Find the particular solution given that $\mathbf{r} = 5\mathbf{i} - 6\mathbf{j}$ when $t = 0$.

SOLUTION

Integrating $\dfrac{\mathrm{d}\mathbf{r}}{\mathrm{d}t} = 6t^2\mathbf{i} + (3t^2 - 8t)\mathbf{j}$, we obtain

$$\mathbf{r} = 2t^3\mathbf{i} + (t^3 - 4t^2)\mathbf{j} + \mathbf{c}$$

This is the general solution, in which $\mathbf{c}$ is an arbitrary vector constant. To find the particular solution, we need the value of $\mathbf{c}$.

To find the value of $\mathbf{c}$, we use the initial condition, $\mathbf{r} = 5\mathbf{i} - 6\mathbf{j}$ when $t = 0$, which gives

$$5\mathbf{i} - 6\mathbf{j} = 0\mathbf{i} + 0\mathbf{j} + \mathbf{c} \quad \Rightarrow \quad \mathbf{c} = 5\mathbf{i} - 6\mathbf{j}$$

Substituting for $\mathbf{c}$ in the general solution, we obtain the particular solution:

$$\mathbf{r} = 2t^3\mathbf{i} + (t^3 - 4t^2)\mathbf{j} + 5\mathbf{i} - 6\mathbf{j}$$

or $\qquad \mathbf{r} = (2t^3 + 5)\mathbf{i} + (t^3 - 4t^2 - 6)\mathbf{j}$

Notation

In mechanics, we often have to differentiate and integrate functions with respect to time. In fact, since most real-life mechanics problems involve some form of change over a period of time, our mathematical models predominantly involve establishing a differential equation with time as the independent variable and then solving it. Because of this, there is an alternative notation for the derivative which is used **only** when the independent variable is time:

$$\frac{\mathrm{d}x}{\mathrm{d}t} \quad \text{can be written as } \dot{x} \qquad \frac{\mathrm{d}^2x}{\mathrm{d}t^2} \quad \text{can be written as } \ddot{x}$$

This notation saves time, but you should be particularly careful when using it to make sure that you have the correct number of dots in the correct place.

Thus, we have

$$\dot{\mathbf{r}} = \mathbf{v} \qquad \dot{\mathbf{v}} = \mathbf{a} \qquad \ddot{\mathbf{r}} = \mathbf{a}$$

For example, the opening equation in Example 3 can be written as

$$\dot{\mathbf{r}} = 6t\mathbf{i} + (2t - 9t^2)\mathbf{j}$$

Equations like this, containing $\dot{\mathbf{r}}$ or $\ddot{\mathbf{r}}$ (sometimes both), are called **differential equations**. We solve them, as you have seen, by integration. More complicated examples need more sophisticated techniques to solve them (see pages 105–6 and 420–40). All the examples we use in this chapter can be solved by direct integration.

Example 5 Differentiate the vector $\mathbf{r} = (3t^2 + 2t)\mathbf{i} + (2t - t^2)\mathbf{j} + 4t^3\mathbf{k}$ with respect to t.

SOLUTION

Writing $\mathbf{r} = x\mathbf{i} + y\mathbf{j} + z\mathbf{k}$, we have

$$x = 3t^2 + 2t \qquad y = 2t - t^2 \qquad z = 4t^3$$

from which we obtain

$$\dot{x} = 6t + 2 \qquad \dot{y} = 2 - 2t \qquad \dot{z} = 12t^2$$

$$\Rightarrow \quad \dot{\mathbf{r}} = (6t + 2)\mathbf{i} + (2 - 2t)\mathbf{j} + 12t^2\mathbf{k}$$

Exercise 7A

1 Differentiate each of the following vectors once only with respect to t.

a) $\mathbf{r} = 4t\mathbf{i} + (8 + 2t^2)\mathbf{j}$

b) $\mathbf{r} = (t^2 - 4t)\mathbf{i} + (t^3 - 2t^2)\mathbf{j}$

c) $\mathbf{v} = t^3\mathbf{i} + 3t^2\mathbf{j}$

d) $\mathbf{v} = (t - t^2)\mathbf{i} + (3t - 5)\mathbf{j}$

e) $\mathbf{r} = (15t - 5t^2)\mathbf{j}$

f) $\mathbf{v} = 15\mathbf{i} + (20 - 10t)\mathbf{j}$

g) $\mathbf{v} = 3\sin 2t\mathbf{i} + 3\cos 2t\mathbf{j} - 5t\mathbf{k}$

h) $\mathbf{r} = 2\cos^2 t\mathbf{i} + 2\sin^2 t\mathbf{j} + \sqrt{t}\mathbf{k}$

i) $\mathbf{v} = 3t\mathbf{i} + 2\cos t\mathbf{j} - 4\sin \frac{1}{2}t\mathbf{k}$

2 Integrate each of the following vectors once only with respect to t.

a) $\mathbf{a} = 6t\mathbf{i} + (2 - 4t)\mathbf{j}$

b) $\mathbf{a} = -10\mathbf{j}$

c) $\mathbf{v} = 3\mathbf{i} + (4 - 5t)\mathbf{j}$

d) $\mathbf{v} = (2t - 6t^2)\mathbf{i} + (3t^2 - 4t^3)\mathbf{j}$

e) $\mathbf{a} = 2\mathbf{i} + 4t\mathbf{j}$

f) $\mathbf{v} = (3 - 6t)\mathbf{i} + (2t - 9t^2)\mathbf{j}$

g) $\mathbf{v} = 4t\mathbf{i} + 4\cos t\mathbf{j} + 2\sin t\mathbf{k}$

h) $\mathbf{a} = -5\cos t\mathbf{i} - 5\sin t\mathbf{j} + 6t\mathbf{k}$

i) $\mathbf{v} = -8\sin t\mathbf{i} + 8\cos t\mathbf{j} + 4\mathrm{e}^{2t}\mathbf{k}$

3 For each of the following differential equations, find the particular solution giving the expression for **r** consistent with the stated values.

a) $\dot{\mathbf{r}} = 4t\mathbf{i} + 8t^3\mathbf{j}$, given that $\mathbf{r} = 2\mathbf{i} - \mathbf{j}$ when $t = 0$.
b) $\dot{\mathbf{r}} = 4\sin t\mathbf{i} + 2\mathbf{j}$, given that $\mathbf{r} = 3\mathbf{i} + 2\mathbf{j}$ when $t = 0$.
c) $\dot{\mathbf{r}} = -5\sin t\mathbf{i} - 5\cos t\mathbf{j} + 3\cos 2t\mathbf{k}$, given that $\mathbf{r} = \mathbf{j} - 5\mathbf{k}$ when $t = 0$.

4 For each of the following differential equations, find the particular solutions giving the expressions for **v** and **r** consistent with the stated values.

a) $\dot{\mathbf{v}} = (3 - 2t)\mathbf{i} + (2t - 6t^3)\mathbf{j}$, given that $\mathbf{v} = 3\mathbf{i}$ and $\mathbf{r} = \mathbf{i} - 2\mathbf{j}$ when $t = 0$.
b) $\ddot{\mathbf{r}} = 4\cos 2t\mathbf{i} + 8\sin 2t\mathbf{j}$, given that $\mathbf{v} = 2\mathbf{i} + \mathbf{j}$ and $\mathbf{r} = 4\mathbf{j}$ when $t = 0$.
c) $\ddot{\mathbf{r}} = (4t - 3)\mathbf{i} + (6t - 2)\mathbf{k}$, given that $\mathbf{v} = 2\mathbf{i} - 3\mathbf{j}$ and $\mathbf{r} = \mathbf{i} + \mathbf{j} + 2\mathbf{k}$ when $t = 0$.

Motion in two and three dimensions

We can solve kinematics problems by using the calculus techniques developed in the previous section, remembering to use the known values (called **initial conditions** or **boundary conditions**) to find the vector constants of integration.

Example 6 An aircraft is flying in such a way that its position vector relative to a watchtower is given by

$$\mathbf{r} = 150t\mathbf{i} + 200t\mathbf{j} + 600\mathbf{k}$$

where the unit vectors **i**, **j** and **k** are measured in the directions east, north and vertically upwards respectively. All distances are measured in metres and time in seconds.

a) Find expressions for
   i) the aircraft's velocity
   ii) its acceleration.
b) **i)** Find the position of the aircraft when $t = 0$.
   ii) Find the speed and direction of the aircraft.
   iii) What is the significance of the **k**-component of the displacement?

SOLUTION

a) **i)** To find the velocity, **v**, we differentiate

$$\mathbf{r} = 150t\mathbf{i} + 200t\mathbf{j} + 600\mathbf{k}$$

which gives

$$\mathbf{v} = \dot{\mathbf{r}} = 150\mathbf{i} + 200\mathbf{j}$$

ii) To find the acceleration, **a**, we differentiate **v**:

$$\mathbf{a} = \dot{\mathbf{v}} = \mathbf{0} \qquad \text{(Note that } \mathbf{0} \text{ is the zero vector.)}$$

b) i) When $t = 0$, $\mathbf{r} = 600\mathbf{k}$. So, the aircraft is directly above the watchtower.

ii) The speed of the aircraft is given by

$$|\mathbf{v}| = \sqrt{150^2 + 200^2} = 250 \text{ m s}^{-1}$$

The direction is given by the angle θ, where

$$\tan \theta = \frac{200}{150}$$

This gives the direction as $53.12°$ from the **i**-direction, or on a bearing of approximately $037°$.

iii) The **k**-component is a constant, indicating that the aircraft is in level flight at an altitude of 600 m.

Example 7 The unit vectors **i** and **j** are the base vectors for a plane. A particle is moving in the plane with a constant acceleration of $2\mathbf{j}$. At time $t = 0$, the particle is at the point $\mathbf{i} + 4\mathbf{j}$ and has a velocity of $3\mathbf{i} - 4\mathbf{j}$.

a) i) Find an expression for the velocity of the particle at time t.
ii) Find an expression for the position of the particle at time t.
b) Plot the path of the particle over the interval $0 \leqslant t \leqslant 5$, marking on it arrows representing the directions of the velocity and of the acceleration for $t = 0, 1, 2, 3, 4$ and 5.

SOLUTION

a) i) Integrating $\mathbf{a} = 2\mathbf{j}$ this to find the velocity, **v**, we get

$$\mathbf{v} = 2t\mathbf{j} + \mathbf{C}$$

where **C** is an arbitrary constant.

To find **C**, we use the initial condition, $\mathbf{v} = 3\mathbf{i} - 4\mathbf{j}$ when $t = 0$, which gives

$$3\mathbf{i} - 4\mathbf{j} = 0\mathbf{j} + \mathbf{C} \quad \Rightarrow \quad \mathbf{C} = 3\mathbf{i} - 4\mathbf{j}$$

Thus, the velocity at time t is given by

$$\mathbf{v} = 2t\mathbf{j} + 3\mathbf{i} - 4\mathbf{j} \quad \Rightarrow \quad \mathbf{v} = 3\mathbf{i} + (2t - 4)\mathbf{j}$$

which is the required expression.

ii) Integrating again, we have

$$\mathbf{r} = 3t\mathbf{i} + (t^2 - 4t)\mathbf{j} + \mathbf{K}$$

where **K** is an arbitrary constant.

To find $\mathbf{K}$, we use the initial condition, $\mathbf{r} = \mathbf{i} + 4\mathbf{j}$ when $t = 0$, which gives

$$\mathbf{i} + 4\mathbf{j} = 0\mathbf{i} + 0\mathbf{j} + \mathbf{K} \quad \Rightarrow \quad \mathbf{i} + 4\mathbf{j} = \mathbf{K}$$

Thus, the position at time t is given by

$$\mathbf{r} = 3t\mathbf{i} + (t^2 - 4t)\mathbf{j} + \mathbf{i} + 4\mathbf{j}$$

$$\Rightarrow \quad \mathbf{r} = (3t + 1)\mathbf{i} + (t^2 - 4t + 4)\mathbf{j}$$

b) The expressions for $\mathbf{a}$, $\mathbf{v}$ and $\mathbf{r}$ give the following results:

t	0	1	2	3	4	5
$\mathbf{r}$	$\mathbf{i} + 4\mathbf{j}$	$4\mathbf{i} + \mathbf{j}$	$7\mathbf{i}$	$10\mathbf{i} + \mathbf{j}$	$13\mathbf{i} + 4\mathbf{j}$	$16\mathbf{i} + 9\mathbf{j}$
$\mathbf{v}$	$3\mathbf{i} - 4\mathbf{j}$	$3\mathbf{i} - 2\mathbf{j}$	$3\mathbf{i}$	$3\mathbf{i} + 2\mathbf{j}$	$3\mathbf{i} + 4\mathbf{j}$	$3\mathbf{i} + 6\mathbf{j}$
$\mathbf{a}$	$2\mathbf{j}$	$2\mathbf{j}$	$2\mathbf{j}$	$2\mathbf{j}$	$2\mathbf{j}$	$2\mathbf{j}$

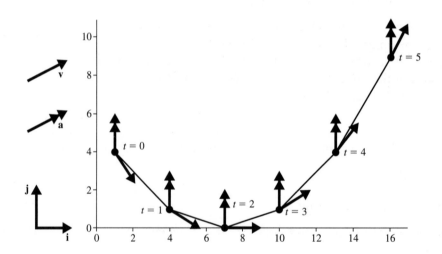

Example 8 Two particles, A and B, leave the origin at the same time. The position vectors of the particles are

$$\mathbf{r}_A = (8t - 2t^2)\mathbf{i} + (4t + t^2)\mathbf{j}$$

$$\mathbf{r}_B = (4t + t^2)\mathbf{i} + (6t + 1\tfrac{1}{2}t^2)\mathbf{j}$$

a) At what times are the particles moving
  i) in the same direction
  ii) in opposite directions
  iii) perpendicular to each other?
b) What are the positions of the particles at these times?

SOLUTION

a) The direction of motion of any body is determined by its velocity.

The velocities of these particles, $\mathbf{v}_A$ and $\mathbf{v}_B$, are found by differentiating $\mathbf{r}_A$ and $\mathbf{r}_B$ respectively:

$$\mathbf{v}_A = (8 - 4t)\mathbf{i} + (4 + 2t)\mathbf{j} \quad \text{and} \quad \mathbf{v}_B = (4 + 2t)\mathbf{i} + (6 + 3t)\mathbf{j}$$

i), ii) Whether they are moving in the same direction or in opposite directions, $\mathbf{v}_A = k\mathbf{v}_B$, where k is some constant. If $k > 0$, they are moving in the same direction. If $k < 0$, they are moving in opposite directions. In either case, we have

$$\frac{4 + 2t}{8 - 4t} = \frac{6 + 3t}{4 + 2t}$$

$$\Rightarrow \quad (4 + 2t)(4 + 2t) = (6 + 3t)(8 - 4t)$$

$$\Rightarrow \quad 4t^2 + 16t + 16 = 48 - 12t^2$$

$$\Rightarrow \quad t^2 + t - 2 = 0 \quad \Rightarrow \quad (t - 1)(t + 2) = 0$$

$$\Rightarrow \quad t = 1 \quad \text{or} \quad t = -2$$

We can reject the negative solution, since it does not satisfy the practical aspects of the problem. Therefore, when $t = 1$, we have

$$\mathbf{v}_A = 4\mathbf{i} + 6\mathbf{j} \quad \text{and} \quad \mathbf{v}_B = 2\mathbf{i} + 3\mathbf{j}$$

These vectors are parallel, since $\mathbf{v}_A = 2\,\mathbf{v}_B$. So, particles A and B are moving in the **same** direction. Since this is the only acceptable solution to the problem, the particles are **never** moving in opposite directions. (In fact, when $t = -2$, $\mathbf{v}_B = 0$.)

iii) When the particles are moving perpendicular to each other, the scalar product $\mathbf{v}_A.\mathbf{v}_B$ of their velocities is zero. This gives

$$(8 - 4t)(4 + 2t) + (4 + 2t)(6 + 3t) = 0$$

$$\Rightarrow \quad (4 + 2t)(14 - t) = 0$$

$$\Rightarrow \quad t = -2 \quad \text{or} \quad t = 14$$

We have already established that $t = -2$ is not an acceptable solution, so the only time at which the particles are moving perpendicular to each other is $t = 14$. This gives

$$\mathbf{v}_A = -48\mathbf{i} + 32\mathbf{j} \quad \text{and} \quad \mathbf{v}_B = 32\mathbf{i} + 48\mathbf{j}$$

b) The positions of the particles at $t = 1$ and $t = 14$ are as follows:

i) When $t = 1$: $\quad \mathbf{r}_A = 6\mathbf{i} + 5\mathbf{j} \quad \text{and} \quad \mathbf{r}_B = 5\mathbf{i} + 7\frac{1}{2}\mathbf{j}$

ii) When $t = 14$: $\quad \mathbf{r}_A = -280\mathbf{i} + 252\mathbf{j} \quad \text{and} \quad \mathbf{r}_B = 252\mathbf{i} + 378\mathbf{j}$

Exercise 7B

1 A particle moves in a plane in such a way that its position at time t is given by

$$\mathbf{r} = (3t - 2)\mathbf{i} + (4t - 2t^2)\mathbf{j}$$

a) Find an expression for the velocity of the particle.
b) Find an expression for the acceleration of the particle.
c) Plot the position of the particle at times $t = 0, 1, 2, 3$. Mark on your diagram arrows representing the directions of the velocity and of the acceleration at the times indicated.

2 The velocity of a particle moving in a plane is given by

$$\mathbf{v} = (4t - 2)\mathbf{i} + (3t^2)\mathbf{j}$$

At time $t = 0$, the particle is at the point whose position vector is $2\mathbf{i} + 3\mathbf{j}$.

a) Find an expression for the acceleration of the particle.
b) Find an expression for the position of the particle.
c) Show that the particle is never stationary.
d) Find the average velocity over the time interval $t = 0$ to $t = 3$.

3 The position vector of a moving particle is given by

$$\mathbf{r} = (2 + 3t + 8t^2)\mathbf{i} + (6 + t + 12t^2)\mathbf{j}$$

a) Find an expression for the velocity of the particle at time t.
b) Find an expression for the acceleration of the particle at time t.

An observer placed at the origin sees the particle in the direction given by $\tan \theta = y/x$, where $\mathbf{r} = x\mathbf{i} + y\mathbf{j}$ is the position vector of the particle at that time. The direction in which the particle is actually moving is given by $\tan \phi = b/a$, where $\mathbf{v} = a\mathbf{i} + b\mathbf{j}$ is the velocity vector of the particle at that time. The particle will be moving directly away from or directly towards the observer when $\tan \theta = \tan \phi$.

c) Find the time(s) when the particle is moving directly away from or directly towards the observer.
d) Find the position of the particle at the time(s) found in part **c**, identifying whether the particle is moving away from or towards the observer.

4 An aircraft is flying at an altitude of 800 m at a speed of 960 km h^{-1} on a bearing of 030°. At time $t = 0$, measured in hours, it passes directly over an observer. Taking the unit vectors, $\mathbf{i}$, $\mathbf{j}$ and $\mathbf{k}$, in the directions east, north and upwards respectively, measured from the observer,

a) write down the velocity vector, $\mathbf{v}$
b) find the position vector, $\mathbf{r}$, of the aircraft t hours after passing the observer.

5 A particle moves so that its displacement is given by

$$\mathbf{r} = 5\cos\left(\frac{\pi t}{3}\right)\mathbf{i} + 5\sin\left(\frac{\pi t}{3}\right)\mathbf{j}$$

a) Find its position when $t = 0, 1, 2, 3, 4, 5, 6$ seconds and plot these points.
b) Find an expression for the velocity of the particle at time t.

c) Find its velocity at the times given in part **a** and draw an arrow on your diagram to represent each velocity.

d) Find the speed of the particle at the times given in part **a**.

e) Find an expression for the acceleration of the particle at time t.

f) Find the acceleration of the particle at the times given in part **a** and draw an arrow on your diagram to represent each acceleration.

6 A particle is moving in a plane so that its acceleration is given by $\mathbf{a} = -2\mathbf{j}$. At time $t = 0$, it is at the point whose position vector is $2\mathbf{i} - 3\mathbf{j}$ and it has a velocity of $2\mathbf{i} + 4\mathbf{j}$.

a) Find expressions for the velocity and displacement of the particle at time t.

b) At what time is the particle moving parallel to the **i**-vector?

c) At what times does the particle cross the **i**-axis?

7 Two particles have velocities given by

$$\mathbf{v}_1 = (2 + 2t)\mathbf{i} + (t + 2)\mathbf{j} \quad \text{and} \quad \mathbf{v}_2 = 3\mathbf{i} - 4\mathbf{j}$$

At time $t = 0$, both particles are at the origin.

a) Find the time at which the particles are moving perpendicular to each other.

b) Find the position of each particle at this time.

8 A particle moves in space in such a way that its displacement from the origin at time t is given by

$$\mathbf{r} = 5\cos\left(\frac{\pi t}{4}\right)\mathbf{i} + 5\sin\left(\frac{\pi t}{4}\right)\mathbf{j} + 2t\,\mathbf{k}$$

a) Find the displacement of the particle at times $t = 1, 2, 3, 4, 5, 6, 7, 8, 9, 10$ seconds.

b) Sketch the displacements found in part **a** and also the path taken by the particle in its motion.

c) Describe the motion of the particle.

d) Find expressions for the velocity and the acceleration of the particle.

Motion with uniform acceleration

Below we recapitulate the equations relating to the motion of a particle in a straight line with uniform acceleration:

$$v = u + at$$
$$s = ut + \tfrac{1}{2}at^2$$
$$v^2 = u^2 + 2as$$
$$s = \tfrac{1}{2}(u + v)t$$
$$s = vt - \tfrac{1}{2}at^2$$

When we are dealing with motion in two or three dimensions in which the acceleration vector is constant, we need to restate these equations in vector terms. Thus the variables become

Initial velocity $\quad \mathbf{u} = u_1\mathbf{i} + u_2\mathbf{j} + u_3\mathbf{k}$
Final velocity $\quad \mathbf{v} = v_1\mathbf{i} + v_2\mathbf{j} + v_3\mathbf{k}$
Acceleration $\quad \mathbf{a} = a_1\mathbf{i} + a_2\mathbf{j} + a_3\mathbf{k}$
Displacement $\quad \mathbf{r} = x\mathbf{i} + y\mathbf{j} + z\mathbf{k}$
Time $\qquad\qquad t$

Each component corresponds to motion in a straight line with uniform acceleration. So, for example, we have

$$v_1 = u_1 + a_1 t$$

$$v_2 = u_2 + a_2 t$$

$$v_2 = u_3 + a_3 t$$

which give

$$v_1\mathbf{i} + v_2\mathbf{j} + v_3\mathbf{k} = (u_1 + a_1 t)\mathbf{i} + (u_2 + a_2 t)\mathbf{j} + (u_3 + a_3 t)\mathbf{k}$$

$$= (u_1\mathbf{i} + u_2\mathbf{j} + u_3\mathbf{k}) + (a_1\mathbf{i} + a_2\mathbf{j} + a_3\mathbf{k})t$$

or

$$\mathbf{v} = \mathbf{u} + \mathbf{a}t \qquad [1]$$

Similarly, we have

$$\mathbf{r} = \mathbf{u}t + \tfrac{1}{2}\mathbf{a}t^2 \qquad [2]$$

$$\mathbf{r} = \tfrac{1}{2}(\mathbf{u} + \mathbf{v})t \qquad [3]$$

$$\mathbf{r} = \mathbf{v}t - \tfrac{1}{2}\mathbf{a}t^2 \qquad [4]$$

The remaining formula requires a more subtle change.

$v^2 = u^2 + 2as$ involves the product of a and s. Since these are now the vectors $\mathbf{a}$ and $\mathbf{r}$, we need to consider the scalar product of $\mathbf{a}$ and $\mathbf{r}$. From [1], we get

$$\mathbf{a} = \frac{\mathbf{v} - \mathbf{u}}{t}$$

and from [3], we get

$$2\mathbf{r} = (\mathbf{u} + \mathbf{v})t$$

These give

$$2\mathbf{a}.\mathbf{r} = (\mathbf{v} - \mathbf{u}).(\mathbf{u} + \mathbf{v})$$

$$\Rightarrow \quad 2\mathbf{a}.\mathbf{r} = \mathbf{v}.\mathbf{v} - \mathbf{u}.\mathbf{u} = v^2 - u^2$$

$$\Rightarrow \quad v^2 = u^2 + 2\mathbf{a}.\mathbf{r} \qquad [5]$$

which, unlike the previous four equations, is a scalar equation and gives the **speed** v.

Projectiles

A situation we frequently need to model is that of a projectile. That is, an object moving freely through the air under gravity. Unless the object is buffeted by strong side winds, it is effectively moving in two dimensions. An example of a projectile is the shot which is putt in athletics meetings. (The sport originated when soldiers competed to see who could throw a 16 lb shot the furthest.) We are going to investigate the problem of how a shot putter can get the maximum distance from a putt.

If you have access to the spreadsheet SHOT 1, you may wish to explore this simulation before continuing. (SHOT 1 is available on the Oxford University Press website: http://www.oup.co.uk./mechanics)

1 The real problem

In a shot competition, the athlete must putt the shot as far as possible. At what angle should the shot be projected in order to achieve this?

2 Setting up the model

What factors will affect the throw?

There are several factors which may affect the distance the shot is putt. Some of them are:

- Angle of projection
- Height of the shot putter
- Physical build of the shot putter
- Mass of the shot
- Wind direction and strength
- Air resistance
- Speed of projection

Simplifying assumptions

In order to simplify the problem, we assume the following:

- The shot is a particle.
- There is no air resistance.
- The shot putter has zero height.
- The acceleration due to gravity is constant.
- The initial position of the shot is at the origin.

Variables and parameters

Initially, we use the following symbols:

$\mathbf{r}$	Displacement of the shot at time t, where $\mathbf{r} = x\mathbf{i} + y\mathbf{j}$	θ	Angle of protection
$\mathbf{i}$	Horizontal unit vector	m	Mass of the shot
$\mathbf{j}$	Vertical unit vector	g	Acceleration due to gravity
U	Speed of the shot at the point of projection		

3 Formulating the mathematical model

During flight, the only force acting on the shot is its weight, $-mg\mathbf{j}$. Hence, the equation of motion (Newton's second law) is

$$-mg\mathbf{j} = m\mathbf{a}$$
$$\Rightarrow \quad \mathbf{a} = -g\mathbf{j} \qquad [1]$$

Thus, the acceleration of the shot is constant and is vertically downwards.

4 Solving the mathematical model

We can use the equations of motion with uniform acceleration.

We know that the initial velocity, $\mathbf{u}$, is given by

$$\mathbf{u} = U\cos\theta\,\mathbf{i} + U\sin\theta\,\mathbf{j}$$

Using $\mathbf{v} = \mathbf{u} + \mathbf{a}\,t$, we get

$$\mathbf{v} = U\cos\theta\,\mathbf{i} + U\sin\theta\,\mathbf{j} - gt\,\mathbf{j}$$
$$\Rightarrow \quad \mathbf{v} = U\cos\theta\,\mathbf{i} + (U\sin\theta - gt)\mathbf{j} \qquad [2]$$

Equation [2] gives us the horizontal and vertical components of velocity at any time.

Using $\mathbf{r} = \mathbf{u}\,t + \frac{1}{2}\,\mathbf{a}\,t^2$, we get

$$\mathbf{r} = (U\cos\theta\,\mathbf{i} + U\sin\theta\,\mathbf{j})t - \tfrac{1}{2}gt^2\mathbf{j}$$
$$\Rightarrow \quad \mathbf{r} = (Ut\cos\theta)\mathbf{i} + (Ut\sin\theta - \tfrac{1}{2}gt^2)\mathbf{j} \qquad [3]$$

Equation [3] gives us the horizontal and vertical components of displacement at any time.

The task we set ourselves is to find the value of θ which produces the maximum range. We therefore need an expression for the horizontal displacement, x, of the shot at the moment it hits the ground.

From [3], the vertical displacement of the shot at time t is

$$y = Ut\sin\theta - \tfrac{1}{2}gt^2$$

Because we assume that the shot putter has zero height, the shot hits the ground when $y = 0$. So, we have

$$Ut\sin\theta - \tfrac{1}{2}gt^2 = 0$$
$$\Rightarrow \quad t(U\sin\theta - \tfrac{1}{2}gt) = 0$$
$$\Rightarrow \quad t = 0 \quad \text{or} \quad t = \frac{2U\sin\theta}{g}$$

The value of $t = 0$ is the time of projection, so the shot hits the ground when

$$t = \frac{2U \sin \theta}{g}$$

From [3], the horizontal displacement of the shot is $x = Ut \cos \theta$. Hence, when the shot hits the ground, we have

$$\text{Range} = \frac{2U^2 \sin \theta \cos \theta}{g}$$

Using the double-angle formula $\sin 2\theta \equiv 2 \sin \theta \cos \theta$, this becomes

$$\text{Range} = \frac{U^2 \sin 2\theta}{g} \qquad [4]$$

The maximum range occurs when $\sin 2\theta = 1$,

$$\Rightarrow \quad 2\theta = 90° \quad \Rightarrow \quad \theta = 45°$$

The model therefore predicts that the maximum range is $\dfrac{U^2}{g}$, which is attained when the angle of projection is 45°.

5 Comparing with reality

It would be quite difficult to test our conclusion experimentally with an actual shot putter. But if we examine our initial assumptions, we can see that one of them – 'the shot putter has zero height' – is implausible, and so we explore how the model could be adjusted on pages 132–3.

In fact, for other situations where an object **is** projected from ground level, such as a ball being kicked, it can be shown experimentally that, for small objects and low speeds, the model developed above provides a realistic solution.

Example 9 A golf ball is hit towards the pin with a velocity of $50 \, \text{m s}^{-1}$ at an angle of 30° to the horizontal.

a) If the pin is 180 m away, how far from the pin will the ball land?
b) What is the furthest the ball could be hit with this initial speed?

SOLUTION

We assume that the ball is a particle, that there is no air resistance and that $g = 10 \, \text{m s}^{-2}$. We take unit vectors as shown in the diagram.

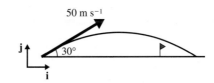

a) The initial velocity, **u**, of the ball is given by

$$\mathbf{u} = 50 \cos 30° \mathbf{i} + 50 \sin 30° \mathbf{j} = 25\sqrt{3}\mathbf{i} + 25\mathbf{j}$$

The acceleration of the ball is $-10\mathbf{j}$.

Using $\mathbf{r} = \mathbf{u}t + \frac{1}{2}\mathbf{a}t^2$, we get

$$\mathbf{r} = (25\sqrt{3}\mathbf{i} + 25\mathbf{j})t - 5t^2\mathbf{j}$$

$$\Rightarrow \quad \mathbf{r} = 25\sqrt{3}\,t\mathbf{i} + (25t - 5t^2)\mathbf{j}$$

The vertical displacement of the ball is $y = 25t - 5t^2$, and the ball hits the ground when $y = 0$. Therefore, we have

$$25t - 5t^2 = 0$$

$$\Rightarrow \quad 5t(5 - t) = 0$$

$$\Rightarrow \quad t = 0 \quad \text{or} \quad t = 5$$

The value $t = 0$ corresponds to the instant the ball was struck, so the ball lands when $t = 5$.

The horizontal displacement of the ball is $x = 25\sqrt{3}\,t$. So, the distance travelled by the ball is

$$25\sqrt{3} \times 5 = 216.5\,\text{m}$$

Hence, the ball lands $36.5\,\text{m}$ beyond the pin.

b) The maximum range is achieved when the angle of projection is $45°$. This gives an initial velocity of

$$\mathbf{u} = 50\cos 45°\mathbf{i} + 50\sin 45°\mathbf{j} = 25\sqrt{2}\mathbf{i} + 25\sqrt{2}\mathbf{j}$$

Using $\mathbf{r} = \mathbf{u}t + \frac{1}{2}\mathbf{a}t^2$, the displacement of the ball at time t is

$$\mathbf{r} = (25\sqrt{2}\mathbf{i} + 25\sqrt{2}\mathbf{j})t - 5t^2\mathbf{j}$$

$$\Rightarrow \quad \mathbf{r} = 25\sqrt{2}\,t\mathbf{i} + (25\sqrt{2}\,t - 5t^2)\mathbf{j}$$

The ball lands when the vertical displacement $25\sqrt{2}\,t - 5t^2 = 0$. That is, when $t = 0$ or $t = 5\sqrt{2}$.

The value $t = 0$ is the projection time, so the value $t = 5\sqrt{2}$ is the landing time. The maximum range is, therefore,

$$25\sqrt{2} \times 5\sqrt{2} = 250\,\text{m}$$

Note   In Example 9, part **b**, we could have used the formula we derived for the maximum range, but for examination purposes it is better that you derive the result each time from the basic equations.

Exploring the model further

On page 124, we derived the following equations for the velocity and displacement of a projectile:

$$\mathbf{v} = U\cos\theta\,\mathbf{i} + (U\sin\theta - gt)\mathbf{j} \qquad [2]$$

$$\mathbf{r} = (Ut\cos\theta)\mathbf{i} + (Ut\sin\theta - \tfrac{1}{2}gt^2)\mathbf{j} \qquad [3]$$

These can be used to solve other related problems. We consider three of them: achieving maximum height, hitting a target on the ground and finding the equation of the flight path.

Maximum height

The projectile is at the highest point of its flight when the vertical component of velocity is zero. That is,

$$U\sin\theta - gt = 0$$

$$\Rightarrow \quad t = \frac{U\sin\theta}{g}$$

Notice that, according to our model, this is exactly half the time the projectile spends in the air.

The vertical displacement is $y = Ut\sin\theta - \tfrac{1}{2}gt^2$. So, substituting the value of t, we have

$$y = \frac{U^2\sin^2\theta}{g} - \frac{U^2\sin^2\theta}{2g}$$

which gives

$$\text{Maximum height} = \frac{U^2\sin^2\theta}{2g}$$

Example 10 A boy kicks a ball from the ground with a velocity of $12\,\mathrm{m\,s^{-1}}$ at an angle of $60°$ to the horizontal. Can the ball clear a fence $5\,\mathrm{m}$ high?

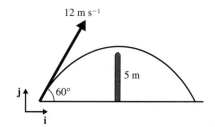

SOLUTION

We make the following assumptions:

- There is no air resistance
- The position of the fence corresponds to the highest point in the ball's trajectory
- $g = 9.8\,\mathrm{m\,s^{-2}}$

Taking unit vectors as shown in the diagram, the initial velocity, **u**, is given by

$$\mathbf{u} = 12\cos 60°\,\mathbf{i} + 12\sin 60°\,\mathbf{j} = 6\mathbf{i} + 6\sqrt{3}\,\mathbf{j}$$

The acceleration is $\mathbf{a} = -9.8\,\mathbf{j}$. Using $\mathbf{v} = \mathbf{u} + \mathbf{a}t$, the velocity of the ball at time t is given by

$$\mathbf{v} = (6\mathbf{i} + 6\sqrt{3}\,\mathbf{j}) - 9.8t\,\mathbf{j}$$
$$\Rightarrow \quad \mathbf{v} = 6\mathbf{i} + (6\sqrt{3} - 9.8t)\,\mathbf{j}$$

The ball reaches maximum height when the vertical velocity component is zero. That is, when

$$6\sqrt{3} - 9.8t = 0 \quad \Rightarrow \quad t = \frac{3\sqrt{3}}{4.9}$$

Using $\mathbf{r} = \mathbf{u}t + \frac{1}{2}\mathbf{a}t^2$, the displacement of the ball at time t is given by

$$\mathbf{r} = (6\mathbf{i} + 6\sqrt{3}\,\mathbf{j})\,t - 4.9t^2\,\mathbf{j}$$
$$\Rightarrow \quad \mathbf{r} = 6\mathbf{i} + (6\sqrt{3}\,t - 4.9t^2)\,\mathbf{j}$$

The maximum height is the vertical component of displacement when $t = \dfrac{3\sqrt{3}}{4.9}$. Therefore, we have

$$\text{Maximum height} = 6\sqrt{3} \times \left(\frac{3\sqrt{3}}{4.9}\right) - 5 \times \left(\frac{3\sqrt{3}}{4.9}\right)^2 = 5.4\,\text{m}$$

Thus, the ball can clear the fence, provided that it is at or near its maximum height as it passes over the fence.

Hitting a target

Suppose that an object is projected from the ground with initial speed U and angle of projection θ, and that we require it to hit a target on the ground at a distance R, which is less than the maximum range. On page 125, we derived the following expression for the range:

$$\text{Range} = \frac{U^2 \sin 2\theta}{g}$$

So, we have

$$R = \frac{U^2 \sin 2\theta}{g}$$

$$\Rightarrow \quad \sin 2\theta = \frac{Rg}{U^2}$$

In general, solving this equation gives two complementary values $\theta = \alpha$ and $\theta = (90° - \alpha)$.

(You may have noticed this if you tried the spreadsheet investigation SHOT 1.)

Example 11 A shell is fired with a velocity of $200 \, \text{m s}^{-1}$ at an angle θ to the horizontal, to hit a target on the same level 3000 m away. At what angle of projection should the shell be fired?

SOLUTION

Making the usual assumptions and taking $g = 9.8 \, \text{m s}^{-2}$, we have

Initial velocity of the shell: $\mathbf{u} = 200 \cos \theta \, \mathbf{i} + 200 \sin \theta \, \mathbf{j}$

Acceleration: $\mathbf{a} = -9.8 \mathbf{j}$

Using $\mathbf{r} = \mathbf{u} t + \frac{1}{2} \mathbf{a} t^2$, the displacement of the shell at time t is given by

$$\mathbf{r} = (200 \cos \theta \, \mathbf{i} + 200 \sin \theta \, \mathbf{j}) \, t - 4.9 t^2 \mathbf{j}$$

$$\Rightarrow \quad \mathbf{r} = 200 \cos \theta \, t \mathbf{i} + (200 \sin \theta \, t - 4.9 t^2) \mathbf{j}$$

The shell lands when the vertical component of displacement is zero.

$$200 \sin \theta \, t - 4.9 t^2 = 0$$

which gives

$$t = 0 \quad \text{or} \quad t = \frac{200 \sin \theta}{4.9}$$

The value $t = 0$ is the time of projection, so the shell lands when

$$t = \frac{200 \sin \theta}{4.9}$$

The range is given by the horizontal component of displacement at this time. So, we have

$$\text{Range} = 200 \cos \theta \times \frac{200 \sin \theta}{4.9}$$

Using the double-angle formula $\sin 2\theta \equiv 2 \sin \theta \cos \theta$, this becomes

$$\text{Range} = \frac{20\,000 \sin 2\theta}{4.9}$$

The required range is 3000 m. So, we have

$$\frac{20\,000 \sin 2\theta}{4.9} = 3000$$

$$\Rightarrow \quad \sin 2\theta = 0.735$$

$$\Rightarrow \quad 2\theta = 47.3° \quad \text{or} \quad 132.7°$$

$$\Rightarrow \quad \theta = 23.7° \quad \text{or} \quad 66.3° \text{ (to 3 sf)}$$

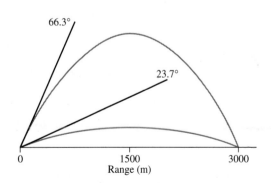

Note that the two possible angles of projection are complementary (add up to 90°).

Equation of the path of a projectile

On page 124, we derived the following expression for the displacement of a projectile at time t:

$$\mathbf{r} = (Ut\cos\theta)\mathbf{i} + (Ut\sin\theta - \tfrac{1}{2}gt^2)\mathbf{j} \qquad [3]$$

where $\qquad \mathbf{r} = x\,\mathbf{i} + y\,\mathbf{j}$

Equating components, we have

$$x = Ut\cos\theta \qquad\qquad [5]$$

$$y = Ut\sin\theta - \tfrac{1}{2}gt^2 \qquad [6]$$

From [5], we have

$$t = \frac{x}{U\cos\theta}$$

Substituting for t in [6], we get

$$y = U\left(\frac{x}{U\cos\theta}\right)\sin\theta - \frac{g}{2}\left(\frac{x}{U\cos\theta}\right)^2$$

$$\Rightarrow \quad y = x\tan\theta - \left(\frac{g\sec^2\theta}{2U^2}\right)x^2$$

This equation gives y as a quadratic function of x and so, according to our model, the path of the projectile is a parabola.

Note As with the range, it is better to learn how to derive the equation of the path of the projectile than to try to memorise it.

Example 12 A projectile is launched with a velocity of $40\,\mathrm{m\,s^{-1}}$ at an angle of $30°$ to the horizontal. What is the equation of the path of the projectile? (Take $g = 10\,\mathrm{m\,s^{-2}}$.)

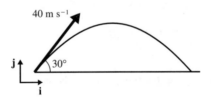

SOLUTION

With the usual assumptions and taking unit vectors as shown, we have:

Initial velocity: $\quad \mathbf{u} = 40\cos 30°\,\mathbf{i} + 40\sin 30°\,\mathbf{j}$

$$= 20\sqrt{3}\,\mathbf{i} + 20\,\mathbf{j}$$

Acceleration: $\quad \mathbf{a} = -10\,\mathbf{j}$

Using $\mathbf{r} = \mathbf{u}t + \tfrac{1}{2}\mathbf{a}t^2$, the displacement of the projectile at time t is given by

$$\mathbf{r} = (20\sqrt{3}\,\mathbf{i} + 20\,\mathbf{j})t - 5t^2\,\mathbf{j}$$

$$\Rightarrow \quad \mathbf{r} = 20\sqrt{3}\,t\,\mathbf{i} + (20\,t - 5t^2)\,\mathbf{j}$$

But, we have

$$\mathbf{r} = x\mathbf{i} + y\mathbf{j}$$

So, equating components, we get

$$x = 20\sqrt{3}\,t \qquad [1]$$

and $\quad y = 20t - 5t^2 \qquad [2]$

From [1], we get

$$t = \frac{x}{20\sqrt{3}}$$

Substituting for t in [2], we obtain the equation of the projectile's path:

$$y = 20\left(\frac{x}{20\sqrt{3}}\right) - 5\left(\frac{x}{20\sqrt{3}}\right)^2$$

$$\Rightarrow \quad y = \frac{x\sqrt{3}}{3} - \frac{x^2}{240}$$

So, the path is a parabola.

Note Knowing the equation of the projectile's path could be used to solve other problems, but in practice it is usually better to solve problems starting from the basic equations of motion.

Example 13 A ball is kicked with a velocity of $10\,\text{m s}^{-1}$ at an angle of $40°$ to the horizontal towards a wall which is $7\,\text{m}$ away.

a) How far up the wall does the ball hit?
b) What is the speed of the ball when it hits the wall?
c) In what direction is the ball moving when it hits the wall?

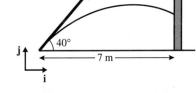

SOLUTION

We make the usual assumptions and take $g = 9.8\,\text{m s}^{-2}$.

With unit vectors as shown in the diagram, we have

Initial velocity: $\quad \mathbf{u} = 10\cos 40°\,\mathbf{i} + 10\sin 40°\,\mathbf{j}$

Acceleration: $\quad \mathbf{a} = -9.8\,\mathbf{j}$

Using $\mathbf{v} = \mathbf{u} + \mathbf{a}t$, the velocity at time t is given by

$$\mathbf{v} = (10\cos 40°\,\mathbf{i} + 10\sin 40°\,\mathbf{j}) - 9.8\,t\,\mathbf{j}$$

$$\Rightarrow \quad \mathbf{v} = 7.660\,\mathbf{i} + (6.428 - 9.8\,t)\,\mathbf{j}$$

Using $\mathbf{r} = \mathbf{u}\,t + \frac{1}{2}\mathbf{a}\,t^2$, the displacement at time t is given by

$$\mathbf{r} = (10\cos 40°\,\mathbf{i} + 10\sin 40°\,\mathbf{j})\,t - 4.9\,t^2\,\mathbf{j}$$
$$\Rightarrow \quad \mathbf{r} = 7.660\,t\,\mathbf{i} + (6.428\,t - 4.9\,t^2)\,\mathbf{j}$$

a) The horizontal displacement of the ball at time t is $7.660t$.

When the ball has travelled $7\,m$ horizontally to hit the wall, we have

$$7.660t = 7 \quad \Rightarrow \quad t = 0.9138$$

The height of the ball at time t is $6.428t - 4.9t^2$. So, when the ball hits the wall

$$\text{Height} = 6.428 \times 0.9138 - 4.9 \times 0.9138^2 = 1.78\,m \quad \text{(to 3 sf)}$$

b) The velocity of the ball at this point is

$$\mathbf{v} = 7.660\mathbf{i} + (6.428 - 9.8 \times 0.9138)\mathbf{j}$$
$$= 7.66\mathbf{i} - 2.53\mathbf{j}$$

The speed at impact is the magnitude of this velocity. So, we have

$$v = \sqrt{7.66^2 + 2.53^2} = 8.07\,m\,s^{-1}$$

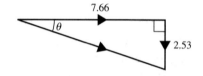

c) The direction of the motion is given by the angle θ in the diagram, where

$$\tan\theta = \frac{2.53}{7.66} \quad \Rightarrow \quad \theta = 18.3°$$

So, when the ball hits the wall, it is travelling in a direction $18.3°$ below the horizontal.

Revising the model

Finally, let us return to the problem of the shot putter. We can now adjust our model to allow for the height of the putter.

The equation $\mathbf{r} = \mathbf{u}\,t + \frac{1}{2}\mathbf{a}\,t^2$ assumed that the shot started at the origin. If, in fact, it starts from position vector $\mathbf{s}$, the equation becomes

$$\mathbf{r} = \mathbf{s} + \mathbf{u}\,t + \frac{1}{2}\mathbf{a}\,t^2 \qquad [7]$$

In our case, $\mathbf{s} = h\mathbf{j}$, where h is the height of the shot above the ground when it is released. As before, $\mathbf{u} = U\cos\theta\,\mathbf{i} + U\sin\,\mathbf{j}$ and the acceleration is $-g\mathbf{j}$, so [7] becomes

$$\mathbf{r} = h\mathbf{j} + (U\cos\theta\,\mathbf{i} + U\sin\theta\,\mathbf{j})t - \tfrac{1}{2}gt^2\mathbf{j}$$

$$\Rightarrow \quad \mathbf{r} = Ut\cos\theta\,\mathbf{i} + (h + Ut\sin\theta - \tfrac{1}{2}gt^2)\mathbf{j}$$

The shot lands when

$$h + Ut\sin\theta - \tfrac{1}{2}gt^2 = 0$$

This is a quadratic equation in t whose roots are

$$t = \frac{U\sin\theta \pm \sqrt{U^2\sin^2\theta + 2gh}}{g}$$

One of the roots is negative and can be rejected.

The horizontal displacement of the shot is $Ut\cos\theta$. Substituting the positive value of t gives

$$\text{Range} = \frac{U^2\sin\theta\cos\theta + U\cos\theta\sqrt{U^2\sin^2\theta + 2gh}}{g}$$

This expression is more complex than that produced by the previous, simpler model, and maximising it is not easy. If you have access to the spreadsheet SHOT 2, you can explore the model. You will find that the maximum range occurs when the angle of projection is slightly less than 45°, with the exact value depending on the height of the point of projection.

(SHOT 2 is available on the Oxford University Press website: http://www.oup.co.uk/mechanics)

Exercise 7C

1 A projectile is launched from ground level with a speed of $15\,\mathrm{m\,s^{-1}}$ at an angle of 35° to the horizontal.

   a) For how long is the projectile in the air?
   b) What is the horizontal range of the projectile?
   c) Find the time taken to reach maximum height.
   d) What is the greatest height reached?

2 A projectile is launched from ground level with a speed of $10\,\mathrm{m\,s^{-1}}$ at an angle of 60° to the horizontal.

   a) For how long is the projectile in the air?
   b) What is the horizontal range of the projectile?
   c) Find the time taken to reach maximum height.
   d) What is the greatest height reached?

3 A particle is projected from a point at ground level with a speed of $24\,\mathrm{m\,s^{-1}}$ at an angle of 50° to the horizontal. A wall is situated 30 m away from the projection point.

a) Find how far up the wall the particle hits it.

b) What is the speed of the particle when it hits the wall?

c) Find the direction of motion of the particle when it hits the wall.

4 James is throwing pebbles into the water. He finds that the greatest distance that he can throw them is 80 m. Assuming that his height is negligible, how fast does he throw them and how long do they take to hit the water?

5 The maximum range of a projectile, fired with speed U, is R. A target is placed h m above the landing point. Find the speed with which it must be projected if it is to hit this target without changing the angle of projection.

6 Two projectiles are released simultaneously from the same point with the same speed, one at an angle of elevation θ, and the other at an angle of elevation α. Show that, during flight,

a) the line joining the two particles has a constant gradient

b) the distance between them is increasing at a constant rate.

7 A projectile, released with a velocity U at an angle of elevation θ to the horizontal, just clears two obstacles, both of height h m, whose distances from the projection point are b m and $3b$ m respectively. Find the range of the projectile.

8 A ball is projected with a velocity of $(30\mathbf{i} + 40\mathbf{j})$ from a point on the ground.

a) Find the position of the ball 1 s and 2 s later.

b) How long would the ball stay in the air if the ground were level?

c) Find the range of the ball.

d) Find the equation of the path of the ball.

Take the value of g to be $10\,\mathrm{m\,s^{-2}}$.

9 A ball was projected at an angle of 60° to the horizontal. One second later another ball was projected from the same point at an angle of 30° to the horizontal. One second after the second ball was released, the two balls collided. Show that the velocities of the balls were $12.99\,\mathrm{m\,s^{-1}}$ and $15\,\mathrm{m\,s^{-1}}$. Take the value of g to be $10\,\mathrm{m\,s^{-2}}$.

10 A netball is projected with a velocity of $2\sqrt{5}(\mathbf{i} + 2\mathbf{j})\,\mathrm{m\,s^{-1}}$ towards a net. Take the point of projection as the origin and ignore air resistance. Assume $g = 10\,\mathrm{m\,s^{-2}}$.

a) Draw a diagram showing the forces acting on the ball whilst in flight.

b) Write down the equation of motion of the ball and from it find expressions for the velocity, **v**, and the position, **r**, of the ball at a subsequent time, t seconds after projection.

c) Find the equation of the path of the ball.

The net is 5 m away from the point of projection.

d) If the ball is to pass through the net, how high above the point of projection should the net be?

e) By considering the velocity of the ball, show that it is moving downwards as it passes through the net.

Examination questions

Chapters 5 to 7

Chapter 5

1 The radius of the earth is R and the acceleration due to gravity at the earth's surface is g. The gravitational force acting on a particle P of mass m, when it is at a distance x, $x \geqslant R$, from the centre of the earth, has magnitude $\dfrac{km}{x^2}$ where k is a constant.

   a) Find k in terms of R and g.

   At time $t = 0$, P is projected vertically upwards from the surface of the earth with speed $\sqrt{(1.6gR)}$. At time t, it is at a distance x from the centre of the earth and is moving with velocity v.

   Ignoring the effects of air resistance, and the motion of the earth,

   b) show that $v\dfrac{\mathrm{d}v}{\mathrm{d}x} = -\dfrac{R^2 g}{x^2}$

   c) Hence find, in terms of R, the greatest height of P above the earth's surface.   (EDEXCEL)

2 A lift cage in a mine shaft is carrying a passenger of mass 80 kg. The cage is being pulled up the mine shaft by a cable and is accelerating at a rate of $3\,\mathrm{m\,s^{-2}}$.

   a) Taking $g = 10\,\mathrm{m\,s^{-2}}$, find the reaction of the floor of the cage on the passenger.
   b) State the magnitude and direction of the reaction of the passenger on the floor of the cage.   (NEAB)

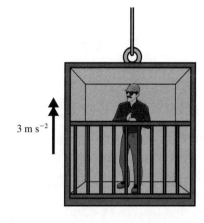

3 A tractor of mass 1600 kg is pulling a trailer of mass 800 kg along a level muddy field. The driving force of the tractor is 7000 N and the resistances to motion acting on the tractor and trailer are 1000 N and 1200 N respectively.

   i) Show that the common acceleration of the tractor and trailer is $2\,\mathrm{m\,s^{-2}}$ in the direction of their motion.

The tractor and the trailer are connected by a light, horizontal coupling, as shown in the diagram.

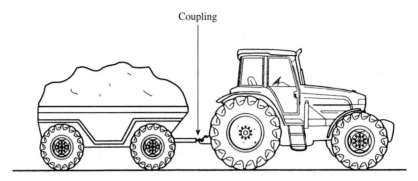

Coupling

ii) Calculate the tension in the coupling.

The coupling between the tractor and the trailer breaks at the time when the speed is $2.7\,\mathrm{m\,s^{-1}}$. Assume that the resistances to motion and the driving force of the tractor do not change.

iii) a) Show that the deceleration of the trailer is $1.5\,\mathrm{m\,s^{-2}}$.
b) Calculate the distance moved by the trailer as it comes to rest.
c) Calculate the distance between the tractor and trailer when the trailer comes to rest.

(MEI)

4

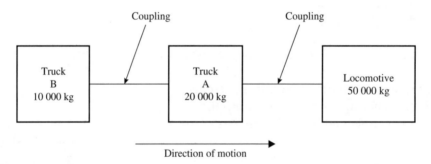

Direction of motion

A train made up of a locomotive and two trucks is travelling along a straight, level track. There is a driving force of $10\,000\,\mathrm{N}$ from the locomotive. The resistances to forward motion are $500\,\mathrm{N}$ on each of the trucks and $1000\,\mathrm{N}$ on the locomotive. The masses of the locomotive and the two trucks are shown in the diagram.

i) Draw a diagram showing all the horizontal forces acting on the locomotive and on each of the trucks, including the forces in the couplings.
ii) Show that the acceleration of the train is $0.1\,\mathrm{m\,s^{-2}}$. Calculate the forces in each of the two couplings.

With the first locomotive still exerting a driving force of $10\,000\,\mathrm{N}$, a second locomotive is added to the train. This locomotive is behind truck B and has the effect of applying a forward force of $4000\,\mathrm{N}$ to it.

iii) Show that the coupling between trucks A and B is now under a compression of $2000\,\mathrm{N}$.

The second locomotive has a mass of $40\,000\,\mathrm{kg}$ and resistance to forward motion of $800\,\mathrm{N}$.

iv) Calculate the total driving force of the second locomotive.   (MEI)

5 Two particles of mass $4m$ and m respectively are joined by a light inextensible string passing over a smooth peg. The particles are held at rest with the string taut and then released from rest.

a) State which of the underlined words enables you to assume that
   i) the tension is constant in the string on either side of the peg
   ii) the tension is the same on both sides of the peg at the points of contact with the string.
b) Write down the equation of motion of each particle and determine the acceleration of the particles   (WJEC)

6

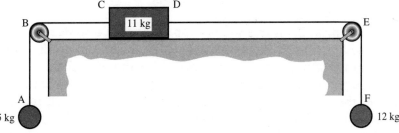

Light, inextensible strings AC and DF are attached one to each side of a block of mass 11 kg which is on a rough horizontal table. The string sections BC and DE are parallel to the table and the strings pass over smooth pulleys at B and E. Objects of mass 5 kg and 12 kg are attached to the free ends A and F respectively of the strings and hang freely. The diagram shows this system. A constant friction force of 35 N opposes the motion of the 11 kg block and all other resistances to motion are negligible. The system is released and moves from rest.

i) Draw three separate diagrams showing all the forces acting on the block and on the two hanging objects.
ii) Show that the magnitude of the acceleration of the system is $1.2\,\mathrm{m\,s^{-2}}$. Calculate also the tensions in the strings AC and DF.

When the system reaches a speed of $4\,\mathrm{m\,s^{-1}}$, the string DF breaks at D. The friction force on the 11 kg block is unchanged.

iii) Calculate the total distance travelled by the block, after its release from rest, before it comes instantaneously to rest.   (MEI)

7 A trolley of mass 2 kg can move on a horizontal table. One end of a light inextensible string is fixed to the trolley. The string passes over a smooth pulley at the edge of the table, and a wooden block B of mass 0.5 kg hangs freely at the other end of the string. The part of the string between the trolley and the pulley is horizontal (see diagram). Resistances to the motion of the system, from all causes, are modelled as a constant horizontal force of magnitude F newtons acting on the trolley.

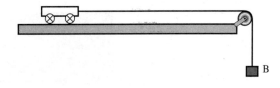

i) The system is released from rest with B at a height of 1 m above the floor, and B hits the floor 2.5 s later. Use this information to calculate the acceleration of B while it is falling, and the speed with which it hits the floor.
ii) Hence find the value of F and the tension in the string while B is falling.   (OCR)

8

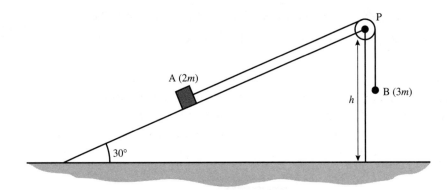

A small parcel A of mass $2m$ is free to move on the face of a wedge which is inclined at an angle of 30° to the horizontal. The wedge is fixed on horizontal ground. One end of a string is attached to A and the string passes over a small smooth pulley fixed at the point P, which is at the top of the wedge, the string remaining parallel to the line of greatest slope of the wedge. The other end of the string is attached to a ball B of mass $3m$ which can move freely in a vertical line below P, as shown in the diagram. The pulley is at a height h above the ground. The parcel and the ball are modelled as particles, the inclined face of the wedge is assumed to be smooth, and the string is assumed to be light and inextensible. The system is released from rest with the string taut and B initially at the level of P. In the period before B hits the ground, the magnitude of the acceleration of B is f.

a) State which assumption in the modelling of the situation implies that the magnitude of the acceleration of A is also f.
b) Write down an equation of motion for A and an equation of motion for B.
c) Hence show that $f = \frac{2}{5}g$.

When B hits the ground, it rebounds vertically in such a way that its initial speed on leaving the ground is the same as its speed just before it hit the ground. The string then becomes slack and A continues to move up the wedge. Given that, when B reaches the highest point on its initial rebound from the ground, A has just reached the pulley P,

d) find the length of the string in terms of h.   (EDEXCEL)

9

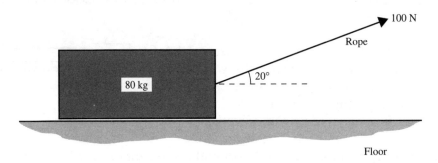

A box of mass 80 kg is to be pulled along a horizontal floor by means of a light rope. The rope is pulled with a force of 100 N and the rope is inclined at 20° to the horizontal, as shown in the diagram.

i) Explain briefly why the box cannot be in equilibrium if the floor is smooth.

In fact the floor is not smooth and the box is in equilibrium.

ii) Draw a diagram showing all the external forces acting on the box.

iii) Calculate the frictional force between the box and the floor and also the normal reaction of the floor on the box, giving your answers correct to three significant figures.

The maximum value of the frictional force between the box and the floor is 120 N and the box is now pulled along the floor with the rope always inclined at 20° to the horizontal.

iv) Calculate the force with which the rope must be pulled for the box to move at a constant speed. Give your answer correct to three significant figures.

v) Calculate the acceleration of the box if the rope is pulled with a force of 140 N.   (MEI)

10 A skier, of mass 60 kg, is being pulled up a slope by a rope as shown in the first diagram. The slope is at 20° to the horizontal and the rope is at 10° to the slope. The skier is to be modelled as a particle with forces acting as shown in the second diagram. Assume that the magnitude, F, of the resistance force is 80 N.

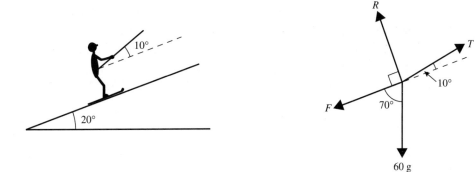

a) Show that when travelling at a constant speed, the tension in the rope T is approximately 290 N and find the magnitude, R, of the normal reaction.

b) Initially the skier accelerates at 0.1 m s^{-2}. Find the magnitude of the tension in the rope while the skier is accelerating.   (AEB 97)

Chapters 6 and 7

11 The acceleration $\mathbf{a}$ m s^{-2} of a particle P at time t seconds is given by

$$\mathbf{a} = 3t\,\mathbf{i} + 2\,\mathbf{j}$$

When $t = 0$ the velocity of P is $(2\mathbf{i} + \mathbf{j})$ m s^{-1}.

a) Find the velocity of P when $t = 2$.

When $t = 2$, the direction of motion of P makes an angle θ with the vector $\mathbf{j}$.

b) Find, to the nearest degree, the value of θ.   (EDEXCEL)

12 At time t seconds the position vector, $\mathbf{r}$ metres, of a moving particle P is given by

$$\mathbf{r} = (2t - 1)\mathbf{i} + 4\cos 3t\,\mathbf{j} + 4\sin 3t\,\mathbf{k}$$

Find
a) the velocity $\mathbf{v}$ m s^{-1} of P at time t seconds
b) the time when $\mathbf{v}$ and $\mathbf{r}$ are perpendicular to each other.   (WJEC)

13 A golf ball is struck from a point O on horizontal ground so that initially it is moving with speed $25\,\text{m s}^1$ at an angle $\tan^{-1}(\frac{3}{4})$ to the horizontal.

  a) Write down expressions for the horizontal and vertical components of its displacement from O at any subsequent time t.

  b) State **two** physical assumptions that you have made in determining the displacements.

  c) Determine the time to reach maximum height and find the range as predicted by your model.   (WJEC)

14 Darts are thrown at a vertical dartboard. The trajectory of each dart thrown is in a vertical plane.

  i) One dart is projected horizontally from a point 1.7 m above a horizontal floor and 3 m away from the dartboard. This dart strikes the board at a point which is 1.6 m above the floor.

   a) Show that the time taken for the dart to reach the dartboard is $\frac{1}{7}$th of a second.

   b) Show that the speed with which the dart was projected was $21\,\text{m s}^{-1}$.

   c) Find the angle, in degrees correct to one decimal place, which the trajectory of the dart makes with the vertical when it strikes the dartboard.

  ii) A second dart is projected from the same position at $20\,\text{m s}^{-1}$. It strikes the board at a point which is 1.8 m above the floor.

   a) Find the angles, in degrees correct to one decimal place, to the horizontal at which the dart could be thrown.

   b) If there is a horizontal ceiling 2.8 m above the floor, state which is the appropriate angle of projection.   (NICCEA)

15 The diagram shows a stone projected horizontally with speed $5\,\text{m s}^{-1}$ from the top of a vertical cliff at a height of 19.6 m above sea level. The stone strikes the sea at the point Q.

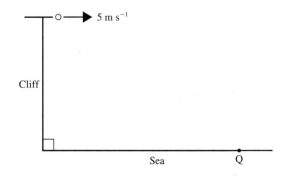

  a) Find the time taken for the stone to reach sea level and the distance of Q from the base of the cliff.

  b) Find the tangent of the angle between the horizontal and the direction of motion of the stone just as it strikes the sea at Q.   (WJEC)

16 A ball is thrown with speed $7\sqrt{10}\,\text{m s}^{-1}$ from the top of a vertical cliff at an angle of θ above the horizontal. The top of the cliff is 50 m above a horizontal seashore. The ball first strikes the seashore at a horizontal distance $50\sqrt{3}\,\text{m}$ from the point of projection.

  i) Find the only possible angle of projection.

  ii) Find the direction of motion of the ball just before it strikes the seashore.

  iii) Give one assumption that you have made in answering parts **i** and **ii** of this question.   (NICCEA)

17

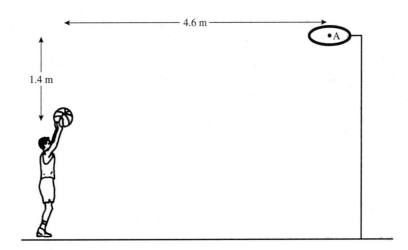

During a practice session, a basketball player throws a ball towards a horizontal ring of centre A. In a simple model this ball is treated as a particle.

a) The ball is first projected at a speed of $8\,\text{m s}^{-1}$ and at an angle of $40°$ to the horizontal, from a point at a horizontal distance of $4.6\,\text{m}$ from A and $1.4\,\text{m}$ below A.
  i) Find the time taken for the ball to travel a horizontal distance of $4.6\,\text{m}$, giving your answer correct to two significant figures.
  ii) Taking $g = 10\,\text{m s}^{-2}$, show that the ball passes below A.
b) The player throws again from the same point as before. He projects the ball at an angle of $40°$ to the horizontal but increases the speed of projection to $V\,\text{m s}^{-1}$.
  i) Determine the value of V for which the ball passes through A.
  ii) Show that, for this value of V, the ball is descending as it passes through A.   (NEAB)

18

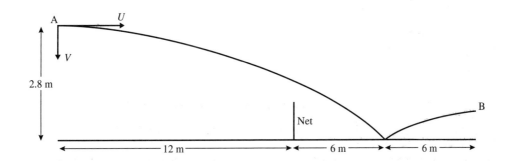

The diagram shows the trajectory of a tennis ball during a serve. The server's racket hits the ball at a point A which is $2.8\,\text{m}$ above the ground, and projects the ball towards the receiver with initial velocity components $U\,\text{m s}^{-1}$ horizontally and $V\,\text{m s}^{-1}$ vertically downwards. The server and the receiver are each at a distance of $12\,\text{m}$ horizontally from the net, and the ball bounces halfway between the net and the receiver. The receiver's racket hits the ball $0.6\,\text{s}$ after the serve, at the point B. Assume that the ball may be treated as a particle, that air resistance may be neglected, and that the ball's horizontal speed is unaffected by the bounce.

i) Show that $U = 40$.
ii) Find the value of V, and show that the ball clears the net, which has a height of $0.91\,\text{m}$, by approximately $0.24\,\text{m}$.

iii) The point B is 0.75 m above the ground. Calculate the direction in which the ball is travelling when the receiver's racket hits it at B.

Suppose now that air resistance is taken into account, but that the other assumptions and all the given distances and times remain unchanged. State, with a reason, whether the value of U is larger or smaller than 40.   (OCR)

19 A particle P, of mass 0.2 kg, moves under the action of a resultant force, **F** newtons, which at time t seconds is given by

$$\mathbf{F} = \frac{2t}{5}\mathbf{i} + \frac{2}{5}\mathbf{j} \qquad t \geq 0$$

a) i) The velocity of P when $t = 0$ is $2\mathbf{i}$ m s^{-1}. Find an expression for the velocity of P at time t.
ii) Find the time at which the direction of P is momentarily parallel to $\mathbf{j}$.
iii) Show that the speed of P at time t is given by $\sqrt{(t^4 + 4)}$ m s^{-1}.
b) The particle passes through the points A and B at times $t = 1$ and $t = 4$, respectively. Calculate the distance AB, giving your answer in the form $n\sqrt{2}$, where n is a positive integer.
    (NEAB)

20 A submarine fires a torpedo, of mass m, horizontally into the water at a speed of 30 m s^{-1}. It continues to move horizontally subject to a resistance force of magnitude mkv^3, where k is a constant and v is the speed of the torpedo at time t. The torpedo hits a target, which is 100 m away from the submarine, at a speed of 10 m s^{-1}.

a) Find an expression for $\dfrac{dv}{dt}$ and show that

$$\frac{1}{v^2} = 2kt + \frac{1}{900}$$

b) Also show that

$$\frac{1}{v} = kx + \frac{1}{30}$$

where x is the distance of the torpedo from the submarine at time t.

c) Find the time the torpedo takes to reach the target.   (AEB 97)

21 A luggage transporter at an airport is designed so that luggage is moved smoothly from one location to another taking a total time of 3 minutes. The luggage starts from rest and moves in a straight line. For the first 60 seconds, the acceleration of the luggage is constant and is equal to f m s^{-2}. For the last 60 seconds, there is a constant deceleration of f m s^{-2}. For the 60 seconds between, the acceleration changes from $+f$ m s^{-2} to $-f$ m s^{-2}. A model of the motion makes the assumption that this change in acceleration is uniform with respect to time, so that the acceleration a m s^{-2} of the luggage, t seconds after it starts to move, $60 \leq t \leq 120$, is given by

$$a = 3f - \frac{ft}{30}$$

a) Find, in terms of f, the maximum speed reached by the luggage.

Given that the maximum speed reached by the luggage is $54 \, \mathrm{km \, h^{-1}}$,

b) show that $f = \frac{1}{5}$.

c) Find the distance travelled by the luggage in the first 90 seconds for this value of f.

(EDEXCEL)

22

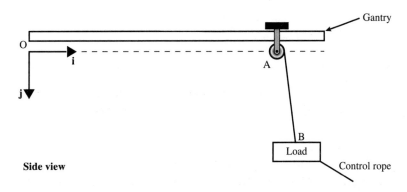

Side view

A horizontal overhead gantry is used for unloading lorries. A light wire is wound round a drum at A and is attached to a hook at B, which supports a load. A further rope attached to the load may be pulled to help control its motion. The drum at A may be moved along the gantry. The position vectors of points A and B are $\mathbf{r}_A$ and $\mathbf{r}_B$ respectively, where the unit of length is the metre. O is the origin, the unit vector $\mathbf{i}$ is horizontal in the direction OA and the unit vector $\mathbf{j}$ is vertically downwards, as shown in the diagram.

During the time interval $0 \leqslant t \leqslant 2.5$, where t is the time in seconds, the motions of A and B are modelled as

$$\mathbf{r}_A = \frac{t}{2}\mathbf{i} \quad \text{and} \quad \mathbf{r}_B = \frac{t^2}{8}\mathbf{i} + \left(\frac{3t^3}{4} + 1\right)\mathbf{j}$$

i) Calculate the distance between A and B when
    a) $t = 0$ **b)** $t = 1$

ii) Find expressions for the velocity and acceleration of B at time t.

iii) What is the direction of motion of B when $t = 1.5$?

iv) By considering the acceleration of B, show that, for some values of t, the controlling rope must exert a downward force on the load.   (MEI)

8 Relative motion

'Will you walk a little faster?' said a whiting to a snail,
'There's a porpoise close behind us, and he's treading on my tail.'
LEWIS CARROLL

Imagine that you are in a train. You roll a ball backwards down the carriage to a friend at the other end of the carriage. Which way does the ball travel – forwards or backwards?

From your point of view, it obviously travels backwards, but what would a track-side observer conclude? This depends on the speed of the train and how fast you roll the ball.

Case 1

Let us assume that the train is travelling at $15\,m\,s^{-1}$ and that you roll the ball so that it takes 2 seconds to reach your friend. Let us also assume that it is 20 m from you to your friend.

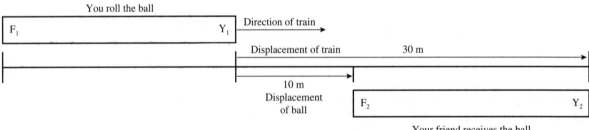

The ball takes 2 seconds to travel from you to your friend. Suppose, when you start the ball rolling, the track-side observer fixes your position at the point Y_1 and that of your friend at the point F_1 20 m away. Then, 2 seconds later, the observer would fix your position at the point Y_2 and your friend's at the point F_2, both of you having travelled 30 m.

The ball, which started at the point Y_1, finishes at the point F_2. It therefore has travelled a total distance of 10 m forwards, according to the observer. Since this took 2 seconds, he would say that the velocity of the ball is $5\,m\,s^{-1}$ **forwards**.

Case 2

Let us assume that the train is travelling at $5\,m\,s^{-1}$ and that you roll the ball in the same way.

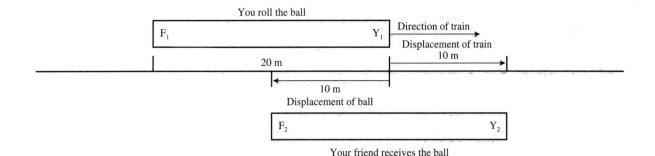

The ball again takes 2 seconds to travel from you to your friend. Suppose, when you start the ball rolling, the observer fixes your position at the point Y_1 and that of your friend at the point F_1. Then as before, 2 seconds later, the observer would fix your position at the point Y_2 and your friend's at the point F_2, both of you having travelled 10 m.

The ball, which started at the point Y_1, again finishes at the point F_2. It therefore has travelled a total distance of 10 m backwards, according to the observer. Since this took 2 seconds, he would say that the velocity of the ball is $5 \, \text{m s}^{-1}$ **backwards**.

Thus, according to the observer, the velocity of the ball has changed, but according to you and your friend, it is exactly the same as before. You would say that the ball has travelled backwards 20 m in 2 seconds, a speed of $10 \, \text{m s}^{-1}$.

The example of the ball and the train illustrates a general principle. Every observer measures velocity relative to some frame of reference in which the observer is stationary, hence the term **relative velocity**. In the example, the train provided the frame of reference for you and your friend. The frame of reference for the track-side observer was the Earth.

Almost all the velocities you encounter are relative to the surface of the Earth. When this is the case, we do not use the term 'relative', although it is implied. However, when observers are moving relative to the Earth, and perhaps relative to each other, we need to be clear what frame of reference is being used.

In the same way, displacement is measured relative to a frame of reference. In the example, you and your friend measured the displacement of the ball as 20 m backwards relative to the train. The track-side observer measured the ball's displacement as 10 m forwards in Case 1, and 10 m backwards in Case 2, relative to the Earth. Strictly, we should, therefore, talk about **relative displacement**, although in cases where all the displacements are relative to the Earth we do not bother to state this explicitly.

Notation

We use the following notation:

$$_A\mathbf{v}_B = \text{Velocity of A relative to B}$$

$$_A\mathbf{r}_B = \text{Displacement of A relative to B}$$

Let us now examine the ball-on-the-train situation using this notation, and find the relationship between the velocities involved.

We have

Velocity of the train relative to the Earth	$_T\mathbf{v}_E$
Velocity of the ball relative to the train	$_B\mathbf{v}_T$
Velocity of the ball relative to the Earth	$_B\mathbf{v}_E$

which give

Case 1: $_T\mathbf{v}_E = 15\,\mathrm{m\,s^{-1}}$ $_B\mathbf{v}_T = -10\,\mathrm{m\,s^{-1}}$ $_B\mathbf{v}_E = 5\,\mathrm{m\,s^{-1}}$

Case 2: $_T\mathbf{v}_E = 5\,\mathrm{m\,s^{-1}}$ $_B\mathbf{v}_T = -10\,\mathrm{m\,s^{-1}}$ $_B\mathbf{v}_E = -5\,\mathrm{m\,s^{-1}}$

It is evident that in each case

$$_B\mathbf{v}_E = {_B\mathbf{v}_T} + {_T\mathbf{v}_E}$$

This illustrates a general relationship between relative velocity vectors:

$$_A\mathbf{v}_C = {_A\mathbf{v}_B} + {_B\mathbf{v}_C}$$

Notice the order of the references. This will help you to remember the relationship.

To show that the relationship holds in general, consider observers on two vehicles, A and B, moving relative to a point C, with velocities $_A\mathbf{v}_C$ and $_B\mathbf{v}_C$ respectively.

To the observer on B, B appears to be stationary and A to be moving with velocity $_A\mathbf{v}_B$. This is equivalent to adding velocity vector $-_B\mathbf{v}_C$ to the whole system so that B becomes stationary. A now moves with velocity given by the vector sum of $_A\mathbf{v}_C$ and $-_B\mathbf{v}_C$, as shown in the right hand diagram on page 147.

This gives

$$_A\mathbf{v}_B = {_A}\mathbf{v}_C - {_B}\mathbf{v}_C$$

$$\Rightarrow \quad _A\mathbf{v}_C = {_A}\mathbf{v}_B + {_B}\mathbf{v}_C$$

as was stated on page 146.

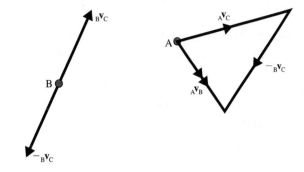

Notice, in particular, that if A and B have velocities $_A\mathbf{v}_E$ and $_B\mathbf{v}_E$ relative to the Earth, then

$$_A\mathbf{v}_B = {_A}\mathbf{v}_E - {_B}\mathbf{v}_E$$

However, as we rarely refer explicitly to velocity relative to the Earth, we would more usually label the velocities of A and B as $\mathbf{v}_A$ and $\mathbf{v}_B$ respectively, giving

$$_A\mathbf{v}_B = \mathbf{v}_A - \mathbf{v}_B$$

Solving relative velocity problems

There are three possible approaches we can use when solving relative velocity problems:

- Use a scale drawing of the velocity triangle (in practice this is rarely done).
- Use trigonometry to solve the velocity triangle.
- Combine the velocity vectors in component form.

Whichever method you use, you should **always** first sketch the situation. This often involves two diagrams: a space diagram to show the physical situation, and a vector diagram to show the velocities.

Current and wind effects

You have met similar problems on pages 13–15. Examples 1 and 2 illustrate them in the context of the relative velocity notation.

Example 1 A canoe, which can be paddled at $4\,\mathrm{m\,s^{-1}}$ in still water, is launched on a river flowing at $3\,\mathrm{m\,s^{-1}}$. The river is $24\,\mathrm{m}$ wide. We make the modelling assumption that the flow of the river is the same at every point across its width.

a) If the canoe is steered directly across the river, where on the far bank will it land?

b) In which direction must it be steered to land directly opposite its starting point?

SOLUTION

We solve the problem in two ways: first, by solving the velocity triangle trigonometrically; second, by expressing the vectors in component form.

Method 1

a) The velocity of the canoe relative to the water, $_C\mathbf{v}_W = 4\,\mathrm{m\,s^{-1}}$ perpendicular to the bank.

The velocity of the water relative to the bank, $_W\mathbf{v}_B = 3\,\mathrm{m\,s^{-1}}$ parallel to the bank.

We need to find $_C\mathbf{v}_B$.

The relationship $_C\mathbf{v}_B = {_C\mathbf{v}_W} + {_W\mathbf{v}_B}$ gives the vector triangle shown on the right. From this triangle, we have

$$\tan\theta = \frac{4}{3}$$

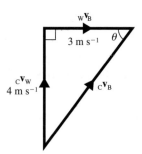

From the diagram of the river (space diagram), we see that

$$\frac{24}{x} = \tan\theta$$

$$\Rightarrow \quad x = \frac{24}{\tan\theta} = 18$$

So, the canoe lands at a point 18 m downstream from its starting point.

Velocity diagram

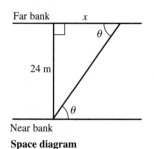

Space diagram

b) We are given the velocity of the canoe relative to the water, $_C\mathbf{v}_W = 4\,\mathrm{m\,s^{-1}}$ in a direction to be found.

We know the velocity of the water relative to the bank, $_W\mathbf{v}_B = 3\,\mathrm{m\,s^{-1}}$ parallel to the bank.

The velocity of the canoe relative to the bank, $_C\mathbf{v}_B$, must be perpendicular to the bank.

Using $_C\mathbf{v}_B = {_C\mathbf{v}_W} + {_W\mathbf{v}_B}$, we get the vector triangle shown on the right. From this triangle, we have

$$\cos\phi = \frac{3}{4} \quad \Rightarrow \quad \phi = 41.41°$$

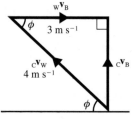

Velocity diagram

The canoe lands directly opposite its starting point if it sets a course upstream at an angle of 41.41° to the river bank.

Method 2

a) Taking unit vectors as shown in the diagram, we have

$$_C\mathbf{v}_W = 4\mathbf{j} \quad \text{and} \quad _W\mathbf{v}_B = 3\mathbf{i}$$

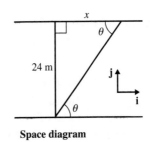

Using $_C\mathbf{v}_B = {_C}\mathbf{v}_W + {_W}\mathbf{v}_B$, we get

$$_C\mathbf{v}_B = 3\mathbf{i} + 4\mathbf{j}$$

Space diagram

The direction of $_C\mathbf{v}_B$ is given by θ, where $\tan\theta = \frac{4}{3}$.

From the diagram of the river, we have

$$\frac{24}{x} = \tan\theta \quad \Rightarrow \quad x = \frac{24}{\tan\theta} = 18$$

So, the canoe lands at a point 18 m downstream from its starting point.

b) Taking unit vectors as shown in the diagram, we have

$$_C\mathbf{v}_W = -4\cos\phi\,\mathbf{i} + 4\sin\phi\,\mathbf{j}$$

$$_W\mathbf{v}_B = 3\mathbf{i}$$

and if the magnitude of $_C\mathbf{v}_B$ is v, we have

$$_C\mathbf{v}_B = v\mathbf{j}$$

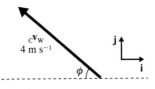

Using $_C\mathbf{v}_B = {_C}\mathbf{v}_W + {_W}\mathbf{v}_B$, we obtain

$$v\mathbf{j} = -4\cos\phi\,\mathbf{i} + 4\sin\phi\,\mathbf{j} + 3\mathbf{i}$$
$$= (-4\cos\phi + 3)\mathbf{i} + 4\sin\phi\,\mathbf{j}$$

Comparing **i**-components, we have

$$-4\cos\phi + 3 = 0$$
$$\Rightarrow \quad \cos\phi = 0.75 \quad \Rightarrow \quad \phi = 41.41°$$

The canoe lands directly opposite its starting point if it sets a course upstream at an angle of 41.41° to the river bank.

Example 2 An aircraft can fly at $200\,\text{km h}^{-1}$ in still air. The pilot wishes to set a course so that he can fly from town A to town B, which is 500 km from A on a bearing of 030°. There is a wind blowing from the south-east at a speed of $40\,\text{km h}^{-1}$.

a) What course should the pilot set?
b) How long will it take to travel from A to B?
c) What course should he set for the return journey and how long will it take?

SOLUTION

We solve the problem in two ways: first, by solving the velocity triangle trigonometrically; second, by expressing the vectors in component form.

Note the following:
- The velocity of the aircraft **relative to the air** is $_P\mathbf{v}_A$. The magnitude of this is often called its **air speed**.
- The wind velocity is the velocity of the air **relative to the ground**, $_A\mathbf{v}_G$.
- The velocity of the aircraft **relative to the ground** is $_P\mathbf{v}_G$. The magnitude of this is often called its **ground speed**.

Method 1

a) We know the following:

$$_P\mathbf{v}_A = 200\,\text{km h}^{-1} \text{ in some direction to be determined}$$
$$_A\mathbf{v}_G = 40\,\text{km h}^{-1} \text{ from the south-east}$$
$$_P\mathbf{v}_G \text{ is on a bearing of } 030° \text{ with magnitude, } x, \text{ to be determined}$$

The relation $_P\mathbf{v}_G = {}_P\mathbf{v}_A + {}_A\mathbf{v}_G$ gives the velocity triangle shown.

We require the angle θ, which we can find using the sine rule:

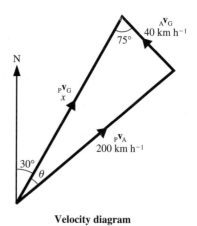

$$\frac{\sin\theta}{40} = \frac{\sin 75°}{200}$$

$$\Rightarrow \quad \sin\theta = \frac{40\sin 75°}{200} = 0.1932$$

$$\Rightarrow \quad \theta = 11.14° \quad \text{or} \quad 168.86°$$

As the second solution is not valid in this case, the course the pilot should set is

$$030° + 11.14° = 041.14°$$

Velocity diagram

b) To find how long the journey will take, we need to find the magnitude of $_P\mathbf{v}_G$, represented by the length x in the velocity diagram.

The third angle in the triangle is

$$180° - (75° + 11.14°) = 93.86°$$

We can now use the sine rule:

$$\frac{x}{\sin 93.86°} = \frac{200}{\sin 75°}$$

$$\Rightarrow \quad x = \frac{200\sin 93.86°}{\sin 75°} = 206.59\,\text{km h}^{-1}$$

So, the journey will take $\dfrac{500}{206.59} = 2.42\,\text{h}$ (2 h 25 min) to complete.

c) The diagram for the return journey is shown on the right. Notice that this triangle is not congruent to the previous one.

We again use the sine rule to find the bearing of $_P\mathbf{v}_A$:

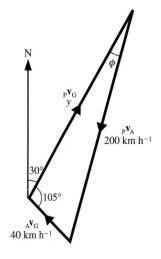

$$\frac{\sin \phi}{40} = \frac{\sin 105°}{200}$$

$$\Rightarrow \quad \sin \phi = \frac{40 \sin 105°}{200} = 0.1932$$

$$\Rightarrow \quad \phi = 11.14° \quad \text{or} \quad 168.86°$$

Again, the second value is not a valid solution.

So, the required bearing is

$$180° + (30° - 11.14°) = 198.86°$$

To find how long the return journey takes, we need to find the magnitude of $_P\mathbf{v}_G$, represented by the length y in the velocity diagram.

The third angle of the triangle is

$$180° - (105° + 11.14°) = 63.86°$$

We can now use the sine rule:

$$\frac{y}{\sin 63.86°} = \frac{200}{\sin 105°}$$

$$\Rightarrow \quad y = \frac{200 \sin 63.86°}{\sin 105°} = 185.88 \, \text{km h}^{-1}$$

So, the return journey will take $\dfrac{500}{185.88} = 2.69\,\text{h}$ (2 h 41 min) to complete.

Method 2

Note This approach requires the solution of trigonometric equations of the form $a \cos x + b \sin x = c$. If you have not encountered this, you may wish to skip what follows.

a) Putting $|_P\mathbf{v}_G| = v$ and taking unit vectors as shown in the diagram, we have

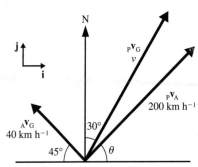

$$_P\mathbf{v}_A = 200 \cos \theta \, \mathbf{i} + 200 \sin \theta \, \mathbf{j}$$

$$_A\mathbf{v}_G = -40 \cos 45° \, \mathbf{i} + 40 \sin 45° \, \mathbf{j}$$

$$= -20\sqrt{2}\,\mathbf{i} + 20\sqrt{2}\,\mathbf{j}$$

$$_P\mathbf{v}_G = v \sin 30° \, \mathbf{i} + v \cos 30° \, \mathbf{j}$$

$$= \frac{v}{2}\,\mathbf{i} + \frac{v\sqrt{3}}{2}\,\mathbf{j}$$

Using $_P\mathbf{v}_G = {}_P\mathbf{v}_A + {}_A\mathbf{v}_G$, we have

$$\frac{v}{2}\mathbf{i} + \frac{v\sqrt{3}}{2}\mathbf{j} = (200\cos\theta - 20\sqrt{2})\mathbf{i} + (200\sin\theta + 20\sqrt{2})\mathbf{j}$$

Comparing components, we have

$$\frac{v}{2} = 200\cos\theta - 20\sqrt{2} \qquad [1]$$

and $\qquad \dfrac{v\sqrt{3}}{2} = 200\sin\theta + 20\sqrt{2} \qquad [2]$

From [1] and [2], we obtain

$$\sqrt{3}(200\cos\theta - 20\sqrt{2}) = 200\sin\theta + 20\sqrt{2}$$

Dividing through by 20 and rearranging, we get

$$10\sqrt{3}\cos\theta - 10\sin\theta = \sqrt{2} + \sqrt{6}$$

Rewriting the left-hand side in the form $R\cos(\theta + \alpha)$, we get

$$20\cos(\theta + 30°) = \sqrt{2} + \sqrt{6}$$

$$\Rightarrow \quad \cos(\theta + 30°) = 0.1932$$

$$\Rightarrow \quad \theta + 30° = 78.86° \quad \Rightarrow \quad \theta = 48.86°$$

The course the pilot should set is, therefore,

$$90° - 48.86° = 041.14°$$

b) Substituting for θ in [1], we have

$$\frac{v}{2} = 200\cos 48.86° - 20\sqrt{2} \quad \Rightarrow \quad v = 206.59\,\text{km h}^{-1}$$

So, the journey will take $\dfrac{500}{206.59} = 2.42\,\text{h}$ (2 h 25 min) to complete.

c) Putting $|_P\mathbf{v}_G| = V$ and taking unit vectors as shown, we obtain

$$_P\mathbf{v}_A = -200\sin\phi\,\mathbf{i} - 200\cos\phi\,\mathbf{j}$$

$$_A\mathbf{v}_G = -40\cos 45°\,\mathbf{i} + 40\sin 45°\,\mathbf{j}$$

$$= -20\sqrt{2}\,\mathbf{i} + 20\sqrt{2}\,\mathbf{j}$$

$$_P\mathbf{v}_G = -V\cos 60°\,\mathbf{i} - V\sin 60°\,\mathbf{j}$$

$$= -\frac{V}{2}\mathbf{i} - \frac{V\sqrt{3}}{2}\mathbf{j}$$

Using $_P\mathbf{v}_G = {}_P\mathbf{v}_A + {}_A\mathbf{v}_G$, we get

$$-\frac{V}{2}\mathbf{i} - \frac{V\sqrt{3}}{2}\mathbf{j} = (-200\sin\phi - 20\sqrt{2})\mathbf{i} + (-200\cos\phi + 20\sqrt{2})\mathbf{j}$$

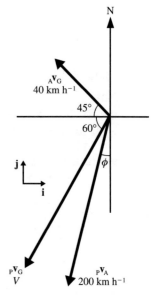

Comparing components, we have

$$\frac{V}{2} = 200 \sin \phi + 20\sqrt{2} \qquad [3]$$

and $\qquad \dfrac{V\sqrt{3}}{2} = 200 \cos \phi - 20\sqrt{2} \qquad [4]$

From [3] and [4], we have

$$\sqrt{3}(200 \sin \phi + 20\sqrt{2}) = 200 \cos \phi - 20\sqrt{2}$$

Dividing through by 20 and rearranging, we get

$$10 \cos \phi - 10\sqrt{3} \sin \phi = \sqrt{2} + \sqrt{6}$$

Rewriting the left-hand side in the form $R \cos(\theta + \alpha)$, we get

$$20 \cos(\phi + 60°) = \sqrt{2} + \sqrt{6}$$

$$\Rightarrow \quad \cos(\phi + 60°) = 0.1932$$

$$\Rightarrow \quad \phi + 60° = 78.86°$$

$$\Rightarrow \quad \phi = 18.86°$$

The course the pilot should set is, therefore,

$$180° + 18.86° = 198.86°$$

Substituting for ϕ in [3], we have

$$\frac{V}{2} = 200 \sin 18.86° + 20\sqrt{2} \quad \Rightarrow \quad V = 185.88 \, \text{km h}^{-1}$$

So, the return journey will take $\dfrac{500}{185.88} = 2.69 \, \text{h} \, (2 \, \text{h} \, 41 \, \text{min})$ to complete.

Example 3 Two fish, Angel and Barb, pass close to each other. In relation to a fixed frame of reference, they have velocities $\mathbf{v}_A = (2\mathbf{i} + \mathbf{j} - \mathbf{k}) \, \text{m s}^{-1}$ and $\mathbf{v}_B = (-\mathbf{i} + 2\mathbf{j} + 2\mathbf{k}) \, \text{m s}^{-1}$ respectively. From Angel's point of view, how fast does Barb appear to be moving?

SOLUTION

We need to find $|\,_B\mathbf{v}_A\,|$.

We use the relationship $_B\mathbf{v}_A = \mathbf{v}_B - \mathbf{v}_A$, which gives

$$_B\mathbf{v}_A = -3\mathbf{i} + \mathbf{j} + 3\mathbf{k}$$

$$\Rightarrow \quad |\,_B\mathbf{v}_A\,| = \sqrt{(-3)^2 + 1^2 + 3^2} = 4.36 \, \text{m s}^{-1}$$

So Barb appears to Angel to be moving at $4.36 \, \text{m s}^{-1}$

Exercise 8A

1 A ferry can travel at $10\,\mathrm{m\,s^{-1}}$ in still water. It is crossing a river that is 40 m wide.

    a) If the river is flowing at $6\,\mathrm{m\,s^{-1}}$ parallel to the banks and the ferry steers perpendicular to them, how far downstream from its starting point will it drift?

    b) If the river is flowing at $6\,\mathrm{m\,s^{-1}}$ parallel to the banks, at what angle to the near bank should the boat be steered in order to cross the river to reach a point on the far bank directly opposite the starting point? How long does the crossing take?

    c) If the river is flowing at $12\,\mathrm{m\,s^{-1}}$, can the ferry reach the point directly opposite its starting point?

    d) If the river is flowing at $12\,\mathrm{m\,s^{-1}}$ and the ferry steers upstream at an angle of 30° to the near bank, how far downstream does it drift before reaching the opposite bank?

2 In order to travel due north at $25\,\mathrm{km\,h^{-1}}$, a ship with a cruising speed of $20\,\mathrm{km\,h^{-1}}$ has to steer a course of 320°. At what speed is the current flowing?

3 Rain is falling with a speed of $24\,\mathrm{m\,s^{-1}}$. It is being blown by a wind so that it appears to be falling at an angle of 67.38° to the ground. What is the speed of the wind?

4 A power boat has a cruising speed of $60\,\mathrm{km\,h^{-1}}$. It is taking part in a race around three buoys arranged in an equilateral triangle. The second buoy, B, is 5 km due east of the starting/finishing buoy, A. The third buoy, C, is to the north of the line joining A and B. There is a current flowing at $10\,\mathrm{km\,h^{-1}}$ from the south-west. In what direction should be boat be steered on each leg of the race and how long will it take the boat to complete one circuit?

5 Microlight aircraft A and B pass each other. A, which has speed of $55\,\mathrm{km\,h^{-1}}$, is travelling on a bearing of 040° and climbing at an angle of 20° to the horizontal. B, which has speed of $70\,\mathrm{km\,h^{-1}}$, is travelling on a bearing of 110° and descending at an angle of 25° to the horizontal.

    a) Taking east, north and upwards as the **i**-, **j**- and **k**-directions respectively, express the velocities of the two aircraft in component form. Hence find in the component form the velocity of A relative to B.

    b) From B's point of view, what are A's apparent speed and angle of ascent?

Example 4 To a cyclist travelling east along a straight road at a speed of $15\,\mathrm{km\,h^{-1}}$, the wind appears to be coming **from** the direction 120°. When she increases her speed to $20\,\mathrm{km\,h^{-1}}$, the wind appears to be coming **from** the direction 110°. What is the true velocity of the wind?

SOLUTION

First part

We have $_C\mathbf{v}_G$ is $15\,\mathrm{km\,h^{-1}}$ due east, and $_W\mathbf{v}_C$ has a direction **from** 120° but an unknown speed.

We need to find $_W\mathbf{v}_G$, the velocity of the wind relative to the ground.

Using $_w\mathbf{v}_G = {_w}\mathbf{v}_C + {_c}\mathbf{v}_G$, we obtain Diagram 1.

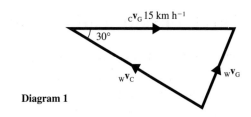

Diagram 1

Second part

We have $_c\mathbf{v}_G$ is $20\,\text{km h}^{-1}$ due east, and $_w\mathbf{v}_C$ has a direction **from** $110°$ but an unknown speed.

Again using $_w\mathbf{v}_G = {_w}\mathbf{v}_C + {_c}\mathbf{v}_G$, we obtain Diagram 2.

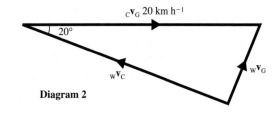

Diagram 2

In these two diagrams, $_w\mathbf{v}_G$ is the same and so we can superimpose the two diagrams to give Diagram 3.

We obtain the velocity of the wind by drawing Diagram 3 to scale and measuring it, or by calculating it using trigonometry.

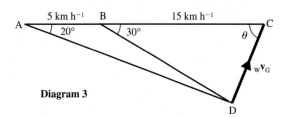

Diagram 3

In triangle ABD, $\widehat{ADB} = 10°$. So, using the sine rule, we have

$$\frac{BD}{\sin 20°} = \frac{5}{\sin 10°}$$

$$\Rightarrow \quad BD = \frac{5\sin 20°}{\sin 10°} = 9.848$$

In triangle BCD, using the cosine rule, we have

$$|_w\mathbf{v}_G|^2 = 9.848^2 + 15^2 - 2 \times 9.848 \times 15 \times \cos 30°$$

$$\Rightarrow \quad |_w\mathbf{v}_G| = 8.132$$

And, using the sine rule, we get

$$\frac{\sin \theta}{9.848} = \frac{\sin 30°}{8.132}$$

$$\Rightarrow \quad \sin \theta = \frac{9.848 \sin 30°}{8.132}$$

$$\Rightarrow \quad \theta = 37.27°$$

The velocity of the wind is, therefore, $8.13\,\text{km h}^{-1}$ **from** a direction whose bearing is $232.73°$. It is blowing **towards** the direction whose bearing is $52.73°$.

Interception, collision and closest approach

The English Channel between Dover and Calais is one of the busiest shipping routes in the world. Ships travelling north-east are separated from those travelling south-west by two 'lanes' marked on navigation charts. However, the routes of cross-channel ferries cut across these lanes, leading to the possibility of collisions.

If you are aboard a ship and trying to decide how close to you a second ship will come, it is necessary to find the velocity of the second vessel relative to you. Effectively, you imagine that you are stationary and plot the relative course of the other vessel. This will indicate whether there will be a collision or, if not, how close the ships will be when they pass each other.

Example 5 A ferry is travelling due south at 15 knots. (A knot is one nautical mile per hour, where one nautical mile = 1.852 km). The captain notices a tanker, initially 5 nautical miles away and on a bearing of 210° from the ferry. The tanker signals that it is travelling due east at 10 knots. Will the two ships collide if they do not take avoiding action? If they are not on a collision course, how far apart will they be at their closest point? (Make the modelling assumption that any wind or current affects both vessels equally and so can be ignored.)

SOLUTION

Using the subscripts E, F and T to stand for Earth, ferry and tanker respectively, we have

$$_T\mathbf{v}_E = 10 \text{ knots due east}$$

$$_F\mathbf{v}_E = 15 \text{ knots due south}$$

We need to find the direction of $_T\mathbf{v}_F$.

We sketch a space diagram showing the initial positions, F and T, of the ferry and the tanker.

The relative velocities are connected by the relationship

$$_T\mathbf{v}_E = {}_T\mathbf{v}_F + {}_F\mathbf{v}_E$$

This leads to the velocity diagram on the right.

From this triangle, we have

$$\tan\theta = 1.5 \quad \Rightarrow \quad \theta = 56.3°$$

The direction of $_T\mathbf{v}_F$ is a bearing of 033.7°.

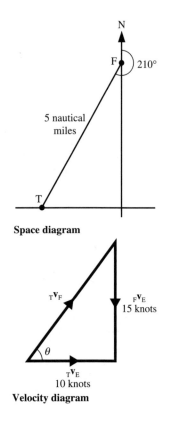

Space diagram

Velocity diagram

To the captain of the ferry, the ferry appears stationary and the tanker appears to be travelling on a bearing of 033.7°. We can now superimpose this relative course on the space diagram as the line TS.

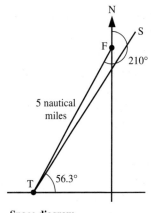

Space diagram

This indicates that the vessels will not collide. Since the ferry is actually travelling south, the tanker will pass in front of it. The shortest distance between the vessels is the perpendicular distance, d, from F to the line TS.

We calculate d from the triangle shown below right:

$$d = 5 \sin 3.7° = 0.322\,66 \text{ nautical miles}$$

The vessels pass within 0.322 66 nautical miles (about 600 m) of each other.

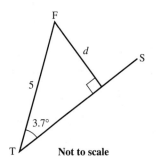

Not to scale

Example 6 The captain of a patrol boat, capable of a speed of 50 km h^{-1}, spots a smuggler's craft 800 m away on a bearing of 060°. Radar indicates that the smuggler's craft is travelling at 30 km h^{-1} due west. Assuming there is no current flowing, what course should the patrol boat's captain set in order to intercept the smuggler in the shortest possible time? How long before they meet?

SOLUTION

Using the subscripts P, S and W to stand for patrol, smuggler and water respectively, we have

$$_S\mathbf{v}_W = 30 \text{ km h}^{-1} \text{ due west}$$

$$|_P\mathbf{v}_W| = 50 \text{ km h}^{-1}$$

In order to intercept the smuggler's craft, the direction of $_S\mathbf{v}_P$ must be 240°. This is because the smuggler's craft must appear, to someone on the patrol boat, to be coming straight towards them.

We need to find the direction of $_P\mathbf{v}_W$. We also need $|_S\mathbf{v}_P|$ in order to find the time taken.

We sketch a space diagram showing the initial positions, P and S, of the patrol boat and the smuggler's craft.

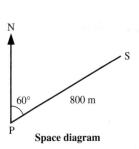

Space diagram

The relative velocities are connected by the relationship

$$_S\mathbf{v}_W = {}_S\mathbf{v}_P + {}_P\mathbf{v}_W$$

This leads to the velocity diagram on the right.

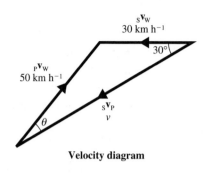

Velocity diagram

Using the sine rule, we obtain

$$\frac{\sin\theta}{30} = \frac{\sin 30°}{50}$$

$$\Rightarrow \quad \sin\theta = \frac{30\sin 30°}{50} = 0.3$$

$$\Rightarrow \quad \theta = 17.46° \quad \text{or} \quad 162.54°$$

In this case, the second value does not give a viable triangle. The course that the patrol needs to set is therefore on a bearing of

$$060° - 17.46° = 042.54°$$

To find $|{}_S\mathbf{v}_P|$, labelled v in the diagram, we use the sine rule again.

The third angle in the triangle is

$$180° - (30° + 17.46°) = 132.54°$$

which gives

$$\frac{v}{\sin 132.54°} = \frac{50}{\sin 30°}$$

$$\Rightarrow \quad v = \frac{50\sin 132.54°}{\sin 30°} = 73.68\,\text{km}\,\text{h}^{-1}$$

This gives an interception time of $\dfrac{0.8}{73.68} = 0.010\,86\,\text{h} = 39.1\,\text{s} \quad \text{(to 1 dp)}$

Note If the speed of the patrol boat had been less than $30\,\text{km}\,\text{h}^{-1}$, both values of θ would have given viable triangles. This is illustrated in Example 7. There are also situations in which the pursuing vessel is unable to catch its quarry. In such a situation, it is impossible to draw the velocity triangle.

Example 7 A motor boat is travelling at a speed of $16\,\text{km}\,\text{h}^{-1}$. A rower, who can row at $10\,\text{km}\,\text{h}^{-1}$, wishes to intercept the motor boat. Initially, the rower is $500\,\text{m}$ from the motor boat on a bearing of $120°$.

a) If the motor boat is travelling due east, which direction should the rower take in order to intercept it? Find the corresponding travel time.

b) If the motor boat is travelling due south, can the rower intercept it?

SOLUTION

a) Using subscripts M, R and W for the motor boat, the rower and the water respectively, we have

$$_M\mathbf{v}_W = 16\,\text{km h}^{-1} \text{ due east}$$

$$|_R\mathbf{v}_W| = 10\,\text{km h}^{-1}$$

If the rower is to intercept the motor boat, the direction of $_M\mathbf{v}_R$ must be 120°, because the motor boat must appear to the rower to be travelling straight towards her.

We need to find the direction of $_R\mathbf{v}_W$.

We draw a space diagram to show the initial positions, M and R, of the motor boat and the rower.
The relative velocities are connected by the relationship

$$_M\mathbf{v}_W = {_M\mathbf{v}_R} + {_R\mathbf{v}_W}$$

When we sketch the velocity diagram, we find that there are two possible triangles, ABC_1 and ABC_2, which fit the facts, as shown on the right. This is because, having drawn the line AB representing $_M\mathbf{v}_W$, an arc of radius 10 centred on B cuts the desired direction of $_M\mathbf{v}_R$ in two places, C_1 and C_2.

Using the sine rule, we have

$$\frac{\sin\theta}{16} = \frac{\sin 30°}{10}$$

$$\Rightarrow \quad \sin\theta = \frac{16\sin 30°}{10} = 0.8$$

$$\Rightarrow \quad \theta = 53.13° \quad \text{or} \quad 126.87°$$

This time, each value is a valid solution to the problem.

So, there are two possible directions for the rower to take: namely, a bearing of 353.13° or of 066.87°.

To find the travel times, we need to find the possible values of $|_M\mathbf{v}_R|$, which are given by the lengths AC_1 and AC_2 in the velocity diagram.

For triangle ABC_1, the third angle is $180° - (30° + 53.13°) = 96.87°$

For triangle ABC_2, the third angle is $180° - (30° + 126.87°) = 23.13°$

Space diagram

Velocity diagram

Using the sine rule for triangle ABC_1, we have

$$\frac{AC_1}{\sin 96.87°} = \frac{10}{\sin 30°}$$

$$\Rightarrow \quad AC_1 = \frac{10 \sin 96.87°}{\sin 30°} = 19.86 \, \text{km h}^{-1}$$

Using the sine rule for triangle ABC_2, we have

$$\frac{AC_2}{\sin 23.13°} = \frac{10}{\sin 30°}$$

$$\Rightarrow \quad AC_2 = \frac{10 \sin 23.13°}{\sin 30°} = 7.856 \, \text{km h}^{-1}$$

The first course gives an interception time of

$$\frac{0.5}{19.86} = 0.025 \, 18 \, \text{h} = 1 \, \text{min} \, 31 \, \text{s}$$

The second course gives an interception time of

$$\frac{0.5}{7.856} = 0.0636 \, \text{h} = 3 \, \text{min} \, 49 \, \text{s}$$

Notice that these calculations give the possible courses for **interception**, in which the motor boat and the rower arrive at the same point at the same time. If the rower were to set any course between the two, she could reach a point on the motor boat's course, and then sit and wait for it to arrive.

b) The details here are exactly the same as for part **a**, except that the motor boat is now travelling south.

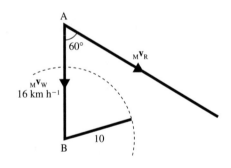

When we try to draw the velocity diagram, we find that there is no triangle which fits the facts. This is because, having drawn the line AB representing $_M v_W$, an arc of radius 10 centred on B does not cut the desired direction of $_M v_R$.

It is, therefore, impossible for the rower to intercept the motor boat in this case.

Note It is possible to solve problems of interception and closest approach by using vectors in component form, as in Examples 8 and 9.

Example 8 Two flies, Buz and Nat, have position vectors $r_B = 2i - 6j + 9k$ and $r_N = 10i + 10j - 7k$ respectively. They are flying with constant velocities $v_B = i - 2j + 3k$ and $v_N = -i - 6j + 7k$ respectively. Show that, unless they alter course or speed, they will collide. Find the position vector of the point where they meet.

SOLUTION

The displacement of Nat relative to Buz is

$$_N\mathbf{r}_B = \mathbf{r}_N - \mathbf{r}_B = 8\mathbf{i} + 16\mathbf{j} - 16\mathbf{k}$$

The velocity of Nat relative to Buz is

$$_N\mathbf{v}_B = \mathbf{v}_N - \mathbf{v}_B = -2\mathbf{i} - 4\mathbf{j} + 4\mathbf{k}$$

We can see that

$$_N\mathbf{v}_B = -\tfrac{1}{4} \, _N\mathbf{r}_B$$

As the multiplier here is negative, $_N\mathbf{v}_B$ is in exactly the opposite direction to $_N\mathbf{r}_B$. So, from Buz's point of view, Nat appears to be heading straight towards him. They will therefore collide.

As $|_N\mathbf{v}_B| = \tfrac{1}{4}|_N\mathbf{r}_B|$, the collision will happen after $4\,\text{s}$.

In this time, Buz will have moved to the point with position vector

$$(2\mathbf{i} - 6\mathbf{j} + 9\mathbf{k}) + 4(\mathbf{i} - 2\mathbf{j} + 3\mathbf{k}) = 6\mathbf{i} - 14\mathbf{j} + 21\mathbf{k}$$

Similarly, Nat will have moved to the point with position vector

$$(10\mathbf{i} + 10\mathbf{j} - 7\mathbf{k}) + 4(-\mathbf{i} - 6\mathbf{j} + 7\mathbf{k}) = 6\mathbf{i} - 14\mathbf{j} + 21\mathbf{k}$$

and so this is the point where they meet.

Example 9 Two birds, A and B, are initially at points with position vectors $(5\mathbf{i} + 8\mathbf{j} + 12\mathbf{k})\,\text{m}$ and $(2\mathbf{i} - 4\mathbf{j} + 15\mathbf{k})\,\text{m}$ respectively. They are flying with constant velocities of $(2\mathbf{i} + \mathbf{j} + 3\mathbf{k})\,\text{m s}^{-1}$ and $(\mathbf{i} + 2\mathbf{j} + 2\mathbf{k})\,\text{m s}^{-1}$ respectively. Find the time at which they are closest together, and the distance that then separates them.

SOLUTION

After t seconds, bird A has a displacement of $t(2\mathbf{i} + \mathbf{j} + 3\mathbf{k})$ from its initial position. Hence, after t seconds, bird A is at the point with position vector

$$\mathbf{r}_A = (5\mathbf{i} + 8\mathbf{j} + 12\mathbf{k}) + t(2\mathbf{i} + \mathbf{j} + 3\mathbf{k})$$

Similarly,

$$\mathbf{r}_B = (2\mathbf{i} - 4\mathbf{j} + 15\mathbf{k}) + t(\mathbf{i} + 2\mathbf{j} + 2\mathbf{k})$$

The displacement of A relative to B at time t seconds is given by

$$_A\mathbf{r}_B = \mathbf{r}_A - \mathbf{r}_B$$
$$\Rightarrow \quad _A\mathbf{r}_B = (3 + t)\mathbf{i} + (12 - t)\mathbf{j} + (t - 3)\mathbf{k}$$
$$\Rightarrow \quad |_A\mathbf{r}_B|^2 = (3 + t)^2 + (12 - t)^2 + (t - 3)^2$$
$$= 3t^2 - 24t + 162 \qquad\qquad [1]$$

The closest approach occurs when $|_A\mathbf{r}_B|^2$ is a minimum. So, differentiating, we get

$$\frac{d(|_A\mathbf{r}_B|^2)}{dt} = 6t - 24$$

For a minimum, we have

$$6t - 24 = 0 \quad \Rightarrow \quad t = 4$$

So the birds are at their closest point after 4 s.

Substituting for t in [1], we have

$$|_A\mathbf{r}_B|^2 = 114 \quad \Rightarrow \quad _A\mathbf{r}_B = 10.68$$

So, the minimum distance between the birds is 10.68 m.

Exercise 8B

1 Two trains leave a station. One travels west along the straight main line at 50 km h^{-1} and the other travels at 30 km h^{-1} along a branch line making an angle of 30° with the main line, as shown. What is the velocity of the branch-line train relative to the main-line train?

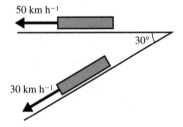

2 A horse rider feels that the wind is blowing at 12 km h^{-1} directly across her path from left to right. If she is travelling from west to east at 16 km h^{-1}, what is the velocity of the wind?

3 A cyclist, travelling due north at 20 km h^{-1}, feels that the wind is coming from due west. When travelling in the opposite direction at the same speed, the wind appears to be coming from a bearing of 210°. What is the velocity of the wind?

4 A baseball player has just hit the ball, which is fielded by a member of the opposing team. The fielder is at position $-9\mathbf{i} + 12\mathbf{j}$, the batter being at the origin. The batter runs with a constant velocity of $2\mathbf{i} + 3\mathbf{j}$ and the fielder throws the ball with a velocity of $5\mathbf{i} - \mathbf{j}$. Assume that the ball is not affected by gravity.

a) Show that the ball will hit the batter if he does not take avoiding action.
b) Where would the batter be if he were hit by the ball?

5 A boat, A, is 10 km south of another boat, B. A is travelling at 24 km h^{-1} due east, whilst B can travel at 26 km h^{-1}. What course should B set in order to intercept A and how long will it take to do so?

6 A ferry, travelling at $35\,\text{km}\,\text{h}^{-1}$, is about to cross a sea lane perpendicularly. It observes a tanker in the lane travelling at $20\,\text{km}\,\text{h}^{-1}$. The tanker is $5\,\text{km}$ away from the ferry in a direction making an angle of $35°$ with the direction of the ferry. Will the two ships collide? If not, how far apart are the ships when they are at their point of closest approach?

7 A patrol boat, travelling at $40\,\text{km}\,\text{h}^{-1}$ on a bearing of $030°$, sees a suspect craft travelling at $15\,\text{km}\,\text{h}^{-1}$ due north. It intercepts the craft 1 hour later. What was the position of the craft relative to the patrol boat when it was first sighted?

8 An enemy aircraft, flying at a height of $1000\,\text{m}$ and at a speed of $600\,\text{km}\,\text{h}^{-1}$, passes directly over a missile site. A missile is fired at $800\,\text{km}\,\text{h}^{-1}$ to intercept the aircraft. Assuming that the missile is not of the heat seeking type and must travel in a straight line, at what angle should the missile be fired? How long would it take to hit the plane?

9 A motor cyclist is travelling at a steady speed of $60\,\text{km}\,\text{h}^{-1}$ due south along a straight road. The wind appears to be blowing in a direction with bearing $040°$. On her return journey, at the same speed and with the same wind conditions, the wind appears to be blowing in a direction with bearing $150°$. What is the velocity of the wind?

10 Particles A and B start at points with position vectors $(2\mathbf{i}+\mathbf{j})\,\text{m}$ and $(3\mathbf{i}+4\mathbf{j})\,\text{m}$ respectively. They have constant velocities of $(-\mathbf{i}+2\mathbf{j})\,\text{m}\,\text{s}^{-1}$ and $(p\mathbf{i}+q\mathbf{j})\,\text{m}\,\text{s}^{-1}$ respectively.

a) Show that, if the particles are to collide at some time after the start,

$$p < -1, q < 2 \quad \text{and} \quad q = 3p+5$$

b) Show that, if the particles have their closest approach to one another at some time after the start, $p+3q < 5$.

c) If $p = -2$ and $q = 2$, find the time taken for the particles to reach their point of closest approach, and find the distance that then separates them.

11 Particles A and B start at points with position vectors $(16\mathbf{i}+3\mathbf{j}+4\mathbf{k})\,\text{m}$ and $(6\mathbf{j}+2\mathbf{k})\,\text{m}$ respectively. They have constant velocities of $(\mathbf{i}+2\mathbf{j})\,\text{m}\,\text{s}^{-1}$ and $(2\mathbf{i}-\mathbf{j}-2\mathbf{k})\,\text{m}\,\text{s}^{-1}$ respectively. Find the time which elapses before the particles are at their point of closest approach, and find the distance which then separates them.

9 Friction

Ay, there's the rub.
WILLIAM SHAKESPEARE

So far, either we have assumed that friction is negligible or we have talked about 'frictional resistance forces' without discussing how they might be modelled.

To get a feel for the problems of modelling friction, you should ideally do some practical investigation along the lines suggested in Experiments 1 to 3. Before starting, though, there are things we can reasonably agree upon.

- Friction occurs where we are sliding, or attempting to slide, one surface over another in contact with it. (The word derives from the Latin *fricare*, which means to rub).

- Friction always tries to prevent movement. The direction of the friction force is always opposite to that in which the object is moving or would move if there were no friction.

- Friction is a passive force. That is, it happens as a response to an attempt to slide surfaces. A block placed on a table has no friction force acting on it unless we try to push it along. When we push gently, the block stays still, meaning that the friction force 'adjusts' to exactly match the applied force, up to a limit beyond which the block starts to move.

We will investigate the simplest friction situation: a block sliding on a plane surface.

Experiments

Experiment 1

You need a plank, string, some weights, a pulley and three blocks.

Two of the blocks must be of the same material but their corresponding faces must have different areas. (Or you could have one block with faces of different areas.) The third block should have a markedly different type of surface: for example you might try sticking sandpaper to it. You need to know the mass of each block.

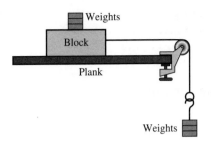

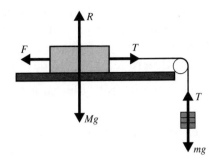

The block plus some weights, total mass M, are placed on the horizontal plank, and the block is connected to the suspended load, mass m, by means of string passing over the pulley at the end of the plank, as shown. You assume that the string is light and the pulley is smooth in comparison with the other masses and forces involved, so that the tension T is the same throughout the string. The contact forces between the block and the plank are the normal reaction, R, and the friction force, F.

As the suspended mass, m, is increased, the friction increases to keep the block stationary, until the point is reached where an increase in the load causes the block to move. At this point, the friction force is at its maximum value, called **limiting friction**.

For this experiment, you find the value of m corresponding to limiting friction for ten different values of M. In practice, you may find it more satisfactory to fix the value of m and alter the value of M until the block is just on the point of sliding.

You find the values of F and R as follows.

Resolving vertically for the suspended mass gives	$T - mg = 0$
Resolving horizontally for the block gives	$F - T = 0$
from which you obtain	$F = mg$
Resolving vertically for the block gives	$R - Mg = 0$
from which you obtain	$R = Mg$

Analysis and interpretation

You analyse the results as a table showing F, R and the ratio F/R. You can also draw a scatter graph of the various pairs of values and draw a line of best fit. This can be done by hand, but if you have access to a spreadsheet you can use the sheet FRIC1 available on the Oxford University Press website (http://www.oup.co.uk/mechanics). Alternatively, if you have a graphics calculator with list functions, you can analyse the results more conveniently.

The table on page 166 shows sample results for one run of the experiment. The results you obtain should be similar and you will probably notice three things:

- The values for F/R are approximately the same for different loadings of each particular block. This corresponds to a scatter graph with a convincing linear pattern and a line of best fit passing through, or near to, the origin (as expected, because a block with zero mass would have zero friction).

- The two blocks of the same material but different areas have similar values of F/R.

- The value of F/R changes for the block of different material.

Mass on block, M (kg to nearest 0.01)	Suspended mass, m (kg to nearest 0.01)	Reaction R (newtons)	Friction F (newtons)	$\dfrac{F}{R}$
0.14	0.03	1.372	0.294	0.214
0.19	0.04	1.862	0.392	0.211
0.24	0.05	2.352	0.490	0.208
0.29	0.06	2.842	0.588	0.207
0.35	0.07	3.430	0.686	0.200
0.41	0.08	4.018	0.784	0.195
0.46	0.09	4.508	0.882	0.196
0.51	0.10	4.998	0.980	0.196
0.56	0.11	5.488	1.078	0.196
0.60	0.12	5.880	1.176	0.200

Mean F/R	0.202

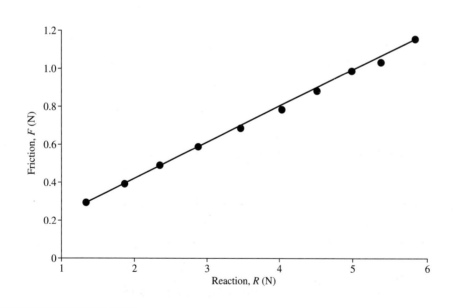

Experiment 2

Experiment 1 suggests that the friction force depends on the reaction between the surfaces. However, as R was given by Mg, it could be that the friction depends on the mass. To test this, you re-run the experiment using only one of the blocks but with the plank inclined at an angle α to the horizontal, so that R is no longer equal to Mg.

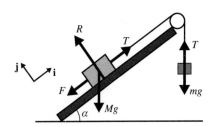

You find the values of F and R as follows.

Resolving vertically for the suspended mass gives $\qquad T - mg = 0$
Resolving in the **i**-direction for the block gives $\qquad T - F - Mg\sin\alpha = 0$
from which you obtain $\qquad F = mg - Mg\sin\alpha$
Resolving in the **j**-direction for the block gives $\qquad R - Mg\cos\alpha = 0$
from which you obtain $\qquad R = Mg\cos\alpha$

Analysis and interpretation

Once again, you tabulate the values of F, R and F/R and draw a scatter graph. Spreadsheet FRIC 2 (also available at http://www.oup.co.uk/mechanics) may be used for this purpose.

The table shows sample results for this experiment, using $\alpha = 30°$.

Mass on block, M (kg to nearest 0.01)	Suspended mass, m (kg to nearest 0.01)	Reaction R (newtons)	Friction F (newtons)	$\dfrac{F}{R}$
0.11	0.07	0.934	0.147	0.157
0.12	0.08	1.018	0.196	0.192
0.14	0.09	1.188	0.196	0.165
0.15	0.10	1.273	0.245	0.192
0.16	0.11	1.358	0.294	0.217
0.18	0.12	1.528	0.294	0.192
0.20	0.13	1.697	0.294	0.173
0.21	0.14	1.782	0.343	0.192
0.23	0.15	1.952	0.343	0.176
0.24	0.16	2.037	0.392	0.192

Mean F/R	0.185

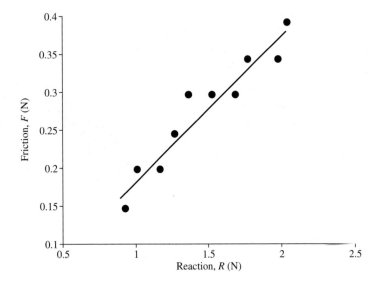

You should find that the value of F/R remains approximately constant, as before, and that this constant value is similar to that obtained when the plank is horizontal. This shows that it is the reaction between the surfaces, rather than the mass of the object, which determines the friction.

Coefficient of friction

Let us now summarise the results so far.

- When we try to move the block by applying an increasing force, the friction force increases to exactly match the applied force up to a maximum, called the **limiting friction**, and the system is then said to be in **limiting equilibrium**. Applying a force greater than this causes the block to move.

- The value of limiting friction depends on the nature of the surfaces in contact but is independent of the area of contact.

- For a given pair of surfaces, the ratio between the limiting friction and the normal reaction is constant. This constant is called the **coefficient of friction** for the surfaces and is denoted by μ. As the friction force can never be greater than the limiting friction, we have

$$\frac{F}{R} \leqslant \mu \quad \text{or} \quad F \leqslant \mu R$$

Experiment 3

You can try to confirm your value for the coefficient of friction between your blocks and plank by taking a different approach.

Place one of the blocks on the plank and gradually raise one end of the plank until the block just starts to move. Measure the angle of inclination of the plank. (You may, in practice, find it easier to measure the length of the plank and the height to which the end is raised, using these to find the sine of the required angle.)

Friction is now directed up the slope to try to stop the block sliding down, as shown.

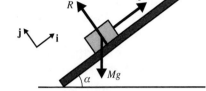

Suppose the block is on the point of moving when the angle is α.

Resolving parallel to the slope gives you

$$F - Mg\sin\alpha = 0 \qquad [1]$$

Resolving perpendicular to the slope gives you

$$R - Mg\cos\alpha = 0 \qquad [2]$$

From [1], you have: $F = Mg\sin\alpha$ [3]

and from [2], you have: $R = Mg\cos\alpha$ [4]

Dividing [3] by [4] gives you

$$\frac{F}{R} = \frac{\sin\alpha}{\cos\alpha} = \tan\alpha$$

But $F/R = \mu$, the coefficient of friction, so

$$\mu = \tan\alpha$$

You will need to repeat the process several times and average your results to obtain a reasonable value for α and therefore for μ.

Using our block on a 39 cm plank, we obtained heights (to the nearest 0.5) of 8.5 cm, 7.5 cm, 8 cm, 9.5 cm and 8.5 cm, giving an average height of 8.4 cm and an angle of 12.4°. This estimates the coefficient of friction as

$$\mu = \tan 12.4° = 0.22$$

Note   Do **not** worry if in these experiments you obtain results which vary quite widely. It is notoriously difficult to get accurate measurements of friction using simple apparatus. The best you can hope for is to get a general feel for the way friction behaves.

Static and dynamic friction

In Experiments 1 to 3, you were dealing with **static friction**. That is, the friction available to hold the object stationary. You may have noticed that, once the block started to move, it tended to accelerate, which would indicate that the force exerted by friction when the block is moving is less than that exerted when it is still.

When you place your block on the plank and raise the end until the block is close to moving, a slight nudge will get it started, and once it is moving it will continue to do so.

The friction force operating between two moving surfaces is called **dynamic** (or **kinetic) friction**, which is usually slightly less than the static friction between those surfaces. Dynamic friction is also encountered in situations where two surfaces in relative motion are brought into contact: for example, when the brakes of a car are applied. For our purposes, we usually make the modelling assumption that the coefficients of static and dynamic friction are equal, but you should be aware that the difference can have a significant effect on experimental results.

It is possible to show experimentally that the ratio F/R for dynamic friction is approximately constant and is **independent** of the speed. This ratio is called the **coefficient of dynamic friction**.

It should be noted that the situation is markedly different when the surfaces are lubricated. When there is a film of oil between the surfaces, F/R varies in proportion to speed.

Example 1 A block of mass 4 kg rests on a rough horizontal surface. The coefficient of friction between the block and the surface is 0.35. A horizontal force P is applied to the block so that it is just on the point of moving. Find the value of P.

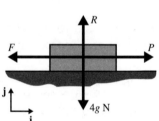

SOLUTION

Resolving in the **i**-direction, we get

$$P - F = 0 \qquad [1]$$

Resolving in the **j**-direction, we get

$$R - 4g = 0 \qquad [2]$$

As friction is limiting, we have

$$\frac{F}{R} = 0.35 \quad \Rightarrow \quad F = 0.35R \qquad [3]$$

Substituting from [2] into [3], we have

$$F = 0.35 \times 4g = 13.72 \, \text{N}$$

And so from [1] we get $P = 13.72 \, \text{N}$.

Example 2 Using the same block and surface as in Example 1, we now apply the force P at an angle of 20° to the horizontal. Find the value of P when the block is about to move.

SOLUTION

Resolving in the **i**-direction, we get

$$P \cos 20° - F = 0 \qquad [1]$$

Resolving in the **j**-direction, we get

$$R + P \sin 20° - 4g = 0 \qquad [2]$$

As friction is limiting, we have

$$F = 0.35R \qquad [3]$$

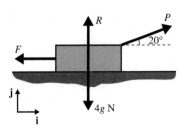

From [2] and [3], we obtain

$$F = 0.35(4g - P \sin 20°)$$

Substituting in [1], we have

$$P \cos 20° = 0.35(4g - P \sin 20°)$$

$$\Rightarrow \quad P = \frac{0.35 \times 4g}{\cos 20° + 0.35 \times \sin 20°} = 12.95 \, \text{N}$$

Example 3 A block of mass m is on a rough plane inclined at $30°$ to the horizontal. The coefficient of friction between the block and the plane is μ. A horizontal force P acts on the block. μ is sufficiently small for the block to slide down the slope if P does not act. Find the range of possible values of P if the block remains stationary.

SOLUTION

The block moves in one of two ways:

a) If P is too small, the block slides down the slope.

b) If P is too large, the block slides up the slope.

a) Suppose the block is in limiting friction and about to slide down the slope. The friction force is directed up the slope to oppose this.

Limiting friction means

$$F = \mu R \qquad [1]$$

Resolving in the **i**-direction, we get

$$F + P \cos 30° - mg \sin 30° = 0 \qquad [2]$$

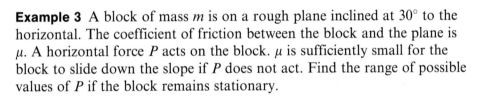

Resolving in the **j**-direction, we get

$$R - P \sin 30° - mg \cos 30° = 0 \qquad [3]$$

Substituting from [1] into [2] and doubling both sides, we obtain

$$2\mu R + P\sqrt{3} - mg = 0 \qquad [4]$$

Doubling both sides of [3], we have

$$2R - P + mg\sqrt{3} = 0 \qquad [5]$$

Multiplying [5] by μ and then subtracting it from [4], we get

$$P(\mu + \sqrt{3}) - mg(1 + \mu\sqrt{3}) = 0$$

$$\Rightarrow \quad P = \frac{mg(1 + \mu\sqrt{3})}{(\mu + \sqrt{3})}$$

b) Now suppose the block is in limiting friction and about to move up the slope. The friction force is now directed down the slope to oppose this.

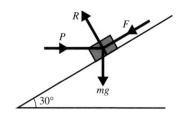

Resolving as before, we obtain

$$P\cos 30° - F - mg\sin 30° = 0 \qquad [1]$$

$$R - P\sin 30° - mg\cos 30° = 0 \qquad [2]$$

Substituting $F = \mu R$ and doubling as before, we get

$$P\sqrt{3} - 2\mu R - mg = 0 \qquad [3]$$

$$2R - P - mg\sqrt{3} = 0 \qquad [4]$$

Multiplying [4] by μ and then adding it to [3], we get

$$P(\sqrt{3} - \mu) - mg(1 + \mu\sqrt{3}) = 0$$

$$\Rightarrow \quad P = \frac{mg(1 + \mu\sqrt{3})}{(\sqrt{3} - \mu)}$$

So, the range of values of P is

$$\frac{mg(1 + \mu\sqrt{3})}{(\mu + \sqrt{3})} \leqslant P \leqslant \frac{mg(1 + \mu\sqrt{3})}{(\sqrt{3} - \mu)}$$

Exercise 9A

1 Each of the following diagrams shows a block of mass 5 kg resting on a rough horizontal surface. If the block is in limiting equilibrium, find the coefficient of friction.

a)

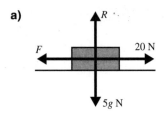

b)

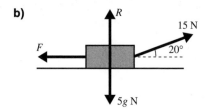

c)

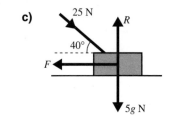

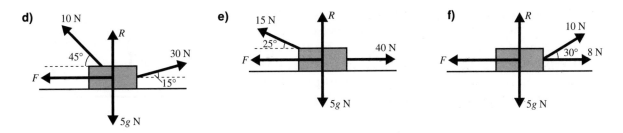

2 Each of the following diagrams shows a block of mass 8 kg resting on a rough horizontal surface. If the block is in limiting equilibrium and the coefficient of friction is as stated, find the force *P*.

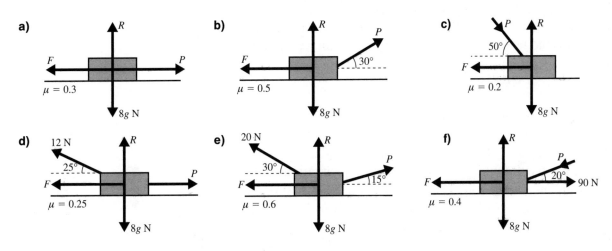

3 A block of mass 5 kg rests in limiting equilibrium on a rough plane inclined at 27° to the horizontal. Find the magnitude of the friction force acting on the block and the coefficient of friction between the block and the plane.

4 A block of mass *m* rests in limiting equilibrium on a rough plane inclined at an angle α to the horizontal. Show that the coefficient of friction between the block and the plane is tan α.

5 A block of mass 4 kg rests on a rough plane inclined at 10° to the horizontal. The coefficient of friction between the block and the plane is 0.3. A force *P* acts on the block parallel to the plane. Find the magnitude and direction of *P* if the block is about to move **a)** up the plane, **b)** down the plane.

6 A block of mass 6 kg rests in limiting equilibrium on a rough plane inclined at 20° to the horizontal. Find the horizontal force which would have to be applied to the block to cause it to be on the point of sliding up the plane.

7 An object of mass 50 kg rests on a rough plane inclined at an angle α to the horizontal. It is supported in this position by a light string parallel to the plane which is attached to the object and fixed to a point at the top of the plane. The string has a breaking strain of 200 N, and the coefficient of friction between the object and the plane is 0.2. Find the largest value of α for the string to remain intact.

8 The diagram shows two blocks resting on rough planes inclined to the horizontal at 40°. The blocks are connected by means of a light string passing over a smooth pulley at the top of the slopes. The coefficient of friction at each block is μ. If the blocks have masses of 10 kg and 4 kg and the system is in limiting equilibrium, find the value of μ.

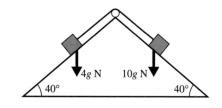

9 The diagram shows a block of mass km resting on a rough horizontal table. It is connected by means of a light string passing over a smooth pulley at the edge of the table to a particle off mass m. This particle is acted upon by a horizontal force P so that the string is inclined at an angle θ to the vertical. If the coefficient of friction between the block and the table is μ and the system is in limiting equilibrium, show that

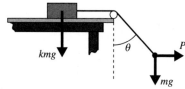

$$\cos \theta = \frac{1}{k\mu}$$

and find P in terms of k, m and μ.

10 The diagram shows particles of mass 2 kg and 1 kg placed on a fixed double inclined plane with inclinations of 60° and 30° respectively. The particles are connected by a light string passing over a smooth pulley at the vertex. The coefficient of friction between the particles and the planes is μ. Show that if the heavier particle is on the point of slipping then $\mu = 5\sqrt{3} - 8$.

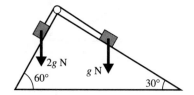

11 A block of mass m rests on a rough plane inclined at an angle α to the horizontal. A force P_1 acting up the plane causes the block to be on the point of moving in that direction. A force P_2 acting down the plane causes the block to be on the point of moving in that direction. Show that $P_1 - P_2$ is independent of the coefficient of friction between the block and the plane.

12 The diagram shows particles of mass m and km (where $k > 1$) resting on a double inclined plane with each part inclined at 30° to the horizontal. The coefficient of friction on each plane is μ. The particles are connected by a light string passing over a smooth pulley at the vertex. If the system is about to move, show that

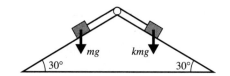

$$k = \frac{1 + \mu\sqrt{3}}{1 - \mu\sqrt{3}}$$

What does this tell you about the value of μ?

13 A particle of mass m can just rest on a rough plane inclined at 30° to the horizontal without slipping down. Show that the least horizontal force needed to maintain its position if the inclination is increased to 45° is $mg(2 - \sqrt{3})$.

14 Two rough rings, each of mass m, are threaded onto a horizontal rod, the coefficient of friction between the rings and the rod being μ. Two light strings, each of length a, connect the rings to a particle of mass m which hangs freely below the rod. Show that the maximum distance between the rings on the rod for equilibrium to be maintained is

$$\frac{6a\mu}{\sqrt{9\mu^2 + 1}}$$

15 The diagram shows a block of mass km resting on a rough horizontal table, the coefficient of friction being μ. A light string attached to the block passes over a smooth pulley at a hole in the table and its other end is fixed to a point on the underside of the table. A smooth ring of mass m is threaded on the string. Find the minimum value of μ for equilibrium to be possible with the string through the ring inclined at $30°$ to the vertical.

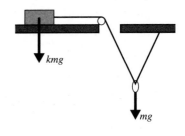

Situations involving acceleration

In Examples 4 to 6, the coefficient of friction, μ, is assumed to be that for dynamic friction.

Example 4 A block of mass $5\,\text{kg}$ moves on a rough horizontal plane with coefficient of (dynamic) friction 0.2 under the action of a horizontal force of $30\,\text{N}$. If the block starts from rest, find the distance it travels in the first 3 seconds of motion.

SOLUTION

Let the block have acceleration a.

From the laws of friction, we have

$$F = 0.2R \qquad [1]$$

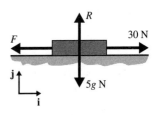

Applying Newton's laws and resolving, we obtain:

In the **j**-direction: $\quad R - 5g = 0 \qquad [2]$

In the **i**-direction: $\quad 30 - F = 5a \qquad [3]$

Substituting from [2] into [1], we have

$$F = 0.2 \times 5g = 9.8\,\text{N}$$

Substituting in [3], we get

$$20.2 = 5a \quad \Rightarrow \quad a = 4.04\,\text{m s}^{-2}$$

Using $s = ut + \frac{1}{2}at^2$, where $u = 0$, $t = 3\,\text{s}$ and $a = 4.04\,\text{m s}^{-2}$, we obtain

$$s = \tfrac{1}{2} \times 4.04 \times 9 = 18.18\,\text{m}$$

So, the block travels 18.18 m in the first 3 s of motion.

Example 5 A particle of mass 6 kg, moving at $8\,\text{m s}^{-1}$ on a smooth horizontal surface, goes onto a rough horizontal surface with a coefficient of friction 0.25. Find the distance it moves across the rough surface before coming to rest.

SOLUTION

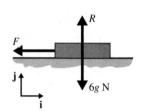

Let the particle have acceleration a.

From the laws of friction, we have

$$F = 0.25R \qquad\qquad\qquad [1]$$

Applying Newton's laws and resolving, we obtain:

In the **j**-direction: $\quad R - 6g = 0 \qquad [2]$

In the **i**-direction: $\quad -F = 6a \qquad [3]$

Substituting from [2] into [1], we get

$$F = 0.25 \times 6g = 14.7\,\text{N}$$

Substituting in [3], we get

$$-14.7 = 6a \quad \Rightarrow \quad a = -2.45\,\text{m s}^{-2}$$

Using $v^2 = u^2 + 2as$, where $u = 8\,\text{m s}^{-1}$, $v = 0$ and $a = -2.45\,\text{m s}^{-2}$, we obtain

$$0 = 8^2 - 2 \times 2.45 \times s$$
$$\Rightarrow \quad s = 13.06\,\text{m}$$

So, the particle travels a distance of 13.06 m across the rough surface.

Example 6 The diagram shows a small block of mass 2 kg able to move on a rough plane of length 8 m inclined at 20° to the horizontal. The block is attached by means of a light string passing over a smooth pulley at the top of the plane to a particle of mass 5 kg hanging freely. The coefficient of friction between the block and the plane is 0.2. The system is released from rest with the block at the bottom of the plane. Find the time which elapses before it reaches the top.

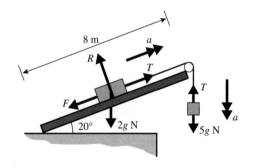

SOLUTION

From the laws of friction, we have

$$F = 0.2R \qquad [1]$$

Resolving vertically for the suspended particle, we get

$$5g - T = 5a \qquad [2]$$

Resolving perpendicular to the plane for the block, we get

$$R - 2g \cos 20° = 0 \qquad [3]$$

Resolving parallel to the plane for the block, we get

$$T - F - 2g \sin 20° = 2a \qquad [4]$$

Adding [2] and [4], we obtain

$$5g - 2g \sin 20° - F = 7a \qquad [5]$$

Substituting from [3] into [1], we have

$$F = 0.2 \times 2g \cos 20° = 3.684 \, \text{N}$$

Substituting in [5], we have

$$38.61 = 7a \quad \Rightarrow \quad a = 5.52 \, \text{m s}^{-2}$$

Using $s = ut + \frac{1}{2}at^2$, where $s = 8 \, \text{m}$, $u = 0$ and $a = 5.52 \, \text{m s}^{-2}$, we obtain

$$8 = \tfrac{1}{2} \times 5.52 \times t^2$$
$$\Rightarrow \quad t = 1.7 \, \text{s}$$

So, the block reaches the top of the plane in 1.7 s.

Exercise 9B

1 A block of mass 3 kg is being towed across a horizontal surface with coefficient of friction 0.2 by a horizontal force of 18 N. Find the acceleration of the block.

2 A block of mass 5 kg is being towed across a horizontal surface with coefficient of friction μ by a horizontal force of 40 N. If the acceleration of the block is $5 \, \text{m s}^{-2}$, find the value of μ.

3 A block of mass 3 kg is moving at $10 \, \text{m s}^{-1}$ on a smooth horizontal surface when it moves onto a rough horizontal surface with coefficient of friction 0.35. Find the distance which it travels on the rough surface before coming to rest.

4 A block of mass 6 kg moves on a rough horizontal surface (coefficient of friction 0.25) under the action of a horizontal force. It accelerates from rest to a speed of $4 \, \text{m s}^{-1}$ in a distance of

12 m, continues for a time at this speed and then decelerates to rest in a distance of 2 m. Find the magnitude and direction of the horizontal force required during each stage of the journey.

5 A particle, moving at $6\,\mathrm{m\,s^{-1}}$ on a smooth horizontal surface, goes onto a rough horizontal surface and is brought to rest in a distance of 20 m. Find the coefficient of friction involved.

6 A particle moving on a smooth horizontal surface encounters two rough areas, each 10 m wide. The coefficients of friction for the two areas are 0.2 and 0.4 respectively. Find the minimum initial speed of the particle if it just makes it across the two areas.

7 A box of mass 20 kg rests on a rough horizontal floor, the coefficient of friction being 0.3. A light string is attached to the box and a tension T is exerted with the string inclined upwards at 30° to the horizontal. If the resulting acceleration of the box is $0.5\,\mathrm{m\,s^{-2}}$, find the value of T.

8 Find the force needed to accelerate a 2 kg block at $3\,\mathrm{m\,s^{-2}}$ up a rough plane (coefficient of friction 0.2) inclined at 25° to the horizontal if the force is

a) parallel to the slope   b) horizontal   c) at 45° to the upward vertical

9 A particle of mass 4 kg is being towed at constant speed up a rough plane inclined at 30° to the horizontal by a force of $4g$ N acting parallel to the slope. At the top of the slope the particle moves onto a rough horizontal plane with the same coefficient of friction. If the towing force continues to act in the same direction, show that the particle undergoes an acceleration of

$$\frac{g\sqrt{3}}{6}\,\mathrm{m\,s^{-2}}$$

10 A particle of mass 5 kg is being towed at a constant speed of $6\,\mathrm{m\,s^{-1}}$ on a rough horizontal plane with coefficient of friction 0.2. At a certain point the towing force is reversed in direction. Find the distance that the particle will travel before coming to rest and explain what will happen after it does so.

11 The diagram shows particles A and B, each of mass m, resting on rough planes inclined at angles α and β to the horizontal. The coefficients of friction at A and B are μ_A and μ_B respectively. The particles are connected by a light string passing under a smooth pulley at the point where the planes meet. It is found that a force P applied to particle A directly up its slope causes the system to accelerate with acceleration a_1 with the string taut. The same force P applied to particle B directly up its slope causes the system to accelerate with acceleration a_2 with the string taut. Show that

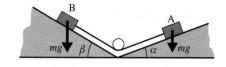

$$a_1 - a_2 = g(\sin\beta - \sin\alpha)$$

Angle of friction

In many situations, it is convenient to replace the friction and normal reaction forces F and R with a single **total reaction force**, S say.

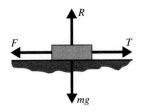
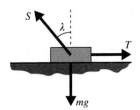

The advantage of this approach is that four-force problems may be reduced to three-force problems, enabling us to use triangles of forces. We should stress that you should use one method or the other – do **not** put both sets of forces on one diagram.

The angle between the total reaction, S, and the vertical is usually represented by λ. Resolving horizontally and vertically, we get

$$F = S\sin\lambda \quad \text{and} \quad R = S\cos\lambda$$

which give

$$\frac{F}{R} = \tan\lambda$$

When friction is limiting, $F/R = \mu$ and λ has its maximum possible value. In this case,

Coefficient of friction $\mu = \tan\lambda$

In this limiting-friction case, λ is called the **angle of friction**.

Example 7 A block of mass 10 kg rests on a rough plane inclined at $20°$ to the horizontal. The coefficient of friction between the block and the plane is 0.3. A force P acts directly up the plane and the block is about to move in that direction. Find the value of P.

SOLUTION

If λ is the angle of friction, we have

$$\tan\lambda = 0.3 \quad \Rightarrow \quad \lambda = 16.7°$$

As friction is limiting, the total reaction, S, is inclined at $16.7°$ to the perpendicular to the plane, as shown.

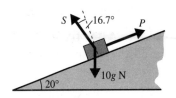

We draw a triangle of forces. Then, applying the sine rule, we get

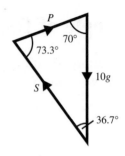

$$\frac{P}{\sin 36.7°} = \frac{10g}{\sin 73.3°}$$

$$\Rightarrow \quad P = \frac{10g \sin 36.7°}{\sin 73.3°} = 61.1 \, \text{N}$$

We would, of course, have achieved the same result using Lami's theorem (see pages 59–60).

Example 8 Show that for a particle to rest in limiting equilibrium on a rough inclined plane, the angle of inclination of the plane must equal the angle of friction between the particle and the plane.

SOLUTION

The diagram shows a particle of mass m resting on a plane inclined at θ to the horizontal. The total reaction, S, of the plane must act vertically to maintain equilibrium, as there are only two forces acting on the particle.

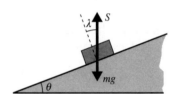

If equilibrium is limiting, the angle between S and the normal to the plane is the angle of friction, λ. It is evident from simple geometry that λ equals θ.

Example 9 A particle of mass 6 kg rests on a rough horizontal plane with coefficient of friction 0.35. A force P is exerted on the particle so that it is on the point of moving. Find the least magnitude of P for this to happen.

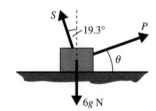

SOLUTION

If λ is the angle of friction, we have

$$\tan \lambda = 0.35 \quad \Rightarrow \quad \lambda = 19.3°$$

So, the total reaction, S, of the plane on the block is at an angle of 19.3° to the vertical, as shown.

Let the force P act at an angle θ to the horizontal.

We draw a triangle of forces, ABC, as shown.

As AB is fixed and the direction of BC is fixed, the minimum length of BC is clearly when the angle at C is 90°, which means $\theta = 19.3°$. This gives

$$\text{Minimum } P = 6g \sin 19.3° = 19.4 \, \text{N}$$

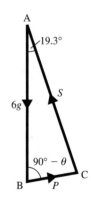

Example 10 A particle of mass 2 kg rests at point P inside a rough sphere, centre O. The coefficient of friction is 0.27. The particle is just kept in place by a force T along AP where A is the lowest point of the sphere. Angle AOP is 40°. Find the value of T.

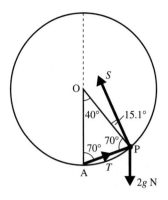

SOLUTION

If the angle of friction is λ, we have

$$\tan \lambda = 0.27 \quad \Rightarrow \quad \lambda = 15.1°$$

As the system is in limiting equilibrium, the total reaction, S, makes an angle of 15.1° with the radius OP, as shown.

We draw a triangle of forces, as shown. Then, applying the sine rule, we have

$$\frac{T}{\sin 24.9°} = \frac{2g}{\sin 85.1°}$$

$$\Rightarrow \quad T = \frac{2g \sin 24.9°}{\sin 85.1°} = 8.28 \text{ N}$$

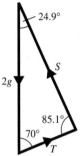

Exercise 9C

1 A particle of mass 3 kg rests on a rough horizontal plane with coefficient of friction 0.2. The particle is being pulled by a light string inclined at an angle of 30° to the horizontal. The system is in limiting equilibrium. Use a triangle of forces to find the tension in the string.

2 A particle of mass 5 kg rests on a rough horizontal plane with coefficient of friction 0.3. The particle is being pulled by a light string inclined at an angle θ to the horizontal. The tension in the string is 40 N and the system is in limiting equilibrium. Use a triangle of forces to find the value of θ.

3 A particle of mass 8 kg rests on a rough horizontal plane with coefficient of friction μ. The particle is being pulled by a light string inclined at an angle of 25° to the horizontal. The tension in the string is 30 N and the system is in limiting equilibrium. Use a triangle of forces to find the value of μ.

4 A particle of mass 4 kg rests on a rough plane inclined at 30° to the horizontal. The coefficient of friction is 0.25. The particle is supported by a light string.

   a) Use a triangle of forces or Lami's theorem to find the tension in the string under each of the conditions **i)** to **iv)**.
   i)   The string is parallel to the slope and the particle is about to slide down the slope.
   ii) The string is parallel to the slope and the particle is about to slide up the slope.
   iii) The string is at 20° to the plane and the particle is about to slide down the slope.
   iv) The string is at 10° to the plane and the particle is about to slide up the slope.

b) Find the least tension in the string when the particle is about to slide down the slope.

c) Find the least tension in the string when the particle is about to slide up the slope.

5 A particle of mass 3 kg rests at point P on the inside surface of a rough hollow sphere, centre O, with coefficient of friction 0.35. OP makes an angle of 50° with the vertical. The particle is acted on by a horizontal force which is on the point of moving it up the surface of the sphere. Find the magnitude of this force.

6 A particle of mass 6 kg rests at point P on the outer surface of a rough sphere, centre O, with coefficient of friction 0.4. OP makes an angle of 30° with the vertical. Find the least force needed to keep the particle in this position.

7 A particle of mass m rests on a rough plane inclined at an angle α to the horizontal. The angle of friction is λ. A horizontal force of mg N is just sufficient to prevent the particle from sliding down the slope. A horizontal force of $3mg$ N causes the particle to be on the point of sliding up the slope. Find the values of α and λ.

8 Two particles of mass m are connected by a light inextensible string of length a and are placed on the rough outer surface of a sphere of radius r with one of the particles at the highest point of the sphere. The string is taut and the second particle is as low as possible on the sphere. The angle of friction is λ. Show that the maximum possible length of the string is $2r\lambda$, with λ given in radians.

10 Turning effects of forces

Give me somewhere to stand, and I will move the Earth.
ARCHIMEDES

If a strip of cardboard is pinned to a table top by a single drawing pin so that it is free to rotate, any horizontal force applied to the strip makes it rotate (unless the force acts along a line which passes through the drawing pin). How strong this turning effect is depends not only on the magnitude of the force applied, but also on where it is applied and in which direction.

Experiment

You need a rigid uniform rod (for example, a piece of wood) at least 1 metre long. Fix the rod so that it is free to rotate in a vertical plane about its centre. (Ideally, drill a hole in the rod and then nail it to a vertical post through the hole.) If this is done accurately, the rod should stay in a horizontal position when placed there.

Hang two 40-gram masses equal distances on either side of the pivot – 48 cm each way is convenient. The rod should remain horizontal because its tendency to turn anticlockwise is being exactly balanced by its tendency to turn clockwise: the two applied forces have equal and opposite turning effects.

Now increase one of the masses (m) so that the rod no longer balances. Move the larger mass towards the pivot to restore the balance. Record its distance (x) from the pivot. Repeat this so that there is data for eight different masses including the original 40 grams.

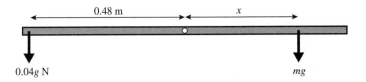

Analysis and interpretation

The relationship between the force and distance values should be fairly obvious, but you can examine it in tabular and graphical form. See the table and graphs on page 184 for an example of this. The spreadsheet MOM 1, available on the Oxford University Press website (http://www.oup.co.uk/mechanics), enables you to reproduce this for your data from the experiment.

Distance (m)	Mass (g)	Force (N)	Force × Distance (N m)	1/Distance (m⁻¹)
0.48	40	0.392	0.188	2.083 333 333
0.32	60	0.588	0.188	3.125
0.24	80	0.784	0.188	4.166 666 667
0.19	100	0.98	0.186	5.263 157 895
0.16	120	1.176	0.188	6.25
0.14	140	1.372	0.192	7.142 857 143
0.12	160	1.568	0.188	8.333 333 333
0.11	180	1.764	0.194	9.090 909 091

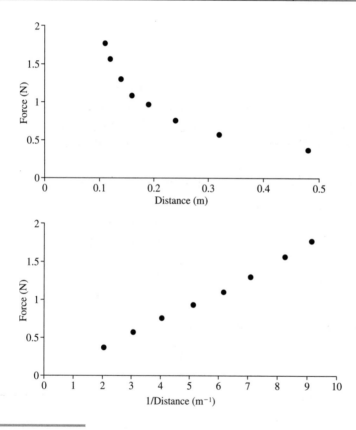

Moment of a force

It is clear from the experiment that the relation is

Force × Distance = Constant

(In the experiment, the constant is 0.188.) Each pairing of force and distance achieved the same turning effect, which leads to the following definition.

When a force **F** acts in a plane at a perpendicular distance d from an axis which is perpendicular to the plane, the turning effect of the force about the axis has magnitude $|\mathbf{F}| \times d$.

This turning effect is called the **moment of the force** or the **torque**.

As $|\mathbf{F}|$ is in newtons and d is in metres, the moment of a force is measured in **newton metres** (N m), which is the SI unit for this quantity.

Although we should strictly always talk about 'moment about an axis', all the situations we consider involve forces acting in a single plane. It therefore makes sense to talk about **moment about a point**, meaning about an axis through the point and perpendicular to the plane of the forces.

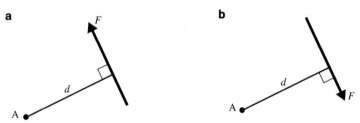

We also need to be clear about the **sense** of the turning effect. In the two situations shown above, the magnitude of the turning effect is the same. However, in case **a**, the tendency is anticlockwise rotation, whereas in case **b** the tendency is clockwise rotation. It is usual to regard **anticlockwise** as the **positive** rotational direction, so the moment in case **a** is $+Fd$ and in case **b** is $-Fd$.

Note   Strictly speaking, the moment of a force is a vector quantity because it has magnitude Fd and positive or negative direction along the axis. When the force acts in the x–y plane, the direction of the moment is the positive or negative z-direction.

When several forces act in the same plane, their **total turning effect** about a given point is the **sum of the moments of the individual forces**.

Example 1   Forces of 10 N, 15 N and 18 N act as shown on a rectangular lamina ABCD with AB = 6 m and BC = 4 m. Find the total moment of the forces about A.

(If you have not met the term *lamina* before, it means an idealised, infinitely thin, plane figure. The plural is *laminae*, sometimes *laminas*.)

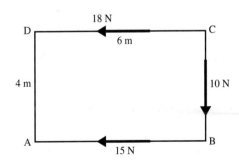

The 18 N force acts 4 m from A and turns in an anticlockwise sense. Therefore, its moment is

$$+18 \times 4 = +72 \, \text{N m}$$

The 10 N force acts 6 m from A and turns in a clockwise sense. Therefore, its moment is

$$-10 \times 6 = -60 \, \text{N m}$$

The 15 N force acts directly through A and so has no turning effect. Its moment is 0 N m.

The total moment of the forces is, therefore,

$$72 - 60 + 0 = 12 \, \text{N m}$$

Exercise 10A

1 Find the moment of each of the following forces about the point A, making sure you indicate whether it is a positive or a negative moment.

a)

4 m   A

14 N

b)

A   3.5 m

25 N

c)

A

6.3 m

16 N

d)

A

7.5 N

4.2 m

e)

48 N   9.4 m

A

f)

60 N

A   12 m

2 Find the total moment of the forces shown in each of the following diagrams about **i)** A and **ii)** B.

a)

12 N   9 N

A   B

2.3 m   1.6 m

7 N

b)

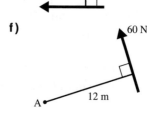

A

6 N

1.6 m

4 N   B

1.4 m

10 N

2.2 m

8 N

c)

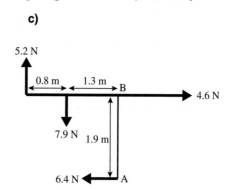

5.2 N

0.8 m   1.3 m   B

4.6 N

7.9 N   1.9 m

6.4 N   A

3 The rectangular lamina ABCD has AB = 4.8 m and BC = 3.6 m. M is the mid-point of AB and O is the centre of the rectangle. Forces 2 N, 3 N, 4 N and 6 N act as shown in the diagram. Find the sum of the moments of these forces about

a) A   **b)** B   **c)** M   **d)** O

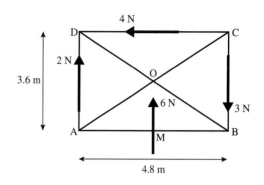

4 The diagram shows an aerial view of a revolving door. Four people are exerting forces of 40 N, 60 N, 80 N and 90 N as shown. Find the distance x if the total moment of the forces about O is

a) 12 N m   **b)** −8 N m   **c)** 0 N m

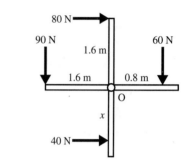

Forces at an angle

The diagram shows a rod AB of length a. A force of magnitude F is applied at B at an angle θ to the rod.

There are two ways of treating the moment of F about A.

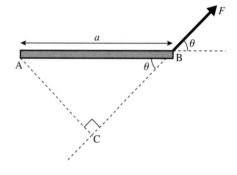

Method 1   The perpendicular distance from A to the line of action F is AC, so

$$\text{Moment of } F \text{ about A} = F \times \text{AC}$$

$$= Fa \sin \theta$$

Method 2   The force F can be resolved into two components:

$F \cos \theta$ in the direction AB

$F \sin \theta$ perpendicular to AB

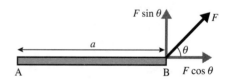

The first component is along a line through A and so has no turning effect about A. The moment of F is produced by the second component. So,

$$\text{Moment of } F \text{ about A} = F \sin \theta \times a$$

$$= Fa \sin \theta$$

You should become familiar with both approaches, though you will probably find the second method to be more useful.

Example 2 Find the total moment about the point A of the forces shown in the diagram.

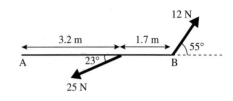

SOLUTION

The components of the forces perpendicular to AB are $25 \sin 23°$ and $12 \sin 55°$ respectively. Therefore,

$$\text{Total moment} = -25 \sin 23° \times 3.2 + 12 \sin 55° \times 4.9$$

$$= 16.9 \,\text{N}\,\text{m}$$

Example 3 The diagram shows a rectangular lamina ABCD. A force of 20 N is applied at C, as shown. Find the moment of this force about A.

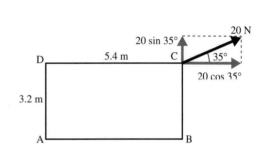

SOLUTION

The force has a component $20 \sin 35°$ in the direction BC. This component has a moment about A of

$$20 \sin 35° \times 5.4 = 61.95 \,\text{N}\,\text{m}$$

The force has a component $20 \cos 35°$ in the direction DC. This component has a moment about A of

$$-20 \cos 35° \times 3.2 = -52.43 \,\text{N}\,\text{m}$$

Therefore, the total moment about A is

$$61.95 - 52.43 = 9.52 \,\text{N}\,\text{m}$$

Forces in vector component form

Sometimes, forces may be given in the vector component form $a\mathbf{i} + b\mathbf{j}$, and points as either cartesian coordinates (x, y) or position vectors $x\mathbf{i} + y\mathbf{j}$.

Example 4 Find the moment about the point P, with position vector $(\mathbf{i} + 3\mathbf{j})\,\text{m}$, of the force $(5\mathbf{i} + 2\mathbf{j})\,\text{N}$ acting at the point Q, with position vector $(4\mathbf{i} + 5\mathbf{j})\,\text{m}$.

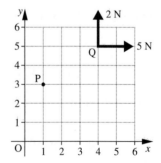

SOLUTION

From the diagram, we can see that

The 5 N component has a clockwise moment

$$-5 \times 2 = -10 \,\text{N}\,\text{m}$$

The 2 N component has an anticlockwise moment

$$2 \times 3 = 6\,\text{N}\,\text{m}$$

Therefore, the total moment of the force about P is

$$-10 + 6 = -4\,\text{N}\,\text{m}$$

Exercise 10B

1 Find the moment of each of the following forces about the point A, making sure you indicate whether it is a positive or a negative moment.

a)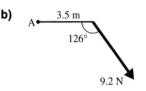

b)

c)

d)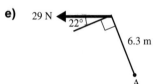

e)

f)

2 Find the total moment of the forces shown in each of the following diagrams about **i)** A and **ii)** B.

a)

b)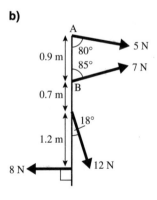

c)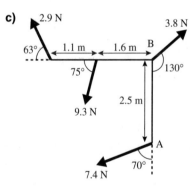

3 A force of 12 N acts along the diagonal AC of a rectangular lamina ABCD, as shown. Find the moment of force about B.

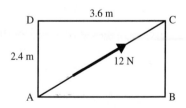

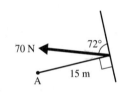

4 Find the moment about the point P, with position vector $(2\mathbf{i} + \mathbf{j})\,\text{m}$, of a force $(4\mathbf{i} + 6\mathbf{j})\,\text{N}$ acting at the point Q, with position vector $(4\mathbf{i} + 7\mathbf{j})\,\text{m}$.

5 Find the moment about the point P, with position vector $(\mathbf{i} - 3\mathbf{j})\,\text{m}$, of a force $(2\mathbf{i} - 3\mathbf{j})\,\text{N}$ acting at the point Q, with position vector $(2\mathbf{i} + 4\mathbf{j})\,\text{m}$.

6 Find the moment about the point P, with position vector $(-5\mathbf{i} + \mathbf{j})\,\text{m}$, of a force $(-3\mathbf{i} + 5\mathbf{j})\,\text{N}$ acting at the point Q, with position vector $(-3\mathbf{i} - 5\mathbf{j})\,\text{m}$.

Parallel forces: resultants and couples

Principle of moments

We know that any system of forces is equivalent to a single force, called the **resultant**. We also know that the combined **translational** effect of the forces is the same, both in magnitude and direction, as that of their resultant.

We now state that the same holds good for their **rotational** effect. This is, in fact, an assumption, although a plausible one, and those of you who like the mathematical niceties may wish to turn to the appendix to this chapter for a proof (see pages 209–11). For now, though, we apply this idea to parallel forces and state the **principle of moments**.

The **total moment of a pair of parallel forces** about any point is equal to the **moment of their resultant** about that point.

Resultant of parallel forces

When two forces act along parallel lines, they are called **like** forces when they act in the same direction, and **unlike** forces when they act in opposite directions.

The magnitude of the resultant of a pair of parallel forces is obviously the sum (for like forces) or the difference (for unlike forces) of the two forces. The problem now is to decide the line of action of this resultant, and for this we use the principle of moments.

Consider like forces of $P\,\text{N}$ and $Q\,\text{N}$ acting at points A and B and at right angles to the line AB. Let the length of AB be a metres. Their resultant has magnitude $(P + Q)\,\text{N}$ and acts through a point C on AB, where $\text{AC} = x$ metres.

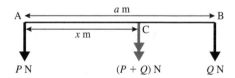

Now consider moments about A.

Total moment of the pair of like forces $= 0 - Qa\ \text{N}\,\text{m}$

Moment of the resultant $= -(P + Q)x\ \text{N}\,\text{m}$

By the principle of moments these are equal. Therefore, we have

$$(P + Q)x = Qa$$

$$\Rightarrow \quad x = \frac{Qa}{P + Q}$$

This means that the point C is between A and B such that

$$AC : BC = Q : P$$

That is, the line of action divides the line AB **internally** in the ratio $Q : P$.

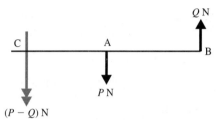

If the forces P N and Q N are unlike forces, as shown on the right, their resultant is $(P - Q)$ N. Its line of action is through the point C, where

$$AC : BC = Q : P$$

as before, but C now divide AB **externally** in the ratio $Q : P$. (You might like to show this for yourself.)

To summarise:

- The resultant of a pair of parallel forces P N and Q N, passing through points A and B respectively, cuts the line AB at C, where

$$AC : BC = Q : P$$

- When the forces are **like** forces, the point C is **internal** to AB.

- When the forces are **unlike** forces, the point C is **external** to AB.

Couples

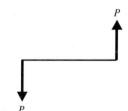

When a pair of unlike forces have equal magnitude their resultant is zero. But unless they act along the same line, they still have a turning effect. Such a pair of forces is called a **couple**.

Moment of a couple

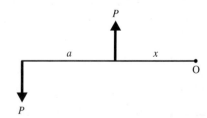

The diagram on the right shows a couple formed by unlike forces of magnitude P N acting a metres apart. We choose a general point, O, at a distance of x metres, as shown.

The total moment of the forces about O is

$$P(a + x) - Px = Pa \, \text{N m}$$

This means that the turning effect of a couple is the **same** about any point in the plane. We can, therefore, talk about the **moment of a couple** without specifying a point.

A couple is not confined to just two forces. Any number of forces may together form a couple provided the following two conditions obtain.

- Their resultant has magnitude zero.
- Their total moment is non-zero.

Example 5 Show that the forces given in the diagram form a couple and find the moment of that couple.

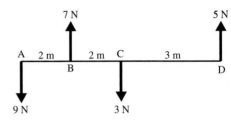

SOLUTION

Resolving perpendicular to AD, we get

$$\text{Resultant} = 7 + 5 - 9 - 3 = 0$$

As the resultant is zero, the forces are either in equilibrium or form a couple. So, taking moments about A, we have

$$\text{Total moment} = 7 \times 2 + 5 \times 7 - 3 \times 4 = 37\,\text{N m}$$

The forces therefore form a couple with a moment of $37\,\text{N m}$.

We can check that the moment is the same about other points by taking moments about C, say, which gives

$$\text{Total moment} = 9 \times 4 + 5 \times 3 - 7 \times 2 = 37\,\text{N m}$$

Equilibrium of parallel forces

We have seen that parallel forces may reduce to a single resultant force or to a couple. When the resultant is zero and there is no overall turning effect, the forces are in equilibrium. This means that for parallel forces in equilibrium:

- if we resolve in any direction, the resultant is zero
- if we take moments about any point, the total moment is zero.

Example 6 A uniform beam AB of mass $10\,\text{kg}$ and length $4\,\text{m}$ rests in a horizontal position on a single support at C, 1 metre from A. The other end of the beam is supported by a vertical string, as shown. Find the reaction, R, at the support and the tension, T, in the string.

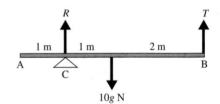

SOLUTION

Resolving vertically, we have

$$R + T - 10g = 0 \qquad [1]$$

Taking moments about C, we get

$$3T - 10g = 0 \qquad [2]$$

$$\Rightarrow \quad T = 3\tfrac{1}{3}g = 32.7\,\text{N}$$

Substituting into [1], we get: $\quad R = 6\tfrac{2}{3}g = 65.3\,\text{N}$

Note We choose to take moments about C rather than A because R has zero moment about C and so only one unknown (T) appears in Equation [2]. On a similar basis, instead of resolving, we could take moments about B to find the value of R.

We could also solve the problem without taking moments by noting that the resultant of R and T must be equal and opposite to the weight of the beam. This means that $R + T = 10g$ and that $R = 2T$, because the resultant divides CB in the ratio $T : R$, which is here $1 : 2$. Solving these two equations would give the required result.

Exercise 10C

1 The following diagrams show a light rod AB, of length 4 m, acted upon by parallel forces perpendicular to it. Decide whether the forces are in equilibrium, form a couple or have a single resultant. If they form a couple, find the moment. If they have a resultant, find its magnitude and the distance of its line of action from A and B.

a)

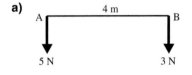

b)

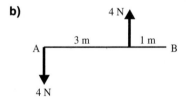

c)

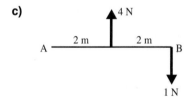

d)

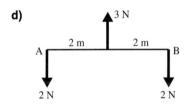

e)

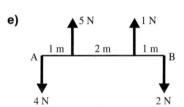

f)

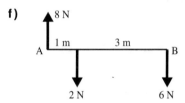

2 The diagram on the right shows forces acting on a rod AB of length 5 m. Find the values of R and x when the forces

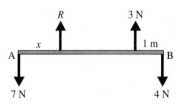

   a) are in equilibrium
   b) reduce to a couple of moment 4 N m
   c) reduce to a resultant of 2 N downwards passing through the mid-point of AB.

3 ABCD is a rectangle with $AB = 6$ m and $BC = 4$ m. Forces of 8 N, 8 N, P N and P N act along AB, CD, CB and AD respectively in the directions indicated by the order of the letters. Find the value of P when

   a) the forces are in equilibrium
   b) the forces form a couple with moment 8 N m in the sense ABCD.

4 Forces of $(5\mathbf{i} - 4\mathbf{j})\,\mathrm{N}$, $(2\mathbf{i} + \mathbf{j})\,\mathrm{N}$ and $(-7\mathbf{i} + 3\mathbf{j})\,\mathrm{N}$ act at points P, Q and R with position vectors $(-\mathbf{i} + 2\mathbf{j})\,\mathrm{m}$, $(4\mathbf{i} + 5\mathbf{j})\,\mathrm{m}$ and $(2\mathbf{i} - 3\mathbf{j})\,\mathrm{m}$ respectively. Show that the system is a couple and find its moment.

5 Forces of $(3\mathbf{i} + 2\mathbf{j})\,\mathrm{N}$, $(-2\mathbf{i} + \mathbf{j})\,\mathrm{N}$ and $(a\mathbf{i} + b\mathbf{j})\,\mathrm{N}$ act at points P(2, 5), Q(−3, 1) and R(−2, −4) respectively (distances in metres). If the system reduces to a couple, find the values of a and b, and find the moment of the couple.

6 A light rod AB of length 2 m is suspended by two vertical strings at A and B.

  a) An object of weight 200 N is suspended from the rod at C, where AC = 0.5 m. Calculate the tensions in the strings.
  b) The strings have a breaking strain of 180 N. The object is gradually moved along the rod towards A. How close can it get to A before the string breaks?
  c) A couple is applied to the rod so that the object can just be moved to A. Find the moment of the couple.

7 A uniform beam AB of length 4 m and mass 50 kg rests on supports at A and B. Objects of mass 20 kg and 40 kg are hung on the beam at C and D respectively, where AC = 1.4 m and AD = 3.2 m. Find the reactions at the supports.

8 A heavy beam AB rests on two supports at points C and D, where CD = a. An object of weight W rests on the beam. If the object is moved a distance b in the direction DC, show that, provided equilibrium is maintained, the reaction at C will be increased by Wb/a.

9 A rod AB of length $12a$ and weight $4W$ is suspended by strings attached at C and D, where AC = $3a$ and BD = $4a$. The breaking strain of the string at C is $3W$ and that at D is $3.8W$. An object of weight W is attached to the beam at a distance x from A. Find the range of values of x if neither string is to break.

10 A beam AB of weight W and length a has its centre of gravity a distance b from A. It is placed symmetrically on two supports a distance c apart. Find, in terms of W, a, b and c, the reactions at the supports.

11 A uniform plank is 12 m long and has mass 100 kg. It is placed on horizontal ground at the edge of a cliff, with 4 m of the plank projecting over the edge.

  a) How far out from the cliff can a man of mass 75 kg safely walk?
  b) The man wishes to walk to the end of the plank. What is the minimum mass he should place on the other end of the plank so he may safely do this?

12 A rectangular lamina ABCD is free to rotate in a vertical plane about its centre O. Weights of W, $5W$, $2W$ and $3W$ are attached at A, B, C and D respectively. If the system is in equilibrium, find the inclination of AC to the horizontal. Show that, by changing the order in which the weights are attached, a second equilibrium position is possible and find the inclination of the diagonal in this case.

Non-parallel forces: resultants and couples

On page 190, we met the principle of moments for parallel forces. In fact, this principle is true in general for **any** system of forces:

The **total moment of a system of forces** about any point is the same as the **moment of their resultant** about that point.

The proof of this can be found in the appendix at the end of this chapter (see pages 209–11).

Couples

It is possible for a system of non-parallel forces to form a couple.

Example 7 Forces of 10 N, 10 N and 12 N act along the sides of an isosceles triangle ABC, where AC = BC = 5 m and AB = 6 m, as shown. Show that the forces form a couple and find the moment of the couple.

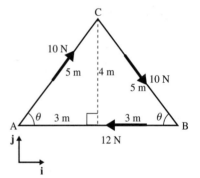

SOLUTION

From the diagram, we have

$$\cos \theta = \tfrac{3}{5} \quad \text{and} \quad \sin \theta = \tfrac{4}{5}$$

Taking the **i**- and **j**-directions as indicated in the diagram, the resultant force is

$$(10 \cos \theta \, \mathbf{i} + 10 \sin \theta \, \mathbf{j}) + (10 \cos \theta \, \mathbf{i} - 10 \sin \theta \, \mathbf{j}) - 12 \mathbf{i}$$

$$= (6\mathbf{i} + 8\mathbf{j}) + (6\mathbf{i} - 8\mathbf{j}) - 12\mathbf{i} = 0$$

As the resultant is zero, the forces are either in equilibrium or else they form a couple.

Taking moments about C, the only force with a turning effect is the 12 N force. Thus we have

$$\text{Total moment} = -12 \times 4 = -48 \, \text{N m}$$

The forces therefore form a couple with a moment of $-48 \, \text{N m}$.

We can check that this turning effect is the same about other points by taking moments about A, say. The only force involved now is that along CB. Therefore, we have

$$\text{Total moment} = -10 \sin \theta \times 6 = -8 \times 6 = -48 \, \text{N m}$$

Equilibrium of non-parallel forces

We have seen that a system of non-parallel forces can reduce to a single resultant force (a translational effect) or to a couple (a rotational effect). To be in equilibrium, both the translational and rotational effects must be zero. This will be the case provided the following two conditions obtain.

- The resultant force in each of two directions is zero.

- The total moment about any chosen point is zero.

These are the most commonly used **conditions for equilibrium**.

Example 8 A ladder AB of mass 20 kg rests on smooth horizontal ground and leans against a smooth vertical wall. The inclination of the ladder to the horizontal is 60°. The ladder is kept in position by a horizontal force P N applied to the bottom of the ladder. Find the value of P and the reactions at the wall and the ground.

SOLUTION

Suppose the ladder is of length $2a$. The reactions R and S at the ground and the wall are normal because the contacts are smooth. The system is in equilibrium.

Resolving vertically, we get

$$R - 20g = 0 \quad \Rightarrow \quad R = 196\,\text{N}$$

Resolving horizontally, we get

$$P - S = 0 \qquad\qquad [1]$$

Taking moments about B, we have

$$20g \cos 60° \times a - S \sin 60° \times 2a = 0 \qquad [2]$$

From [2]: $\quad 10g = S\sqrt{3} \quad \Rightarrow \quad S = 56.6\,\text{N}$

From [1]: $\quad P = 56.6\,\text{N}$

Note You should always be careful in your choice of the directions for resolving and of the point for taking moments. For instance, in Example 8 we could resolve parallel and perpendicular to the ladder and take moments about the middle of the ladder. The resulting equations would be

$$S \cos 60° + 20g \sin 60° - R \sin 60° - P \cos 60° = 0$$

$$S \sin 60° + R \cos 60° - 20g \cos 60° - P \sin 60° = 0$$

$$(R \cos 60° \times a) - (S \sin 60° \times a) - (P \sin 60° \times a) = 0$$

These equations would still yield the correct values, but the algebra involved is more tedious.

Alternative conditions for equilibrium

We have seen on page 196 that we can obtain three independent equations if we equate to zero the total components in two chosen directions and the total moment about one chosen point. We can achieve a similar result in two other ways, which can occasionally prove more convenient.

- **Method 1** Resolve in one direction and take moments about two points.

- **Method 2** Take moments about three points.

In Method 1, if we take moments about P and Q, say, we must **not resolve perpendicular to PQ**. (Otherwise a system comprising a single force along PQ would appear to be in equilibrium.)

In Method 2, the three points must **not be collinear** (for a similar reason).

Example 9 A uniform rod AB, of length 2 m and mass 5 kg, rests with A on smooth horizontal ground and B on a rough peg 1 m above the ground. Find the reaction at A and the normal reaction and friction forces at B.

SOLUTION

From the dimensions given, the angle at A is 30°.

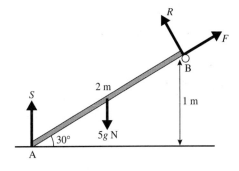

Taking moments about B, we have

$$5g\cos 30° \times 1 - S\cos 30° \times 2 = 0$$
$$\Rightarrow \quad S = 2.5g = 24.5 \, \text{N}$$

Taking moments about A, we have

$$R \times 2 - 5g\cos 30° \times 1 = 0$$
$$\Rightarrow \quad R = 2.5g\cos 30° = 21.2 \, \text{N}$$

Resolving along AB, we get

$$F + S\sin 30° - 5g\sin 30° = 0$$
$$\Rightarrow \quad F = 2.5g\sin 30° = 12.25 \, \text{N}$$

Example 10 A fixed smooth cylinder, radius a and centre O, rests on a smooth horizontal surface with its axis horizontal. A rod AB of weight W rests with A on the horizontal surface and B on the cylinder such that AB is inclined at 60° to the horizontal and is a tangent to the cylinder. The rod is held in place by a light string, AP, attached to the cylinder at P so that APO is a straight line. Find the tension in the string and the reactions at A and B.

SOLUTION

In the diagram, C is the point where the lines of action of the reactions R and S meet, and E is the point where the line of action of the weight meets the surface. By simple geometry and trigonometry, we can show that

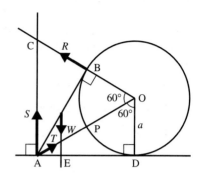

CP is perpendicular to AO

$$AB = AD = CP = a\sqrt{3} \quad \text{and} \quad AE = \tfrac{1}{4}AD$$

Taking moments about A, we have

$$Ra\sqrt{3} - \tfrac{1}{4}Wa\sqrt{3} = 0 \quad \Rightarrow \quad R = \tfrac{1}{4}W$$

Taking moments about O, we have

$$\tfrac{3}{4}Wa\sqrt{3} - Sa\sqrt{3} = 0 \quad \Rightarrow \quad S = \tfrac{3}{4}W$$

Taking moments about C, we have

$$Ta\sqrt{3} - \tfrac{1}{4}Wa\sqrt{3} = 0 \quad \Rightarrow \quad T = \tfrac{1}{4}W$$

Reactions in hinges and joints

Some problems require us to find the reaction in a hinge or a joint. This is often best achieved by finding the horizontal and vertical components of the reaction.

Example 11 A rod AB of length a and weight W is hinged to a vertical wall at A and is held at an angle of 30° above the horizontal by a light string BC, also of length a, which is fixed to the wall at C, a distance a vertically above A. Find the reaction in the hinge at A.

SOLUTION

Suppose the reaction has horizontal and vertical components, X and Y, as shown.

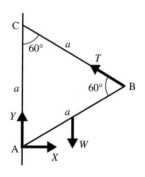

Taking moments about C, we have

$$Xa - \tfrac{1}{2}Wa\sin 60° = 0 \quad \Rightarrow \quad X = \tfrac{1}{4}W\sqrt{3}$$

Taking moments about B, we have

$$\tfrac{1}{2}Wa\sin 60° + Xa\cos 60° - Ya\sin 60° = 0$$

$$\Rightarrow \quad Y\sqrt{3} = X + \tfrac{1}{2}W\sqrt{3} \quad \Rightarrow \quad Y = \tfrac{3}{4}W$$

We can now combine X and Y to find the reaction R:

$$R = \sqrt{X^2 + Y^2} = \tfrac{1}{2}W\sqrt{3}$$

The reaction makes an angle θ to the horizontal, where

$$\tan\theta = \frac{Y}{X} = \sqrt{3} \quad \Rightarrow \quad \theta = 60°$$

Three forces in equilibrium

We have already looked at problems involving three concurrent forces. We are now in a position to make a stronger statement:

> When a system of three forces is in equilibrium, the **lines of action** of the forces must all **pass through a single point**.

The reasoning is as follows.

If the lines of action of two of the forces meet at point A, the resultant of those two forces is a force **R** passing through A. For equilibrium, the third force in the system must be $-\mathbf{R}$, and its line of action must also pass through A. (If it did not, the system would be a couple.) The three forces are therefore concurrent.

We make use of this property to reduce some problems to simple geometry.

Example 12 A uniform ladder AB rests with A on rough horizontal ground and B against a smooth vertical wall. The ladder is inclined at 45° to the horizontal and is on the point of slipping. Find the coefficient of friction μ between the ladder and the ground.

SOLUTION

If the ladder is on the point of slipping, the reaction S at A makes an angle λ with the vertical, where $\mu = \tan \lambda$.

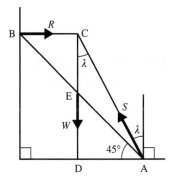

The lines of action of R, the normal reaction at B, and of W, the weight, meet at C. As there are only three forces, the line of action of S must also pass through C.

We can see that BCE and ADE are congruent right-angled isosceles triangles, which means that CD = 2AD. Therefore,

$$\tan \lambda = \frac{AD}{CD} = 0.5 \quad \Rightarrow \quad \mu = 0.5$$

An unreal problem

The following is a question from an old mechanics book.

> A uniform rod AB of length $2a$ and weight W rests with A in contact with a smooth vertical wall. B is attached by means of a light inextensible string of length $2b$ to a point C on the wall, a distance x vertically above A. If the system is in equilibrium, find the length AC in terms of a and b.

The solution to this problem uses the principle that three forces in equilibrium are concurrent.

The wall is smooth, so the reaction, R, is normal. W, R and the tension, T, are concurrent at D, the mid-point of BC.

From triangle ACD, we have

$$x = b\cos\theta \qquad [1]$$

From triangle ABC, we also have by the cosine rule

$$4a^2 = x^2 + 4b^2 - 4xb\cos\theta \qquad [2]$$

Substituting from [1] into [2], we obtain

$$4a^2 = x^2 + 4b^2 - 4x^2$$
$$\Rightarrow \quad 3x^2 = 4(b^2 - a^2)$$
$$\Rightarrow \quad x = 2\sqrt{\frac{b^2 - a^2}{3}}$$

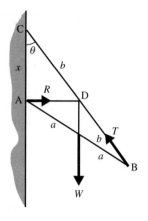

The difficulty with this problem is that if the wall were really frictionless, it would in practice be impossible to get the rod to remain in equilibrium. The equilibrium is unstable, which means that once the system has made the slightest move from the equilibrium position (and any vibration or air current would achieve this), it would continue to move until both A and B were against the wall.

A real problem

Experiment

You are going to try to find a model that is more closely related to the real world. You are still going to make assumptions, such as ignoring any weight or stretch in the string, but clearly you cannot ignore friction. If the wall contact is rough, there exists a range of values of x for which the system is in equilibrium. There is an upper limiting position in which A is about to slide **up** the wall, and a lower limiting position in which it is about to slide **down**. You are going to examine this both theoretically and practically.

The problem

A uniform rod AB of length $2a$ and weight W rests with A in contact with a rough vertical wall, the coefficient of friction being μ. B is attached by means of a light inextensible string of length $2b$ to a point C on the wall, a distance x vertically above A. The angle BCA is θ. Find expressions for μ for both the upper and lower limiting equilibrium positions.

Carry out an experiment with a suitable rod and wall. Calculate the value of μ from both positions and compare the results.

Now using a different vertical surface, calculate μ from the upper limiting equilibrium position. Use your result to predict the value of x corresponding to the lower limiting equilibrium position and confirm this experimentally.

The theory

As A is about to slide up the wall, the total reaction acts at an angle λ to the normal to the wall, where $\mu = \tan \lambda$, as shown.

If you set up the experiment and measure x, you can then calculate θ from triangle ABC using the cosine rule:

$$\cos \theta = \frac{x^2 + 4b^2 - 4a^2}{4bx}$$

From triangle CDE, $DE = b \sin \theta$ and $CE = b \cos \theta$.

In triangle ADE, angle D is λ. Therefore, you have

$$\tan \lambda = \frac{AE}{DE} \quad \Rightarrow \quad \mu = \frac{(b \cos \theta) - x}{b \sin \theta}$$

You should draw the diagram for the lower limiting equilibrium position and confirm that in this case

$$\mu = \frac{x - b \cos \theta}{b \sin \theta}$$

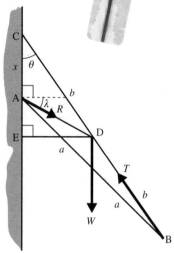

The practical

When you come to do the experiment, you will have more success if you replace the rod with a strip of wood 2 or 3 cm wide, and the string with a fairly wide, light tape which can be stuck to the underside of the strip at B, passing round the end of the strip, and pinned to the wall at C. This helps to prevent the tendency of the rod to slip sideways on the wall.

You should find it relatively easy to gradually manoeuvre the rod into a position where A is about to slide upwards. Measure the distance AC. You need to be as accurate as possible, because the formula for μ is quite sensitive to small changes in the input values.

Now move the strip (rod) until A is about to slide downwards. Again measure the distance AC.

You should now be able to calculate the estimated values of μ given by your results. This can be done directly from the formulae above. Alternatively, you may wish to make use of spreadsheet MOM 2, which is available on the Oxford University Press website (http://www.oup.co.uk/mechanics). This will be particularly helpful for the second part of the problem when you try to find x from a given value of μ.

Our attempt used a 40 cm rod and a 54 cm tape, so $a = 0.2$ m and $b = 0.27$ m.

Upper position: $\quad x = 0.194$ m $\quad \Rightarrow \quad \theta = 36.13°$

$$\Rightarrow \quad \text{Estimated } \mu = \frac{0.27 \cos 36.13° - 0.194}{0.27 \sin 36.13°} = 0.15$$

Lower position: $\quad x = 0.233$ m $\quad \Rightarrow \quad \theta = 42.38°$

$$\Rightarrow \quad \text{Estimated } \mu = \frac{0.233 - 0.27 \cos 42.38°}{0.27 \sin 42.38°} = 0.18$$

Calculating x

Instead of finding a second estimate of μ using the lower position, we could have used our estimate $\mu = 0.15$ from the upper position to predict x for the lower position. If you are using the spreadsheet, this can be done by entering the value for μ and using the 'goal seek' (EXCEL) or 'backsolver' (1–2–3) commands to obtain x. Otherwise, we need to solve the equations

$$\cos \theta = \frac{x^2 + 4b^2 - 4a^2}{4bx} \quad [1] \quad \text{and} \quad \mu = \frac{x - b \cos \theta}{b \sin \theta} \quad [2]$$

Substituting known values, we obtain

Eqn [1] $\quad \Rightarrow \quad x^2 - 1.08x \cos \theta + 0.1316 = 0 \quad$ [3]

Eqn [2] $\quad \Rightarrow \quad x = 0.0405 \sin \theta + 0.27 \cos \theta \quad$ [4]

Substituting from [4] into [3], we have

$$0.001\,640\,025 \sin^2 \theta - 0.021\,87 \sin \theta \cos \theta - 0.2187 \cos^2 \theta + 0.1316 = 0$$

Dividing through by $\cos^2 \theta$ and using $\sec^2 \theta = 1 + \tan^2 \theta$, we get

$$0.133\,240\,25 \tan^2 \theta - 0.02187 \tan \theta - 0.0871 = 0$$

Solving this quadratic equation, we obtain

$$\tan \theta = 0.8947 \quad (\text{or} \ -0.7306)$$

$$\Rightarrow \quad \theta = 41.82°$$

Substituting in [4], we get $x = 0.228$ m, which is in good experimental agreement with the recorded value of 0.233 m.

Exercise 10D

1 The forces $(3\mathbf{i} + 2\mathbf{j})\,\text{N}$, $(-5\mathbf{i} - 4\mathbf{j})\,\text{N}$ and $(2\mathbf{i} + 2\mathbf{j})\,\text{N}$ act at points with position vectors $(\mathbf{i} + 3\mathbf{j})\,\text{m}$ and $(4\mathbf{i} - \mathbf{j})\,\text{m}$ and $(-2\mathbf{i} - 5\mathbf{j})\,\text{m}$ respectively. Show that these forces form a couple, and find the moment of the couple.

2 Triangle ABC has $AB = 16\,\text{m}$, $AC = BC = 17\,\text{m}$. Forces of $8\,\text{N}$, $8.5\,\text{N}$ and $8.5\,\text{N}$ act along AB, BC and CA respectively, with the direction given by the order of the letters. Show that these forces form a couple, and find the moment of the couple.

3 Triangle ABC has sides of length a, b and c labelled according to the usual convention. Forces of magnitude ka, kb and kc act along BC, CA and AB respectively, with the direction given by the order of the letters. By considering the vector sum of the forces, or otherwise, show that these forces form a couple, and find the moment of the couple in terms of the area of the triangle ABC.

4 Forces of $1\,\text{N}$, $6\,\text{N}$, $2\,\text{N}$, $4\,\text{N}$, $3\,\text{N}$ and $5\,\text{N}$ act in that order along the sides of a regular hexagon of side length a. All the forces act in the same direction around the hexagon. Show that the system is a couple and find the magnitude of its moment.

5 A uniform rod AB of length 3 metres and mass $15\,\text{kg}$ is hinged at A. A light string is attached to B and holds the rod in equilibrium at an angle of $60°$ to the upward vertical through A. Find the tension in the string when

a) the string is at right angles to AB
b) the string is vertical
c) the string is horizontal.

6 The diagram shows a horizontal uniform rod AB of length $2a$ and weight W. A light string is attached to A and B and passes through a smooth ring at C, vertically above A, so that angle ABC is $30°$. A horizontal force P is applied to A.

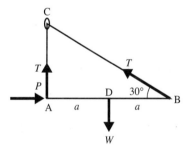

a) Show that the system cannot be in equilibrium.
b) A weight W is attached at a point D on AB so that equilibrium is maintained. Find the distance AD and the force P.

7 The diagram shows a cross-section of a uniform horizontal shelf hinged to a vertical wall. The length AB is $2a$ and the shelf has weight W. The shelf is supported by a light string CD connecting a point D on the shelf, where $AD = x$, to a point C on the wall, a distance a vertically above A. The breaking strain of the string is $4W$. Find the minimum value of x.

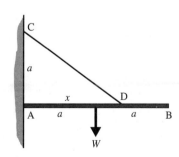

8 A uniform rod AB of mass 10 kg and length 2 m rests with A on smooth horizontal ground and B on a smooth peg 1 m above the ground. The rod is held in position by a horizontal force of P N at A. Find the value of P and the magnitude of the reactions at A and B.

9 A uniform ladder of mass 20 kg and length 3 m rests against a smooth wall with the bottom of the ladder on smooth horizontal ground and attached by means of a light inextensible string, 1 m long, to the base of the wall.

a) Find the tension in the string.

b) If the breaking strain of the string is 250 N, find how far up the ladder a man of mass 80 kg can safely ascend.

10 A uniform ladder of weight W rests at an angle α to the horizontal with its top against a smooth vertical wall and its base on rough horizontal ground with coefficient of friction 0.25. Find the minimum value of α if the ladder does not slip.

11 A uniform ladder of weight W rests at an angle α to the horizontal with its top against a rough vertical wall and its base on rough horizontal ground with coefficient of friction 0.25 at each contact. Find the minimum value of α if the ladder does not slip.

12 The diagram shows a cross-section ABCD of a uniform rectangular block of mass 20 kg. AB is 0.75 m and BC is 1 m. The block rests with A on rough horizontal ground and AB at 20° to the horizontal. It is held in place by a horizontal force P N applied at C. The block is on the point of slipping. Find the value of P and the coefficient of friction between the block and the ground.

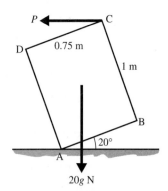

13 The diagram shows a uniform rod AB of weight W and length $2a$. The rod rests with A on rough horizontal ground and leans against a rough fixed prism of semicircular cross-section of radius a. The coefficient of friction at both contacts is μ. When friction is limiting the rod makes an angle θ with the horizontal. Show that

$$\sin\theta = \sqrt{\frac{\mu}{1 + \mu^2}}$$

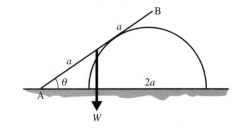

14 A uniform rod AB of length $2a$ and weight W is hinged to a horizontal ceiling at A and is suspended by a light inextensible string BC of length a connecting B to a point C on the ceiling such that angle ABC is 90°. Show that the tension in the string is $W/\sqrt{5}$, and find the horizontal and vertical components of the reaction of the hinge on the rod.

15 A uniform ladder of length $2a$ and weight W leans at an angle θ to the vertical against a rough vertical wall. The bottom of the ladder rests on rough horizontal ground. The angle of friction is λ at both contacts and the ladder is on the point of slipping. By using the fact that three forces in equilibrium must be concurrent, or otherwise, show that $\theta = 2\lambda$.

Equivalent force systems

In general, a system of forces is either equivalent to a couple or it has a resultant force. We have seen how to find the line of action of this resultant force in the case of parallel forces. Now consider the following system of forces.

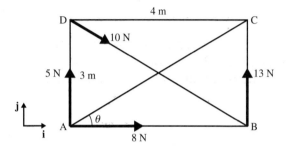

We can see that ABC is a 3–4–5 triangle, giving $\sin\theta = 0.6$ and $\cos\theta = 0.8$.

We find the vector sum, **R**, of the forces, taking the **i**- and **j**-directions as shown in the diagram:

$$\mathbf{R} = 8\mathbf{i} + 5\mathbf{j} + 13\mathbf{j} + (10\cos\theta\,\mathbf{i} - 10\sin\theta\,\mathbf{j}) = 16\mathbf{i} + 12\mathbf{j}$$

We can see that the resultant has magnitude 20 N and acts at an angle ϕ to the **i**-direction, where $\tan\phi = 0.75$.

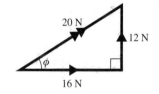

This means that $\phi = \theta$, so the system of forces is equivalent to a single force of 20 N acting in a direction parallel to AC.

We now need to find the line of action of this resultant force.

Suppose it acts through a point E on AB, where $AE = x$ m. The moment of the resultant about any point must be the same as the total moment of the original system of forces.

Taking the moments about A, we have

$$12x = 13 \times 4 - 10\cos\theta \times 3$$

$$\Rightarrow \quad 12x = 28$$

$$\Rightarrow \quad x = 2\tfrac{1}{3} \text{ m}$$

The original system of forces is therefore equivalent to a single force of 20 N acting along a line through E and parallel to AC.

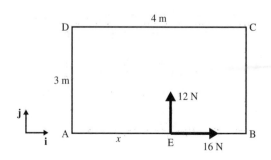

We could also replace the system of forces with a single force of 20 N acting along any other line parallel to AC, provided we also apply a couple so that the total moment is unchanged. For example, if we used a force of 20 N acting along AC, this would have zero moment about A. We would, therefore, have to apply a couple of moment 28 N m.

Again, if we used a force of 20 N parallel to AC but acting through B, this would have a moment about A of $12 \times 4 = 48$ N m. We would, therefore, have to apply a couple of moment -20 N m for the system to be equivalent to the original set of forces.

Similarly, the forces are equivalent to a single force of 20 N parallel to AC but acting through D, together with a couple of moment 76 N m.

Example 13 Forces of $(3\mathbf{i} + 13\mathbf{j})$ N, $(2\mathbf{i} - \mathbf{j})$ N and $(-\mathbf{i} - 4\mathbf{j})$ N act at points with position vectors $(\mathbf{i} + \mathbf{j})$ m, $(3\mathbf{i} + 2\mathbf{j})$ m and $(-3\mathbf{i} + 5\mathbf{j})$ m respectively. Find the magnitude of the resultant force and the equation of its line of action.

SOLUTION

The resultant, **R**, is given by

$$\mathbf{R} = (3\mathbf{i} + 13\mathbf{j}) + (2\mathbf{i} - \mathbf{j}) + (-\mathbf{i} - 4\mathbf{j}) = 4\mathbf{i} + 8\mathbf{j}$$

We can see that

$$|\mathbf{R}| = \sqrt{4^2 + 8^2} = 8.94 \text{ N}$$

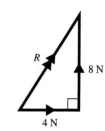

and that the line of action of the resultant has a gradient of 2.

The moment about the origin of the original system of forces is

$$-3 \times 1 + 13 \times 1 - 2 \times 2 - 1 \times 3 + 1 \times 5 + 4 \times 3 = 20 \text{ N m}$$

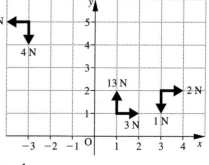

The moment of the resultant about the origin must therefore be 20 N m.

Suppose the line of action of **R** cuts the y-axis at A$(0, c)$. Then, the moment of **R** about the origin is $-4c$ N m, which gives

$$-4c = 20 \quad \Rightarrow \quad c = -5 \text{ m}$$

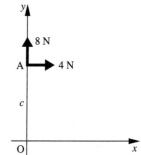

The equation of the line of action is, therefore,

$$y = 2x - 5$$

Example 14 The diagram shows a system of forces acting along the sides of a regular hexagon of side length $2a$ m.

a) Show that the system is equivalent to a force of magnitude 18 N acting along BA together with a couple, and find the magnitude of the couple.

b) Show also that the system reduces to a single force without a couple and find the line of action of this force.

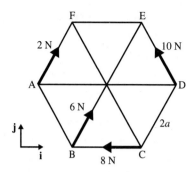

SOLUTION

Taking the **i**- and **j**-directions as shown, the resultant is

$$\mathbf{R} = (\mathbf{i} + \sqrt{3}\mathbf{j}) + (3\mathbf{i} + 3\sqrt{3}\mathbf{j}) + (-5\mathbf{i} + 5\sqrt{3}\mathbf{j}) + (-8\mathbf{i})$$
$$= -9\mathbf{i} + 9\sqrt{3}\mathbf{j}$$

We can see that

$$|\mathbf{R}| = \sqrt{81 + 243} = 18 \text{ N}$$

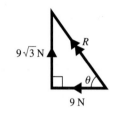

and that the direction is given by θ, as shown, where

$$\tan\theta = \sqrt{3} \quad \Rightarrow \quad \theta = 60°$$

which is the direction of BA.

Suppose the system is equivalent to a force of 18 N along BA together with a couple of moment M, as shown. This should have the same moment about any point as the original system.

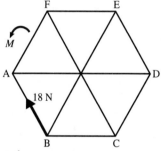

Consider moments about A.

The total moment of the system is

$$10 \times 2a\sqrt{3} + 6 \times a\sqrt{3} - 8 \times a\sqrt{3} = 18a\sqrt{3} \text{ N m}$$

In the equivalent system, the single force passes through A, so the total moment is M. Therefore, we have

$$M = 18a\sqrt{3} \text{ N m}$$

Now suppose the system is equivalent to a single force of 18 N acting along a line cutting AD at G, where AG = x, as shown.

The moment of this about A is

$$18 \sin 60° \times x = 9x\sqrt{3}$$
$$\Rightarrow \quad 9x\sqrt{3} = 18a\sqrt{3}$$
$$\Rightarrow \quad x = 2a$$

This means that G is at the centre of the hexagon, so the system is equivalent to a single force of 18 N acting along CF.

Exercise 10E

1 ABC is an equilateral triangle. D and E are the mid-points of AB and BC respectively. A force of 2 N acts along AC in the direction indicated by the order of the letters, and forces **P** and **Q** act along AB and BC respectively.

a) Find the magnitude and sense of **P** and **Q** if the system reduces to a couple.

b) Find the magnitude and sense of **P** and **Q** if the system reduces to a single force along DE.

2 Relative to axes with origin O, points A and B have coordinates (12, 5) and (6, 8) respectively. Forces of magnitude 6.5 N, 10 N and $4\sqrt{5}$ N act along OA, OB and AB respectively in the directions indicated by the order of the letters.

a) Show that the line of action of the resultant is $29x - 8y = 176$.

b) Find the magnitude and sense of the couple which must be added to the system for it to reduce to a single force passing through A.

3 ABCD is a square. Forces 5 N, 2 N, 3 N, 1 N and Q N act along AB, CB, CD, AD and DB respectively, in the directions indicated by the order of the letters. Find the value of Q if the system reduces to a single force passing through A. For this value of Q, find the magnitude of the resultant and the angle which it makes with AB.

4 A system of forces has a clockwise moment of $22Pa$ about the point A(2a, 2a) and anticlockwise moments of $7Pa$ and $4Pa$ about B(a, 6a) and C(−2a, 3a) respectively. If the system is equivalent to a force $X\mathbf{i} + Y\mathbf{j}$ acting through the point (0, c), find the values of X, Y and c.

5 The diagram shows a lamina ABCDE comprising a rectangle ABCE with AB = 2a and BC = a, connected to an equilateral triangle CDE. Forces of 2 N, 2 N, 3 N, 3 N and 4 N act along ED, DC, BC, BA and EA respectively in the directions indicated. The lamina is pinned to a fixed surface by a pin at point F on CE, where EF = x, so that the system is then in equilibrium. Find the magnitude and direction of the force exerted by the system on the pin and find the value of x. If it were desired to pin the lamina to the surface using a pin at B, what couple would have to be added to the system for there still to be equilibrium?

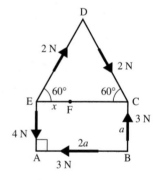

6 The diagram shows a parallelogram ABCD in which angle BAD is 45° and AD = DB = BC. Forces of $2W$ N, $2W\sqrt{2}$ N, $5W\sqrt{2}$ N, P N and Q N act along AB, AD, CB, CD and BD respectively. Find the values of P and Q if the system reduces to

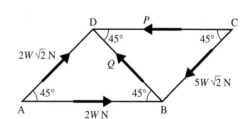

a) a couple

b) a single force acting along BA.

7 ABC is a triangular lamina with AB = 4 cm, BC = 5 m and AC = 3 m. D, E and F are the mid-points of BC, AC and AB respectively. Forces of 4 N, 3 N, 10 N, P N and Q N act along AB, AC, CB, DE and AD respectively, in the directions indicated by the order of the letters.

a) Show that, unless the system is in equilibrium, it reduces to a single force through D.

b) Find the values of P and Q if the resultant is a force of 6 N acting along FD.

c) Find the value of P if the resultant acts along AD and find the magnitude of this resultant in terms of Q.

8 The diagram shows a circular lamina of radius a. Forces of 9 N, 8 N, 2 N and P N act at the ends of two perpendicular diameters, as shown.

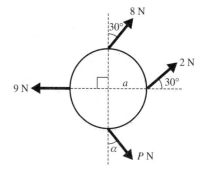

a) Show that the system cannot be in equilibrium.

b) Show that if the system reduces to a couple, the magnitude of the couple is $(2 - \sqrt{3})a$.

c) If the value of P is 10 and the system reduces to a single force through the centre of the circle, find the value of α.

Appendix

Principle of moments for parallel forces

First, we find the line of action of the resultant of a pair of like parallel forces.

Suppose we have like forces of P N and Q N acting at A and B. The resultant of the forces has a magnitude of $(P + Q)$ N and acts at some point C, as shown.

We can add a pair of equal and opposite forces to the system without changing the overall situation. Suppose we add forces of PQ N at A and B, as shown below.

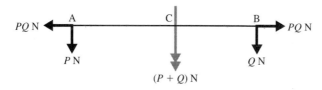

The forces which now act at A have a resultant of magnitude $P\sqrt{Q^2 + 1}$ N acting at an angle θ, where $\tan \theta = Q$, as shown.

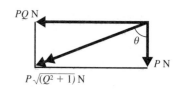

Similarly, the forces which now act at B have a resultant of magnitude $Q\sqrt{P^2 + 1}$ N acting at an angle ϕ, where $\tan \phi = P$, as shown.

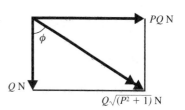

The pairs forces at A and B can be replaced by these resultant forces. The lines of action of these forces intersect at D, as shown. The resultant of the whole system is still $(P + Q)\,\text{N}$ and we can now see that its line of action must pass through D.

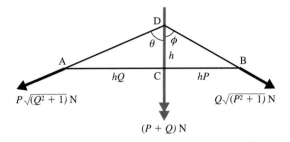

If the length $CD = h$, then we have

$$\tan\theta = Q \quad \Rightarrow \quad AC = hQ$$

$$\tan\phi = P \quad \Rightarrow \quad BC = hP$$

So, the line of action of the resultant divides the line AB **internally** in the ratio $Q:P$.

Note If we have a pair of unlike forces of magnitude $P\,\text{N}$ and $Q\,\text{N}$, where $P > Q$, we can follow a similar procedure to the one above. (You might like to try this for yourself.) The resultant is now a force of magnitude $(P - Q)\,\text{N}$. Its line of action is through the point C, where $AC:BC = Q:P$ as before, but C now divides AB **externally** in the ratio $Q:P$.

Proving the principle of moments

Suppose we have like forces of magnitude $P\,\text{N}$ and $Q\,\text{N}$ and any point O. The line through O at right angles to the direction of the forces cuts the lines of action of P and Q at A and B respectively, and the line of action of their resultant at C, as shown.

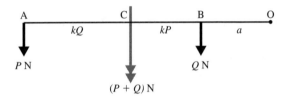

C divides AB in the ratio $Q:P$, so we can put $AC = kQ$ and $BC = kP$. Let $BO = a$.

The total moment of the two forces about O is

$$P(kQ + kP + a) + Qa = kPQ + kP^2 + Pa + Qa$$

The moment of the resultant about O is

$$(P + Q)(kP + a) = kPQ + kP^2 + Pa + Qa$$

So, the moment of the resultant is the **same** as the total moment of the forces. (A similar argument will establish this for unlike parallel forces. Why not check this for yourself.)

Principle of moments for non-parallel forces

We can now establish that the principle holds for non-parallel forces.

We have already seen that the moment about point A of a force F acting at point B is given by the moment about A of the component of the force perpendicular to AB.

For example, in the diagram, the moment of F about A is $Fa\sin\theta$.

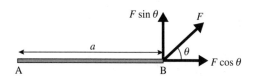

Now consider the moments about point O of forces P and Q whose resultant is R, as shown. We draw the perpendicular OA from O to the line of action of R, cutting the lines of action of P and Q at B and C.

The components $P\sin\theta$ and $Q\sin\phi$ are equal and opposite (because the total of components at right angles to the resultant must be zero). So, R is the resultant of the components $P\cos\theta$ and $Q\cos\phi$. These are parallel forces, therefore, by the principle of moments for parallel forces,

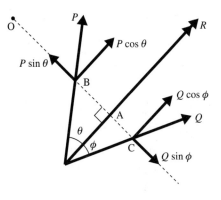

$$P\cos\theta \times \mathrm{OB} + Q\cos\phi \times \mathrm{OC} = R \times \mathrm{OA}$$

But the left-hand side of this equation is the total moment of P and Q about O, and the right-hand side is the moment of their resultant about O. So, the principle of moments holds for non-parallel forces.

Examination questions

Chapters 8 to 10

Chapter 8

1 A teacher, on playground duty, notices Jill 5 m due north of Susan. Jill and Susan are running with velocities of $4\,\text{m s}^{-1}$ due south and $4\sqrt{3}\,\text{m s}^{-1}$ due east respectively.

i) Find the magnitude and direction of the velocity of Jill relative to Susan.

ii) If they maintain these velocities, find the shortest distance between Jill and Susan.

(NICCEA)

2 A bird wishes to fly to a tree 2 km due east of its present position. It can fly at $3\,\text{m s}^{-1}$ in still air. There is a wind of speed $1\,\text{m s}^{-1}$ blowing from the north-east.

i) Find the direction in which the bird should fly.

ii) Find its speed towards the tree.

The wind abruptly changes direction t seconds after the bird has started its flight and blows from the east at the same speed. The total time for the bird to reach the tree is 16 minutes.

iii) Find the value of t.   (NICCEA)

3 A rally car is travelling at $24\sqrt{2}\,\text{m s}^{-1}$ due east. A marshal is running at $3\,\text{m s}^{-1}$ in a south-easterly direction.

i) Find the magnitude and direction of the velocity of the rally car relative to the marshal.

At a certain instant, the marshal is 25 m north-east of the rally car. In order to read the number on the side of the car, the marshal needs to be within 20 m of it. Assume the car and marshal maintain the velocities given above.

ii) Find how long the marshal will have to read the number. Give your answer in seconds.

(NICCEA)

4 The figure on the right shows two jetties P and Q on the shore of a calm lough. The bearing of Q from P is 120°. A motorboat leaves P and travels directly towards Q at $8\,\text{m s}^{-1}$. A police launch which is patrolling the lough is travelling on a bearing of 210° at $10\,\text{m s}^{-1}$.

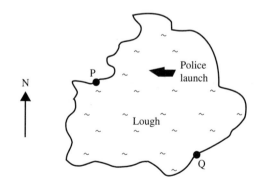

i) Find the velocity of the police launch relative to the motorboat.

At 1.00 pm, when the police launch is 2.5 km due east of the motorboat, it alters course in order to intercept the motorboat.

ii) Find the course on which the police launch should be steered.

Shortly after 1.00 pm the crew of the motorboat sight the police launch at a distance of 0.568 km.

iii) At what time does this sighting occur?　　(NICCEA)

5 Two particles A and B have constant velocities $4\mathbf{i} - 3\mathbf{j}$ and $5.25\mathbf{i} + 7\mathbf{j}$, respectively. Initially A has position vector $4\mathbf{i} + 10\mathbf{j}$ and B has position vector $-\mathbf{i} - 30\mathbf{j}$.

a) Show that the directions of motion of A and B are perpendicular.

b) i) Show that at time t the position vector of A is given by

$$\mathbf{r} = (4\mathbf{i} + 10\mathbf{j}) + t(4\mathbf{i} - 3\mathbf{j})$$

　　ii) Write down, in a similar form, the position vector of B at time t.

c) Given that the particles collide, find the time at which the collision occurs.　　(NEAB)

6 A football pitch is a horizontal plane and O is a fixed point on the pitch. The vectors $\mathbf{i}$ and $\mathbf{j}$ are perpendicular unit vectors in this horizontal plane. Colin and David are two players on the pitch. At time $t = 0$, David kicks the ball from the origin O with a constant velocity $8\mathbf{i}\,\mathrm{m\,s^{-1}}$ and runs thereafter with constant velocity $(3\mathbf{i} + 5\mathbf{j})\,,\mathrm{s^{-1}}$. When David kicks the ball, Colin is at the point with position vector $(10\mathbf{i} + 8\mathbf{j})\,\mathrm{m}$ and starts running with constant velocity $(3\mathbf{i} - 4\mathbf{j})\,\mathrm{m\,s^{-1}}$.

a) Write down the position vectors of Colin and David at time t seconds.

b) Verify that Colin intercepts the ball after 2 seconds.

As soon as Colin intercepts the ball he kicks it, giving it a constant velocity of $(\lambda\mathbf{i} + \mu\mathbf{j})\,\mathrm{m\,s^{-1}}$. He aims to pass it to David who maintains his constant velocity. Given that David intercepts the ball 2 seconds after Colin has kicked it,

c) find the values of λ and μ.　　(EDEXCEL)

7 Two motorboats A and B are moving with constant velocities. The velocity of A is $30\,\mathrm{km\,h^{-1}}$ due north, and B is moving at $20\,\mathrm{km\,h^{-1}}$ on a bearing of $060°$. The unit vectors $\mathbf{i}$ and $\mathbf{j}$ are directed due east and north respectively. At 10 am the position vector of B is $70\mathbf{j}\,\mathrm{km}$ relative to a fixed origin O and A is at the point O; t hours later, the position vectors of A and B are $\mathbf{r}\,\mathrm{km}$ and $\mathbf{s}\,\mathrm{km}$ respectively.

a) Find the velocity of B in the form $(p\mathbf{i} + q\mathbf{j})\,\mathrm{km\,h^{-1}}$.

b) Find expressions for $\mathbf{r}$ and $\mathbf{s}$ in terms of t.

The boats can maintain radio contact with each other, provided that the distance between them is no more than 70 km.

c) Find the time at which the boats are again at the maximum distance at which they can maintain radio contact with each other.　　(EDEXCEL)

Chapter 9

8

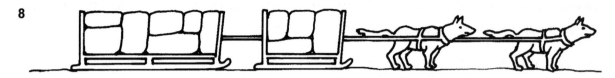

The diagram shows two sledges being pulled by a team of dogs at a steady speed over rough level ground. The ropes connecting the sledges and the dogs are horizontal. The first sledge has mass m, the second sledge has mass $2m$, and the coefficient of friction between each sledge and the ground is μ.

a) Show, on a sketch, the main horizontal and vertical forces acting on each sledge.
b) Find expressions in terms of μ, m and g for
  i) the tension in the rope connecting the two sledges
  ii) the total horizontal force exerted by the dogs on the first sledge. (NEAB)

9 A man irons a shirt by pressing down on an iron, applying a force of magnitude R newtons at an angle $\alpha°$ to the vertical. He moves the iron slowly at constant speed over the shirt which is spread out on a horizontal ironing board. The iron has mass $\frac{1}{2}$ kg, and the coefficient of friction between the iron and the shirt is $\frac{1}{4}$. To iron out the creases on the shirt successfully, the magnitude of the total vertical force exerted on the shirt by the iron must be 20 N. By modelling the iron as a particle, and ignoring any effect of air resistance, find, to three significant figures,

a) the value of α
b) the value of R.
c) Explain why, in the situation described, it is a reasonable assumption to ignore air resistance. (EDEXCEL)

10 In an exercise, a gymnast assumes a position in which her body is in a straight line and her hands are resting flat against a horizontal circular rail, as shown in Fig. 1. It is required to determine the least coefficient of friction between her shoes and the ground which will prevent her from slipping. Friction between her hands and the rail is considered negligible.

Fig. 1   Fig. 2

To model this problem, the gymnast is considered as a uniform rod AB, of length $2l$ and weight W, in equilibrium with the end A resting against a smooth horizontal circular rail and the end B resting on a horizontal floor, as shown in Fig. 2. The rod is assumed to be tangential to the rail and the angle between the rod and the downward vertical through A is denoted by α.

a) i) Explain why the force on the rod at A must be perpendicular to AB.
  ii) Explain why the lines of action of the force on the rod at A and the force on the rod at B must meet at a point vertically above the mid-point of AB.

iii) Draw a diagram to illustrate the forces acting on the rod.

b) In the case when $\tan \alpha = 2$, the force on the rod at B makes an angle β with AB. The angle of friction at B is λ.

   i) Show that $\beta = 45°$.

   ii) Show that $\lambda \geqslant \alpha - \beta$.

   iii) Deduce that the coefficient of friction between the rod and the floor must be at least $\frac{1}{3}$.

c) Give **one** criticism of the mathematical model which has been used.   (NEAB)

11 The figure shows a particle A, of mass $3m$, on a plane inclined at an angle α to the horizontal, and a particle B, of mass $2m$, on a plane inclined at an angle $90° - \alpha$ to the horizontal, where $\tan \alpha = \frac{4}{3}$. The plane on which A moves is rough, the coefficient of friction between A and the plane being $\frac{1}{3}$. The plane on which B moves is smooth. A

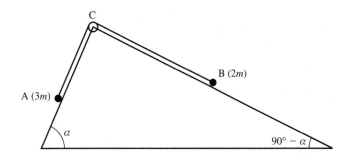

and B are connected by a string passing over a small smooth pulley fixed at C, the highest point of the two planes. The string is assumed to be light and inextensible. The particles are released from rest with the string taut and the sections AC and BC of the string parallel to lines of greatest slope in the respective planes, and A moves down the plane. Given that neither particle has reached the top or the bottom of its respective plane,

a) show that the frictional force acting on A has magnitude $\frac{3}{5}mg$

b) find

   i) the magnitude of the acceleration of the particles in terms of g

   ii) the magnitude of the tension in the string in terms of m and g.

c) State where you have used the assumption that

   i) the string is inextensible

   ii) the pulley is smooth.   (EDEXCEL)

12 A particle of mass $0.5\,\text{kg}$ is at rest on a rough horizontal table. The coefficient of friction is 0.3. A horizontal force of magnitude $1.47t^2\,\text{N}$, where t denotes the time in seconds, is applied to the particle, starting at time $t = 0$.

a) Show that the particle will not move until $t = 1$.

b) Show that for $t \geqslant 1$ the acceleration of the particle at time t is $2.94t^2 - 2.94$.

c) Find the speed of the particle when $t = 2$.   (WJEC)

13 A sledge of mass $40\,\text{kg}$ moves in a straight line on a horizontal snow surface and is attached to one end of a rope. The other end of the rope is attached to a towing hook on a motorised snowmobile. The snowmobile pulls the sledge, giving it an acceleration of $0.12\,\text{m\,s}^{-2}$. The coefficient of friction between the sledge and the snow surface is 0.1.

In an initial modelling of the situation, the sledge is modelled as a particle and the rope is assumed to be horizontal. Using this model,

a) find the tension in the rope.

In a more refined model, account is taken of the fact that the towing hook on the snowmobile is above the level of the sledge. The rope is now assumed to make an angle of 20° to the horizontal. Using this model,

b) find, in N to one decimal place, a revised value for the tension in the rope.   (EDEXCEL)

14 A uniform ladder rests with its lower end on a rough horizontal path and its upper end against a smooth vertical wall. The ladder rests in a vertical plane perpendicular to the wall. A woman stands on the top of this ladder, and the ladder is in limiting equilibrium. The weight of the woman is twice the weight of the ladder, and the coefficient of friction between the path and the ladder is $\frac{5}{12}$. By modelling the ladder as a uniform rod and the woman as a particle, find, to the nearest degree, the angle between the ladder and the horizontal.   (EDEXCEL)

15 The diagram on the right shows a uniform rod AB, of weight W and length $2a$, resting in equilibrium in a vertical plane with its end A on a rough horizontal plane and the point C on the rod in contact with a smooth fixed peg. The length AC is $\frac{3}{2}a$. Given that the rod makes an angle θ with the horizontal, show that the force exerted by the peg on the rod is $\frac{2}{3}W\cos\theta$.

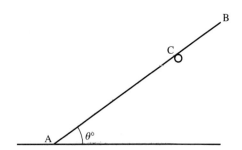

Find, in terms of W and θ, the normal and frictional forces exerted by the plane on the rod at A.

The coefficient of friction between the rod and the plane at A is μ. Deduce that, for equilibrium to be possible,

$$\mu \geqslant \frac{2\sin\theta\cos\theta}{3 - 2\cos^2\theta} \qquad \text{(OCR)}$$

Chapter 10

16

The diagram shows a horizontal light rod AB resting on smooth supports at P and Q where $AP = 0.3\,\text{m}$ and $PQ = 0.9\,\text{m}$. A particle of weight 12 N is placed at A and a second particle, of weight 27 N, is placed at the point C on the rod where $PC = 0.5\,\text{m}$.

Given that the system is in equilibrium, find the reactions at P and Q.   (WJEC)

17

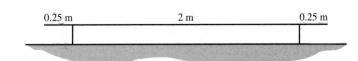

The diagram above shows a gym bench of length 2.5 m, which stands on horizontal ground.

The two supports of the bench are of the same height; they are 2 m apart and each is 0.25 m from an end of the bench. The centre of mass of the bench is equidistant from its ends. When a girl of weight 376 N stands on one end, the bench is on the point of toppling. Calculate the weight of the bench.   (OCR)

18 A uniform beam, AB, of length 4 m and mass 30 kg rests horizontally on two supports which are 1 m from each end. A man of mass 75 kg stands on the beam directly above one support.

   i) Using a suitable model, draw a diagram which shows all the external forces acting on the beam.

   ii) Give **two** assumptions that you have made in your model.

   iii) Find the reaction at each support.   (NICCEA)

19

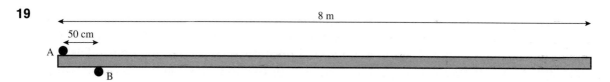

A large uniform plank of wood of length 8 m and mass 30 kg is held in equilibrium by two small steel rollers A and B, ready to be pushed into a saw-mill. The centres of the rollers are 50 cm apart. One end of the plank presses against roller A from underneath, and the plank rests on top of roller B, as shown in the figure. The rollers are adjusted so that the plank remains horizontal and the force exerted on the plank by each roller is vertical.

   a) Suggest a suitable model for the plank to determine the forces exerted by the rollers.

   b) Find the magnitude of the force exerted on the plank by the roller at B.

   c) Find the magnitude of the force exerted on the plank by the roller at A.   (EDEXCEL)

20 A pole of mass m and length $2a$ is used to display a light banner. The pole is modelled as a uniform rod AB, freely hinged to a vertical wall at the point A. It is held in a horizontal position by a light wire. One end of the wire is attached to the end B of the rod and the other end is attached to the wall at a point C which is vertically above A such that $\angle$ ABC is θ, where $\tan \theta = \frac{1}{2}$, as shown in the figure.

   a) Show that the tension in the wire is

$$\frac{mg}{2 \sin \theta}$$

   b) Find, in terms of m and g, the magnitude of the force exerted by the wall on the rod at A.

   c) State, briefly, where in your calculation you have used the modelling assumption that the pole is a rod.   (EDEXCEL)

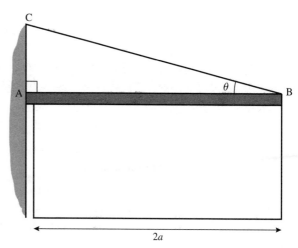

21

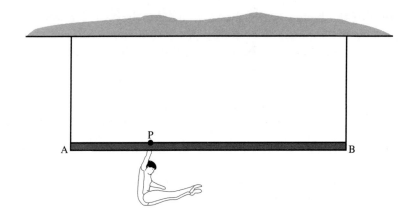

A gymnast of mass 36 kg hangs by one hand from the point P on a bar AB of length 3 m and mass 12 kg. The bar is suspended by two vertical cables which are attached to the ends A and B, and it is hanging in equilibrium in a horizontal position, as shown in the figure. The tension in the cable at A is twice the tension in the cable at B. By modelling the bar as a uniform rod, and the gymnast as a particle,

a) find the distance AP.

b) State two ways in which, in your calculation, you have used the model of the bar as a 'uniform rod'.   (EDEXCEL)

22

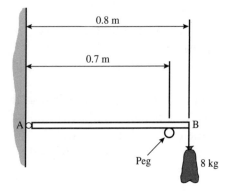

Fig. 1

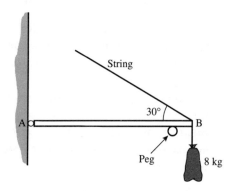

Fig. 2

A uniform, rigid rod AB of mass 4 kg and length 0.8 m is freely pivoted to a wall at A. A sack of mass 8 kg hangs from B. The rod is horizontal and rests on a small smooth peg 0.7 m from A, as shown in Fig. 1.

i) Calculate the reaction of the peg on the rod.

A light string inclined at an angle of 30° to the horizontal is now attached to the rod at B, as shown in Fig. 2. The tension in the string is 98 N.

ii) Calculate the moment about A of the tension in the string.

iii) Show that the reaction of the peg on the rod is half the value found in part **i**.

iv) Calculate the horizontal and vertical components of the force on the rod from the pivot at A.   (MEI)

23

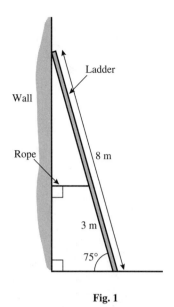

Fig. 1

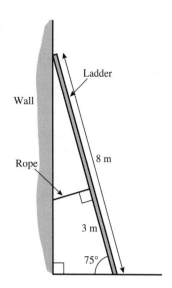

Fig. 2

A uniform ladder is standing in equilibrium on horizontal ground and leaning against a vertical wall. The ladder has length 8 m and mass 15 kg. A light rope is attached to the ladder 3 m from its bottom end and tied to the wall so that the rope is horizontal, as shown in Fig. 1. The rope and ladder are in the same vertical plane. There is negligible friction between the ladder and the ground and between the ladder and the wall

The ladder is at an angle of 75° to the horizontal.

i) Draw a diagram showing all the forces acting on the ladder. Show that the tension in the rope is about 31.5 N.

The rope will break when the tension in it reaches 300 N. A man of mass 80 kg climbs the ladder gently.

ii) How far up the ladder can the man climb before the rope breaks?

The ladder, without the man, is once again set up inclined at 75° to the horizontal. The rope is attached to the ladder 3 m from its bottom end, as before, but is now fixed so that it is perpendicular to the ladder, as shown in Fig. 2. The rope and the ladder are in the same vertical plane.

iii) Calculate the tension in the rope.   (MEI)

24 The diagram on the right shows a light rod AB of length $4a$ rigidly joined at B to a light rod BC of length $2a$ so that the rods are perpendicular to each other and in the same vertical plane. The centre O of AB is fixed and the rods can rotate freely about O in a vertical plane. A particle of mass $4m$ is attached at A and a particle of mass m is attached at C. The system rests in equilibrium with AB inclined at an acute angle θ to the vertical as shown. By taking moments about O, find the value of θ.   (WJEC)

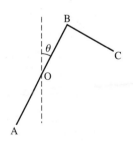

25 A rhombus ABCD has side d centimetres and the angle $\widehat{ABC} = 120°$.

Four forces whose magnitudes are b newtons, $2b$ newtons, b newtons and $\dfrac{4b}{3}$ newtons act along the edges AB, BC, DC and DA respectively, as shown in the figure on the right. This system of forces can be replaced by a single force acting through D and a couple.

i) Find the magnitude and direction of the single force.
ii) Find the magnitude and the sense of the couple.

(NICCEA)

26 A playground roundabout ABCDEF is in the shape of a regular hexagon of side 1.5 m. It rotates about a vertical axis through its centre O. Three children push this roundabout with horizontal forces of magnitudes 10 N, 5 N and P newtons, acting along the sides AF, ED and CB respectively. The horizontal reaction at the axis O has magnitude R newtons and acts at 45° to OC. This system of forces is illustrated in the figure on the right.

The system of forces reduces to a couple.

i) Show that $P = 6.34$.
ii) Find the value of R.
iii) Find the magnitude and sense of the couple. (NICCEA)

27 The figure below shows a rectangle ABCD, where AB and AD have lengths $6l$ metres and $4l$ metres respectively. O is the centre of the rectangle and E is the point on BC such that CE is x metres.

The following four forces are applied to this rectangle:

$10W$ newtons at B parallel to BA
$20W$ newtons at E parallel to BA
$5W$ newtons at D making an angle
$\quad 2\alpha$ with CD
R newtons at A making an angle
$\quad \alpha$ with BA

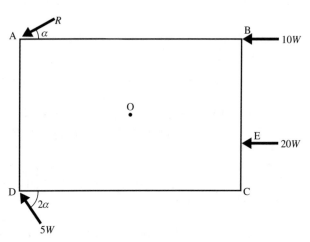

The system of forces reduces to a single force P newtons acting through O parallel to BA.

i) Obtain an expression, in terms of W and α, for R.
ii) Hence show that $P = 5W(7 + 2\cos 2\alpha)$.

iii) Prove that $x = \dfrac{l}{2}$. (NICCEA)

11 Centre of mass

A wonderful bird is the pelican,
His bill will hold more than his belican.
DIXON LANIER MERRITT

Experiment

Take a sheet of card and cut it to an irregular shape. Make a
number of holes at random positions around its edge.

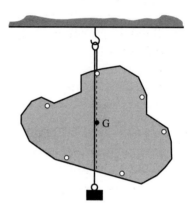

Take a length of string. Tie a loop about two thirds of the way
along. Tie a small weight to the end of the longer portion. Tie
the other end to one of the holes in the card.

Suspend the assembly by means of the loop. When it is hanging
at rest, mark on the card the line indicated by the string.

Repeat the process with the card attached by a different hole.
The two lines you have drawn will cross at a point G.

Now suspend the card from the other holes in turn. You will
find that the line of the string always passes through G.

Interpretation

The string with the weight indicates a vertical line through the
point of suspension. The line you drew was, therefore, the line of
action of the weight of the card, which must pass through the
point of suspension as the card is in equilibrium.

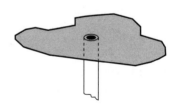

As this line of action always passes through G, the weight of the
card behaves as if the card were a single point mass positioned
at G.

You should be able to balance the card on the flat end of a pencil
placed at G.

Centre of gravity, centre of mass and centroid

You will often meet these three terms used almost interchangeably. Strictly speaking, the point you found in the experiment is the **centre of gravity**, which can be defined as follows:

> When a system consisting of one or more bodies having a total mass M is acted upon by gravity, there is a point G such that the magnitude and line of action of the weight of the system are the same as those of a particle of mass M placed at G. G is the centre of gravity of the system.

Provided the object under consideration is small, the centre of gravity is independent of the orientation of the object and coincides with a point called the **centre of mass**. This depends on the distribution of mass within the object. Even in the weightless conditions of space, where centre of gravity is meaningless, the centre of mass is still important as it affects, for example, the behaviour of the object in collision with other objects.

The centre of gravity and the centre of mass coincide provided the object is small enough for the acceleration due to gravity to have effectively the same magnitude and direction at each point of the object.

Consider a simple system consisting of two point masses of weight w_1 and w_2. They are placed at points A and B on a horizontal line through an origin O so that $OA = x_1$ and $OB = x_2$.

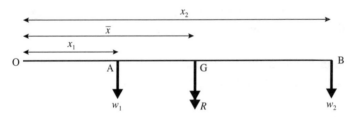

We suppose that the system is small enough for the weights to be considered as parallel forces. As shown on pages 190–1, they are then equivalent to a resultant force, R, acting through some point G on AB, where $OG = \bar{x}$.

The moment of R about O is the same as the total moments of w_1 and w_2 about O. That is,

$$R\bar{x} = w_1x_1 + w_2x_2$$

$$\Rightarrow \quad \bar{x} = \frac{w_1x_1 + w_2x_2}{R}$$

But $R = w_1 + w_2$, which gives

$$\bar{x} = \frac{w_1x_1 + w_2x_2}{w_1 + w_2}$$

When the line OAB is at an angle θ to the horizontal, the above analysis is modified as follows.

$$R\bar{x}\cos\theta = w_1 x_1 \cos\theta + w_2 x_2 \cos\theta$$

$$\Rightarrow \quad \bar{x} = \frac{w_1 x_1 + w_2 x_2}{R}$$

But $R = w_1 + w_2$, which gives

$$\Rightarrow \quad \bar{x} = \frac{w_1 x_1 + w_2 x_2}{w_1 + w_2}$$

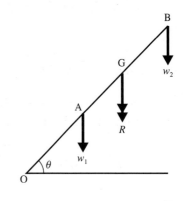

G is therefore independent of the orientation of the system.

G is both the **centre of gravity** and the **centre of mass** of the system.

For a very large object, the centre of gravity would not usually coincide with the centre of mass. Consider, for example, a system consisting of two equal masses m placed at A and B as shown, where A is twice as far as B from E, the centre of the Earth. The centre of mass of the system is at M, the mid-point of AB. By the inverse square law for gravity, the weight of particle A is a quarter of that of particle B and it acts in a different direction.

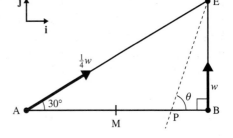

The resultant of the two weights is

$$R = \frac{1}{4}w\cos 30° \, \mathbf{i} + \left(w + \frac{1}{4}w\sin 30°\right)\mathbf{j}$$

$$= \frac{w\sqrt{3}}{8}\mathbf{i} + \frac{9w}{8}\mathbf{j}$$

The line of action of R is PE, where $\tan\theta = 9/\sqrt{3}$. This means that BP is a ninth of BA and so the resultant weight does **not** act through the centre of mass. The centre of gravity of this system would be that point on PE at which a mass of $2m$ would have a weight of R. This centre of gravity would move in relation to A and B if AB were moved in relation to E.

Situations such as the one above are unlikely to be encountered in practice. The objects we deal with are small enough for the centre of gravity and centre of mass to coincide. We can define the centre of mass of an object as follows:

The centre of mass of a system is the centre of gravity of that system when it is placed in a gravitational field such that each part of the system is subject to the same gravitational acceleration.

For a solid or a lamina (plane shape) there is a third centre, the **centroid**, which is a geometrical centre. For example, the centroid of a rectangular lamina is at the intersection of its diagonals. The centroid coincides with the centre of mass when the object is made of a uniformly dense material.

The centroid would assume importance if, for example, a plane surface were subjected to the pressure of a uniform air flow. Regardless of the mass distribution of the object, the resultant force caused by the pressure would act through the centroid of the surface.

Finding the centre of mass

We need to be able to find the centre of mass of a system independent of a particular gravitational field.

Suppose we have a system consisting of particles m_1, m_2, ..., m_n placed at points (x_1, y_1), (x_2, y_2), ..., (x_n, y_n) in a plane, as shown.

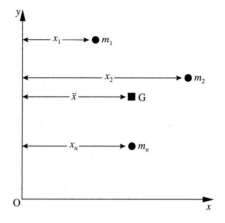

If the system were placed in a uniform gravitational field perpendicular to the plane, the resultant weight would be the total of the weights of the individual particles and would act through the centre of mass, $G(\bar{x}, \bar{y})$.

Taking moments about the y-axis, we obtain

$$\left(\sum_{i=1}^{n} m_i g \right) \bar{x} = \sum_{i=1}^{n} m_i g x_i$$

$$\Rightarrow \quad \bar{x} = \frac{\displaystyle\sum_{i=1}^{n} m_i g x_i}{\displaystyle\sum_{i=1}^{n} m_i g}$$

Cancelling by g, we get

$$\bar{x} = \frac{\displaystyle\sum_{i=1}^{n} m_i x_i}{\displaystyle\sum_{i=1}^{n} m_i}$$

Similarly, we can find the y-coordinate of G.

Taking moments about the x-axis, we obtain

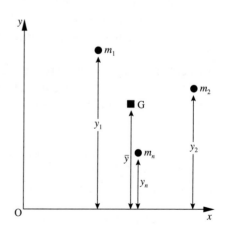

$$\left(\sum_{i=1}^{n} m_i g \right) \bar{y} = \sum_{i=1}^{n} m_i g y_i$$

$$\Rightarrow \quad \bar{y} = \frac{\displaystyle\sum_{i=1}^{n} m_i g y_i}{\displaystyle\sum_{i=1}^{n} m_i g} = \frac{\displaystyle\sum_{i=1}^{n} m_i y_i}{\displaystyle\sum_{i=1}^{n} m_i}$$

It is usual to combine the above formulae into a vector equation:

$$\begin{pmatrix} \bar{x} \\ \bar{y} \end{pmatrix} = \frac{\displaystyle\sum_{i=1}^{n} m_i \begin{pmatrix} x_i \\ y_i \end{pmatrix}}{\displaystyle\sum_{i=1}^{n} m_i}$$

This can be extended to three dimensions as

$$\begin{pmatrix} \bar{x} \\ \bar{y} \\ \bar{z} \end{pmatrix} = \frac{\displaystyle\sum_{i=1}^{n} \begin{pmatrix} x_i \\ y_i \\ z_i \end{pmatrix}}{\displaystyle\sum_{i=1}^{n} m_i}$$

This vector form can be expressed more neatly. If each mass m_i has position vector $\mathbf{r}_i$ and the centre of mass, G, has position vector $\bar{\mathbf{r}}$, then

$$\bar{\mathbf{r}} = \frac{\displaystyle\sum_{i=1}^{n} m_i \mathbf{r}_i}{\displaystyle\sum_{i=1}^{n} m_i}$$

Example 1 Masses of 2 kg, 3 kg and 5 kg are placed at A(3, 1), B(5, 7) and C(1, −4) respectively. Find the position of the centre of mass.

SOLUTION

$$\bar{\mathbf{r}} = \frac{2\begin{pmatrix} 3 \\ 1 \end{pmatrix} + 3\begin{pmatrix} 5 \\ 7 \end{pmatrix} + 5\begin{pmatrix} 1 \\ -4 \end{pmatrix}}{2 + 3 + 5}$$

$$= \begin{pmatrix} 2.6 \\ 0.3 \end{pmatrix}$$

So, the centre of mass is G(2.6, 0.3)

Example 2 Masses of 2 kg, 4 kg, 5 kg and 3 kg are placed respectively at the vertices A, B, C and D of a light rectangular framework ABCD, where AB = 3 m and BC = 2 m. Further masses of 1 kg and 5 kg are placed at E and F, the mid-points of BC and CD respectively. If the framework is suspended from A, find the angle which AB makes with the vertical.

SOLUTION

The mass of the rods forming the framework is assumed to be negligible in comparison with the masses attached to it.

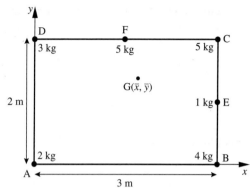

Take AB and AD to be the *x*- and *y*-axes, as shown.

Let G($\bar{x}$, $\bar{y}$) be the centre of mass of the system. Therefore, we have

$$\begin{pmatrix} \bar{x} \\ \bar{y} \end{pmatrix} = \frac{2\begin{pmatrix} 0 \\ 0 \end{pmatrix} + 4\begin{pmatrix} 3 \\ 0 \end{pmatrix} + 1\begin{pmatrix} 3 \\ 1 \end{pmatrix} + 5\begin{pmatrix} 3 \\ 2 \end{pmatrix} + 5\begin{pmatrix} 1.5 \\ 2 \end{pmatrix} + 3\begin{pmatrix} 0 \\ 2 \end{pmatrix}}{2 + 4 + 1 + 5 + 5 + 3}$$

$$= \begin{pmatrix} 1.875 \\ 1.35 \end{pmatrix}$$

So, the centre of mass is at G(1.875, 1.35) m.

When the framework is suspended from vertex A, the line AG is vertical, as shown.

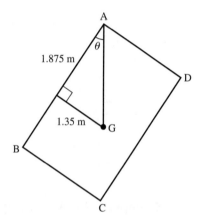

If θ is the angle between AB and the vertical, then we have

$$\tan \theta = \frac{1.35}{1.875} \quad \Rightarrow \quad \theta = 35.75°$$

Exercise 11A

1 Find the coordinates of the centre of mass of each of the following systems of masses placed respectively at the given points.

  a) 3 kg, 5 kg and 7 kg at A(2, 5), B(3, 1) and C(4, 9)
  b) 9 kg, 4 kg, 2 kg and 5 kg at A(4, 8), B(−2, 6), C(4, −4) and D(−2, −5)
  c) 6 kg, 12 kg and 15 kg at A(0, −8), B(6, −3) and C(−4, −9)
  d) 2 kg, 1 kg, 5 kg and 3 kg at A(2, 1, 6), B(3, 2, 0), C(5, −2, −8) and D(−1, 2, 3)

2 Masses of 3 kg, 8 kg and 5 kg are placed at points A, B and C with position vectors $3\mathbf{i} + 6\mathbf{j}$, $4\mathbf{i} - 2\mathbf{j}$ and $6\mathbf{i} - 8\mathbf{j}$ respectively. Find the position vector of the centre of mass.

3 Masses of 5 kg, 7 kg and 6 kg are placed at points A, B and C with position vectors $2\mathbf{i} - 7\mathbf{j} + 4\mathbf{k}$, $-3\mathbf{i} - 5\mathbf{j} + 8\mathbf{k}$ and $\mathbf{i} - 12\mathbf{k}$ respectively. Find the position vector of the centre of mass.

4 Masses of 4 kg, 9 kg and 6 kg are placed at A(5, 3), B(6, −2) and C(−1, 4) respectively. Where should a mass of 5 kg be placed so that the centre of mass of the whole system is at G(0, −1)?

5 A light rectangular framework ABCD has AB = 4 m and BC = 3 m. Masses of 5 kg, 4 kg, 2 kg and 3 kg are placed at A, B, C and D respectively. A fifth mass m kg is placed at a point E on CD so that the centre of mass of the system is at the centre of the rectangle. Find the value of m and the position of E.

6 A cuboidal framework of light rods has a rectangular base ABCD and vertices E, F, G and H vertically above A, B, C and D respectively. AB = 4 m, AD = 3 m and AE = 2 m. Masses of 4 kg, 6 kg, 2 kg, 2 kg and 5 kg are placed at B, C, F, G and H respectively. Taking AB, AD and AE to be the x-, y- and z-axes, find the coordinates of the centre of mass of the system.

7 Masses of 2 kg, 4 kg, 6 kg and 9 kg are placed respectively at the vertices A, B, C and D of a light rectangular framework ABCD, where AB = 5 cm and BC = 3 m. Find the angle which AB makes with the vertical when the framework is suspended from A.

8 A light triangular framework ABC has AB = 4.6 m, AC = 6.3 m and angle BAC = 68°. Masses of 3 kg, 6 kg and 8 kg are placed at A, B and C respectively. The framework is suspended from A. Find the angle which AB makes with the vertical.

9 ABCDE is a light framework consisting of a square ABCE and an equilateral triangle CDE, as shown. Masses of 2 kg, 1 kg, 4 kg, 5 kg and m kg are attached to A, B, C, D and E respectively. The framework is then suspended from A. Find the value of m if the diagonal AC makes an angle of 20° with the vertical.

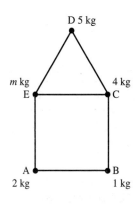

Centre of mass of a rigid body

We can find the centre of mass of some common shapes by considering their symmetry, provided that the bodies are uniformly dense. (Objects with variable density fall outside the scope of this book.)

One dimension

In one dimension, we have a **uniform rod** whose thickness is assumed to be negligible compared with its length.

By symmetry, the centre of mass, G, of a rod AB lies at the mid-point of AB.

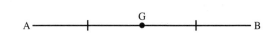

Two dimensions

Any plane figure whose thickness is negligible compared with its other dimensions is called a **lamina** (plural *laminae*, sometimes *laminas*.)

• Uniform rectangular lamina

By symmetry, the centre of mass, G, of a uniform rectangular lamina ABCD is at the intersection of its diagonals, as shown.

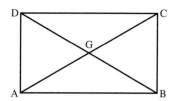

• Uniform circular lamina

By symmetry, the centre of mass, G, of a uniform circular lamina is at the centre of the circle.

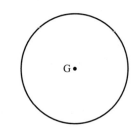

• Uniform triangular lamina

A triangle can be regarded as being made up of a large number of strips of negligible thickness (that is, uniform rods) parallel to one of its sides, as shown. The centres of mass (G_1, G_2, G_3, etc) of these strips lie at their mid-points.

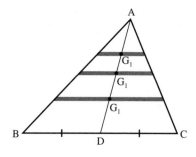

The centre of mass, G, of the triangle must therefore lie on the line formed by G_1, G_2, G_3, etc. This is the line AD in the diagram, joining A to the mid-point D of BC. This line is called a **median** of the triangle.

By considering the triangle divided into strips parallel to AC, we can see that G lies also on the median BE.

By considering the triangle divided into strips parallel to AB, we can see that G lies also on the median CF.

The medians of a triangle meet at the point which divides each median in the ratio $2:1$. So, in this diagram, we have

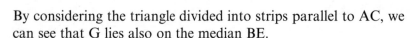

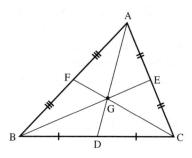

$$AG:GD = BG:GE = CG:GF = 2:1$$

(You may not have encountered this standard geometrical result. A proof is given on page 249.)

• Uniform semicircular lamina

The centre of mass of a uniform semicircular lamina of radius r lies on the line of symmetry, as shown, at a distance h from the straight edge (diameter).

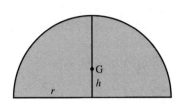

We show by calculus methods on pages 237–8 that

$$h = \frac{4r}{3\pi}$$

Three dimensions

• Uniform solid sphere

The centre of mass of a uniform sphere must, by symmetry, lie at the centre of the sphere.

• Uniform solid cylinder

The centre of mass of a uniform cylinder lies on its axis, halfway along the cylinder.

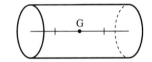

This is a special case of the more general result relating to a uniform solid prism.

• Uniform solid prism

A prism is a solid with a uniform cross-section, as shown in the diagram. G_1 and G_2 are the centroids of the laminae which correspond to the two ends of the prism. The centre of mass, G, of the prism is at the mid-point of G_1G_2.

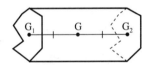

• Uniform tetrahedron

A tetrahedron, ABCD, can be regarded as made up of a series of triangular laminae parallel to BCD. The centre of mass of the tetrahedron lies on the line, AP, formed by the centres of mass of these triangular laminae, as shown in the diagram. This can be repeated with laminae parallel to ABC, ABD and ACD. The four lines so generated intersect at G, the centre of mass of the tetrahedron, as shown.

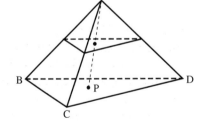

It can be shown that G divides AP, BQ, CR and DS in the ratio 3 : 1.

The result can be extended to establish the centre of mass of a cone.

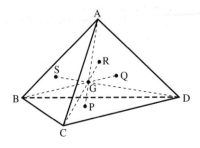

- **Uniform cone**

Given a cone whose vertex is V and the centre of whose base is O, then its centre of mass, G, lies on VO, where $VG : GO = 3 : 1$.

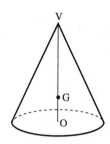

- **Uniform hemisphere**

The centre of mass of a uniform hemisphere lies on its axis of symmetry, as shown.

We show on page 239 that the distance h of the centre of mass from the plane face of the hemisphere is given by

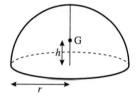

$$h = \frac{3r}{8}$$

Composite bodies

Many complex shapes are made up of several components, each of which is one of the standard shapes just described. Each of these components may be regarded as a point mass located at its centre of mass. The centre of mass of the overall shape may then be found from these point masses.

Example 3 The diagram shows an L-shaped lamina ABCDEF with uniform density. Find its centre of mass.

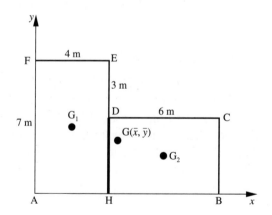

SOLUTION

We can regard this lamina as being composed of two rectangles, AHEF and HBCD, as shown. If we take the density to be $\rho\,\text{kg}\,\text{m}^{-2}$, the mass of AHEF is $28\rho\,\text{kg}$ and that of HBCD is $24\rho\,\text{kg}$.

Taking AB and AF to be the x- and y-axes respectively, the centre of mass of the body is that of a mass of $28\rho\,\text{kg}$ at $G_1(2, 3.5)$ and another of $24\rho\,\text{kg}$ at $G_2(7, 2)$.

Therefore, we have

$$\begin{pmatrix} \bar{x} \\ \bar{y} \end{pmatrix} = \frac{28\rho \begin{pmatrix} 2 \\ 3.5 \end{pmatrix} + 24\rho \begin{pmatrix} 7 \\ 2 \end{pmatrix}}{(28 + 24)\rho} = \begin{pmatrix} 4.31 \\ 2.81 \end{pmatrix}$$

So, the centre of mass is at G(4.31, 2.81) m.

Example 4 The lamina in Example 3 is folded along DH so that the angle AHB is 90°, as shown. Find the centre of mass.

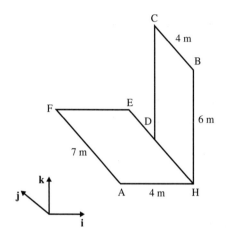

SOLUTION

Taking A as the origin and directions **i**, **j** and **k** as shown, we have:

Centre of mass of AHEF is $G_1(2, 3.5, 0)$

Centre of mass of HBCD is $G_2(4, 2, 3)$

Therefore, the position, **r̄**, of the centre of mass of the whole body is given by

$$\bar{\mathbf{r}} = \frac{28\rho \begin{pmatrix} 2 \\ 3.5 \\ 0 \end{pmatrix} + 24\rho \begin{pmatrix} 4 \\ 2 \\ 3 \end{pmatrix}}{(28 + 24)\,\rho} = \begin{pmatrix} 2.92 \\ 2.81 \\ 1.38 \end{pmatrix}$$

So, the centre of mass is at G(2.92, 2.81, 1.38) m.

Example 5 A rectangular lamina ABCD of uniform density $1\,\text{kg}\,\text{m}^{-2}$ has a hole cut in it, consisting of a rectangle PQRS and a triangle RST, as shown. AD = 3 m and AB = 4 m. S is at the centre of the rectangle ABCD, and T and Q lie on the diagonals so that

$$ST = \tfrac{1}{2}SC \quad \text{and} \quad SQ = \tfrac{1}{2}SB$$

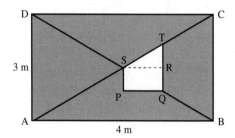

Find the centre of mass.

SOLUTION

Take AB and AD to be the x- and y-axes respectively.

Original rectangle ABCD has mass 12 kg and centre of mass S(2, 1.5). The pieces removed are:

Rectangle PQRS, having mass 0.75 kg and centre of mass $G_1(2.5, 1.125)$.

Triangle RST, having mass 0.375 kg and centre of mass $G_2(2\frac{2}{3}, 1.75)$. (That is, one third of the way along from R towards S and towards T; see page 228.)

The required shape has mass $12 - 0.75 - 0.375 = 10.875$ kg and centre of mass $G(\bar{x}, \bar{y})$.

Taking moments about the axes, we obtain

Moment of ABCD = Moment of shaded lamina + Moment of PQTS

$$12\binom{2}{1.5} = 10.875\binom{\bar{x}}{\bar{y}} + 0.75\binom{2.5}{1.125} + 0.375\binom{2\frac{2}{3}}{1.75}$$

$$\Rightarrow \quad 10.875\bar{x} = 24 - 1.875 - 1 = 21.125$$

$$\Rightarrow \quad \bar{x} = 1.943 \text{ m}$$

and $\Rightarrow \quad 10.875\bar{y} = 18 - 0.843\,75 - 0.656\,25 = 16.5$

$$\Rightarrow \quad \bar{y} = 1.517 \text{ m}$$

So, the centre of mass is G(1.943, 1.517) m.

Example 6 The diagram shows an object comprising a solid cylinder of radius 0.1 m and length h m attached to a solid, right circular cone of radius 0.1 m and height 0.12 m. Both objects have a uniform density ρ kg m^{-3}. Find the position of the centre of mass in terms of h. Hence find the maximum value of h for which the object would remain with the surface of the cone resting on a horizontal surface.

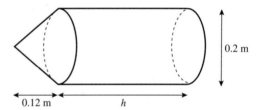

SOLUTION

Because the centre of mass must lie on the axis of symmetry, this is essentially a one-dimensional problem.

Volume of cone $= \frac{1}{3}\pi \times 0.1^2 \times 0.12$

$$= 0.0004\pi \text{ m}^3$$

So, mass of cone $= 0.0004\pi\rho$ kg

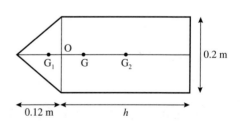

Volume of cylinder $= \pi \times 0.1^2 \times h$

$$= 0.01h\pi \text{ m}^3$$

So, mass of cylinder $= 0.01h\pi\rho$ kg

Taking O in the diagram to be the origin, we have

$$OG_1 = \tfrac{1}{4} \times 0.12 = 0.03\,\text{m} \quad [\text{See page 230.}]$$

$$OG_2 = \tfrac{1}{2}h\,\text{m}$$

$$OG = \bar{x}$$

which give

$$\bar{x} = \frac{0.01h\pi\rho \times \tfrac{1}{2}h - 0.0004\pi\rho \times 0.03}{0.01h\pi\rho + 0.0004\pi\rho}$$

Cancelling by $\pi\rho$ and multiplying top and bottom by $250\,000$, we get

$$\bar{x} = \frac{1250h^2 - 3}{2500h + 100}$$

For the object to balance on the surface of the cone, as shown, $\bar{x}$ must not exceed OA in the diagram.

Triangles OAB and OBC are similar. Therefore, we have

$$OA:OB = OB:OC$$

$$\Rightarrow \quad OA:0.1 = 0.1:0.12$$

$$\Rightarrow \quad OA = \tfrac{1}{12}\,\text{m}$$

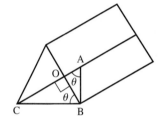

So, for the cylinder to balance in this position, we have

$$\frac{1250h^2 - 3}{2500h + 100} \leqslant \frac{1}{12}$$

$$\Rightarrow \quad 15000h^2 - 2500h - 136 \leqslant 0$$

$$\Rightarrow \quad -0.04 \leqslant h \leqslant 0.21$$

So, the length of the cylinder cannot exceed $0.21\,\text{m}$

Exercise 11B

1 Find the centre of mass of each of the following laminae relative to the origin O and the axes shown. You may assume a uniform density of $1\,\text{kg}\,\text{m}^{-2}$ in each case.

a)

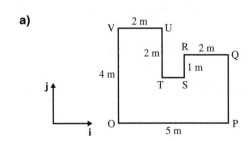

b)

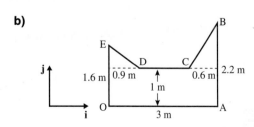

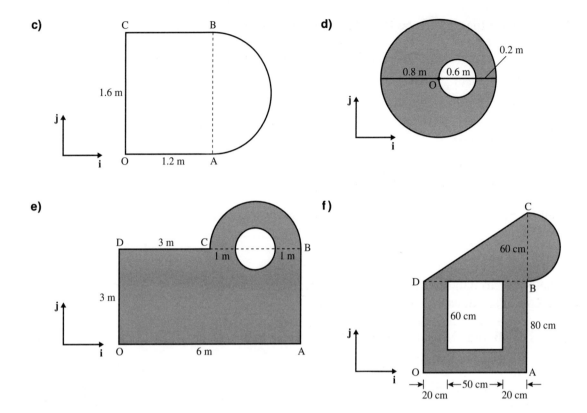

2 The diagram shows a triangular lamina ABC, in which angle ABC is 90°, AB = 0.6 m and BC = 0.9 m. The triangle is attached to a second, rectangular, lamina PQRS, where PQ = 1.2 m and PS = 0.8 m, so that BC lies on PQ and PB = 0.2 m, as shown. Assume that both laminae have uniform densities of $1 \, \text{kg m}^{-2}$. Taking PQ and PS to be the x- and y-axes respectively, find the position of the centre of mass of the combined object.

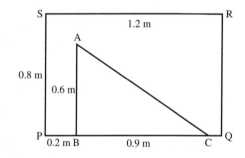

3 A uniform rectangular card ABCD of density $\rho \, \text{kg m}^{-2}$ is folded along OF and BE, as shown in the diagram. AB = 120 cm, AD = 40 cm, AO = 20 cm and CE = 40 cm. Taking O as the origin and axes as shown, find the centre of mass of the folded card.

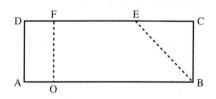

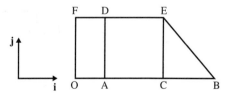

4 This lamina is folded so that each rectangle is perpendicular to its neighbour, as shown. The lamina has a uniform density of ρ kg m^{-2}. Taking O as the origin and axes as shown, find the coordinates of the centre of mass of the object.

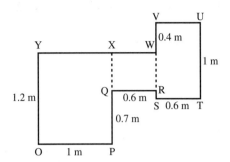

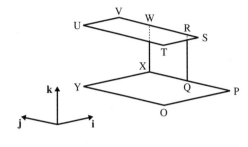

5 A uniform rod AB of mass 5 kg and length 2 m is attached at a point B on the rim of a uniform disc, centre C, of radius 0.6 m and mass 10 kg, so that the rod is perpendicular to the plane of the disc.

a) Taking BC and BA as the x- and y-axes respectively, find the position of the centre of mass of the object.

b) If the object is suspended from A, find the angle between the rod AB and the vertical.

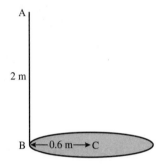

6 The diagram shows the cross-section of a prism of uniform density. Find the minimum length of AB if the prism is to rest with AB in contact with a horizontal surface.

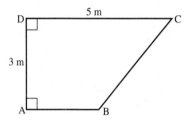

7 The diagram shows a cool-box consisting of a cylinder of diameter 60 cm and height 80 cm, with a hollow cylindrical interior and a hollow hemispherical cap. The thickness of the wall, cap and base is 10 cm throughout. Find the height of the centre of mass of the empty box above its base.

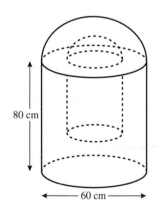

8 The diagram shows a primitive flushing device. A conical vessel (outer dimensions: diameter 60 cm, height 80 cm; inner dimensions: diameter 57 cm, height 76 cm) is pivoted symmetrically at a height h cm above its vertex, as shown. As the vessel fills with water, the centre of mass rises above the level of the pivot, the system becomes unstable and tips, causing the flush. The vessel is made of a material which is three times as dense as water. Find the range of values of h for which the device will work.

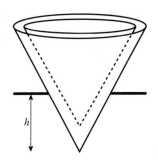

9 A Tippee-Toy comprises a hemisphere of diameter 16 cm topped by a cone of diameter 16 cm and height h cm. The toy is intended to return to an upright position when tipped.

a) If the toy is made of a uniformly dense material,
   i) find the maximum value of h for which the toy will work at all.
   ii) Show that, whatever the value of h, if the toy is tipped too far it will rest in equilibrium with the surface of the cone on the ground.

b) If $h = 30$ cm and the hemisphere is made of a material k times as dense as the cone's, find the minimum value for k for which the toy cannot rest in equilibrium with the cone in contact with the ground.

16 cm

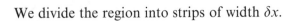

Calculus methods: laminae and solids of revolution

Centre of mass of a uniform lamina

Consider a uniform lamina comprising the region bounded by $y = f(x)$, the x-axis, and the lines $x = a$ and $x = b$. Let the density (mass per unit area) of the lamina be ρ, and the centre of mass be $G(\bar{x}, \bar{y})$.

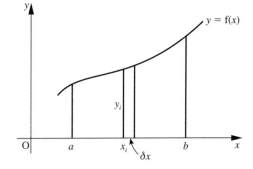

We divide the region into strips of width δx.

The diagram shows the ith strip. The area of this strip is approximately $y_i \delta x$, and so its mass is

$$m_i \approx \rho y_i \delta x$$

The centre of mass of the ith strip is approximately $(x_i, \frac{1}{2} y_i)$.

We can therefore think of the lamina as being equivalent to a system of masses m_i at points with position vectors $\begin{pmatrix} x_i \\ \frac{1}{2} y_i \end{pmatrix}$. We can find the centre of mass of the system in the usual way:

$$\begin{pmatrix} \bar{x} \\ \bar{y} \end{pmatrix} \approx \frac{\sum m_i \begin{pmatrix} x_i \\ \frac{1}{2} y_i \end{pmatrix}}{\sum m_i} = \frac{\sum \rho y_i \begin{pmatrix} x_i \\ \frac{1}{2} y_i \end{pmatrix} \delta x}{\sum \rho y_i \, \delta x}$$

Cancelling by ρ and separating the components, we have

$$\bar{x} \approx \frac{\sum x_i y_i \, \delta x}{\sum y_i \, \delta x} \quad \text{and} \quad \bar{y} \approx \frac{\frac{1}{2} \sum y_i^2 \, \delta x}{\sum y_i \, \delta x}$$

When $\delta x \to 0$, the limits of the summations are given by the corresponding integrals. This gives

$$\bar{x} = \frac{\int_a^b xy \, dx}{\int_a^b y \, dx} = \frac{\int_a^b x f(x) \, dx}{\int_a^b f(x) \, dx}$$

and

$$\bar{y} = \frac{\frac{1}{2} \int_a^b y^2 \, dx}{\int_a^b y \, dx} = \frac{\frac{1}{2} \int_a^b f(x)^2 \, dx}{\int_a^b f(x) \, dx}$$

Example 7 Find the centre of mass of the uniform lamina comprising the region enclosed by $y = \sqrt{x}$, the x-axis and the lines $x = 1$, $x = 4$.

SOLUTION

$$\bar{x} = \frac{\int_1^4 x\sqrt{x} \, dx}{\int_1^4 \sqrt{x} \, dx} = \frac{\left[\frac{2}{5} x^{\frac{5}{2}}\right]_1^4}{\left[\frac{2}{3} x^{\frac{3}{2}}\right]_1^4} = 2.657$$

$$\bar{y} = \frac{\frac{1}{2} \int_1^4 x \, dx}{\int_1^4 \sqrt{x} \, dx} = \frac{\left[\frac{1}{4} x^2\right]_1^4}{\left[\frac{2}{3} x^{\frac{3}{2}}\right]_1^4} = 0.8036$$

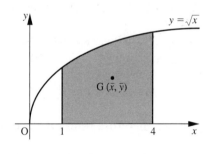

So, the centre of mass is G(2.657, 0.8036).

Example 8 Show that the centre of a uniform, semicircular lamina of radius r is $\dfrac{4r}{3\pi}$ from the straight edge.

SOLUTION

The semicircular lamina, as illustrated, has centre of mass $G(0, \bar{y})$, by symmetry.

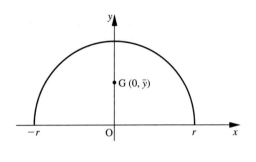

The circle has equation $x^2 + y^2 = r^2$.

We have
$$\bar{y} = \frac{\frac{1}{2}\int_{-r}^{r} y^2 \, dx}{\int_{-r}^{r} y \, dx}$$

The denominator of this expression is the area of the lamina, which we know to be $\frac{1}{2}\pi r^2$. Substituting for y^2 from the circle equation, we have

$$\bar{y} = \frac{\frac{1}{2}\int_{-r}^{r} (r^2 - x^2) \, dx}{\frac{1}{2}\pi r^2} = \frac{\left[r^2 x - \frac{1}{3}x^3\right]_{-r}^{r}}{\pi r^2}$$

$$\Rightarrow \quad \bar{y} = \frac{(r^3 - \frac{1}{3}r^3) - (-r^3 + \frac{1}{3}r^3)}{\pi r^2} = \frac{4r}{3\pi}$$

Centre of mass of a uniform solid of revolution

Consider the solid formed by rotating, about the x-axis, the region bounded by $y = f(x)$, the x-axis, and the lines $x = a$ and $x = b$. Let the density (mass per unit volume) of the solid be ρ. By symmetry, the centre of mass is $G(\bar{x}, 0)$.

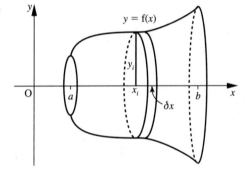

We divide the solid into slices of thickness δx.

The diagram shows the ith slice. The volume of this slice is approximately $\pi y_i^2 \, \delta x$, and so its mass is $m_i \approx \pi \rho y_i^2 \, \delta x$.

The centre of mass of the ith slice is approximately $(x_i, 0)$.

We can therefore think of the solid as being equivalent to a system of masses m_i at points with position vectors $\begin{pmatrix} x_i \\ 0 \end{pmatrix}$. We can find the centre of mass of the system in the usual way:

$$\bar{x} \approx \frac{\sum m_i x_i}{\sum m_i} = \frac{\sum \pi \rho x_i y_i^2 \delta x}{\sum \pi \rho y_i^2 \delta x}$$

Cancelling by $\pi \rho$, we have

$$\bar{x} \approx \frac{\sum x_i y_i^2 \, \delta x}{\sum y_i^2 \, \delta x}$$

When $\delta x \to 0$, the limit of the summation is given by the corresponding integral. This gives

$$\bar{x} = \frac{\displaystyle\int_a^b xy^2 \, dx}{\displaystyle\int_a^b y^2 \, dx} = \frac{\displaystyle\int_a^b x\mathrm{f}(x)^2 \, dx}{\displaystyle\int_a^b \mathrm{f}(x)^2 \, dx}$$

Example 9 The region enclosed by the curve $y = \sqrt{x}$, the x-axis and the lines $x = 1$, $x = 4$, is rotated about the x-axis to form a uniform solid of revolution. Find the coordinates of the centre of mass of this solid.

SOLUTION

The centre of mass is $G(\bar{x}, 0)$, where

$$\bar{x} = \frac{\displaystyle\int_1^4 x(\sqrt{x})^2 \, dx}{\displaystyle\int_1^4 (\sqrt{x})^2 \, dx} = \frac{\displaystyle\int_1^4 x^2 \, dx}{\displaystyle\int_1^4 x \, dx}$$

$$\Rightarrow \quad \bar{x} = \frac{\left[\frac{1}{3}x^3\right]_1^4}{\left[\frac{1}{2}x^2\right]_1^4} = 2.8$$

So, the centre of mass is $G(2.8, 0)$.

Example 10 Show that the centre of mass of a uniform hemisphere of radius r is at a distance $\dfrac{3r}{8}$ from the plane face of the hemisphere.

SOLUTION

The hemisphere is formed by rotating the quadrant enclosed by the circle $x^2 + y^2 = r^2$ and the axes about the x-axis, as shown.

The centre of mass is $G(\bar{x}, 0)$, where

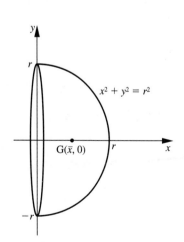

$$\bar{x} = \frac{\displaystyle\int_0^r xy^2 \, dx}{\displaystyle\int_0^r y^2 \, dx} = \frac{\displaystyle\int_0^r x(r^2 - x^2) \, dx}{\displaystyle\int_0^r (r^2 - x^2) \, dx}$$

$$\Rightarrow \quad \bar{x} = \frac{\left[\frac{1}{2}r^2 x^2 - \frac{1}{4}x^4\right]_0^r}{\left[r^2 x - \frac{1}{3}x^3\right]_0^r} = \frac{(\frac{1}{2}r^4 - \frac{1}{4}r^4) - (0)}{(r^3 - \frac{1}{3}r^3) - (0)} = \frac{3r}{8}$$

So, the centre of mass is a distance $\dfrac{3r}{8}$ from the plane face of the hemisphere.

Calculus methods: arcs and shells

When dealing with a lamina or a solid of revolution, we derive the formulae by considering an elemental section of width δx, which we then approximate as a rectangle or a disc respectively. This approach lets us down when we come to consider the arc of a curve or the surface of revolution (shell) generated by rotating the arc about the x-axis.

In these cases, we need to consider an element of arc of length δs.

If we are using cartesian coordinates, we have

$$\delta s \approx \sqrt{(\delta x)^2 + (\delta y)^2} = \sqrt{1 + \left(\frac{\delta y}{\delta x}\right)^2}\, \delta x \qquad [1]$$

In some cases, the work is simplified by using polar coordinates. The expression for δs is then

$$\delta s \approx \sqrt{r^2 + \left(\frac{\delta r}{\delta \theta}\right)^2}\, \delta\theta \qquad [2]$$

Centre of mass of an arc of uniform density

Suppose the density (mass per unit length) of the arc is ρ. Then the mass of an element of arc length $\delta s \approx \rho\delta s$.

Considering the arc as the sum of point masses, we have

$$\begin{pmatrix} \bar{x} \\ \bar{y} \end{pmatrix} \approx \frac{\sum \rho \begin{pmatrix} x_i \\ y_i \end{pmatrix} \delta s}{\sum \rho \delta s}$$

Cancelling by ρ and separating the components, we obtain

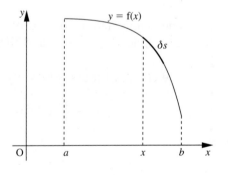

$$\bar{x} \approx \frac{\sum x_i\, \delta s}{\sum \delta s} \quad \text{and} \quad \bar{y} \approx \frac{\sum y_i\, \delta s}{\sum \delta s}$$

Letting $\delta x \to 0$, we have

$$\bar{x} = \frac{\int x\, ds}{\int ds} \qquad [3]$$

and

$$\bar{y} = \frac{\int y\, ds}{\int ds} \qquad [4]$$

The denominator is the total arc length, which in simple cases is already known.

If we are working in cartesian coordinates, with $y = f(x)$, equations [3] and [4] become

$$\bar{x} = \frac{\int_a^b x \sqrt{1 + \left(\dfrac{dy}{dx}\right)^2} \, dx}{\int_a^b \sqrt{1 + \left(\dfrac{dy}{dx}\right)^2} \, dx} \qquad [5]$$

and

$$\bar{y} = \frac{\int_a^b f(x) \sqrt{1 + \left(\dfrac{dy}{dx}\right)^2} \, dx}{\int_a^b \sqrt{1 + \left(\dfrac{dy}{dx}\right)^2} \, dx} \qquad [6]$$

If we are using polar coordinates, with $r = f(\theta)$, we have

$$x = r \cos \theta \quad \text{and} \quad y = r \sin \theta$$

and so equations [3] and [4] become

$$\bar{x} = \frac{\int_{\theta_1}^{\theta_2} r \cos \theta \sqrt{r^2 + \left(\dfrac{dr}{d\theta}\right)^2} \, d\theta}{\int_{\theta_1}^{\theta_2} \sqrt{r^2 + \left(\dfrac{dr}{d\theta}\right)^2} \, d\theta} \qquad [7]$$

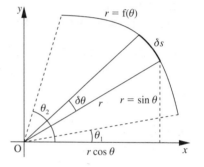

and

$$\bar{y} = \frac{\int_{\theta_1}^{\theta_2} r \sin \theta \sqrt{r^2 + \left(\dfrac{dr}{d\theta}\right)^2} \, d\theta}{\int_{\theta_1}^{\theta_2} \sqrt{r^2 + \left(\dfrac{dr}{d\theta}\right)^2} \, d\theta} \qquad [8]$$

These can easily lead to quite complex integrals, which can be solved analytically in only the simplest of cases. You will usually only meet the situation in which the curve is an arc of a circle centred on the origin. In this case, $\dfrac{dr}{d\theta}$ is zero and $\delta s = r \delta \theta$, leading to a much simplified integral.

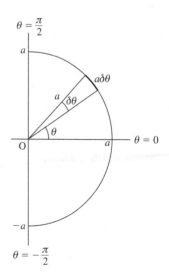

Example 11 Find the coordinates of the centre of mass of a uniform wire bent into a semicircle of radius a.

SOLUTION

Taking the semicircle as shown, the polar equation is $r = a$ and the arc we require is for $-\dfrac{\pi}{2} \leqslant \theta \leqslant \dfrac{\pi}{2}$.

By symmetry, the y-coordinate of the centre of mass is zero.

As $r = a$, we have $\dfrac{dr}{d\theta} = 0$, and equation [7] gives

$$\bar{x} = \frac{\displaystyle\int_{-\frac{\pi}{2}}^{\frac{\pi}{2}} a^2 \cos\theta \, d\theta}{\displaystyle\int_{-\frac{\pi}{2}}^{\frac{\pi}{2}} a \, d\theta}$$

$$= \frac{[a^2 \sin\theta]_{-\frac{\pi}{2}}^{\frac{\pi}{2}}}{[a\theta]_{-\frac{\pi}{2}}^{\frac{\pi}{2}}} = \frac{2a^2}{\pi a} = \frac{2a}{\pi}$$

Hence, the centre of mass has coordinates $\left(\dfrac{2a}{\pi}, 0\right)$.

Centre of mass of a shell of uniform density

The shell, or surface of revolution, is formed by rotating a curve $y = f(x)$ about the x-axis.

By symmetry $\bar{y} = 0$.

Let the density of the shell be ρ.

Consider an elemental section, shown shaded in the diagram. The area of this is approximately that of a cylindrical shell of radius y and length δs.

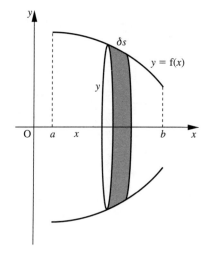

Hence, the mass of the ith element $\approx 2\pi \rho y_i \delta s$. Considering the shell as being the sum of such elements, we have

$$\bar{x} \approx \frac{\sum 2\pi \rho x_i y_i \, \delta s}{\sum 2\pi \rho y_i \, \delta s}$$

Cancelling by $2\pi\rho$ and letting $\delta s \to 0$, we obtain

$$\bar{x} = \frac{\displaystyle\int xy \, ds}{\displaystyle\int y \, ds} \qquad\qquad [9]$$

If we are using cartesian coordinates, with $y = f(x)$, equation [9] becomes

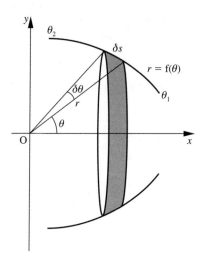

$$\bar{x} = \frac{\displaystyle\int_a^b x\, f(x) \sqrt{1 + \left(\dfrac{dy}{dx}\right)^2} \, dx}{\displaystyle\int_a^b f(x) \sqrt{1 + \left(\dfrac{dy}{dx}\right)^2} \, dx} \qquad\qquad [10]$$

If we are using polar coordinates, with $r = f(\theta)$, we have

$$x = r\cos\theta \quad \text{and} \quad y = r\sin\theta$$

and so equation [9] becomes

$$\bar{x} \approx \frac{\displaystyle\int_{\theta_1}^{\theta_2} r^2 \cos\theta \sin\theta \sqrt{r^2 + \left(\frac{dr}{d\theta}\right)^2} \, d\theta}{\displaystyle\int r\sin\theta \sqrt{r^2 + \left(\frac{dr}{d\theta}\right)^2} \, d\theta}$$

which can be written as

$$\bar{x} = \frac{\displaystyle\int_{\theta_1}^{\theta_2} r^2 \sin 2\theta \sqrt{r^2 + \left(\frac{dr}{d\theta}\right)^2} \, d\theta}{\displaystyle\int_{\theta_1}^{\theta_2} 2r\sin\theta \sqrt{r^2 + \left(\frac{dr}{d\theta}\right)^2} \, d\theta} \qquad [11]$$

Again, the integrals are potentially complex, and you will usually only meet situations in which $\dfrac{dy}{dx}$ is constant or $\dfrac{dr}{d\theta}$ is zero.

Example 12 Find the coordinates of the centre of mass of a hemispherical shell of radius a.

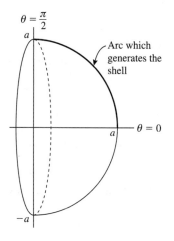

$\theta = \dfrac{\pi}{2}$

Arc which generates the shell

$\theta = 0$

SOLUTION

As in Example 11, we have polar equation $r = a$, but this time for $0 \leqslant \theta \leqslant \dfrac{\pi}{2}$.

As $r = a$ and $\dfrac{dr}{d\theta} = 0$, we have from equation [11]

$$\bar{x} = \frac{\displaystyle\int_0^{\frac{\pi}{2}} a^3 \sin 2\theta \, d\theta}{\displaystyle\int_0^{\frac{\pi}{2}} 2a^2 \sin\theta \, d\theta}$$

$$= \frac{\left[-\frac{1}{2}a^3 \cos 2\theta\right]_0^{\frac{\pi}{2}}}{\left[-2a^2 \cos\theta\right]_0^{\frac{\pi}{2}}}$$

$$= \frac{a^3}{2a^2} = \frac{a}{2}$$

So, the coordinates of the centre of mass are $\left(\dfrac{a}{2}, 0\right)$.

Exercise 11C

1 The region enclosed by the curve $y = x^2 + 1$, the x-axis and the lines $x = 2$, $x = 4$ is rotated about the x-axis. Find the centre of mass of the uniform solid formed.

2 Confirm by integration that the centre of mass of the uniform triangular lamina with vertices $(0, 0)$, $(3a, 0)$ and $(3a, 3b)$ lies at the point $(2a, b)$.

3 The triangle whose vertices are $(0, 0)$, $(h, 0)$ and (h, r) is rotated about the x-axis. Show by integration that the centre of mass of the resulting uniform cone is at the point $(\frac{3}{4}h, 0)$.

4 Find the centre of mass of the uniform lamina formed by the x-axis and the loop of the curve $y = \sin x$ between $x = 0$ and $x = \pi$. Make use of symmetry where possible.

5 a) Find the centre of mass of the uniform lamina formed by the curve $y = e^x$, the axes and the line $x = 2$.
   b) If the region in part **a** is rotated about the x-axis, find the centre of mass of the uniform solid formed.

6 a) Find the centre of mass of the uniform lamina formed by the y-axis and the curves $y = x^2$, $y = 8 - x^2$.
   b) Find the centre of mass of the uniform solid formed when the region in part **a** is rotated about the x-axis.

7 The curve $y^2 = 4ax$ is a parabola. The point $S(a, 0)$ is called the focus of the parabola. The region bounded by the parabola and the line $x = k$ is rotated about the x-axis. Find the value of k if the centre of mass of the resulting uniform solid lies at the focus.

8 A uniform lamina comprises the region in the first quadrant enclosed by the ellipse $\dfrac{x^2}{a^2} + \dfrac{y^2}{b^2} = 1$. Given that the area of the ellipse is πab, find the coordinates of the centre of mass of the lamina. Find also the centre of mass of the uniform solid generated when the region is rotated about the x-axis.

9 A uniform lamina comprises the region enclosed by the curve $y = x^3$, the y-axis and the line $y = 8$. Find

   a) the centre of mass of the lamina
   b) the centre of mass of the solid formed by rotating the lamina about the y-axis.
   [You will need to modify the integrals involved so that the roles of x and y are reversed.]

10 A uniform solid is formed by rotating about the x-axis the region under curve $y = 2 - \frac{3}{4}x^{\frac{1}{2}}$ about the x-axis. Show that the centre of mass of the solid is at the point $(1.42, 0)$.

11 Find the position of the centre of mass of a uniform wire of length $2a\alpha$ which is bent to form an arc of a circle of radius a.

12 Find the position of the centre of mass of a uniform cap of a sphere of radius a which subtends an angle of 2α at the centre of the sphere.

13 Show that the centre of mass of the uniform conical shell formed by rotating the line $x + y = 6$ for $0 \leqslant x \leqslant 6$ about the x-axis is at $(2, 0)$.

14 The portion of the curve $y = 2\sqrt{x}$ between $x = 0$ and $x = 3$ is rotated about the x-axis to form a uniform shell. Find the coordinates of its centre of mass.

Sliding and toppling

Consider an object resting in equilibrium on a rough surface. If we gradually change the forces acting on the object, equilibrium will be broken in one of two ways.

- **Sliding** When the resultant force on the object parallel to the plane of contact becomes non-zero (that is, the limiting friction force is exceeded by the other forces), the object will slide.

- **Toppling** When the total moment of the forces acting on the object becomes non-zero, the object will topple over.

One of these situations will come about before the other, so we may decide whether the object will slide or topple.

Example 11 A uniform cubical block of mass 8 kg and side length 0.4 m rests on a rough horizontal surface. A gradually increasing horizontal force P is applied to the mid-point of an upper edge, as shown. The coefficient of friction between the block and the surface is 0.6. Does the block slide or topple?

SOLUTION

First, we find the value of P needed to make the block slide.

Resolving in the **j**-direction (see the diagram), we obtain

$$R - 8g = 0 \quad \Rightarrow \quad R = 8g \, \text{N}$$

When the block is on the point of sliding, $F = \mu R = 4.8g \, \text{N}$.

Resolving in the **i**-direction, we obtain

$$F - P = 0 \quad \Rightarrow \quad P = F = 4.8g \, \text{N}$$

So, for the block to slide, P must exceed $4.8g \, \text{N}$.

We now find the value of P needed to make the block topple.

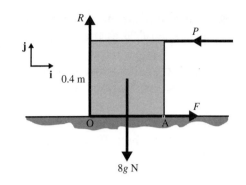

When the block is on the point of toppling, the normal reaction of the surface will act at O, as shown, because the corner A is about to lift off the surface.

Taking moments about O, we get

$$0.4P - 8g \times 0.2 = 0 \quad \Rightarrow \quad P = 4g\,\text{N}$$

So, for the block to topple, P must exceed $4g\,\text{N}$.

It requires a smaller force to make the block topple than to make it slide, so equilibrium will be broken by toppling.

Example 12 A prism of mass m, having a cross-section as shown, rests on a rough horizontal plank PQ. The coefficient of friction between the prism and the plank is 0.4. The end Q of the plank is gradually raised until equilibrium is broken. Will the prism slide or topple?

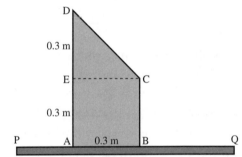

SOLUTION

First, we find the angle needed for the prism to slide.

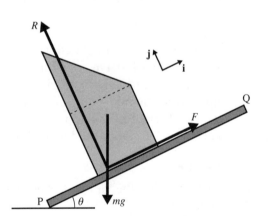

Resolving in the **i**-direction as shown, we obtain

$$F - mg \sin \theta = 0$$

$$\Rightarrow \quad F = mg \sin \theta \qquad [1]$$

Resolving in the **j**-direction, we obtain

$$R - mg \cos \theta = 0$$

$$\Rightarrow R = mg \cos \theta \qquad [2]$$

Dividing [1] by [2], we get

$$\tan \theta = \frac{F}{R}$$

The coefficient of friction is 0.4, so when equilibrium is limiting, $F = 0.4R$. Therefore, we have

$$\tan \theta = 0.4 \quad \Rightarrow \quad \theta = 21.8°$$

So, for the prism to slide, the angle must exceed $21.8°$

We now find the angle needed for the prism to topple.

First, we have to find the centre of mass of the prism.

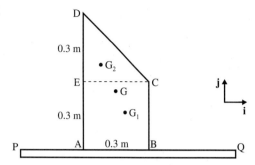

The portion with cross-section ABCE has mass $\frac{2}{3}m$ and centre of mass at $G_1(0.15, 0.15)$

The portion with cross-section CDE has mass $\frac{1}{3}m$ and centre of mass at $G_2(0.1, 0.4)$

With origin A and axes as shown, we have

$$\begin{pmatrix} \bar{x} \\ \bar{y} \end{pmatrix} = \frac{\dfrac{2m}{3}\begin{pmatrix} 0.15 \\ 0.15 \end{pmatrix} + \dfrac{m}{3}\begin{pmatrix} 0.1 \\ 0.4 \end{pmatrix}}{m}$$

$$= \begin{pmatrix} 0.1\dot{3} \\ 0.2\dot{3} \end{pmatrix}$$

So, the centre of mass is at $G(0.1\dot{3}, 0.2\dot{3})$ m.

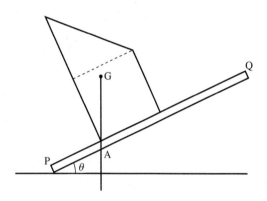

The prism will be on the point of toppling when G is vertically above A. Therefore, we have

$$\tan\theta = \frac{0.1\dot{3}}{0.2\dot{3}} \quad \Rightarrow \quad \theta = 29.7°$$

So, for the prism to topple, the inclination must exceed $29.7°$.

It therefore requires a greater inclination to make the prism topple than is needed to make it slide, so equilibrium will be broken by sliding.

Exercise 11D

1 The diagram shows a uniform rectangular block ABCD resting on a rough horizontal surface. $AB = 1$ m and $BC = 1.2$ m. The mass of the block is 50 kg and the coefficient of friction between the surface and the block is 0.5. A horizontal force P is applied to the mid-point of a top edge, as shown. The magnitude of P is gradually increased until the block breaks from equilibrium. Investigate whether the block will slide or topple.

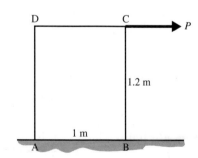

2 The diagram shows a uniform rectangular block ABCD of mass m, resting on a rough horizontal surface. AB = 20 cm and BC = 60 cm. The coefficient of friction between the block and the surface is 0.2. A horizontal force P is applied symmetrically at a point a distance h above A. The magnitude of P is gradually increased until the block breaks from equilibrium. Find the values of h for which equilibrium will be broken by sliding.

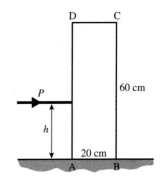

3 A uniform prism of mass 5 kg has cross-section ABC, where AB = 40 cm and BC = AC = 52 cm. It rests with AB in contact with a rough horizontal plank PQ. The coefficient of friction between the prism and the plank is μ. The end Q of the plank is gradually raised until the prism moves. Find the range of values of μ for which the prism will slide.

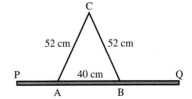

4 The diagram shows a chair made of four rods AP, IQ, JR and DS and two rectangular laminae ABCD and HIJK. AP = DS = AD = 50 cm. IQ = JR = 120 cm. AB = HI = 40 cm. The density of the rods is 600 g m^{-1}. The density of each lamina is 4 kg m^{-2}.

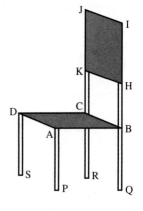

a) Taking P as the origin, with PQ, PS and PA as the x-, y- and z-axes respectively, find the coordinates of the centre of mass of the chair.

b) The chair rests on a horizontal floor and the coefficient of friction between the chair and the floor is μ. A horizontal force T is applied perpendicular to IJ at its mid-point. If it is applied in the positive x-direction, the chair topples. If it is applied in the negative x-direction, the chair slides. Find the range of possible values of μ.

5 A cone of radius r and height h rests on a rough plane. The angle between the plane and the horizontal is gradually increased. Show that the cone will slide before it topples over if the coefficient of friction between the cone and the plane is less than $4r/h$.

6 A cube rests on a rough plane inclined at α to the horizontal. A force P, parallel to the slope, is applied perpendicular to the upper edge of the cube at its mid-point. Find in terms of α the maximum value of μ for which the cube will slide up the slope when P is gradually increased.

7 The diagram shows a rectangular block resting on a rough horizontal plane, the coefficient of friction between the plane and the block being μ. The block has height a and width b, as shown. A gradually increasing force P is applied, as shown, to the mid-point of the top edge of the block so as to make an angle θ with the top face of the block. Show that, if the block topples,

$$\tan \theta > \frac{b - 2a\mu}{b\mu}$$

8 A closed cuboidal box, whose walls are of negligible thickness, has mass 60 g and dimensions as shown in the diagram. The box contains 540 g of washing powder, which occupies the lower $\frac{3}{4}$ of the box. The box rests on a moving conveyor belt, as shown, and is prevented from moving with the belt by a fixed horizontal bar a distance h above the belt. The coefficient of (dynamic) friction between the box and the belt is 0.6. What is the maximum value of h if the box is not to topple?

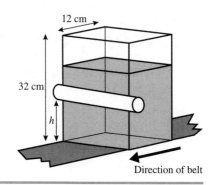

Direction of belt

Appendix

Concurrence of medians

To prove that the medians of a triangle are concurrent and their common point divides each median in the ratio $2:1$.

Let the position vectors of A, B and C relative to some origin be **a**, **b** and **c**.

The mid-point D of BC has position vector

$$\mathbf{d} = \mathbf{b} + \tfrac{1}{2}\overrightarrow{BC}$$
$$= \mathbf{b} + \tfrac{1}{2}(\mathbf{c} - \mathbf{b})$$
$$\Rightarrow \quad \mathbf{d} = \tfrac{1}{2}(\mathbf{b} + \mathbf{c})$$

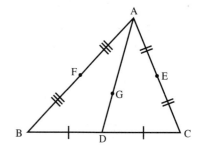

Similarly, the points E and F of AC and AB have position vectors

$$\mathbf{e} = \tfrac{1}{2}(\mathbf{a} + \mathbf{c})$$
and
$$\mathbf{f} = \tfrac{1}{2}(\mathbf{a} + \mathbf{b})$$

If G is the point on AD such that $AG:GD = 2:1$, its position vector is

$$\mathbf{g} = \mathbf{a} + \tfrac{2}{3}\overrightarrow{AD}$$
$$= \mathbf{a} + \tfrac{2}{3}(\mathbf{d} - \mathbf{a})$$
$$= \mathbf{a} + \tfrac{2}{3}(\tfrac{1}{2}\mathbf{b} + \tfrac{1}{2}\mathbf{c} - \mathbf{a})$$
$$\Rightarrow \quad \mathbf{g} = \tfrac{1}{3}(\mathbf{a} + \mathbf{b} + \mathbf{c})$$

But the point which divides BE in the ratio $2:1$ has position vector $\mathbf{b} + \tfrac{2}{3}(\mathbf{e} - \mathbf{b})$, and the point which divides CF in the ratio $2:1$ has position vector $\mathbf{c} + \tfrac{2}{3}(\mathbf{f} - \mathbf{c})$. Each of these reduces to $\tfrac{1}{3}(\mathbf{a} + \mathbf{b} + \mathbf{c})$. This means that G lies on all three medians and divides each in the ratio $2:1$.

(You might try to establish a similar proof regarding the position of the centre of mass of a tetrahedron.)

12 Work, energy and power

I like work: it fascinates me. I can sit and look at it for hours. I love to keep it by me:
the idea of getting rid of it nearly breaks my heart.
JEROME K. JEROME

Work

If you were to lift a heavy object, drag a packing case along or pedal a cycle, you would in each case know that you were doing work. Work is done whenever a force is applied to alter the motion or position of an object. The amount of work done depends on the magnitude of the force needed and the distance through which the point of application of the force moves.

For example, Gladys and Tracy are lifting a 50 kg object using pulleys.

Gladys uses the first arrangement shown. We assume that friction forces in the pulley can be neglected. The object is being raised at a constant rate.

Resolving vertically for the object, we obtain

$$T - 50g = 0 \quad \text{(no acceleration)}$$

So, the tension throughout the rope is $50g$ N.

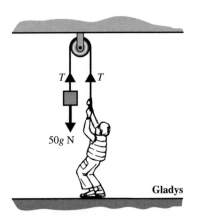

This means that to move the object upwards through a distance of 0.5 m, Gladys has to exert a downward force of $50g$ N on the rope and move it down through a distance of 0.5 m.

Tracy is not as strong as Gladys, but more ingenious. She uses the second arrangement shown. We assume that the additional pulley is smooth and has negligible mass.

Resolving vertically for the object, we obtain

$$T_1 - 50g = 0$$
$$\Rightarrow \quad T_1 = 50g \, \text{N}$$

Resolving vertically for the small pulley, we obtain

$$2T_2 - T_1 = 0$$
$$\Rightarrow \quad T_2 = 25g \, \text{N}$$

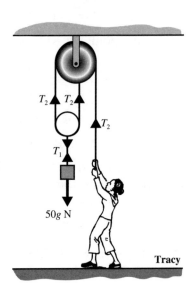

This means that to move the object at a constant speed, Tracy has to exert a force of only $25g$ N. However, because the rope is 'shared' between the two sides of the small pulley, to raise the object through 0.5 m, Tracy would have to pull 1 m of rope through the pulley.

Both Gladys and Tracy do the **same amount of work**: that is, they raise a mass of 50 kg through a distance of 0.5 m. Tracy exerts only half the force that Gladys does, but she exerts it through twice the distance to achieve the same effect.

This leads us to the definition of work.

When the point of application of a force F undergoes a displacement s in the **direction of the force**, the work done is $F \times s$.

The SI unit of work is the **Joule** (J), which is defined as the amount of work done when a force of 1 newton moves a distance of 1 metre.

It should be stressed that the displacement must take place in the **direction of the force**. For example, when a block is dragged along a horizontal surface by means of a rope, work is done by the tension in the rope and by the friction force. No work is done by the reaction of the surface on the block or by the block's weight, since these forces are perpendicular to the direction of the displacement.

Example 1 A block of mass 60 kg is dragged a distance of 3 m at constant speed across a horizontal rough plane by means of a horizontal rope. The coefficient of friction between the block and the plane is 0.4. Find the work done by each of the forces acting on the block.

SOLUTION

Resolving vertically, we obtain

$$R - 60g = 0$$
$$\Rightarrow \quad R = 60g \text{ N}$$
$$\Rightarrow \quad F = 24g \text{ N} \quad \text{(since } \mu = 0.4\text{)}$$

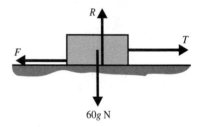

Resolving horizontally, we obtain

$$T - F = 0 \quad \text{(no acceleration)}$$
$$\Rightarrow \quad T = 24g \text{ N}$$

There is no vertical displacement of the block, so the work done by the weight and by R is zero.

Displacement in the direction of T is 3 m, so we have

$$\text{Work done by the tension} = 24g \times 3 = 72g \text{ J}$$

Displacement in the direction of F is -3 m, so we have

$$\text{Work done by the friction} = 24g \times (-3) = -72g \text{ J}$$

Notice the negative work related to friction in Example 1. We can say either that $-72g$ J of work was done by friction, or, more commonly, that $72g$ J of work was done **against** friction.

Work done against gravity

Suppose an object of mass m is raised at constant speed by a force T.

Resolving vertically, we obtain

$$T - mg = 0$$
$$\Rightarrow \quad T = mg$$

Therefore, if the object is raised through a distance h,

Work done by $T = mgh$

Work done by the object's weight $= -mgh$

We say that the **work done against gravity** is mgh.

Example 2 A plank of length 5 m is inclined so that the higher end is 3 m above the lower. A block of mass 40 kg is dragged at constant speed up the plank against a friction force of 120 N. Find the total work done.

SOLUTION

The block undergoes a displacement of -5 m in the direction of the friction force. Therefore,

Work done **against** friction $= 120 \times 5 = 600$ J

The block is raised through a height of 3 m. Therefore,

Work done **against** gravity $= 40g \times 3 = 1176$ J

$\Rightarrow$ Total work done in raising block $= 600 + 1176 = 1776$ J

There is an alternative approach to this problem.

Resolving parallel to the slope, we obtain

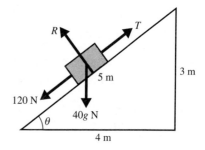

$$T - 120 - 40g \sin\theta = 0$$

But $\sin\theta = \frac{3}{5}$, therefore

$$T = 120 + 24g = 355.2 \text{ N}$$

There is a displacement of 5 m in the direction of T, so

Work done $= 355.2 \times 5 = 1776$ J

Note The concept of work done against gravity is particularly useful when the path of the object is not a straight line.

Example 3 An object of mass 8 kg is dragged at constant speed up a surface forming a quarter of a circle of radius 2 m, against a constant frictional resistance of 45 N. Find the total work done.

SOLUTION

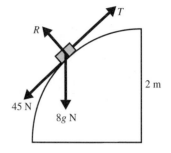

The object is raised through 2 m. Therefore, we have

$$\text{Work done against gravity} = 8g \times 2 = 156.8\,\text{J}$$

The object travels a distance of $-(\frac{1}{4} \times 4\pi) = -\pi\,\text{m}$ in the direction of the friction force. Therefore, we have

$$\text{Work done against friction} = 45\pi = 141.4\,\text{J}$$

So, we have

$$\text{Total work done} = 156.8 + 141.4 = 298.2\,\text{J}$$

Displacement at an angle to the force

Frequently, the force applied is directed at an angle to the direction in which displacement occurs. For example, if a block is dragged along a horizontal surface using a rope, the rope may be inclined to the horizontal.

Suppose a force F is applied to an object which is then displaced by a distance s in a direction making an angle θ to the direction of F. There are two ways in which we can think about this, each leading to the same result.

Method 1 The point of application of the force moves from A to B, but the displacement in the **direction** of F is represented by AC, where

$$AC = s\cos\theta$$

This gives

$$\text{Work done by } F = Fs\cos\theta$$

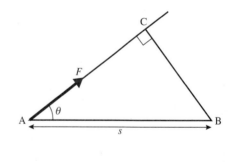

Method 2 The force F can be resolved into two components, parallel and perpendicular to the direction of the displacement, as shown.

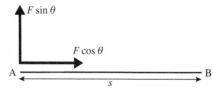

The perpendicular component, $F\sin\theta$, does no work because there is no displacement in that direction.

The parallel component, $F\cos\theta$, is displaced a distance s. Therefore,

$$\text{Work done by } F = (F\cos\theta) \times s = Fs\cos\theta$$

Example 4 A packing case is dragged a distance of 8 m along a horizontal surface by means of a rope inclined at 40° to the horizontal. The tension in the rope is 500 N. Find the work done by the tension.

SOLUTION

The component of the tension in the direction of motion is 500 cos 40°.

The displacement in this direction is 8 m.

Therefore, we have

$$\text{Work done} = 500\cos 40° \times 8 = 3064.2\,\text{J}$$

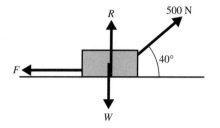

Work in vector terms

On page 26, the scalar product of two vectors **a** and **b** is defined as

$$\mathbf{a.b} = |\mathbf{a}||\mathbf{b}|\cos\theta$$

where θ is the angle between the vectors.

If $\mathbf{a} = a_1\mathbf{i} + a_2\mathbf{j} + a_3\mathbf{k}$, and $\mathbf{b} = b_1\mathbf{i} + b_2\mathbf{j} + b_3\mathbf{k}$, we get the component form

$$\mathbf{a.b} = a_1b_1 + a_2b_2 + a_3b_3$$

Both force **F** and displacement **s** are vectors. So

$$\text{Work done} = |\mathbf{F}||\mathbf{s}|\cos\theta = \mathbf{F.s}$$

Note This means that work is a scalar quantity.

Example 5 A force $\mathbf{F} = (5\mathbf{i} + 2\mathbf{j} - \mathbf{k})\,\text{N}$ is applied to an object which undergoes a displacement $\mathbf{s} = (4\mathbf{i} - \mathbf{j} + 6\mathbf{k})\,\text{m}$. Find the work done by the force.

SOLUTION

$$\begin{aligned}\text{Work done} &= \mathbf{F.s}\\ &= 5 \times 4 + 2 \times (-1) + (-1) \times 6\\ &= 12\,\text{J}\end{aligned}$$

Note Using the component form of the scalar product in this way effectively finds the total of the amounts of work done in the x-, y- and z-directions. In Example 5, the force did 20 J of work in the x-direction, -2 J in the y-direction and -6 J in the z-direction.

Work done by a variable force

If a force which is doing work varies as its point of application moves, we need to use calculus to find the work done. We will only consider the case in which motion takes place along a straight line.

Suppose that force F is a function of the displacement x of its point of application from an origin O. We require a method of finding the work done by the force in moving its point of application from $x = a$ to $x = b$.

A small change, δF, in the force will accompany a small change, δx, in the displacement. During this change, the force lies between F and $(F + \delta F)$ and so the work done, δW, lies between $F\delta x$ and $(F + \delta F)\,\delta x$. So, we have

$$F\delta x \leqslant \delta W \leqslant (F + \delta F)\,\delta x$$

$$\Rightarrow \quad F \leqslant \frac{\delta W}{\delta x} \leqslant (F + \delta F)$$

As $\delta x \to 0$, $\delta F \to 0$ and $\dfrac{\delta W}{\delta x} \to \dfrac{\mathrm{d}W}{\mathrm{d}x}$. So, in the limit, we have

$$F \leqslant \frac{\mathrm{d}W}{\mathrm{d}x} \leqslant F$$

$$\Rightarrow \quad \frac{\mathrm{d}W}{\mathrm{d}x} = F$$

$$\Rightarrow \quad W = \int F \,\mathrm{d}x$$

If x changes from a to b, this becomes

$$W = \int_a^b F \,\mathrm{d}x$$

Example 6 A particle is placed a magnetic field so that it is attracted towards a point O by a force of magnitude $\dfrac{100}{x^2}$ N, where x m is the displacement of the particle from O. Find the work done in moving the particle at constant speed from point A to point B, where OAB is a straight line, OA = 1 m and OB = 4 m.

SOLUTION

The work done is given by

$$W = \int_1^4 \frac{100}{x^2} \,\mathrm{d}x$$

$$= \left[-\frac{100}{x} \right]_1^4 = 75$$

So, the work done by the force is 75 J.

Exercise 12A

1 Find the work done by a crane which raises a load of 250 kg through a distance of 5.6 m.

2 A man of mass 85 kg requires to move a load of mass 30 kg from ground level on to staging 6 m up. How much work will he do if

a) he hoists it up using a single smooth pulley and a light rope
b) he carries it up a ladder?

3 Blocks, each of mass 5 kg and height 10 cm, are lying side by side on the ground. How much work would be involved in making a stack **a)** ten blocks high, **b)** n blocks high?

4 A block of mass 15 kg is pulled a constant speed for a distance of 12 m across a rough horizontal plane. The coefficient of friction between the plane and the block is 0.6. Find the work done.

5 A horizontal force is applied to a 6 kg body so that it accelerates uniformly from rest and moves across a horizontal plane against a constant frictional resistance of 30 N. After it has travelled 16 m, it has a speed of 4 m s^{-1}. Find the applied force and hence the work done.

6 A winch raises an object of mass 20 kg from rest with an acceleration of 0.2 m s^{-2}. How much work is done by the winch in the first 12 seconds?

7 Find the work done in pulling a packing case of mass 80 kg a distance of 15 m against a constant resistance of 150 N **a)** on a horizontal surface, **b)** up an incline whose angle θ to the horizontal is given by $\sin \theta = \frac{1}{8}$.

8 A ramp joining a point A to a point C, which is 8 m higher than A, consists of a straight section and an arc of a circle centre B, as shown. A trolley, which has mass 10 kg and a constant resistance to motion of 140 N, is pulled at a steady speed from A to C. Find the total work done.

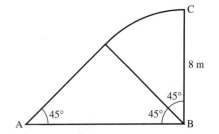

9 A body of mass 10 kg is at rest on a rough horizontal surface. The coefficient of friction between the body and the surface is 0.5. A force of 100 N is applied to the body for a period of 20 seconds in one of three different directions, as shown below.

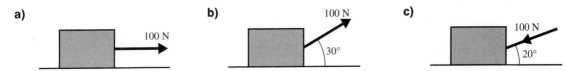

In each case, find the distance travelled by the body and hence the work done by the applied force.

10 Find the work done by a force $\mathbf{F} = (5\mathbf{i} + 3\mathbf{j})$ N whose point of application undergoes a displacement $\mathbf{s} = (3\mathbf{i} + 7\mathbf{j})$ m.

11 A force $\mathbf{F} = (5\mathbf{i} + 4\mathbf{j})$ N acts on a particle which moves from point A, position vector $(\mathbf{i} + 3\mathbf{j})$ m, to point B, position vector $(4\mathbf{i} + 5\mathbf{j})$ m. Find the work done by the force.

12 A force $\mathbf{F} = (2\mathbf{i} - \mathbf{j} + 3\mathbf{k})$ N acts on a particle which moves from point A, position vector $(-2\mathbf{i} + \mathbf{j} - 3\mathbf{k})$ m, to point B, position vector $(5\mathbf{i} - \mathbf{j} + 3\mathbf{k})$ m. Find the work done by the force.

13 The force required to compress a certain spring is proportional to the amount of compression. If it requires a force of 40 N to compress the spring by 2 cm, find the work done in compressing it a further centimetre.

14 A heavy, flexible chain is 5 m long and has a mass of 40 kg. The chain is initially on the ground, and is then gradually raised by one end until it is all hanging vertically. Find by integration the work done, and confirm that it is the same as that required to raise a particle of mass 40 kg to a height of 2.5 m above the ground.

15 The gravitational force acting on an object of mass m kg at a distance h m above the surface of the Earth is

$$\frac{3.99 \times 10^{14} m}{(h + 6.37 \times 10^6)^2} \ \text{N}$$

Find the work done against gravity in raising a rocket payload of 0.5 tonnes from the Earth's surface to a height of 1000 km.

Energy

Gravitational potential energy

Suppose we raise an object of mass 20 kg through a height of 10 m. The work we do against gravity is $20g \times 10 = 1960$ J.

If we now allow the object to sink back to its original level, its weight will do 1960 J of work.

By raising the object we 'stored' 1960 J worth of work, which we 'retrieved' by lowering the object . This principle of storing work is used in many ways. For example, some wall-clocks are powered by means of weights suspended from a chain passing over a cog. The heavier weight is raised manually and then, as it gradually descends, it does the work needed to drive the clock. Similarly, some electricity companies use off-peak electricity to pump water from one reservoir into a higher one. At peak times, the water is allowed to run back through turbines, which 'recover' the electricity.

Stored work is called **energy**. In particular, the energy described above, which depends on the position of an object in a gravitational field, is called **gravitational potential energy** (GPE).

Raising an object increases its GPE, whilst lowering the object decreases its GPE. It is not necessary (or realistic) to talk of the absolute GPE of an object, since the only thing that matters is **change** in its value. It is usual in a given problem to set an arbitrary zero level and measure all GPE in relation to that level.

The work done against gravity in raising an object of mass m kg from the zero level to a height of h metres is mgh J, so

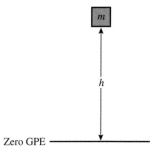

Gravitational potential energy $= mgh$ J

Notice that for an object below the zero level, the GPE would take a negative value, representing the amount of work it would take to raise the object to the zero level. As we are concerned only with changes in energy, this rather curious notion of negative energy is not a problem. You can, if you wish, avoid it completely by setting the zero level at or below the minimum height of all the objects in the problem.

The weight of an object is an example of a **conservative force**. As the object moves from a point A to a point B, the work done by the weight (GPE at B − GPE at A) depends only on the positions of A and B and not on the path taken. If the object follows any closed path, finishing at its starting point, **no** work is done by the weight.

Another example of a conservative force is the tension in an elastic string which is fixed at one end. The work done by the tension when the free end moves from a point A to a point B depends only on the extension of the string in the two positions. Similarly, the force acting on a metal object in a magnetic field is a conservative force.

On the other hand, forces such as friction are not conservative. If friction acts on a body as it moves from a point A to a point B, the work done by friction is greater for a longer path. If the body follows a closed path, friction has done a quantity of work dependent on the arc length of the path. Such a force is called a **dissipative force**.

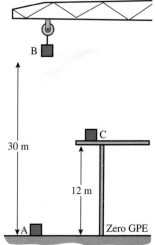

Example 7 A crane carries a load of 400 kg. It raises it from ground level to a height of 30 m, then lowers it on to a platform 12 m above the ground. Find the change in potential energy in each stage.

SOLUTION

Take ground level to be zero GPE.

The diagram shows the initial (A), intermediate (B) and final (C) positions of the load.

At A: GPE $= 0\,$J

At B: GPE $= 400g \times 30 = 117\,600\,$J

At C: GPE $= 400g \times 12 = 47\,040\,$J

During the first stage of motion, GPE increases by $117\,600\,$J.

During the second stage, GPE decreases by $117\,600 - 47\,040 = 70\,560\,$J.

Kinetic energy

Suppose we have an object of mass $20\,$kg at rest on a smooth horizontal surface. We apply a horizontal force of $100\,$N and pull the object for $10\,$m.

The object **accelerates**.

Using $F = ma$, we get

$$100 = 20a \quad \Rightarrow \quad a = 5\,\mathrm{m\,s}^{-2}$$

Using $v^2 = u^2 + 2as$, where $a = 5\,\mathrm{m\,s}^{-2}$, $s = 10\,$m and $u = 0$, we get

$$v^2 = 0^2 + 2 \times 5 \times 10$$
$$\Rightarrow \quad v = 10\,\mathrm{m\,s}^{-1}$$

The work done in giving the object this speed is $100 \times 10 = 1000\,$J.

A moving object has the capacity to do work. We would need to apply a frictional force to stop the above object and work would be done against that frictional force. For example, suppose we applied a frictional force of $200\,$N.

Using $F = ma$, we get

$$-200 = 20a \quad \Rightarrow \quad a = -10\,\mathrm{m\,s}^{-2}$$

Using $v^2 = u^2 + 2as$, where $a = -10\,\mathrm{m\,s}^{-2}$, $u = 10\,\mathrm{m\,s}^{-1}$ and $v = 0$, we have

$$0^2 = 10^2 - 20s \quad \Rightarrow \quad s = 5\,\mathrm{m}$$

So, the work done against friction is $200 \times 5 = 1000\,$J.

By giving the object a speed of $10\,\mathrm{m\,s}^{-1}$ we stored $1000\,$J worth of work, which we 'retrieved' when we brought the object to rest.

The work capacity of an object due to its motion is called **kinetic energy** (KE).

In general, when we apply a force $F\,$N to a stationary object of mass $m\,$kg for a distance $s\,$m, it has acceleration $a\,\mathrm{m\,s}^{-2}$ and final velocity $v\,\mathrm{m\,s}^{-1}$, where

$$v^2 = 2as \quad \Rightarrow \quad as = \tfrac{1}{2}v^2 \qquad [1]$$

Work done $= Fs$ and $F = ma$ $\Rightarrow$ Work done $= mas$ $\qquad [2]$

By substituting from [1] into [2], we get

Work done $= \frac{1}{2}mv^2$ J

So, an object of mass m kg travelling at v m s^{-1} has

Kinetic energy $= \frac{1}{2}mv^2$

Example 8 A particle of mass 4 kg, initially at rest, is acted upon by a force of 60 N for 8 seconds. Find its kinetic energy at the end of this time.

SOLUTION

Using $F = ma$, we obtain

$$60 = 4a \quad \Rightarrow \quad a = 15 \, \text{m s}^{-2}$$

Using $v = u + at$, where $a = 15 \, \text{m s}^{-2}$, $t = 8 \, \text{s}$ and $u = 0$, we obtain

$$v = 0 + 15 \times 8 = 120 \, \text{m s}^{-1}$$

$$\Rightarrow \quad \text{Kinetic energy} = \frac{1}{2} \times 4 \times 120^2$$

$$= 28\,800 \, \text{J} \quad \text{or} \quad 28.8 \, \text{kJ}$$

Alternatively, using $s = ut + \frac{1}{2}at^2$ gives $s = 480 \, \text{m}$. Therefore, we have

$$\text{Work done by force} = 60 \times 480 = 28\,800 \, \text{J}$$

As all of this work went into accelerating the particle, the kinetic energy is 28 800 J.

Other forms of energy

Gravitational potential energy and kinetic energy are two forms of mechanical energy. A third form of mechanical energy – the elastic potential energy of a stretched string or spring – is covered on pages 381–90. There are other, non-mechanical, forms of energy, such as heat, light, electrical, nuclear and chemical energy, which you may meet in your studies in physics. In this book, we consider them only obliquely. For example, a moving object which is slowed by friction has some of its kinetic energy converted into heat energy.

Conservation of mechanical energy

Consider an object suspended by a string and swinging as a pendulum. We treat it as the usual idealised model: a weightless, inextensible string supporting a point mass whose motion is not subject to air resistance.

As the particle swings up from its lowest point, work is done against gravity. As a result, the particle slows down – its kinetic energy reduces – but its height and therefore its gravitational potential energy increases. This continues until the particle stops, at which point it has zero KE and maximum GPE.

The process then reverses. The height of the particle and therefore its GPE decrease, but its speed and therefore its KE increase. This continues until the particle reaches its lowest position, at which point its GPE is at a minimum and its KE is at a maximum.

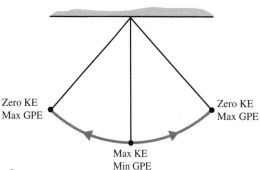

Zero KE
Max GPE

Zero KE
Max GPE

Max KE
Min GPE

In this idealised model, the sequence repeats for ever.

On the way up, KE is being converted into GPE.
On the way down, GPE is being converted into KE.
The total energy of the particle is exactly the same at all points.

The total energy of the system will be altered if one or other of the following happens.

- An external force, other than gravity, acts on the system in such a way that work is done. (Gravity has already been allowed for in the GPE.)

  When an external force does work on the system, the energy of the system is increased.

  When the system does work against an external force (for example, friction), the energy of the system is decreased. Mechanical energy is converted into other forms, such as heat and sound.

- There are sudden changes in the motion of the system.

  This occurs if component particles of the system collide, or if strings connecting component particles are jerked taut. Although the forces involved are internal ones, we see on page 293 that such sudden changes usually involve a loss of energy, again involving a conversion to heat etc.

We can now state the **principle of conservation of mechanical energy**:

The total mechanical energy of a system remains constant provided no external work is done and there are no sudden changes in the motion of the system.

We see on pages 381–2 that this principle includes the mechanical energy stored in a stretched elastic string.

It should be stressed that the conditions necessary for mechanical energy to be conserved are never perfectly realised in practice – it is a model based on the usual assumptions. However, the external forces are often sufficiently small to be neglected in the short term, so the principle can be used to make worthwhile analyses of a situation. We could, for example, make an excellent prediction of the maximum speed of a pendulum by knowing the height from which it was released. It would, however, change its motion over a long time due to the small but cumulative effect of air resistance.

Example 9 A particle of mass 2 kg is released from rest and slides down a smooth plane inclined at 30° to the horizontal. Find the speed of the particle after it has travelled 8 m.

SOLUTION

Let us take the start and finish positions of the particle as A and B, and let the speed of the particle at B be v.

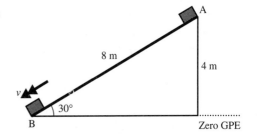

From the diagram, we see that A is at a height of $8 \sin 30° = 4$ m above B.

We have the following energy situations:

$$\text{At A:} \quad KE = 0\,J$$
$$GPE = 2g \times 4 = 78.4\,J$$
$$\text{At B:} \quad KE = \tfrac{1}{2} \times 2 \times v^2 = v^2$$
$$GPE = 0\,J$$

There are no external forces and no sudden changes, so energy is conserved:

$$\text{Total energy at B} = \text{Total energy at A}$$
$$\Rightarrow \quad v^2 = 78.4$$
$$\Rightarrow \quad v = 8.85\,\mathrm{m\,s^{-1}}$$

Note We could, of course, have solved Example 8 using the equations for motion with constant acceleration. Using energy does give us an alternative, convenient way of approaching the problem and has one major advantage: there is no need to assume that AB is a straight line. This means that if the particle follows a curved path, so that acceleration is no longer constant, we can still use energy considerations to find its final speed.

Work–energy principle

If an external force acts on a system so that work is done, mechanical energy is **not** conserved. We can often, however, make use of energy to solve the problem because the total work done on the system equals the change in energy.

We now repeat Example 9 but introduce a friction force.

Example 10 A particle of mass 2 kg is released from rest and slides down a plane inclined at 30° to the horizontal. There is a constant resistance force of 4 N. Find the speed of the particle after it has travelled 8 metres.

SOLUTION

Let us take the start and finish positions of the particle as A and B, and let the speed of the particle at B be v, as before.

From the diagram, we see that A is at a height $8 \sin 30° = 4\,\text{m}$ above B.

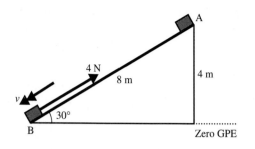

We have the following energy situations:

At A: $\text{KE} = 0\,\text{J}$

   $\text{GPE} = 2g \times 4 = 78.4\,\text{J}$

At B: $\text{KE} = \frac{1}{2} \times 2 \times v^2 = v^2$

   $\text{GPE} = 0\,\text{J}$

The work done against the resistance force is $4 \times 8 = 32\,\text{J}$. Therefore,

Change of energy of system $= -32\,\text{J}$

Total energy at B = Total energy at A − 32

$$\Rightarrow \quad v^2 = 46.4$$

$$\Rightarrow \quad v = 6.81\,\text{m s}^{-1}$$

Example 11 A skateboarder goes down a ramp formed by an arc of a circle of radius 5 m, as shown. She starts from rest at A. The total mass including the board is 50 kg. Find the speed with which she leaves the ramp at B in the following circumstances.

a) There is no appreciable friction.
b) There is a constant resistance force of 10 N.

SOLUTION

Let the speed at B be v m s^{-1}.

The height of B is $CD = 5 - 5 \cos 40° = 1.17\,\text{m}$.

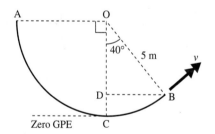

a) We have the following energy situations:

At A: $\text{KE} = 0\,\text{J}$

   $\text{GPE} = 50g \times 5 = 2450\,\text{J}$

At B: $\text{KE} = \frac{1}{2} \times 50 \times v^2 = 25v^2$

   $\text{GPE} = 50g \times 1.17 = 573.2\,\text{J}$

Friction is negligible, hence energy is conserved. Therefore, we have

Total energy at B = Total energy at A

$$\Rightarrow \quad 25v^2 + 573.2 = 2450$$

$$\Rightarrow \quad v = 8.66\,\text{m s}^{-1}$$

b) There is a constant resistance force of 10 N, hence work is done on the system. The distance moved by the resistance force is the arc length of the ramp.

$$\text{Arc length} = 2\pi \times 5 \times \frac{130}{360} = 11.34\,\text{m}$$

$$\text{Work done against the resistance} = 10 \times 11.34 = 113.4\,\text{J}$$

This is the change in energy of the system. Therefore, we have

$$\text{Total energy at B} = \text{Total energy at A} - 113.4$$

$$\Rightarrow \quad 25v^2 + 573.2 = 2450 - 113.4$$

$$\Rightarrow \quad v = 8.40\,\text{m s}^{-1}$$

Example 12 Particles A and B, of mass 2 kg and 5 kg respectively, are connected by a light inextensible string passing over a light smooth pulley. Initially, the particles are held level and at rest, before being released. Find the speed at which they are travelling when they reach 3 m apart.

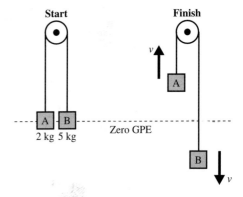

SOLUTION

Let the final speed of the system be v m s^{-1}.

We have the following energy situations:

$$\text{At start:} \quad \text{KE} = 0\,\text{J}$$

$$\text{GPE} = 0\,\text{J}$$

$$\text{At finish:} \quad \text{KE} = \tfrac{1}{2} \times 2 \times v^2 + \tfrac{1}{2} \times 5 \times v^2 = 3.5v^2$$

$$\text{GPE} = 2g \times 1.5 + 5g \times (-1.5) = -44.1\,\text{J}$$

There is no external force so energy is conserved. Therefore, we have

$$\text{Total energy at start} = \text{Total energy at finish}$$

$$\Rightarrow \quad 3.5v^2 - 44.1 = 0$$

$$\Rightarrow \quad v = 3.55\,\text{m s}^{-1}$$

Exercise 12B

1 A ball of mass 0.4 kg is thrown vertically into the air at a speed of 25 m s^{-1}. Assuming that air resistance is negligible, use energy methods to find the speed at which it is moving when it reaches a height of 20 m. Is the mass of the ball a necessary piece of information?

2 A child of mass 25 kg goes down a slide, starting from rest. The total drop in height is 4 metres.

  a) Assuming friction is negligible, find the speed of the child at the bottom of the slide.
  b) In fact, the child reaches the bottom travelling at 6 m s^{-1}. The length of the slide is 6 m. Find the work done against friction and the average friction force.

3 A particle of mass 2 kg starts from rest at A on a smooth 20° slope. A force of 120 N is applied parallel to the slope, moving the particle up the slope to B, where AB = 3 m. The force then stops acting and the particle continues up the slope, coming instantaneously to rest at C before sliding back down to A. Using energy methods:

a) Find the speed of the particle at B.

b) Find the distance BC.

c) Find the speed of the particle when it returns to A.

4 Particles A and B, of mass 0.5 kg and 1.5 kg respectively, are connected by a light inextensible string of length 2.5 m. A rests on a smooth horizontal table at a distance of 1.5 m from its edge. The string passes over the edge of the table and B hangs suspended. The system is held at rest with the string just taut and then released. Find the speed of A when it reaches the edge of the table.

5 Particles A and B, of mass 1 kg and 4 kg respectively, are connected by a light inextensible string of length 4 m. The string passes over a small, light smooth pulley which is 3 m above the ground. The particles are held at rest with A on the ground and B hanging with the string taut. The system is then released.

a) Find the speed of the particles as they pass each other.

b) Explain why you could not have used conservation of energy if the system had started with A on the ground and B held level with the pulley.

6 a) A particle of mass m is attached to the end A of a light rod OA of length a. O is freely hinged to a fixed point and the rod is held in a horizontal position before being released. Find the speed of the particle when the rod makes an angle θ to the downward vertical.

b) If the particle in part **a** had been initially projected downwards with speed u, find the value of u for which the rod will just travel round a complete circle.

7 A light rod AB of length $3a$ is freely hinged at O, where OA = a. Particles each of mass m are attached to the rod at A and B. The rod is held in a horizontal position, then released. Find the maximum speed of B in the subsequent motion.

8 Particles of mass m and $2m$ are connected by a light inextensible string. Initially, the particles lie at opposite edges of a smooth horizontal table with the string just taut. One of the particles is then nudged over the edge of the table. Find the ratio between the two possible speeds of the system when the other particle reaches the edge of the table.

9 A particle is projected with velocity V up a rough plane inclined at an angle θ to the horizontal. The coefficient of friction between the particle and the plane is μ, where μ is sufficiently small so that the particle can slide from rest down the plane. The particle starts at a point A, travels up the plane then slides down through A. At some point B below A on the plane the particle is again travelling with speed V. Show that

$$AB = \frac{\mu V^2 \cos\theta}{\sin^2\theta - \mu^2 \cos^2\theta}$$

Power

My mass is 90 kg. If I were to climb a flight of stairs taking me to a height of 15 m, I would do $90g \times 15 = 13\,230\,\text{J}$ of work.

This would be the same whether I ran up the stairs or walked up slowly. However, the effect on my breathing and heart rate of running up stairs is quite different from that of walking slowly. The rate at which the work is done is clearly important.

The same thing applies in many situations. It requires, for example, a more powerful pump to empty a tank in half an hour than it does to do the same job in five hours.

- The rate at which work is done is called **power**.
- The SI unit of power is the **watt** (W).
- **1 W** is the rate of working of $1\,\text{J\,s}^{-1}$.

If work is done at a variable rate, we can express the relation between work and power in calculus terms. If P is the power and W is the work done, we have

$$P = \frac{\mathrm{d}W}{\mathrm{d}t}$$

and the work done in the time interval from $t = t_1$ to $t = t_2$ is given by

$$W = \int_{t_1}^{t_2} P\,\mathrm{d}t$$

Example 13 A crane lifts a load of 50 kg to a height of 12 m at a steady speed of $0.6\,\text{m\,s}^{-1}$. Find the power required.

SOLUTION

The work done against gravity is $50g \times 12 = 5880\,\text{J}$

Assuming that we can neglect any resistance forces, this is the work done by the crane. Therefore, we have

$$\text{Time taken to lift load} = 12 \div 0.6 = 20\,\text{s}$$

$$\text{Rate of working of crane} = 5880 \div 20 = 294\,\text{W}$$

So, the power required is 294 W.

Example 14 A car of mass 900 kg moves at a steady speed of $15\,\text{m s}^{-1}$ up a slope inclined to the horizontal at an angle whose sine is 0.2. Resistance forces total 400 N. Find the power output of the engine.

SOLUTION

In 1 second, the car travels 15 m on the slope, which raises it through a vertical height of $15 \times 0.2 = 3\,\text{m}$.

$$\text{Work done against gravity} = 900g \times 3 = 26\,460\,\text{J}$$

$$\text{Work done against resistance} = 400 \times 15 = 6000\,\text{J}$$

$$\text{Total work done in each second} = 32\,460\,\text{J}$$

$$\text{Rate of working} = 32\,460\,\text{W} \quad \text{or} \quad 32.46\,\text{kW}$$

So, the power output of the car engine is 32.46 kW.

Example 15 A pump raises water from a tank through a height of 3 m and outputs it through a circular nozzle of radius 3 cm at $8\,\text{m s}^{-1}$. Find the rate at which the pump is working. Ignore any resistance forces.

SOLUTION

In each second, the pump raises and accelerates a 'cylinder' of water 8 m long and with radius 3 cm.

$$\text{Volume of water} = \pi \times 0.03^2 \times 8 = 0.022\,62\,\text{m}^3$$

We will assume the water has a density of $1000\,\text{kg m}^{-3}$, so we have

$$\text{Mass of water} = 0.0226 \times 1000 = 22.62\,\text{kg}$$

The water is raised through 3 m, so we have

$$\text{GPE given to the water} = 22.62\,g \times 3 = 665\,\text{J}$$

The water is accelerated from rest to $8\,\text{m s}^{-1}$, so we have

$$\text{KE given to the water} = \tfrac{1}{2} \times 22.62 \times 8^2 = 723.8\,\text{J}$$

The work done by the pump in each second is $665 + 723.8 = 1388.8\,\text{J}$. Therefore, we have

$$\text{Rate of working} = 1388.8\,\text{W}$$

Exercise 12C

1 A man raises a load of 20 kg through a height of 6 m using a rope and pulley. Assume the rope and pulley are light and smooth.

   a) What would be the man's power output if he completed the task in 30 seconds?

   b) If the man's maximum power output is 180 W, what is the shortest time in which he could complete the task?

2 A crate of mass 80 kg is dragged at a steady speed of $4\,\mathrm{m\,s^{-1}}$ up a slope inclined at an angle θ to the horizontal, where $\sin\theta = 0.4$. The total resistance is 250 N. Find the power required.

3 Arnold filled an empty washing-up liquid bottle with water and squirted a horizontal jet at a friend. The diameter of the nozzle was 4 mm and the water emerged at $10\,\mathrm{m\,s^{-1}}$. Find his rate of working. (Assume density of water to be $1000\,\mathrm{kg\,m^{-3}}$.)

4 Arnold's friend got her own back using a stirrup pump, which is a device for pumping water from a bucket. The pump raised the water through a height of 80 cm and emitted it as a jet with speed $8\,\mathrm{m\,s^{-1}}$ through a circular nozzle of radius 5 mm. Find her rate of working.

5 A winch has a maximum power output of 500 W. It is dragging a 200 kg crate up a slope inclined at 30° to the horizontal. The coefficient of friction between the crate and the slope is 0.6.

   a) Find the work done in dragging the crate a distance a m up the slope.
   b) Hence find the maximum speed at which the winch can drag the crate.

6 A force $\mathbf{F} = (4\mathbf{i} + 5\mathbf{j})\,\mathrm{N}$ acts on a particle, moving it along a straight groove from A to B, where A has position vector $(\mathbf{i} - 2\mathbf{j})\,\mathrm{m}$ and B $(5\mathbf{i} + \mathbf{j})\,\mathrm{m}$. The process takes 6 seconds. Find the rate at which $\mathbf{F}$ is working.

7 A pump, working at 3 kW, raises water from a tank at $1.2\,\mathrm{m^3\,min^{-1}}$ and emits it through a nozzle at $15\,\mathrm{m\,s^{-1}}$. Find the height through which the water is raised.

8 A horse which is capable of a power output of 800 W is able to pull a plough at a constant speed of $1.6\,\mathrm{m\,s^{-1}}$. Find the resistance to the motion of the plough. (In fact, the customary unit of power, 1 horsepower, is equivalent to 746 W).

9 A projectile of mass m kg is accelerated at a constant rate up a vertical tube of height h m. When it emerges it rises a further $3h$ m before coming instantaneously to rest. Show that the average rate of working while the projectile is in the tube is $2m\sqrt{6g^3h}$ W.

10 A device is operated by an electric motor with a rechargeable battery. The motor has power P W which reduces over time as the battery runs down. The power at time t s is given by $P = 1500e^{-0.0001t}$. The battery is fully charged and the device is then run continuously for 20 minutes. Find the total work done.

11 A child pushes a cart for a period of 5 s from a standing start and then jumps aboard. The power she can exert t s after starting is given by $P = (100 - t)\,\mathrm{W}$.

   a) Find the total amount of work she does.
   b) Given that the child plus the cart have a total mass of 50 kg and assuming there is no significant resistance to the motion, find the speed of the cart when she jumps aboard.

Relation between power and velocity

Suppose a constant force F N applied to a particle just balances the resistance forces, so that the particle has a constant velocity v m s^{-1} in the direction of the force.

In 1 second, the particle would travel v m. Therefore, we have

> Work done in 1 second $= Fv$ J
>
> Power exerted by $F = Fv$ W

More generally, when the angle between the force and the velocity is θ:

$$\text{Power} = Fv \cos \theta \text{ W}$$

or in vector terms:

$$\text{Power} = \mathbf{F.v} \text{ W}$$

In situations where the force or the velocity is variable, this expression still gives the power being exerted at a particular instant.

Example 16 A car is being driven at a constant speed of $20\,\text{m s}^{-1}$ on a level road against a constant resistance force of $260\,\text{N}$. Find the power output of the engine.

SOLUTION

Let the applied force of the engine be F N.

Resolving in the direction of travel, we get

$$F - 260 = 0 \quad \text{(no acceleration)}$$
$$\Rightarrow \quad F = 260\,\text{N}$$

Therefore, we have

$$\text{Power} = 260 \times 20 = 5200\,\text{W} \quad \text{or} \quad 5.2\,\text{kW}$$

Example 17 A car of mass $800\,\text{kg}$ is being driven along a level road against a constant resistance of $450\,\text{N}$. The output of the engine is $7\,\text{kW}$.

a) Find the acceleration when the speed is $10\,\text{m s}^{-1}$.
b) Find the maximum speed of the car.

SOLUTION

a) Let the applied force of the engine be F N, and the acceleration be $a\,\text{m s}^{-2}$. We then have

$$\text{Power} = Fv$$
$$\Rightarrow \quad 7000 = 10F \quad \Rightarrow \quad F = 700\,\text{N}$$

Resolving in the direction of travel, we get

$$700 - 450 = 800a$$
$$\Rightarrow \quad a = 0.3125 \, \text{m s}^{-2}$$

So, the acceleration of the car is $0.3125 \, \text{m s}^{-2}$.

b) Resolving in the direction of travel, we get

$$F - 450 = 0 \quad \text{(no acceleration)}$$
$$\Rightarrow \quad F = 450 \, \text{N}$$

which gives

$$\text{Power} = Fv$$
$$\Rightarrow \quad 7000 = 450v \quad \Rightarrow \quad v = 15.56 \, \text{m s}^{-1}$$

So, the maximum speed is $15.56 \, \text{m s}^{-1}$.

As can be seen from Examples 16 and 17, most simple problems involving power can be solved using these two basic equations:

$$\text{Power} = |\,\text{Applied force}\,| \times |\,\text{Velocity}\,|$$

and

$$\text{Resultant force} = \text{Mass} \times \text{Acceleration}$$

It is important to stress that in the first equation the force referred to is that exerted by the engine etc, often called the **tractive force**, whereas in the second equation the force is the component in the direction of motion of the resultant of all the forces.

Example 18 A car of mass 900 kg travels up a hill, inclined at $10°$ to the horizontal, against a constant resistance force of 250 N. Its maximum speed is $45 \, \text{km h}^{-1}$.

a) Find the power output of the engine.
b) Find the initial acceleration when it reaches level road at the top of the hill.

SOLUTION

a) Let the applied force of the engine be F N.

Resolving up the slope, we obtain

$$F - 250 - 900g \sin 10° = 0$$
$$\Rightarrow \quad F = 1781.6 \, \text{N}$$

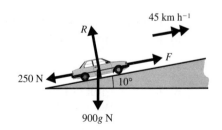

The speed is $45 \, \text{km h}^{-1} = 12.5 \, \text{m s}^{-1}$. Therefore, we have

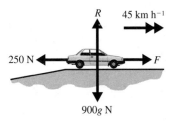

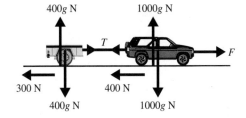

$$\text{Power} = Fv = 1781.6 \times 12.5$$
$$= 22\,269.7\,\text{W} \quad \text{or} \quad 22.3\,\text{kW} \quad \text{(to 3 sf)}$$

b) When the car reaches the level, it has the same power and initially the same speed. So, F is still 1781.6 N.

Let the initial acceleration be $a\,\text{m s}^{-2}$. Then resolving horizontally, we obtain

$$1781.6 - 250 = 900a \quad \Rightarrow \quad a = 1.7\,\text{m s}^{-2}$$

Example 19 A car of mass 1 tonne is towing a trailer of mass 400 kg on a level road. The resistance to motion of the car is 400 N and of the trailer is 300 N. At a certain instant, they are travelling at $10\,\text{m s}^{-1}$ and the power output of the engine is 10.5 kW. Find the tension in the coupling between the car and the trailer.

SOLUTION

Using Power $= |\,\text{Applied force}\,| \times |\,\text{Velocity}\,|$, we get

$$10\,500 = 10F$$
$$\Rightarrow \quad F = 1050\,\text{N}$$

Let the acceleration be $a\,\text{m s}^{-2}$.

Resolving horizontally for the whole system, we get

$$F - 300 - 400 = 1400a$$
$$\Rightarrow \quad a = 0.25\,\text{m s}^{-2}$$

Resolving horizontally for the trailer, we get

$$T - 300 = 400 \times 0.25$$
$$\Rightarrow \quad T = 400\,\text{N}$$

So, the tension in the coupling is 400 N.

Problems with variable resistance

So far, we have made the assumption that resistance to motion is constant. This is never the case in reality, although for a small range of slow speeds it may be approximately true.

In practice, the resistance is variable and depends on the speed of the vehicle. The nature of the relationship may not be a precise one and may itself change as the speed increases. For example, for a small object moving through the air, the air resistance is roughly proportional to the speed v when v is below about $10\,\mathrm{m\,s^{-1}}$, but for higher speeds (up to about $250\,\mathrm{m\,s^{-1}}$) the air resistance is proportional to v^2. Around the speed of sound, there are large changes in air resistance and there is no easy relationship with speed, but once the sound barrier is broken the resistance is again roughly proportional to v.

In a practical situation, we would have to conduct experiments and decide on a relationship which appeared to accord with the experimental data. This would then be one of the assumptions in our model of the situation. You can explore the effect of different models using the spreadsheet POWER, which is available on the Oxford University Press website (http://www.oup.co.uk/mechanics).

Example 20 A car of mass $900\,\mathrm{kg}$ moves against a resistance which is proportional to its speed. Its power output is $6\,\mathrm{kW}$ and on a level road its maximum speed is $40\,\mathrm{m\,s^{-1}}$. Find its maximum speed up an incline whose angle to the horizontal is θ, where $\sin\theta = \frac{1}{30}$.

SOLUTION

Using Power $= |\,\text{Applied force}\,| \times |\,\text{Velocity}\,|$, we get

$$6000 = 40F$$

$$\Rightarrow \quad F = 150\,\mathrm{N}$$

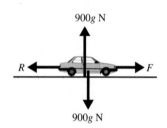

Resistance, R, is proportional to speed, v. So, at maximum speed,

$$R = kv = 40k\,\mathrm{N}$$

Resolving horizontally, we get

$$150 - 40k = 0$$

$$\Rightarrow \quad k = 3.75$$

So, at any speed, $R = 3.75v\,\mathrm{N}$.

Let the maximum speed up the hill be $V\,\mathrm{m\,s^{-1}}$.

From Power $= |\,\text{Applied force}\,| \times |\,\text{Velocity}\,|$, we obtain

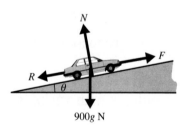

$$F = \frac{6000}{V}$$

Resolving up the slope, we have

$$\frac{6000}{V} - 3.75V - 900g \times \frac{1}{30} = 0$$

$$\Rightarrow \quad 3.75V^2 + 294V - 6000 = 0$$

$$\Rightarrow \quad V = 16.8\,\mathrm{m\,s^{-1}} \quad \text{or} \quad -95.2\,\mathrm{m\,s^{-1}}$$

Clearly, $-95.2\,\mathrm{m\,s^{-1}}$ is inappropriate. So, the maximum speed up the incline is $16.8\,\mathrm{m\,s^{-1}}$.

Exercise 12D

1 A train has a maximum speed of $50\,\mathrm{m\,s^{-1}}$ on the level against a resistance of $40\,\mathrm{kN}$. Find the power output of the engine.

2 A car of mass $800\,\mathrm{kg}$ has a maximum speed of $75\,\mathrm{km\,h^{-1}}$ up an incline against a resistance of $500\,\mathrm{N}$. The incline's angle to the horizontal is θ, where $\sin\theta = \frac{1}{40}$. Find the power output of the engine.

3 A train of mass 40 tonnes has a maximum speed of $15\,\mathrm{m\,s^{-1}}$ up a slope against a resistance of $50\,\mathrm{kN}$. The slope is inclined to the horizontal at an angle θ, where $\sin\theta = \frac{1}{50}$. Assuming the resistance is constant, find the maximum speed of the train down the same slope.

4 The frictional resistances acting on a train are $\frac{1}{100}$ of its weight. Its maximum speed up an incline whose sine is $\frac{1}{80}$ is $48\,\mathrm{km\,h^{-1}}$. Find its maximum speed on the level.

5 An open truck of mass 5 tonnes is carrying a load of $500\,\mathrm{kg}$ of fish up a hill against a constant resistance of $800\,\mathrm{N}$. It is travelling at its maximum speed of $10\,\mathrm{m\,s^{-1}}$. The hill is inclined at θ to the horizontal, where $\sin\theta = \frac{1}{50}$. A flock of gulls, mass $1000\,\mathrm{kg}$, descends on the lorry to eat the fish. Find the initial deceleration of the truck and the new maximum speed. As the truck reaches the top of the hill and moves onto level road, the gulls, having eaten all the fish, fly away. Find the initial acceleration of the truck.

6 A train consists of an engine of mass 50 tonnes and n trucks, each of mass 10 tonnes. The resistance to motion of the engine is $4000\,\mathrm{N}$ and that of each truck is $500\,\mathrm{N}$. The maximum speed of the train on the level when there are five trucks is $120\,\mathrm{km\,h^{-1}}$. Find the power output of the engine and the maximum speed of a train with n trucks going up an incline whose angle is θ, where $\sin\theta = \frac{1}{100}$.

7 a) A car of mass $900\,\mathrm{kg}$ pulls a trailer of mass $200\,\mathrm{kg}$. The resistance to motion of the car is $200\,\mathrm{N}$ and of the trailer is $80\,\mathrm{N}$. Find the power output of the engine if the maximum speed on the level is $40\,\mathrm{m\,s^{-1}}$.

b) The car and trailer are travelling at $8\,\mathrm{m\,s^{-1}}$ on a hill, inclined at θ to the horizontal, where $\sin\theta = \frac{1}{40}$. If the resistance is constant and the engine is exerting full power, find the acceleration and the tension in the coupling between the car and the trailer.

8 The resistance to motion of a car is proportional to its speed. A car of mass $1000\,\mathrm{kg}$ has a maximum speed of $45\,\mathrm{m\,s^{-1}}$ on the level when its power output is $8\,\mathrm{kW}$. Find its acceleration when it is travelling on the level at $20\,\mathrm{m\,s^{-1}}$ and its engine is working at $6\,\mathrm{kW}$.

9 a) A lorry of mass 10 tonnes has a maximum speed of $20\,\mathrm{m\,s^{-1}}$ up an incline when working at $70\,\mathrm{kW}$. The angle of the incline to the horizontal is θ, where $\sin\theta = \frac{1}{100}$. Find the resistance to motion.

b) If the resistance is proportional to the square of the speed, find the maximum speed of the lorry on the level when working at the same rate.

10 A car has a maximum power P. The resistance to motion is kv. Its maximum speed up a certain slope is V and its maximum speed down the same slope is $2V$. Show that $V = \sqrt{P/2k}$.

11 A train of mass 50 tonnes has a maximum speed on the level of $50\,\mathrm{km\,h^{-1}}$ when working at $80\,\mathrm{kW}$. Assuming that resistance is constant, how far would the train travel before coming to rest if, when travelling at maximum speed, the engine were disengaged and the train allowed to coast?

12 A cyclist and her cycle have a combined mass of $80\,\mathrm{kg}$. The resistance to motion is proportional to the speed. On the level, she can travel at a maximum speed of $10\,\mathrm{m\,s^{-1}}$, and she can freewheel down an incline of angle θ at $14\,\mathrm{m\,s^{-1}}$. Find the maximum speed at which she can go up the same incline.

13 A car working at a rate $P\,\mathrm{W}$ has a maximum speed $V\,\mathrm{m\,s^{-1}}$ when travelling on the level against a resistance proportional to the square of its speed. At what rate would the car have to work to double its maximum speed?

13 Momentum and impulse

And there he plays extravagant matches in fitless finger-stalls
On a cloth untrue with a twisted cue and elliptical billiard balls.
W.S. GILBERT

Change of momentum

Suppose that a particle of mass m kg is moving in a straight line with a velocity u m s^{-1}. A constant force F N is applied in the direction of travel for a time t seconds. As a result there is an acceleration a m s^{-2} and the particle's velocity changes to v m s^{-1}.

From Newton's second law, we have

$$F = ma \qquad [1]$$

As acceleration is constant, we have

$$v = u + at$$

$$\Rightarrow \quad a = \frac{v - u}{t} \qquad [2]$$

Substituting from [2] into [1] gives

$$F = \frac{m(v - u)}{t}$$

$$\Rightarrow \quad Ft = mv - mu \qquad [3]$$

The quantity Ft is called the **impulse** of the force. The SI unit of impulse is the **newton second** (N s).

We can see from equation [3] that the impulse is equal to $mv - mu$. That is, it is equal to the change in the value of Mass × Velocity. The product of the mass and the velocity of a body is called its **momentum**. The unit of momentum must obviously be the same as that for impulse, namely N s.

So, in general, we have

Impulse = Change of momentum

Notice that, as force and velocity are both vector quantities, impulse and momentum are also vectors. In general, if $\mathbf{J}$ is the impulse, $\mathbf{F}$ is the force, and $\mathbf{u}$ and $\mathbf{v}$ are the initial and final velocities respectively, then

$$\mathbf{J} = \mathbf{F}t = m\mathbf{v} - m\mathbf{u}$$

The concept of impulse is most useful when the period of time involved is short, although this does not have to be the case. If, for example, a ball is struck with a bat, we would probably be unable to determine the magnitude of the force or the length of time for which it acted. We could, however, more easily measure the velocity of the ball before and after the impact, and so calculate the change of momentum and hence the impulse. Of course, we could not be sure that the force involved was constant. So, even if we knew the duration of the contact, the best we could do would be to find the average force applied.

Example 1 A batsman in the nets hits a ball of mass 0.15 kg at a speed of $40\,\mathrm{m\,s^{-1}}$. The ball strikes the netting and is brought to rest in 0.5 s. Find the average force exerted by the netting on the ball.

SOLUTION

If we take left to right in the diagram as the positive direction, then the net exerts an impulse $+J\,\mathrm{N\,s}$ on the ball, whose velocity changes from $-40\,\mathrm{m\,s^{-1}}$ to zero.

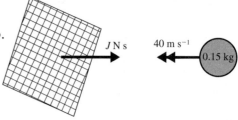

Initial momentum $= 0.15 \times -40 = -6\,\mathrm{N\,s}$

Final momentum $= 0\,\mathrm{N\,s}$

Applying Impulse = Change of momentum, we obtain

$$J = 0 - (-6) = 6\,\mathrm{N\,s}$$

To find the average force, F, we use

Impulse = Average force $\times$ Time

$$\Rightarrow \quad F \times 0.5 = 6 \quad \Rightarrow \quad F = 12\,\mathrm{N}$$

Notice that the ball would have exerted an average force of $-12\,\mathrm{N}$ on the net for 0.5 s, and hence an impulse of $-6\,\mathrm{N\,s}$. This is a consequence of Newton's third law.

Example 2 A particle of mass 3 kg struck a wall and rebounded along the same line. Its speed before the impact was $8\,\mathrm{m\,s^{-1}}$ and afterward was $5\,\mathrm{m\,s^{-1}}$. Find the impulse exerted by the wall on the particle.

SOLUTION

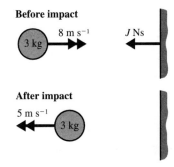

Before impact

8 m s⁻¹   J N s

3 kg

After impact

5 m s⁻¹

3 kg

Taking left to right as the positive direction, we have

Initial momentum of particle $= 3 \times 8 = 24\,\text{N s}$

Final momentum of particle $= 3 \times -5 = -15\,\text{N s}$

To find the impulse of the wall on the particle, $-J\,\text{N s}$, we use

Impulse = Change of momentum

$$\Rightarrow \quad -J = -15 - 24 = -39\,\text{N s}$$

So, the magnitude of the impulse is $39\,\text{N s}$.

The concepts of momentum and impulse also apply to motion in two or three dimensions.

Example 3 A particle of mass 2 kg, travelling with a velocity of $(3\mathbf{i} + 5\mathbf{j})\,\text{m s}^{-1}$, is given an impulse of $(2\mathbf{i} - 4\mathbf{j})\,\text{N s}$. Find its new velocity.

SOLUTION

Let the new velocity of the particle be $\mathbf{v}\,\text{m s}^{-1}$. We then have

Initial momentum $= 2(3\mathbf{i} + 5\mathbf{j}) = (6\mathbf{i} + 10\mathbf{j})\,\text{N s}$

Final momentum $= 2\mathbf{v}\,\text{N s}$

To find $\mathbf{v}$, we use

Impulse = Change of momentum

$$\Rightarrow \quad 2\mathbf{i} - 4\mathbf{j} = 2\mathbf{v} - (6\mathbf{i} + 10\mathbf{j})$$

$$\Rightarrow \quad \mathbf{v} = (4\mathbf{i} + 3\mathbf{j})\,\text{m s}^{-1}$$

Example 4 Steve kicks a ball of mass 0.8 kg along the ground at a velocity of $5\,\text{m s}^{-1}$ towards Monica. She kicks it back towards him but lofts it so that it leaves her foot at $8\,\text{m s}^{-1}$ and with an elevation of $40°$ to the horizontal. Find the magnitude and direction of the impulse from Monica's kick.

SOLUTION

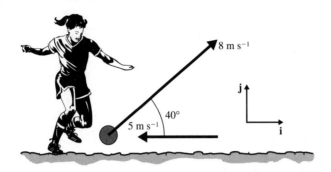

Taking unit vectors **i** and **j** as shown in the diagram, we obtain

$$\text{Initial momentum} = 0.8 \times -5\mathbf{i} = -4\mathbf{i}\,\text{N s}$$

$$\text{Final momentum} = 0.8 \times (8\cos 40°\,\mathbf{i} + 8\sin 40°\,\mathbf{j})\,\text{N s}$$

$$= (4.903\mathbf{i} + 4.114\mathbf{j})\,\text{N s}$$

Applying Impulse **J** = Change of momentum, we obtain

$$\mathbf{J} = (4.903\mathbf{i} + 4.114\mathbf{j}) - (-4\mathbf{i}) = (8.903\mathbf{i} + 4.114\mathbf{j})\,\text{N s}$$

which gives

$$\text{Magnitude of } \mathbf{J} = \sqrt{8.903^2 + 4.114^2} = 9.807\,\text{N s}$$

The direction of **J** is at an angle θ to the horizontal, as shown in the diagram, where

$$\tan\theta = \frac{4.114}{8.903} \quad \text{giving} \quad \theta = 24.8°$$

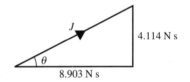

Impulse of a variable force

If the force **F** is a function of t, we have

$$m\frac{d\mathbf{v}}{dt} = \mathbf{F}$$

If the velocity changes from **u** to **v** as t changes from 0 to t, we have

$$m\int_u^v d\mathbf{v} = \int_0^t \mathbf{F}\,dt$$

$$\Rightarrow \quad \int_0^t \mathbf{F}\,dt = m\mathbf{v} - m\mathbf{u}$$

This means that $\int_0^t \mathbf{F}\,dt$ gives the change of momentum, and is therefore the impulse of the force.

Example 5 A particle is acted on by a force $\mathbf{F} = (t^2\mathbf{i} - 2t\mathbf{j})\,$N. The force acts for 3 s on the particle, mass 2 kg, which is initially travelling with velocity $(2\mathbf{i} + \mathbf{j})\,\mathrm{m\,s^{-1}}$. Find the impulse of the force and the final velocity of the particle.

SOLUTION

The impulse $\mathbf{J}$ is given by

$$\mathbf{J} = \int_0^3 (t^2\mathbf{i} - 2t\mathbf{j})\,\mathrm{d}t = \left[\tfrac{1}{3}t^3\mathbf{i} - t^2\mathbf{j}\right]_0^3 = (9\mathbf{i} - 9\mathbf{j})\,\mathrm{N\,s}$$

If the final velocity is $\mathbf{v}$, the change of momentum is $2\mathbf{v} - 2(2\mathbf{i} + \mathbf{j})\,$N s. As Change of momentum = Impulse, we have

$$2\mathbf{v} - 2(2\mathbf{i} + \mathbf{j}) = 9\mathbf{i} - 9\mathbf{j}$$
$$\Rightarrow \quad \mathbf{v} = \tfrac{1}{2}(13\mathbf{i} - 7\mathbf{j})\,\mathrm{m\,s^{-1}}$$

Force exerted by a jet of water

When a jet of water is directed at a surface, it exerts a force. To calculate this force, we make three assumptions.

- The jet strikes the surface at right angles.
- The jet is not slowed down by air resistance after leaving the nozzle.
- The water does not rebound. That is, each particle of water is brought to a standstill when it hits the surface.

We will also assume that the density of the water in the jet is such that 1 litre has a mass of 1 kilogram. This, in general, will depend on the temperature and purity of the water.

Example 6 A jet of water issues from a circular pipe of radius 4 cm at a speed of $8\,\mathrm{m\,s^{-1}}$ and strikes a wall at right angles without rebounding. Find the force exerted on the wall.

SOLUTION

In 1 second, a 'cylinder' of water 8 m long will hit the wall. So,

$$\text{Volume of water} = \pi \times 0.04^2 \times 8 = 0.0402\,\mathrm{m^3}$$

Now, $1\,\mathrm{m^3} = 1000\,$litres, so the volume of water is 40.2 litres, which gives its mass as 40.2 kg. Therefore,

$$\text{Momentum of 'cylinder' of water} = 40.2 \times 8 = 321.7\,\mathrm{N\,s}$$

If the water does not rebound, the momentum after impact is zero. So, in each second, there is a loss of momentum of 321.7 N s.

As impulse equals change of momentum, there must be an impulse on the water of magnitude 321.7 N s, and an equal and opposite impulse on the wall.

To find the force on the wall, F, we use

$$\text{Impulse} = \text{Force} \times \text{Time}$$

where the impulse is 321.7 N s and the time is 1 second, giving

$$F \times 1 = 321.7\,\text{N s}$$
$$\Rightarrow \quad F = 321.7\,\text{N}$$

Exercise 13A

1 An object of mass 4 kg is at rest. It receives an impulse of magnitude 28 N s. With what speed will it commence to move?

2 An object of mass 7 kg is travelling in a straight line at a speed of $4\,\text{m s}^{-1}$. It is acted on by a constant force in the direction of the line and as a result its speed increases to $10\,\text{m s}^{-1}$.

a) Find the impulse exerted on the object.
b) Find the force involved if the process took 0.35 s.

3 A tennis player struck a ball so that its path was exactly reversed. The ball approached the racket at $35\,\text{m s}^{-1}$ and left at $45\,\text{m s}^{-1}$. The mass of the ball was 90 g. Find the magnitude of the impulse exerted on the ball.

4 An engine of mass 20 tonnes is travelling at $54\,\text{km h}^{-1}$. Its brakes are applied for 3 seconds, after which it is travelling at $45\,\text{km h}^{-1}$. Find the change in momentum of the engine and hence the average braking force applied.

5 An object of mass 2 kg has a velocity of $(8\mathbf{i} - 3\mathbf{j})\,\text{m s}^{-1}$. It receives an impulse $\mathbf{J}$ N s, which alters its velocity to $(2\mathbf{i} + 5\mathbf{j})\,\text{m s}^{-1}$. Find $\mathbf{J}$.

6 A hockey ball of mass 200 g, travelling along horizontal ground at $15\,\text{m s}^{-1}$, is struck by a stick, causing it to travel in the opposite direction at the same speed but at an initial angle of 30° to the horizontal. Find the magnitude and direction of the impulse exerted by the stick.

7 A bullet of mass m, travelling at a speed $2u$, strikes an object and ricochets. The effect is that its direction is changed by 60° and its speed is reduced to u. Find, in terms of m and u, the magnitude of the impulse sustained by the bullet.

8 Footballers A, B and C are standing so that the angle ABC is 30°. A kicks a ball of mass 0.5 kg at $8\,\text{m s}^{-1}$ along the ground to B. B kicks it so that it heads directly towards C at $6\,\text{m s}^{-1}$ but with an initial elevation of 40° to the horizontal. Find the magnitude of the impulse which B exerts on the ball.

9 A hose discharges water at the rate of 15 litres per second, with a speed of $20\,\text{m s}^{-1}$. The water strikes a wall at right angles and does not rebound. Find the force exerted on the wall.

10 Water is emitted by a circular nozzle of radius 5 cm at a speed of $25\,\mathrm{m\,s^{-1}}$. It strikes a wall at right angles without rebounding. Find the force exerted on the wall.

11 A rectangular block of mass 4 kg rests on a rough horizontal surface. The coefficient of friction between the block and the surface is 0.4. A jet of water, emerging from a circular nozzle of radius 6 cm with speed $v\,\mathrm{m\,s^{-1}}$, is played directly on one of the vertical faces of the block.

a) Find the value of v for which the block will just commence to move.
b) Find the initial acceleration of the block if the velocity of the jet is twice that in part a.

12 A particle of mass 8 kg, travelling with velocity $(2\mathbf{i} + 3\mathbf{j})\,\mathrm{m\,s^{-1}}$, is acted on for a period of 4 s by a force $\mathbf{F} = [(3t^2 + 1)\mathbf{i} + (2t - t^3)\mathbf{j}]\,\mathrm{N}$. Find

a) the impulse of the force
b) the final velocity of the particle.

Conservation of linear momentum

Consider a system consisting of two bodies, A and B, which may be moving and are certainly free to move.

Suppose that A exerts a force F on B for a time t. This could happen in many ways. There may be a collision between the bodies. The bodies may be connected by a string which becomes taut. There may be magnetic attraction or repulsion between the bodies.

Whatever the reason for the force, B will suffer an impulse $J = Ft$. By Newton's third law, there will be a force $-F$ exerted by B on A, and as a result A will suffer an impulse $-J$.

As impulse equals change of momentum, B's momentum has changed by J and A's momentum has changed by $-J$. This means that the total change of momentum of the system is zero.

This happens because the forces involved are internal to the system. The only way in which the momentum of a system can be changed is if an external force is applied. We can state this as the **principle of conservation of linear momentum**:

The total momentum of a system in a particular direction remains constant unless an external force is applied in that direction.

Another way of arriving at the principle of conservation of momentum is to realise that if no external force acts on a system, the centre of mass of the system will move with uniform velocity, even though individual parts of the system may be moving relative to the centre of mass.

We will analyse the motion in one dimension of a simple two-particle system to illustrate the point.

Suppose we have particles A and B whose masses are m_1 and m_2 respectively. The positions of the particles at some time t relative to an origin O are given by x_1 and x_2 respectively. The centre of mass of the system is at $\bar{x}$.

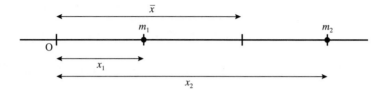

Taking moments about O, we have

$$(m_1 + m_2)\bar{x} = m_1 x_1 + m_2 x_2$$

Differentiating with respect to time, we get

$$(m_1 + m_2)\frac{d\bar{x}}{dt} = m_1\frac{dx_1}{dt} + m_2\frac{dx_2}{dt}$$

The right-hand side of this equation is the total momentum of the system.

If there are no external forces, the centre of mass moves with uniform velocity. Therefore, we have

$$\frac{d\bar{x}}{dt} \text{ is constant}$$

which implies that the total momentum is constant.

Example 7 A body A, of mass 5 kg, is travelling with velocity $6\,\mathrm{m\,s^{-1}}$. It catches up and collides with a body B, of mass 3 kg, which is travelling along the same line with velocity $4\,\mathrm{m\,s^{-1}}$. After collision, the bodies coalesce (merge into a single body). Find the velocity after collision.

SOLUTION

Let v be the velocity of the combined body after impact.

Momentum before collision $= 5 \times 6 + 3 \times 4$

$$= 42\,\mathrm{N\,s}$$

Momentum after collision $= 8v$

There are no external forces, so momentum is conserved. Therefore, we have

$$8v = 42 \quad \Rightarrow \quad v = 5.25\,\mathrm{m\,s^{-1}}$$

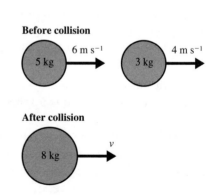

Example 8 Edwina is travelling downhill on a sledge. Edwina plus her sledge have a combined mass of 50 kg. Sarah, whose mass is 30 kg, is standing on the slope when the sledge runs into her at $10\,\text{m s}^{-1}$. She falls on top of Edwina and the sledge continues down the hill with both of them. Find the speed of the sledge immediately after the collision.

SOLUTION

Let the speed after collision be v.

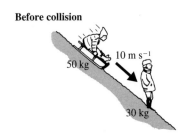

Before collision

$$\text{Momentum before collision} = 50 \times 10 + 30 \times 0$$
$$= 500\,\text{N s}$$

$$\text{Momentum after collision} = 80\,v$$

We assume that there is no friction and that the collision is of such short duration that any acceleration resulting from the component of the weight down the slope is negligible. There is, therefore, effectively no external force involved and so momentum is conserved. So, we have

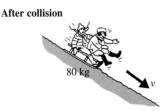

After collision

$$80\,v = 500 \quad \Rightarrow \quad v = 6.25\,\text{m s}^{-1}$$

Example 9 A rail truck, of mass 4 tonnes, is travelling along a straight, horizontal rail at $4\,\text{m s}^{-1}$. It meets another truck, of mass 2 tonnes, travelling in the opposite direction at $5\,\text{m s}^{-1}$. The trucks collide and become coupled together. Find their combined velocity after collision.

SOLUTION

Take left to right to be the positive direction.

Let the combined velocity after collision be v.

Before collision

$$\text{Momentum before} = 4000 \times 4 + 2000 \times (-5)$$
$$= 6000\,\text{N s}$$

After collision

$$\text{Momentum after} = 6000\,v$$

There are no external forces in the direction of motion and so momentum is conserved. Therefore, we have

$$6000\,v = 6000 \quad \Rightarrow \quad v = 1\,\text{m s}^{-1}$$

Example 10 A body of mass 4 kg travelling with velocity $(3\mathbf{i} + 2\mathbf{j})\,\text{m s}^{-1}$ collides and coalesces with a second body of mass 3 kg travelling with velocity $(\mathbf{i} - 3\mathbf{j})\,\text{m s}^{-1}$. Find their common velocity after impact.

SOLUTION

Let the common velocity after impact be $\mathbf{v}$.

Total momentum before collision $= 4(3\mathbf{i} + 2\mathbf{j}) + 3(\mathbf{i} - 3\mathbf{j})$

$$= (15\mathbf{i} - \mathbf{j})\,\mathrm{N\,s}$$

Total momentum after collision $= 7\mathbf{v}$

By the principle of conservation of momentum, we have

$$7\mathbf{v} = 15\mathbf{i} - \mathbf{j}$$

$$\Rightarrow \quad \mathbf{v} = (2\tfrac{1}{7}\mathbf{i} - \tfrac{1}{7}\mathbf{j})\,\mathrm{m\,s}^{-1}$$

Impulsive tension

When two objects are connected by a string and the motion of one or both of them causes the string to become taut, each object experiences a sudden jerk unless the string is elastic. Such a jerk is referred to as an **impulsive tension**.

We consider only those situations in which the string is light and any stretch is so small as to be negligible. Also, we confine ourselves to the case where all the motion takes place in the direction of the line joining the objects.

As the impulsive tension in the string is an internal force, there is no change in the total momentum of the system. Provided the string is not elastic, the objects both have the same velocity after the string has become taut.

Example 11 Two particles, A and B, lie at rest on a smooth horizontal table. They are connected by a light inextensible string which is initially slack. A has a mass of 3 kg and B 2 kg. B is set in motion with velocity $8\,\mathrm{m\,s}^{-1}$ in the direction AB. Find the common velocity of the particles immediately after the string goes taut and the impulsive tension in the string.

SOLUTION

Let the common velocity be v.

Total momentum before $= 2 \times 8 = 16\,\mathrm{N\,s}$

Total momentum after $= 2v + 3v = 5v$

There are only internal forces acting, so momentum is conserved. Therefore, we have

$$5v = 16 \quad \Rightarrow \quad v = 3.2\,\mathrm{m\,s}^{-1}$$

A's momentum before $= 0\,\mathrm{N\,s}$

A's momentum after $= 3 \times 3.2 = 9.6\,\mathrm{N\,s}$

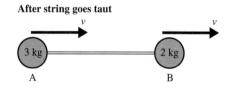

Before string goes taut

After string goes taut

So, A is subject to an impulse of 9.6 N s in the direction AB.

As the total momentum is unchanged, B's momentum is reduced by 9.6 N s. In other words, B is subject to an impulse of 9.6 N s in the direction BA.

The impulsive tension in the string is 9.6 N s.

Explosive forces

Another situation in which internal forces act upon parts of a system is when the parts are affected by an explosive force. For example, when a gun is fired, there is an explosion which exerts a forward force on the bullet and an equal backward force on the gun. If the gun is free to move, it will make a sudden backwards movement – the recoil. No external force is involved, so the momentum of the system is conserved. Usually, the gun is stationary before firing, and so the total momentum of the system is zero before and after the shot is fired.

Example 12 A bullet of mass 50 g is fired horizontally from a gun of mass 1 kg, which is free to move. The bullet is fired with a velocity of 250 m s^{-1}. Find the speed with which the gun recoils.

SOLUTION

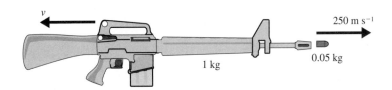

Both the gun and the bullet are stationary before the shot is fired, so the initial momentum is zero.

Take left to right as the positive direction, and let the recoil speed be v.

$$\text{Momentum after firing} = 0.05 \times 250 + 1\,(-v) = 12.5 - v$$

Momentum is conserved, so we have

$$12.5 - v = 0$$
$$\Rightarrow \quad v = 12.5\,\text{m s}^{-1}$$

Example 13 A gun of mass 800 kg fires a shell of mass 4 kg horizontally at $400\,\mathrm{m\,s^{-1}}$. The gun rests on a rough horizontal surface, and the coefficient of friction between the gun and the surface is 0.6. The gun is stationary before the shot is fired. Find the distance which the gun will move as a result of firing.

SOLUTION

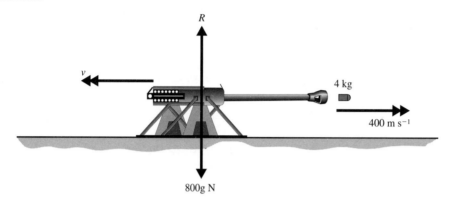

The gun is stationary before firing, so the total initial momentum is zero.

Take left to right to be positive, and the recoil speed to be v.

$$\text{Momentum after firing} = 4 \times 400 + 800\,(-v) = 1600 - 800v$$

If we can assume that the explosion is of sufficiently short duration, the force of friction is negligible compared with that of the explosion. We can, therefore, model the situation as being one in which there is effectively no external force and so momentum is conserved. So, we have

$$1600 - 800v = 0$$
$$\Rightarrow \quad v = 2\,\mathrm{m\,s^{-1}}$$

Resolving vertically, we get

$$R - 800g = 0 \quad \Rightarrow \quad R = 800g\,\mathrm{N}$$

The friction force, F, is given by $F = \mu R$. So, we have

$$F = 0.6R \quad \Rightarrow \quad F = 480g\,\mathrm{N}$$

Resolving horizontally, Newton's second law gives

$$-480g = 800a$$
$$\Rightarrow \quad a = -0.6g\,\mathrm{m\,s^{-2}}$$

The gun has an initial velocity of $2\,\mathrm{m\,s^{-1}}$, a final velocity of zero and an acceleration of $-0.6g\,\mathrm{m\,s^{-2}}$. Using $v^2 = u^2 + 2as$, we have

$$0 = 4 - 1.2gs$$
$$\Rightarrow \quad s = \frac{4}{1.2g} = 0.34\,\mathrm{m}$$

So, the gun moves 0.34 m.

Exercise 13B

1 A bullet of mass 40 grams is fired horizontally with velocity $600 \, \text{m s}^{-1}$ into a block of wood of mass 6 kg, which is resting on a smooth horizontal surface. The bullet becomes embedded in the block. Find the common speed of the bullet and block which results.

2 Two particles, A and B, have masses of 2 kg and 3 kg respectively. They are travelling at speeds of $5 \, \text{m s}^{-1}$ and $2 \, \text{m s}^{-1}$ respectively. They collide and coalesce. Find their common speed after the collision when

   a) they are travelling in the same direction
   b) they are travelling in opposite directions.

3 Arthur balances a box on top of a wall and throws a snowball of mass 0.3 kg at it. The snowball strikes the box at a speed of $10 \, \text{m s}^{-1}$ and sticks to it. Their common speed after impact is $4 \, \text{m s}^{-1}$. Find the mass of the box.

4 A railway truck of mass $3m$, travelling at a speed of $2v$, collides with another of mass $4m$, travelling at a speed of v. The trucks become coupled together. Find, in terms of v, the common speed of the trucks after impact when

   a) they are travelling in the same direction
   b) they are travelling in opposite directions.

5 A sledgehammer of mass 6 kg, travelling at $20 \, \text{m s}^{-1}$, strikes the top of a post of mass 2 kg and does not rebound.

   a) Find the common speed of the hammer and post immediately after impact.
   b) Find the impulse which the hammer gives to the post.

   If the post is driven 15 cm into the ground by the impact, find the average resistance of the ground to the motion of the post.

6 A bullet of mass 50 g is fired horizontally at a wooden block of mass 4 kg, which rests on a rough horizontal surface. The coefficient of friction between the block and the surface is 0.4. As a result of the collision, the block, with the bullet embedded, moves a distance of 10 m along the surface before coming to rest. Find the speed at which the bullet enters the block.

7 A bullet of mass m, travelling at a speed v, strikes a block of mass M horizontally. Prior to impact the block rests on a rough horizontal surface. The coefficient of friction between the block and the surface is μ. The bullet becomes embedded in the block. Find an expression for the distance travelled by the block and the embedded bullet before coming to rest.

8 A block of wood of mass 2 kg rests on top of a vertical wall 3 m high, which stands on horizontal ground. A bullet of mass 50 g strikes the block at a speed of $100 \, \text{m s}^{-1}$ and becomes embedded. How far from the foot of the wall will the block land when

   a) the bullet strikes the block horizontally
   b) the bullet is travelling at an inclination of $60°$ above the horizontal when it strikes the block.

9 Masses of 3 kg and 5 kg are connected by an elastic rope and are held apart on a smooth horizontal surface with the rope stretched. The masses are released from rest and a short time later the smaller mass has a speed of $6 \, \text{m s}^{-1}$. Find the speed of the larger mass at that time.

10 A gun of mass 500 kg, which is free to move, fires a shell of mass 5 kg horizontally at a speed of $200 \, \text{m s}^{-1}$. Find the speed of recoil of the gun.

11 A gun of mass 400 kg fires a shell of mass 8 kg horizontally at a speed of $300 \, \text{m s}^{-1}$. Find the restraining force, assumed constant, which will be needed to bring the gun to rest in a distance of 2 m.

12 An object of mass 2 kg and velocity $(2\mathbf{i} - \mathbf{j}) \, \text{m s}^{-1}$ strikes and coalesces with a second object of mass 3 kg and velocity $(4\mathbf{i} + 6\mathbf{j}) \, \text{m s}^{-1}$. Find their common velocity after impact.

13 An object of mass 4 kg, travelling with velocity $(5\mathbf{i} + 2\mathbf{j}) \, \text{m s}^{-1}$, is struck by a second object of mass 6 kg and velocity $\mathbf{v}$, which sticks to it. Their common velocity afterwards is $(2\mathbf{i} - 4\mathbf{j}) \, \text{m s}^{-1}$. Find $\mathbf{v}$.

14 Particles A and B lie at rest on a smooth horizontal plane and are connected by a light inextensible string which is initially slack. A has mass 4 kg, B 2 kg. A is struck with an impulse of 20 N s so that it moves directly away from B. Find the common speed of the particles after the string has jerked taut.

15 Particles A, mass 2 kg, and B, mass 3 kg, lie on a smooth horizontal plane. They are connected by a light inextensible string which is initially slack. The particles are set in motion directly away from each other, A with speed $4 \, \text{m s}^{-1}$ and B with speed $6 \, \text{m s}^{-1}$. Find the common velocity of the particles after the string goes taut.

16 Particles A, mass 1 kg, and B, mass 2 kg, are connected by a light inextensible string of length 2 m. A rests at the edge of a smooth horizontal table and B rests on the surface of the table so that AB = 1 m and AB is perpendicular to the edge of the table. A is then nudged over the edge so that it falls from rest. With what speed will B be jerked into motion?

17 Particles A, mass 2 kg, and B, mass 1 kg, are connected by a light inextensible string of length 1 m. B is also connected by a similar but longer string to a third particle C, mass 2 kg. The string connecting B and C passes over a smooth pulley. The system is released from rest with A and B on the ground and the string connecting B and C taut. Find the speed with which A leaves the ground and the height to which it rises. (You may assume that C starts sufficiently far up never to reach the ground during the motion.) Investigate and describe the subsequent motion of the system.

Elastic impact

When two objects collide and rebound, there are two unknown quantities – the post-collision velocities of the objects. We therefore need a second equation in order to find these unknowns.

This second equation is provided by an experimental result, credited to Newton and usually called **Newton's law of restitution**. It is important to stress that this 'law' is just a model which appears to conform reasonably well to the results of observation.

Newton observed that when two objects collide, the speed with which they separate after the collision is normally less than the speed with which they approached, and that the ratio between these speeds remains constant for a given pair of objects. This is true when both objects are free to move or when one of them is fixed.

Newton's law of restitution

When two objects collide directly,

$$\frac{\text{Separation speed}}{\text{Approach speed}} = e \quad \text{(a constant)}$$

The constant e is called the **coefficient of restitution** for the objects and takes a value between 0 and 1.

- $e = 0$ corresponds to totally inelastic impact. No rebounding occurs.

- $e = 1$ corresponds to perfectly elastic impact. An object dropped in a vacuum onto a surface giving a perfectly elastic impact would bounce back to the same height, and continue to do so for ever. In reality, perfect elasticity does not occur, so it is more reasonable to say $0 \leqslant e < 1$.

Notice that the law is couched in terms of speed. The approach/separation speed is the rate at which the gap between the objects is diminishing/increasing.

We will be concerned first with **direct impact**, and will investigate **oblique impact** on pages 295–299.

You might like to explore the effects of various masses, speeds and coefficients of restitution on the outcome of a collision, using the spreadsheet MOMENTUM, available on the Oxford University Press website: http://www.oup.co.uk/mechanics.

Example 14 An object of mass $4\,\text{kg}$, travelling at $8\,\text{m s}^{-1}$, strikes a fixed wall at right angles and rebounds. The coefficient of restitution between the object and the wall is 0.3. Find the velocity of the object after impact and the impulse exerted on the wall.

SOLUTION

Let the separation speed be v. Then we have

Before impact

$$\frac{v}{8} = 0.3 \quad \Rightarrow \quad v = 2.4\,\text{m s}^{-1}$$

So, if we take left to right in the diagram as the positive direction, the velocity after impact is $-2.4\,\text{m s}^{-1}$. Therefore, we have

After impact

Momentum of object before impact $= 4 \times 8 = 32\,\text{N s}$

Momentum of object after impact $= 4 \times (-2.4) = -9.6\,\text{N s}$

Change in momentum of object $= -9.6 - 32 = -41.6\,\text{N s}$

The object is subjected to an impulse of $-41.6\,\text{N s}$, and the wall must therefore suffer a corresponding impulse of $41.6\,\text{N s}$.

Example 15 A particle of mass $3\,\text{kg}$, moving on a smooth horizontal plane at $6\,\text{m s}^{-1}$, collides with a stationary particle of mass $5\,\text{kg}$. The coefficient of restitution between the particles is 0.4. Find the velocities of the particles after impact.

SOLUTION

As both particles are free to move, momentum is conserved.

Before collision

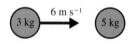

Take the velocities of the particles after impact to be u and v, as shown in the diagram. Therefore, we have

After collision

Momentum before impact $= 3 \times 6 = 18\,\text{N s}$

Momentum after impact $= 3u + 5v$

$\Rightarrow \quad 3u + 5v = 18$ [1]

Approach speed $= 6\,\text{m s}^{-1}$

Separation speed $= v - u$

By Newton's law of restitution, we have

$$\frac{v - u}{6} = 0.4$$

$\Rightarrow \quad v - u = 2.4$ [2]

Solving [1] and [2], we get

$$u = 0.75\,\text{m s}^{-1} \quad \text{and} \quad v = 3.15\,\text{m s}^{-1}.$$

Example 16 A particle A of mass $3\,\text{kg}$, travelling at $6\,\text{m s}^{-1}$, overtakes and collides with a particle B of mass $2\,\text{kg}$, travelling at $2\,\text{m s}^{-1}$. The coefficient of restitution between the particles is 0.8. Find the velocities of the particles immediately after impact.

SOLUTION

Take the velocities of the particles after impact to be u and v.

By the principle of conservation of momentum, we have

$$3u + 2v = 3 \times 6 + 2 \times 2$$

$$\Rightarrow \quad 3u + 2v = 22 \qquad\qquad [1]$$

Approach speed $= 4\,\text{m s}^{-1}$

Separation speed $= v - u$

Before collision

After collision

By Newton's law of restitution, we have

$$\frac{v - u}{4} = 0.8$$

$$\Rightarrow \quad v - u = 3.2 \qquad\qquad [2]$$

Solving [1] and [2], we get

$$u = 3.12\,\text{m s}^{-1} \quad \text{and} \quad v = 6.32\,\text{m s}^{-1}$$

Example 17 Two identical particles collide head-on with speeds $5\,\text{m s}^{-1}$ and $3\,\text{m s}^{-1}$. The coefficient of restitution between the particles is 0.5. Find the speeds of the particles immediately after the impact.

SOLUTION

Let the mass of each particle be m, and their speeds after impact u and v.

By the principle of conservation of momentum, we have

$$mu + mv = 5m - 3m$$

$$\Rightarrow \quad u + v = 2 \qquad\qquad [1]$$

Approach speed $= 8$

Separation speed $= v - u$

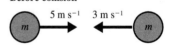

Before collision

After collision

By Newton's law of restitution, we have

$$\frac{v - u}{8} = 0.5$$

$$\Rightarrow \quad v - u = 4 \qquad\qquad [2]$$

Solving [1] and [2], we get

$$u = -1\,\text{m s}^{-1} \quad \text{and} \quad v = 3\,\text{m s}^{-1}$$

So, the particles move in opposite directions with speeds of $1\,\text{m s}^{-1}$ and $3\,\text{m s}^{-1}$.

Some situations involve more than one collision. The principles involved in each collision are the same, but care must be taken to **use clear notation**.

Example 18 Particles A, B and C lie at rest in a straight line in the order stated. Their masses are 3 kg, 2 kg and 4 kg respectively. A is projected towards B with velocity $8\,\mathrm{m\,s^{-1}}$. The coefficient of restitution in each impact is 0.6. Show that there will be a third collision. Find the velocities of the particles after this third collision and hence decide if there will be any further collisions.

SOLUTION

Let u, v and w represent the velocities after impact of A, B and C respectively, with subscripts to show which impact they result from.

First impact
By the principle of conservation of momentum, we have

$$3u_1 + 2v_1 = 24 \qquad [1]$$

By Newton's law of restitution, we have

$$v_1 - u_1 = 4.8 \qquad [2]$$

Solving [1] and [2], we get

$$u_1 = 2.88\,\mathrm{m\,s^{-1}} \quad \text{and} \quad v_1 = 7.68\,\mathrm{m\,s^{-1}}$$

Before first impact

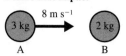

After first impact

Second impact
By the principle of conservation of momentum, we have

$$2v_2 + 4w_2 = 15.36 \qquad [3]$$

By Newton's law of restitution, we have

$$w_2 - v_2 = 4.608 \qquad [4]$$

Solving [3] and [4], we get

$$v_2 = -0.512\,\mathrm{m\,s^{-1}} \quad \text{and} \quad w_2 = 4.096\,\mathrm{m\,s^{-1}}$$

As A has velocity $2.88\,\mathrm{m\,s^{-1}}$ and B has velocity $-0.512\,\mathrm{m\,s^{-1}}$, there is at least one more collision.

Before second impact

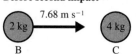

After second impact

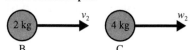

Third impact
By the principle of conservation of momentum, we have

$$3u_3 + 2v_3 = 7.616 \qquad [5]$$

By Newton's law of restitution, we have

$$v_3 - u_3 = 3.392 \qquad [6]$$

Solving [5] and [6], we get

$$u_3 = 0.1664\,\mathrm{m\,s^{-1}} \quad \text{and} \quad v_3 = 3.5584\,\mathrm{m\,s^{-1}}$$

As $w_2 = 4.096\,\mathrm{m\,s^{-1}}$, B will not catch up with C, so there will be no further collisions.

Before third impact

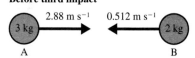

After third impact

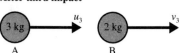

Loss of kinetic energy

Although momentum is conserved in an impact, such sudden changes of velocity lead to a loss of kinetic energy. In Example 18, for instance, we have

$$\text{Initial KE} = \tfrac{1}{2} \times 3 \times 8^2 = 96\,\text{J}$$

$$\text{Total energy after first impact} = \tfrac{1}{2} \times 3 \times 2.88^2 + \tfrac{1}{2} \times 2 \times 7.68^2$$

$$= 71.424\,\text{J}$$

So, the first impact resulted in a loss of kinetic energy of 24.576 J.

This occurs because, in impulse situations, some of the kinetic energy is converted into other forms of energy, such as heat and sound.

Example 19 Particle A, of mass 6 kg and moving with velocity $(3\mathbf{i} + 5\mathbf{j})\,\text{m s}^{-1}$, collides with particle B, of mass 3 kg and moving with velocity $(\mathbf{i} - 2\mathbf{j})\,\text{m s}^{-1}$. Immediately after the impact, A has velocity $(\mathbf{i} + \mathbf{j})\,\text{m s}^{-1}$. Find how much kinetic energy is lost in the impact.

SOLUTION

Let the velocity of B after the impact be $\mathbf{v}$.

$$\text{Momentum before collision} = 6(3\mathbf{i} + 5\mathbf{j}) + 3(\mathbf{i} - 2\mathbf{j})$$

$$= (21\mathbf{i} + 24\mathbf{j})\,\text{N s}$$

$$\text{Momentum after collision} = 6(\mathbf{i} + \mathbf{j}) + 3\mathbf{v}$$

By the principle of conservation of momentum, we have

$$3\mathbf{v} + 6\mathbf{i} + 6\mathbf{j} = 21\mathbf{i} + 24\mathbf{j}$$

$$\Rightarrow \quad \mathbf{v} = (5\mathbf{i} + 6\mathbf{j})\,\text{m s}^{-1}$$

$$\text{Speed of A before impact} = \sqrt{(3^2 + 5^2)} = \sqrt{34}\,\text{m s}^{-1}$$

$$\text{Speed of B before impact} = \sqrt{(1^2 + (-2)^2)} = \sqrt{5}\,\text{m s}^{-1}$$

$$\text{Speed of A after impact} = \sqrt{(1^2 + 1^2)} = \sqrt{2}\,\text{m s}^{-1}$$

$$\text{Speed of B after impact} = \sqrt{(5^2 + 6^2)} = \sqrt{61}\,\text{m s}^{-1}$$

Therefore, we have

$$\text{Total KE before impact} = \tfrac{1}{2} \times 6 \times 34 + \tfrac{1}{2} \times 3 \times 5 = 109.5\,\text{J}$$

$$\text{Total KE after impact} = \tfrac{1}{2} \times 6 \times 2 + \tfrac{1}{2} \times 3 \times 61 = 97.5\,\text{J}$$

So, there has been a loss of KE of 12 J.

Exercise 13C

1 The following table shows information about the masses, velocities and coefficient of restitution of two particles A and B. In each case, calculate the missing quantities.

	A			B			
	Mass (kg)	Initial velocity (m s^{-1})	Final velocity (m s^{-1})	Mass (kg)	Initial velocity (m s^{-1})	Final velocity (m s^{-1})	e
a)	4	8		3	0		0.6
b)	2	6		3	2		0.4
c)	5	4		2	-2		0.8
d)	2	5		5	-4		0.2
e)	m	3		$2m$	1		0.5
f)	3	10	4	5	2		
g)			2	3	6	8	0.4
h)	6	8	0	4			0.3

2 Particles A and B lie at rest on a smooth horizontal plane near a vertical wall so that AB is perpendicular to the wall. The particles have masses 4 kg and 2 kg respectively. The coefficient of restitution between the particles is 0.4 and between the particle B and the wall is 0.6. Find the final velocities of the particles when

a) B is projected directly at the wall with a speed of 8 m s^{-1}
b) A is projected directly at B with a speed of 8 m s^{-1}.

In each case, find the total loss of kinetic energy resulting from the collisions.

3 Particles A and B have masses of 3 kg and 2 kg respectively, and lie at rest on a smooth horizontal plane. The particles are connected together by a light inextensible string, which is initially slack. The coefficient of restitution between the particles is 0.5. A is projected towards B with a velocity of 10 m s^{-1}. Find the common velocity of the particles after the string becomes taut. How much kinetic energy is lost in the process?

4 Particle B has a mass twice that of particle A. The coefficient of restitution between them is 0.5. They lie at rest on a smooth horizontal plane. Show that if A is projected towards B, A will be reduced to rest by the collision and that the kinetic energy is halved by the impact.

5 Particle A, moving at 7 m s^{-1}, overtakes and collides with particle B, which is moving in the same direction at 1 m s^{-1}. The coefficient of restitution is $\frac{3}{4}$. Particle A is reduced to rest by the impact. Find the ratio of the masses of the particles.

6 Particles A, B and C, each of mass m, lie at rest in a straight line in the order stated. A is projected directly towards B with velocity u. The coefficient of restitution is 0.5 in each impact which follows. Show that there will be three impacts and find the final velocities of the particles.

7 A particle A, of mass 5 kg, lies at rest on a smooth horizontal plane at the foot of a smooth plane inclined at 30° to the horizontal. A second particle B, of mass 3 kg, lies on the horizontal plane and is projected directly at A with velocity $8\,\mathrm{m\,s^{-1}}$, so that after the impact A travels up the line of greatest slope of the inclined plane. The coefficient of restitution between the particles is 0.5. Find the time which elapses from the first collision until A overtakes B and collides for a second time, and find the final velocities of the particles.

8 Particles A, B and C, having masses $3m$, m and $2m$ respectively, lie at rest on a smooth horizontal plane in the stated order. The coefficient of restitution between the particles is 0.5. A is projected towards B with a velocity u. Show that there will be four collisions and find the final velocities of the particles.

Oblique impact

When a collision is not in a straight line, the problem is a two-dimensional one. Our choice of unit direction vectors affects the simplicity of our equations. We assume that there is no friction and so the impulse is perpendicular to the plane of contact. It thus makes sense to choose the base directions (unit vectors) parallel and perpendicular to the plane of contact.

There are three basic problems to solve:

- A particle hits a wall (the wall cannot move).
- A particle strikes a second, stationary, particle.
- Two moving particles collide.

Example 20 A particle, of mass 1 kg, slides on the surface of a smooth table with a speed $2\,\mathrm{m\,s^{-1}}$. It collides with a smooth vertical wall, striking the wall at an angle of 45° to its direction of motion. It then rebounds. The coefficient of restitution between the particle and the vertical surface is 0.6. Find

a) the velocity of the particle after impact
b) the kinetic energy lost during the impact.

SOLUTION

a) Let the velocity of the particle after impact be $v\,\mathrm{m\,s^{-1}}$ in the direction making an angle ϕ with the wall, as shown in the diagram.

We choose unit vectors parallel and perpendicular to the plane of contact, as shown.

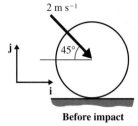

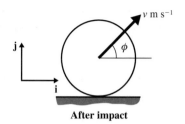

Before impact

After impact

Velocity before impact: $\mathbf{u} = 2\cos 45°\,\mathbf{i} - 2\sin 45°\,\mathbf{j}$ $\Rightarrow$ $\mathbf{u} = \sqrt{2}\,\mathbf{i} - \sqrt{2}\,\mathbf{j}$

Velocity after impact: $\mathbf{v} = v\cos\phi\,\mathbf{i} + v\sin\phi\,\mathbf{j}$

The wall is smooth and so the linear momentum parallel to it (the **i**-direction) will be conserved. So, we have

$$\sqrt{2} = v\cos\phi \qquad\qquad [1]$$

In the **j**-direction, Newton's law of restitution applies. Hence, ww have

Separation speed $= e \times$ Approach speed

$$v\sin\phi = e \times \sqrt{2} = 0.6 \times \sqrt{2} \qquad [2]$$

Dividing [2] by [1], we get

$$\tan\phi = 0.6 \quad\Rightarrow\quad \phi = 30.96°$$

Squaring and adding [1] and [2], we get

$$v^2 = 2 + 2 \times 0.36 = 2.72 \quad\Rightarrow\quad v = 1.649$$

Hence, after impact, the sphere has a speed of $1.65\,\text{m s}^{-1}$, making an angle of $31.0°$ with the wall (both results correct to 3 sf).

b) Before impact: $\text{KE} = \frac{1}{2} \times 1 \times 2^2 = 2\,\text{J}$

After impact: $\text{KE} = \frac{1}{2} \times 1 \times 2.72 = 1.36\,\text{J}$

So, the kinetic energy lost during impact is $(2 - 1.36) = 0.64\,\text{J}$.

Example 21   A small sphere, A, of mass 1 kg, moving with a speed of $4\,\text{m s}^{-1}$, collides with a second, identical, sphere, B, which is initially at rest. At the point of contact, the line joining their centres makes an angle of $60°$ with the direction of motion of sphere A. If the coefficient of restitution between the spheres is 0.5, what are the velocities of the two spheres after the collision?

SOLUTION

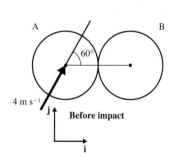

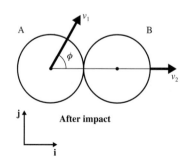

Before impact

After impact

The impact between the spheres is along the line of centres, and so we chose this as one of the base directions. The second base direction is perpendicular to this, as shown.

The second sphere, B, must move in the direction of the impulse acting on it, i.e. along the line of centres.

Before impact, we have

Velocity of A: $\mathbf{u}_1 = 4\cos 60°\,\mathbf{i} + 4\sin 60°\,\mathbf{j} = 2 + 2\sqrt{3}\,\mathbf{j}$
Velocity of B: $\mathbf{u}_2 = 0$

After impact, we have

Velocity of A: $\mathbf{v}_1 = v_1\cos\phi\,\mathbf{i} + v_1\sin\phi\,\mathbf{j}$
Velocity of B: $\mathbf{v}_2 = v_2\,\mathbf{i}$

Momentum is conserved in the **i**-direction. There is no impulse in the **j**-direction and the velocity components in this direction remain unchanged. Hence, we have

In **i**-direction: $2 = v_1\cos\phi + v_2$ [1]

In **j**-direction: $2\sqrt{3} = v_1\sin\phi$ [2]

From Newton's law of restitution we get

$$v_2 - v_1\cos\phi = e \times 2 = 1$$ [3]

We now have the three independent equations needed to find v_1, v_2 and ϕ.

From [1] and [3], we have

$$2v_2 = 3 \quad \Rightarrow \quad v_2 = 1.5$$

Substituting into [1], we obtain

$$v_1\cos\phi = 2 - v_2 = 0.5$$ [4]

Dividing [2] by [4], we get

$$\tan\phi = 4\sqrt{3} \quad \Rightarrow \quad \phi = 81.787°$$

Squaring and adding [2] and [4], we have

$$v_1^2 = (2\sqrt{3})^2 + (0.5)^2 = 12.25$$
$$\Rightarrow \quad v_1 = 3.5$$

Hence, after the collision, sphere A moves with a speed of $3.5\,\text{m}\,\text{s}^{-1}$ in a direction making an angle of $81.8°$ with the line of centres, and sphere B moves along the line of centres with a speed of $1.5\,\text{m}\,\text{s}^{-1}$.

Example 22   Two identical small spheres of mass 2 kg collide. The first, A, moves with a speed of $6\,\text{m s}^{-1}$ and the second, B, with a speed of $8\,\text{m s}^{-1}$. Their directions of motion make angles of 30° and 45° respectively with their line of centres at the instant of contact, as shown. If the coefficient of restitution between the spheres is 0.4, find

a) the velocities of the spheres after the collision
b) the kinetic energy lost during the collision.

SOLUTION

a)

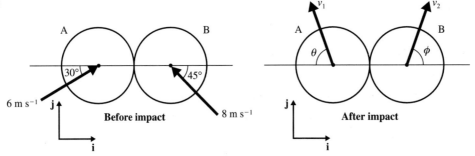

We will assume that the speeds and directions of the spheres after the collision are as shown in the figure. The base directions are as shown.

Before the collision, we have

$$\text{Velocity of A:}\quad \mathbf{u}_1 = 6\cos 30°\,\mathbf{i} + 6\sin 30°\,\mathbf{j} = 3\sqrt{3}\,\mathbf{i} + 3\mathbf{j}$$
$$\text{Velocity of B:}\quad \mathbf{u}_2 = -8\cos 45°\,\mathbf{i} + 8\sin 45°\,\mathbf{j} = -4\sqrt{2}\,\mathbf{i} + 4\sqrt{2}\,\mathbf{j}$$

After the collision, we have

$$\text{Velocity of A:}\quad \mathbf{v}_1 = -v_1\cos\theta\,\mathbf{i} + v_1\sin\theta\,\mathbf{j}$$
$$\text{Velocity of B:}\quad \mathbf{v}_2 = v_2\cos\phi\,\mathbf{i} + v_2\sin\phi\,\mathbf{j}$$

There is no impulse on either sphere in the **j**-direction, so the velocity components in this direction remain unchanged. Hence, we have

$$\text{For A:}\qquad 3 = v_1\sin\theta \qquad [1]$$
$$\text{For B:}\quad 4\sqrt{2} = v_2\sin\phi \qquad [2]$$

In the **i**-direction, linear momentum is conserved, so we have

$$2 \times 3\sqrt{3} + 2 \times (-4\sqrt{2}) = 2 \times (-v_1\cos\theta) + 2v_2\cos\phi$$
$$\Rightarrow\quad 3\sqrt{3} - 4\sqrt{2} = -v_1\cos\theta + v_2\cos\phi \qquad [3]$$

In the **i**-direction, Newton's law of restitution applies, so we have

$$v_2\cos\phi - (-v_1\cos\theta) = 0.4(3\sqrt{3} - (-4\sqrt{2}))$$
$$\Rightarrow\quad v_2\cos\phi + v_1\cos\theta = 1.2\sqrt{3} + 1.6\sqrt{2} \qquad [4]$$

We now have the four independent equations needed to find the unknowns v_1, v_2, θ and ϕ.

Adding [3] and [4], we obtain

$$2v_2 \cos \phi = 4.2\sqrt{3} - 2.4\sqrt{2}$$

$$\Rightarrow \quad v_2 \cos \phi = 2.1\sqrt{3} - 1.2\sqrt{2} \qquad [5]$$

Subtracting [3] from [4], we obtain

$$2v_1 \cos \theta = -1.8\sqrt{3} + 5.6\sqrt{2}$$

$$\Rightarrow \quad v_1 \cos \theta = -0.9\sqrt{3} + 2.8\sqrt{2} \qquad [6]$$

Dividing [1] by [6], we obtain

$$\tan \theta = \frac{3}{-0.9\sqrt{3} + 2.8\sqrt{2}} \quad \Rightarrow \quad \theta = 51.329°$$

Squaring and adding [1] and [6], we have

$$v_1{}^2 = 3^2 + (-0.9\sqrt{3} + 2.8\sqrt{2})^2$$

$$\Rightarrow \quad v_1 = 3.8425$$

Dividing [2] by [5], we obtain

$$\tan \phi = \frac{4\sqrt{2}}{2.1\sqrt{3} - 1.2\sqrt{2}} \quad \Rightarrow \quad \phi = 71.068°$$

Squaring and adding [2] and [5], we obtain

$$v_2{}^2 = (4\sqrt{2})^2 + (2.1\sqrt{3} - 1.2\sqrt{2})^2$$

$$\Rightarrow \quad v_2 = 5.9803$$

Hence, after the collision, A has a speed of $3.84 \, \mathrm{m\,s^{-1}}$ and moves along a line making an angle of $51.3°$ with the line of centres. B has a speed of $5.98 \, \mathrm{m\,s^{-1}}$ and moves along a line making an angle of $71.1°$ with the line of centres. The angles are measured as shown in the diagram.

b) Before the collision, we have

$$\text{Total KE} = \tfrac{1}{2} \times 2 \times 6^2 + \tfrac{1}{2} \times 2 \times 8^2 = 100 \, \mathrm{J}$$

After the collision, we have

$$\text{Total KE} = \tfrac{1}{2} \times 2 \times 14.76 + \tfrac{1}{2} \times 2 \times 35.76 = 50.52 \, \mathrm{J}$$

Thus the KE lost during the collision is $(100 - 50.52) = 49.5 \, \mathrm{J}$ (to 3 sf).

Exercise 13D

1 A snooker ball, moving with a speed of $12 \, \mathrm{m\,s^{-1}}$, collides with the cushion at an angle of $25°$. If the coefficient of restitution between the ball and the cushion is 0.7, what is the velocity of the ball after rebounding?

2 A snooker ball, of mass $0.2 \, \mathrm{kg}$, hits the cushion with a speed of $10 \, \mathrm{m\,s^{-1}}$ so that its direction of motion makes an angle of $60°$ with the cushion. The coefficient of restitution is 0.8. How much kinetic energy is lost during the collision?

3 A ball of mass $0.5\,\text{kg}$ falls from a height of $2\,\text{m}$ onto a surface inclined at an angle $30°$ to the horizontal. If the coefficient of restitution between the ball and the surface is 0.8, what is the velocity of the ball when it rebounds?

4 A ball, A, is projected along a smooth surface towards a second, identical ball, B, which is at rest. The speed of A is $8\,\text{m s}^{-1}$ and it moves in a direction making an angle of $30°$ with the line of centres. After the collision, B has a speed of $4\,\text{m s}^{-1}$. Find

a) value of the coefficient of restitution
b) the velocity of A after the impact.

5 Two smooth spheres of equal radii collide. The mass of A is $3\,\text{kg}$ and that of B is $2\,\text{kg}$. Sphere A has a velocity of $2\mathbf{i} + 3\mathbf{j}$ whilst B has a velocity of $-3\mathbf{i} - 4\mathbf{j}$, where the $\mathbf{i}$-direction is along the line of centres from A to B at the moment of impact. The coefficient of restitution between the two spheres is 0.8. Find

a) the velocities of the spheres after the impact
b) the kinetic energy lost during the collision.

6 An object slides along smooth, horizontal ground towards a wall. It hits the wall and rebounds so that it is now moving perpendicular to its original direction of motion. If it were originally moving at an angle of $60°$ to the wall, find the value of the coefficient of restitution between the object and the wall.

7 A ball, A, is at the origin. A second, identical ball, B, is at rest at the point $3\mathbf{i} + 4\mathbf{j}$. The ball A is projected towards B so that they collide. After the collision, B moves in the $\mathbf{i}$-direction whilst A moves in the direction of $3\mathbf{i} + 16\mathbf{j}$. Show that this can happen whatever the initial speed of A and find the coefficient of restitution between the balls.

8 Two identical balls are moving towards each other, each with a speed of $4\,\text{m s}^{-1}$ but the lines along which they are moving are parallel and separated by the radius of the balls. If the coefficient of restitution between the balls is $\frac{3}{4}$, find the velocity of each ball immediately after impact.

14 Jointed rods and frameworks

Bridges were made for wise men to walk over and fools to ride over.
ENGLISH PROVERB

Many load-bearing structures are made of straight beams and rods. The components of such structures are subject to internal forces, resulting from their own weights and from the applied loads, which tend to stretch, compress, bend and twist them. When a structure is being designed, the analysis of such internal forces is clearly of enormous importance in terms of the safety and the suitability of the structure.

In this chapter, we will confine ourselves to structures in two dimensions and we will make the following simplifying assumptions.

- Rods will be either **light** (of negligible weight compared with the loads applied to the structure) or **uniform**.

- Rods will not stretch, compress or bend under the action of internal forces.

- Rods will be **smoothly jointed**. That is, they will be connected by frictionless hinges of negligible mass.

You will be aware that these are quite sweeping assumptions. Real structures are usually three dimensional and do deform when a load is applied. The components are connected together in a variety of ways, mostly quite different from our assumption of smooth jointing. It does, however, enable us to look at some of the principles used in calculating internal forces.

Jointed rods

The simplest situation is where two rods are smoothly jointed together and no external force acts on the joint. When the system is in equilibrium, the reaction force exerted by the first rod on the second is equal and opposite to that exerted by the second on the first (Newton's third law). We do not, in general, know the direction of this reaction, so it is convenient to represent it by its horizontal and vertical components, as shown below.

 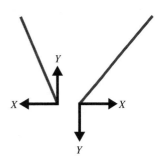

When tackling problems, we can obtain information about external forces by considering the equilibrium of the structure as a whole. When we do this, the internal forces balance out and do not appear in our equations. To get at the internal forces, we consider the equilibrium of some part of the structure. Some of the internal forces are then effectively external to this part of the structure and so appear in the equations.

Example 1 Rods AB and BC, each of length 2 m and mass 8 kg, are smoothly jointed at B and rest with A and C on smooth horizontal ground, 2 m apart. The system is kept in equilibrium by a light, inextensible string of length 2 m connecting A and C.

a) Find the tension in the string.
b) Find the reaction in the joint at B.

SOLUTION

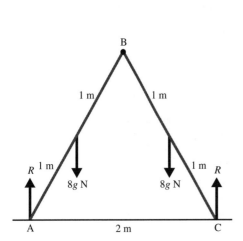
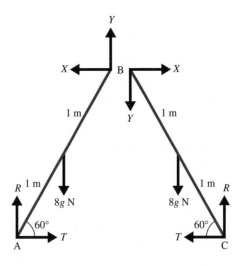

The left-hand diagram shows the external forces acting on the system. The right-hand diagram shows the forces acting on the two individual rods.

The first thing to note is that, because the situation illustrated in the left-hand diagram is symmetrical, the forces in the right-hand diagram should also be symmetrical. This means that the component Y must be zero. In addition, the vertical reactions at A and C are equal, as shown.

a) First, consider the equilibrium of the whole system.

Resolving vertically, we have

$$2R - 16g = 0$$
$$\Rightarrow \quad R = 8g \, \text{N} \qquad\qquad [1]$$

Now consider the equilibrium of the rod AB.

Taking moments about B, we get

$$8g \cos 60° + 2T \sin 60° - 2R \cos 60° = 0$$

Substituting from [1], we obtain

$$2T\sin 60° = 8g\cos 60°$$

$$\Rightarrow \quad T = \frac{4g}{\sqrt{3}}\,\text{N} \quad \text{or} \quad \frac{4g\sqrt{3}}{3}\,\text{N} \qquad [2]$$

b) Resolving horizontally, we have

$$T - X = 0$$

$$\Rightarrow \quad X = \frac{4g}{\sqrt{3}}\,\text{N} \quad \text{or} \quad \frac{4g\sqrt{3}}{3}\,\text{N} \quad \text{(from [2])}$$

Example 2 A rod AB, of length $3a$ and weight W, is smoothly jointed at B to a rod BC, of length $4a$ and weight $2W$. The rods rest with A and C on a rough horizontal surface, $5a$ apart. The coefficient of friction is μ at each point of contact and one of the rods is about to slip.

a) Find the value of μ.

b) Find the reaction in the joint at B.

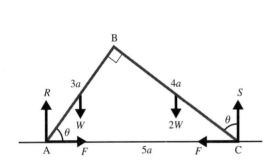 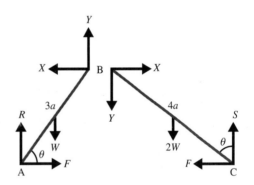

SOLUTION

As the only external horizontal forces acting on the system are those of friction at the two points of contact, these forces must be equal, as shown. Notice that, unlike Example 1, this system is not symmetrical, so the reactions at A and C are **not** equal.

The triangle ABC is a 3–4–5 triangle and so $\sin\theta = \frac{4}{5}$, $\cos\theta = \frac{3}{5}$.

a) Taking moments about A for the whole system, we have

$$S \times 5a - W \times 1.5a\cos\theta - 2W \times (5a - 2a\sin\theta) = 0$$

$$\Rightarrow \quad S = 1.54\,W$$

Resolving vertically for the whole system, we get

$$R + S - 3W = 0$$

$$\Rightarrow \quad R = 1.46W$$

The friction force, F, and the coefficient of friction, μ, are the same at both contact points, and the system is about to slip. As $R < S$, the slipping will take place at A, so $F = 1.46\mu W$.

Taking moments about B for the rod AB, we obtain

$$W \times 1.5a\cos\theta + 1.46\mu W \times 3a\sin\theta - 1.46W \times 3a\cos\theta = 0$$

$$\Rightarrow \quad \mu = 0.493$$

b) Resolving vertically for rod AB, we have

$$1.46W + Y - W = 0$$

$$\Rightarrow \quad Y = -0.46W$$

The fact that Y has proved to be negative shows that the reaction at B exerts a downward force of $0.46W$ on the rod AB and an equivalent upward force on the rod BC.

Resolving horizontally for the rod AB, we obtain

$$1.46\mu W - X = 0$$

$$\Rightarrow \quad X = 1.46\mu W = 0.72W$$

The reaction at B therefore has magnitude

$$W\sqrt{0.72^2 + 0.46^2} = 0.854W$$

and makes an angle ϕ with the horizontal, where

$$\tan\phi = \frac{0.46}{0.72} \quad \text{giving} \quad \phi = 35.6°$$

Example 3 Four equal rods, each of weight W and length $2a$, are smoothly jointed to form a rhombus ABCD. This hangs in a vertical plane suspended by a string attached to A, and is held in the shape of a square by a light rigid rod joining the mid-points of AB and AD. Find the thrust in the rod and the horizontal and vertical reactions in the joint B.

SOLUTION

The tension in the suspending string at A must be $4W$, as shown on page 305.

Because the external forces shown in the left-hand diagram are symmetrical, we can deduce the following.

- The vertical component of the reactions in joints A and C are zero.
- By resolving vertically for rod BC, it follows that the vertical component of the reactions in joints B and D are W, as shown.

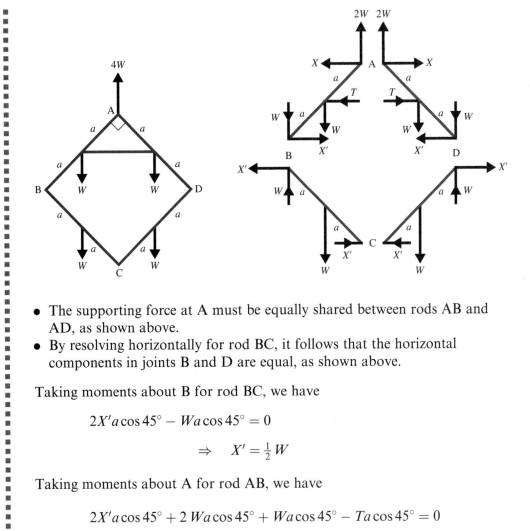

- The supporting force at A must be equally shared between rods AB and AD, as shown above.
- By resolving horizontally for rod BC, it follows that the horizontal components in joints B and D are equal, as shown above.

Taking moments about B for rod BC, we have

$$2X'a\cos 45° - Wa\cos 45° = 0$$

$$\Rightarrow \quad X' = \tfrac{1}{2}W$$

Taking moments about A for rod AB, we have

$$2X'a\cos 45° + 2Wa\cos 45° + Wa\cos 45° - Ta\cos 45° = 0$$

$$\Rightarrow \quad T = 4W$$

Exercise 14A

1 Two equal uniform rods AB and BC are smoothly jointed at B. The ends A and C are hinged to supports so that AC is horizontal and B is above AC. Angle ABC is 60°. The rods each have a mass of 20 kg. Find the reaction in the joint B.

2 Two uniform rods AB and BC, of equal length, hang in a vertical plane from hinges A and C, where AC is horizontal. The rods are smoothly jointed at B. Each rod makes an angle of 40° with the horizontal. AB has a mass of 10 kg and BC has a mass of 20 kg. Find the horizontal and vertical components of the reaction at B.

3 Three equal, uniform rods, each of weight W and length $2a$, are smoothly jointed to form a triangle ABC. The triangle hangs freely in a vertical plane suspended by a string attached to the mid-point of AB. Find the reactions in the joints.

4 A stepladder consists of two equal sections AB and BC, each of length 3 m and mass 8 kg, which are smoothly jointed at B. The ladder stands on smooth horizontal ground with A and C 2 m apart. The ladder is held in position by means of a light cord DE, joining points D and E on AB and BC respectively, where AD and EC are both 1 m. A woman of mass 50 kg climbs to a point two-thirds of the way up AB. Find the tension in the cord, and the horizontal and vertical reactions in the joint B.

5 Four uniform rods, each of length $2a$ and weight W, are connected to form a rhombus ABCD. This hangs in a vertical plane suspended by a string attached to A, and is held in shape by means of a light inextensible string of length $2a\sqrt{3}$ joining A and C. Find the tension in the string, and the horizontal and vertical components of the reaction in the joint B.

6 Rod AB, length a and weight W, and rod AC, length $2a$ and weight $2W$, are smoothly jointed at A. The system hangs from two fixed, smooth hinges at B and C, which are on the same level and are positioned so that angle BAC is 90°. Find, in component form, the reactions at A, B and C.

7 Rod AB, of length $2a$ and weight W, is attached to a fixed smooth hinge at A. A second, identical rod BC is smoothly jointed to AB at B. C is supported by a small ring threaded onto a rough horizontal wire passing through A. The system is in limiting equilibrium when AC = $2a$. Show that the coefficient of friction at C is $\dfrac{1}{2\sqrt{3}}$ and find the magnitude and direction of the reaction in the joint B.

8 A frame is constructed of three rods of length $3a$, $4a$ and $5a$, smoothly jointed. The rods are made of a material such that a length a has weight W. The frame is suspended from a point on its longest side so that that side is horizontal. Find the position of the suspension point and the reactions in the joints.

9 The diagram shows two uniform rods AB and BC, each of length 2 metres and mass 5 kg, which are smoothly jointed at B. The rods are placed symmetrically on the surface of a smooth cylinder of radius 1 metre. Show that for equilibrium angle ABC is 111.4°, and find the reaction in the joint B in that position.

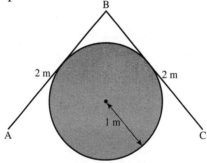

10 The diagram shows two uniform rods AB and BC, each of length
2*a* and weight *W*, which are smoothly jointed at B. C is freely hinged
to a vertical wall. A rests on rough horizontal ground with coefficient
of friction μ. The angles made by the rods with the ground and the
wall are α and β, as shown. If the system is about to slip, show that

$$\mu = \frac{2 \tan \beta}{3 \tan \alpha \tan \beta - 1}$$

and find the components of the reaction in the joint B.

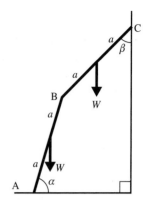

11 AB and BC are two identical uniform rods, each of weight *W* and length 2*a*. They are smoothly
jointed at B. A and C are connected by means of a light inextensible string so that angle ABC is a
right angle. The rods hang in equilibrium from a fixed, smooth hinge at A. Find the angle made by
AB with the vertical and find in component form the reaction in the joint B.

Frameworks

We will now consider structures consisting of light rods. That is, the weights of
the rods are insignificant compared with the loads they bear. The rods will be
smoothly jointed, with all external forces acting at the joints. Such a structure
is referred to as a **framework**.

In a framework, the joints at each end of each rod will be tending either to
move apart or to move towards each other. In the first case, the rod is a **tie**
and the internal force in the rod is a **tension**. In the second case, the rod is a
strut and the force in the rod is a **thrust** or **compression**.

Tie

Strut

Example 4 The diagram shows a framework ABC, with
AB = BC = 5 m and AC = 6 m. The framework is supported
by strings at A and C so that AC is horizontal, and carries a
load of 400 N at B. Find the internal forces in the rods.

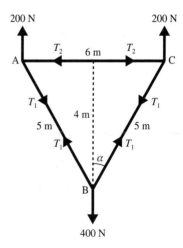

SOLUTION

By symmetry, the supporting forces at A and C are both 200 N,
and the internal forces in AB and BC are equal.

We can see, using symmetry and Pythagoras' theorem, that B
is 4 m vertically below AC, which gives $\sin \alpha = \frac{3}{5}$ and $\cos \alpha = \frac{4}{5}$.

Consider the equilibrium of joint A.

Resolving vertically, we have

$$200 - T_1 \cos \alpha = 0$$
$$\Rightarrow \quad T_1 = 250 \, \text{N}$$

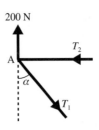

Resolving horizontally, we have

$$T_2 - T_1 \sin \alpha = 0$$
$$\Rightarrow \quad T_2 = 150 \, \text{N}$$

Example 5 The diagram shows a smoothly jointed framework of five equal, light rods AB, BC, CD, AD and BD, each of length $2a$, together with a sixth, DE, of half the length. The framework is attached by smooth hinges to a vertical wall at C and E and carries a load W at A. Find the internal forces in the rods.

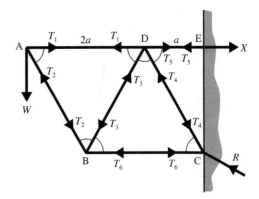

SOLUTION

Taking moments about C for the framework ABCD, we get

$$3Wa - T_5 a \sqrt{3} = 0$$
$$\Rightarrow \quad T_5 = W \sqrt{3}$$

Consider the equilibrium of A.

Resolving vertically, we have

$$T_2 \cos 30° - W = 0$$
$$\Rightarrow \quad T_2 = \frac{2W}{\sqrt{3}}$$

Resolving horizontally, we have

$$T_1 - T_2 \cos 60° = 0$$
$$\Rightarrow \quad T_1 = \frac{W}{\sqrt{3}}$$

Consider the equilibrium of B.

Resolving vertically, we have

$$T_2 \cos 30° + T_3 \cos 30° = 0$$
$$\Rightarrow \quad T_3 = -\frac{2W}{\sqrt{3}} \quad \text{(That is, } T_3 \text{ is a tension)}$$

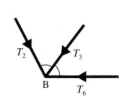

Resolving horizontally, we have

$$T_6 + T_3 \cos 60° = 0$$

$$\Rightarrow \quad T_6 = \frac{W}{\sqrt{3}}$$

Consider the equilibrium of D.

Resolving vertically, we have

$$T_3 \cos 30° + T_4 \cos 30° = 0$$

$$\Rightarrow \quad T_4 = \frac{2W}{\sqrt{3}}$$

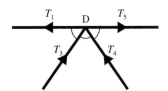

Method of sections

In Examples 4 and 5, we considered the equilibrium of individual joints in the framework. This approach can be extended by dividing the framework into two sections by means of a line passing through at most three rods whose internal forces are not known. We then consider the equilibrium of one of these sections.

Example 6 The framework shown consists of seven identical light rods, each of length $2a$, smoothly jointed. A weight of 400 N is suspended from B and the framework is supported at A and C. Find the tension or compression force acting on each rod.

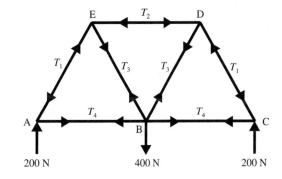

SOLUTION

By symmetry, we can see that the supporting forces are each 200 N and that some of the internal forces are equal, as shown in the diagram.

The second diagram shows the framework divided into two sections by a line through BC, BD and DE. If we consider the left-hand section, we see that the forces in AB, BE and AE are still internal, but that the forces in BC, BD and DE can now be treated as external forces acting on our chosen section.

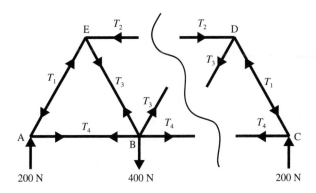

Taking moments about B, we have

$$aT_2\sqrt{3} - 400a = 0$$

$$\Rightarrow \quad T_2 = \frac{400}{\sqrt{3}}$$

Resolving vertically, we have

$$200 + T_3 \cos 30° - 400 = 0$$

$$\Rightarrow \quad T_3 = \frac{400}{\sqrt{3}}$$

Resolving horizontally, we have

$$T_4 + T_3 \cos 60° - T_2 = 0$$

$$\Rightarrow \quad T_4 = \frac{200}{\sqrt{3}}$$

This just leaves T_1 to be found. We can do this by considering the equilibrium of A.

Resolving vertically, we get

$$200 - T_1 \cos 30° = 0$$

$$\Rightarrow \quad T_1 = \frac{400}{\sqrt{3}}$$

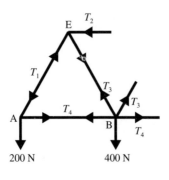

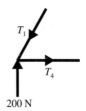

Exercise 14B

1 In each of the following diagrams, a light framework of rigid rods is in equilibrium in a vertical plane under the action of external forces. In some cases, the framework is supported by a smooth, fixed hinge, indicated on the diagram by a small circle. In each case, find the force in each rod.

a)

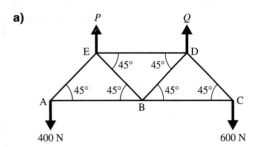

b)

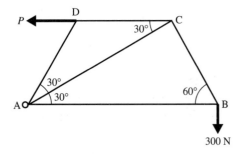

c)

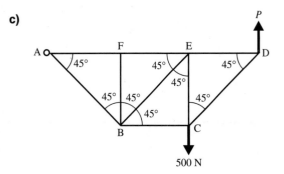

d)

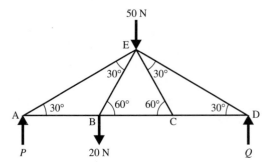

2 A framework consists of five identical light rods forming a regular pentagon ABCDE, together with rods BE and CE which hold it in shape. The framework is suspended from A and carries loads of 50 N at each of C and D. Calculate the forces in the rods BE and CE.

3 The diagram shows a framework consisting of seven identical light rods. The framework is supported by forces P and Q and carries loads of W and $2W$, as shown. The rods are such that a tension or compression force of more than 200 N will cause them to break. Find the largest possible value of W.

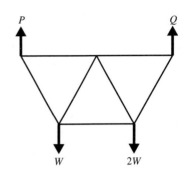

4 The diagram shows a framework consisting of nine identical light rods. The framework is freely hinged at A and is supported by a fixed support at B, so that AB is horizontal. A load of 250 N is suspended as shown. Calculate

a) the pressure on the support at B
b) the forces in the rods CE and CF.

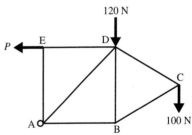

5 The diagram shows a framework ABCDE of light rods resting in a vertical plane on fixed supports at A and B. A load of 80 N is suspended from E and a load of W from C. The value of W is gradually increased until the framework is on the point of tipping. Calculate the forces in the rods at this instant.

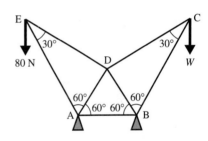

6 The diagram shows a framework consisting of six identical light rods with a seventh rod AD such that ABDE forms a square. The framework is freely hinged at A and carries loads of 100 N and 120 N at C and D respectively. The framework is held with AB horizontal by a horizontal force P at E. Calculate

a) the value of P
b) the magnitude and direction of the reaction in the hinge at A
c) the forces in the rods AD, BD and BC.

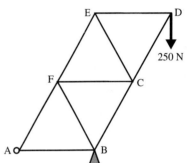

Graphical methods

As the forces acting on a given joint of a framework are in equilibrium, they can be represented by the sides of a closed polygon, which allows us to find the forces by scale drawing and measurement.

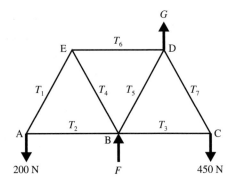

The diagram shows a framework supported by two unknown forces, F and G, and carrying loads of 200 N and 450 N. F and G can be found by taking moments, and we can then find the forces in the rods graphically.

Consider the equilibrium of the joint A. The three forces can be represented by the sides of the triangle PQR below.

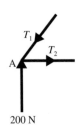

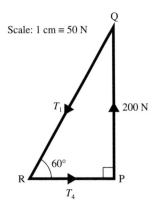

Using the scale indicated, we draw PQ 4 cm long. We can then draw the sides PR and QR at angles of 90° and 30° to PQ and measure PR and QR. We get

$$PR = 2.3\,\text{cm} \quad \Rightarrow \quad T_4 \approx 115\,\text{N}$$

$$QR = 4.6\,\text{cm} \quad \Rightarrow \quad T_1 \approx 230\,\text{N}$$

Because we know that the 200 N force is upwards and that forces 'follow round' a force diagram, we can establish the directions of T_1 and T_2 (T_1 is a thrust and T_2 is a tension) and mark them on our framework diagram.

We could now move on to draw the polygon for the forces in joint B, and so on. This would prove a tedious process, involving as it does a separate diagram for each joint and hence a lot of repetition. This can be avoided by superimposing all the polygons onto a single diagram. We can do this systematically with the help of a method of labelling known as **Bow's notation**.

Bow's notation

When we draw the diagram of a framework, the rods together with the lines of action of the external forces divide the page into regions. We can emphasize this by drawing a boundary round the framework and extending the lines of action of the external forces, as shown. In Bow's notation, each of these regions is labelled with a capital letter. When we draw the force diagram, the line representing a given force is labelled with a pair of lower-case letters which match the regions on either side of the force.

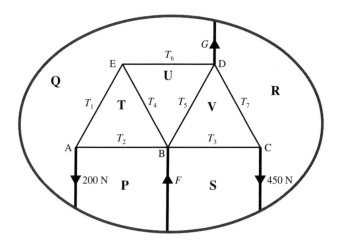

The steps needed to complete the force diagram are:

1 Calculate any unknown external forces.

2 Draw the force diagram for the external forces.

3 Choose a joint where not more than two forces have not been dealt with. Add these forces to the force diagram.

4 By following round the forces in the polygon produced by step **3**, decide on the directions of the new forces. Add appropriate thrust/tension arrows to the framework diagram.

5 Repeat steps **3** and **4** until all forces are included.

6 Find the magnitude of the forces by measurement.

We now apply these six steps to our example.

Step 1

Let the length of each rod be $2a$.

Taking moments about B, we have

$$200 \times 2a + G \times a - 450 \times 2a = 0$$

$$\Rightarrow \quad G = 500\,\text{N}$$

By resolving vertically, we can see that $F = 150\,\text{N}$.

Step 2

Using a scale of 1 cm ≡ 100 N, we start to draw the force diagram.

Regions P and Q are separated by an upward 200 N force, Q and R by a downward 500 N force, R and S by an upward 450 N force, and S and P by a downward 150 N force.

These appear as the line rpsq in the force diagram, as shown on the right.

Step 3

Consider the forces at joint A.

T_1 separates regions Q and T. T_2 separates regions T and P. On the force diagram, we draw a horizontal line from p and a 60° from q to meet at t.

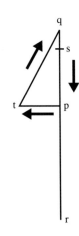

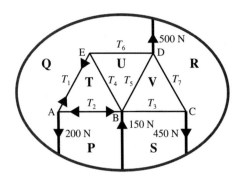

Step 4

By following the forces round the triangle pqt, as shown, we can see that T_1 is a tension and T_2 is a thrust. We indicate this on the framework diagram.

Step 5

We now look at joint E. The forces here separate regions Q, T and U. We already have qt on the force diagram. We draw a horizontal line from q and a 60° line from t to complete this triangle of forces.

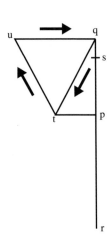

We know the direction of T_1 at E. So, following the forces round, we can see that T_4 is a thrust and T_6 is a tension. We add these to the framework diagram.

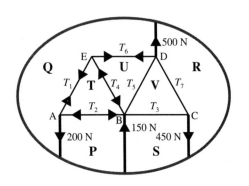

We now look at joint B. The forces here separate regions P, T, U, V and S. We already have pt, tu and sp on the diagram. By drawing a horizontal line from p and a 60° line from u, we complete the force polygon psvut.

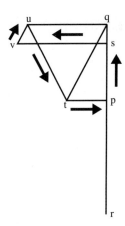

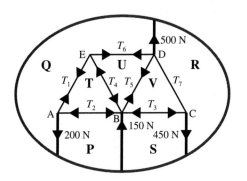

We know that ps is an upward force, so following the forces round the polygon shows us that T_3 is a thrust and T_5 is a tension. We add these to the framework diagram.

Finally, we consider joint C. The triangle of forces for this will be srv, and we already have sr and vs on the diagram. Joining r to v completes the triangle.

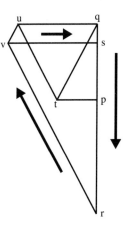

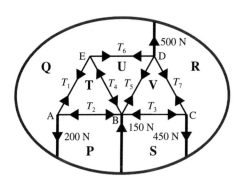

We know that sr is a downward force. Following round as usual shows that T_7 is a tension. We add this to the framework diagram.

Step 6

It now just remains to measure each length in the force diagram and so find the magnitude of each force.

$$qt = 2.3\,\text{cm} \quad \Rightarrow \quad T_1 = 230\,\text{N (tension)}$$

$$pt = 1.15\,\text{cm} \quad \Rightarrow \quad T_2 = 115\,\text{N (thrust)}$$

$$sv = 2.6\,\text{cm} \quad \Rightarrow \quad T_3 = 260\,\text{N (thrust)}$$

$$tu = 2.3\,\text{cm} \quad \Rightarrow \quad T_4 = 230\,\text{N (thrust)}$$

$$uv = 0.6\,\text{cm} \quad \Rightarrow \quad T_5 = 60\,\text{N (tension)}$$

$$qu = 2.3\,\text{cm} \quad \Rightarrow \quad T_6 = 230\,\text{N (tension)}$$

$$vr = 5.2\,\text{cm} \quad \Rightarrow \quad T_7 = 520\,\text{N (tension)}$$

Note   A small scale is used here because of the limitation on space. In practice, you should use the **largest convenient scale** to maximise the accuracy of your results.

Exercise 14C

1 In each of the following diagrams, a framework of light rods is subject to external forces. Use a graphical method to determine the forces in the rods, stating whether they are tensions or thrusts.

a)

60 N   90 N

b)

500 N

c)
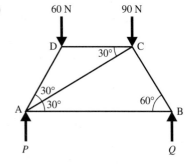

P   90 N

d)
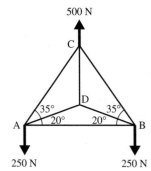

E   D

100 N

e)
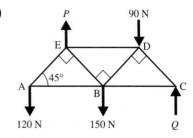

70 N

f)
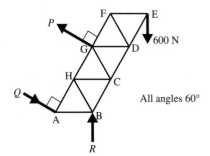

F   E

600 N

All angles 60°

2 The diagram shows a framework ABCD which is smoothly hinged at B. It carries a load W at C and is held with BD vertical by means of a force P at A in the direction DA.

a) Calculate P in terms of W.
b) Find graphically the forces in the rods.
c) Find the maximum load if no rod can withstand a force of more than 500 N.

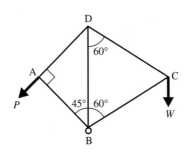

3 The diagram shows a framework ABCDE. It is smoothly hinged at C and carries a load of 400 N at E. The framework is to be held in a vertical plane, with BC horizontal, by means of a horizontal force P which can be applied either at A or at D.

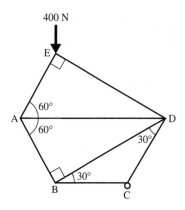

a) Find the required value of P.

b) Investigate graphically whether the two possible positions for P cause different stresses in the rods. If so, is there a preferable option?

4 The framework shown carries a load of 300 N at D. It is supported by a smooth hinge at A and is held in a vertical plane with AB horizontal by means of a horizontal force P at F.

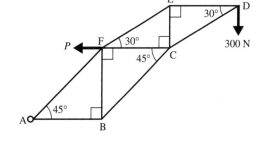

a) Calculate the value of P.

b) Draw a force diagram for the framework and from it find the magnitude and direction of the reaction in the hinge at A and the internal forces in the rods.

5 The diagram shows a framework subject to equal, opposing forces P. By drawing a force diagram, find the internal forces in the rods in terms of P and hence find the largest value of P if the rods break under a strain of more than 80 N.

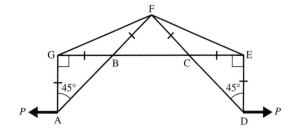

6 The diagram shows a framework smoothly hinged to a vertical wall at A and E. A load of 240 N is suspended from C. Use a graphical method to find the internal forces in the rods.

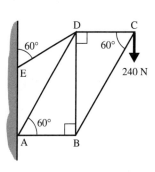

Examination questions

Chapters 11 to 14

Chapter 11

1 Two uniform solid spheres are rigidly joined to the ends of a uniform rod lying along the line containing their centres. One sphere has mass 3 kg and diameter 10 cm; the other has mass 10 kg and diameter 16 cm. The rod has mass 2 kg and length 32 cm (see diagram). The mid-point of the rod is M and G is the centre of mass of the combined body (consisting of the two spheres and the rod). Calculate the distance MG.   (OCR)

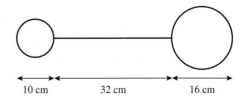

10 cm   32 cm   16 cm

2 The diagram shows a uniform lamina; all the corners are right angles. Using the axes shown, find the coordinates of the centre of mass of the lamina.   (OCR)

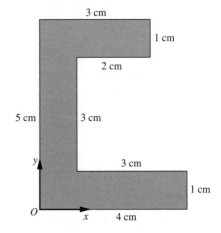

3 A thin uniform wire of length 30 cm is bent to form a framework in the shape of a right-angled triangle PQR, where PQ = 5 cm, QR = 12 cm and RP = 13 cm.

Find the distance of the centre of mass of the framework from

a) PQ   **b)** QR

The framework is freely suspended from the point P and hangs in equilibrium.

c) Find, to the nearest degree, the acute angle that PQ makes with the vertical.   (EDEXCEL)

4 A thin uniform lamina ABCD is formed by taking an equilateral triangle OCD, of side 2 cm, and removing the triangle OAB, where A and B are the mid-points of OC and OD respectively, as shown in the figure on the right.

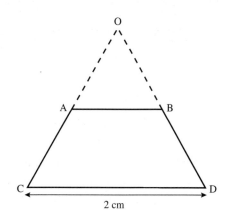

a) Find, in cm, the perpendicular distance of the centre of mass of the lamina from CD, giving your answer in the form $k\sqrt{3}$.

[You may assume that the centre of mass of a uniform equilateral triangular lamina, of height h, is at a perpendicular distance $\frac{1}{3}h$ from each side of the triangle.]

The lamina is freely suspended from A and hangs at rest in a vertical plane.

b) Find, to the nearest degree, the acute angle between AB and the vertical. (EDEXCEL)

5 The area enclosed between the line $y = x + 1$, the x-axis and the lines $x = 0$ and $x = 2$ is reflected in the x-axis to form the trapezium ABCD shown in Fig. 1 on the right.

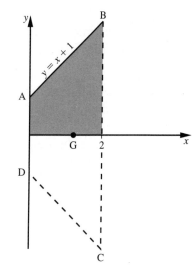

i) Show, by using integration, that the coordinates of the point G, the centre of mass of a uniform lamina in the shape of the trapezium ABCD, are $\left(\frac{7}{6}, 0\right)$.

A uniform prism, whose cross-section is the trapezium ABCD, is placed on a rough plane inclined at an angle α to the horizontal, as shown in Fig. 2 (bottom right).

ii) Find the value of α if the prism is just about to topple. (NICCEA)

Fig. 1

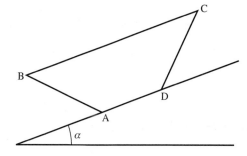

Fig. 2

6 a) The diagram on the right shows a uniform lamina in the shape of a sector of a circle of centre O and radius *a*. The angle AOB $= 2\beta$ radians, and the centre of mass of the lamina is G. By considering the lamina as a system of **either** elementary sectors **or** elementary circular arcs centred at O, show by integration that

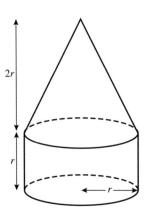

$$OG = \frac{2a \sin \beta}{3\beta}$$

b) Such a sector is to form part of a mobile, which is a household decoration consisting of uniform shapes of thin card suspended from a frame by fine threads. The angle β is to be chosen so that, when the sector is suspended from A and hangs in equilibrium, AB is vertical.

i) Show that β must satisfy the equation $2 \tan \beta = 3\beta$.

ii) Given that $\beta = 1$ is an approximate solution of this equation, use the iterative formula

$$\beta_{n+1} = \tan^{-1}\left(\tfrac{3}{2}\beta_n\right)$$

three times to obtain a better approximation to the value of β, giving two decimal places in your answer.   (NEAB)

7 A simple wooden model of a rocket is made by taking a uniform cylinder, of radius *r* and height 3*r*, and carving away part of the top two-thirds to form a uniform cone of height 2*r*, as shown in the figure on the right. Find the distance of the centre of mass of the model from its plane face.

(EDEXCEL)

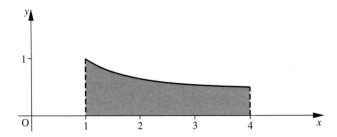

8 A uniform solid is formed by taking the area bounded by the curve $y = \dfrac{1}{\sqrt{x}}$, the *x*-axis and the two lines $x = 1$ and $x = 4$, as shown in the figure below, and rotating it through 2π radians about the *x*-axis.

i) Show that the position of the centre of gravity of the solid is 2.164 from the *y*-axis.

A globe sitting on a table can be modelled as a composite of three parts: a uniform spherical shell of mass 6*M* kilograms; a solid base of mass *M* kilograms, height 3 cm, identical in form to the solid in part **i**; a mount, joining the shell to the base, in the form of a semicircular arc AB of mass 3*M* kilograms, radius 6 cm, with AB inclined at 30° to the vertical. The globe is shown in the diagram on the right.

ii) Find the height of the centre of gravity of the globe above the table. (NICCEA)

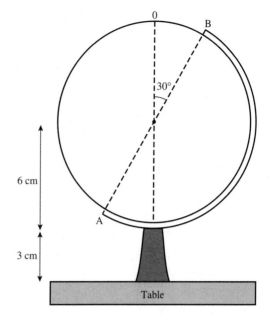

9 A metal toolbox is modelled as a cuboid of length 0.8 m, height 0.4 m and width 0.4 m. The mass of the toolbox and its partitions is 2.5 kg and its centre of mass is at the point on the O*xy* plane with coordinates (0.18, 0.23), referred to the axes shown in the figure below, where the units are metres.

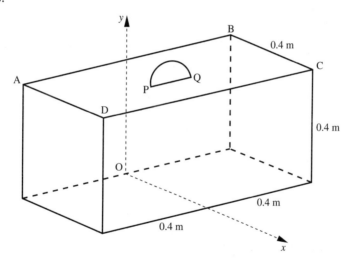

Some tools are put inside on various drawers. The mass of the tools is 7.5 kg and the position of their centre of mass is the point on the O*xy* plane with coordinates (0.34, 0.19).

i) Calculate the position of the centre of mass of the toolbox with the tools.

A light handle is freely pivoted at the points P and Q which are symmetrically placed in the plane ABCD with PQ parallel to AB; the handle PQ is midway between AB and DC and also midway between AD and BC. The box with tools is lifted so that PQ is horizontal and the toolbox is in equilibrium.

ii) Calculate the angle of the plane ABCD with the horizontal.

The toolbox is now returned to its original position with Oy vertical. The uniform lid ABCD of the toolbox has mass 0.5 kg and is hinged to the body of the toolbox along the line AB. The lid is opened through 90° so that the edge DC is vertically above AB (i.e. ABCD now contains the y-axis). The tools remain in their previous positions.

iii) Assuming that the lid has negligible thickness, find the coordinates of the centre of mass of the toolbox and tools with the lid open. (MEI)

Chapter 12

10 A car of mass 1200 kg drives up a straight road inclined at an angle α to the horizontal, where $\sin \alpha = \frac{1}{14}$. It passes a point A on the road with a speed of $20 \, \mathrm{m\,s^{-1}}$, and a point B, higher up the road, with a speed of $15 \, \mathrm{m\,s^{-1}}$, where the distance from A to B is 150 m.

a) Find the change in the **total** energy of the car as it moves from A to B.
b) State whether this change is an increase or a decrease. (EDEXCEL)

11 A particle P, of mass 1.5 kg slides down a smooth plane inclined at an angle of 30° to the horizontal. It starts from rest at a point A, and reaches the bottom of the slope at the point B, where $AB = 3 \, \mathrm{m}$, as shown in the figure on the right.

By using the principle of conservation of energy, find the speed of P when it reaches B, giving your answer in $\mathrm{m\,s^{-1}}$ to three significant figures.

(EDEXCEL)

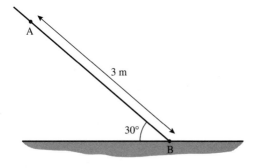

12 A particle P of mass 0.3 kg moves in a straight line on a smooth horizontal surface under the action of a constant horizontal force of magnitude 16 N. The particle starts from rest at the point A and passes the point B with speed $12 \, \mathrm{m\,s^{-1}}$.

a) Find the kinetic energy gained by P in moving from A to B.
b) Write down the work done by the force.
c) Find the distance AB. (EDEXCEL)

13 A block of mass 5 kg is on a plane which is at an angle of 40° to the horizontal. The block is connected by light, inextensible strings to objects A and D which hang vertically and have masses of 4 kg and 15 kg respectively, as shown in the diagram on the right. Between B and C, the strings are parallel to the line of greatest slope of the inclined plane and pass over smooth pegs B and C. The object D, which slides in a groove, is initially 1.5 m above a floor. You may assume that A never reaches the peg B and that the block never reaches the peg C.

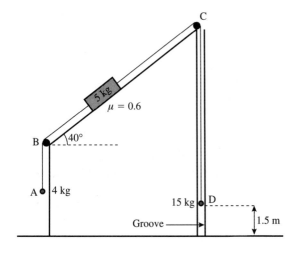

i) Calculate the change in gravitational potential energy of the whole system when D moves from its initial position to the floor. Your answer should specify whether this is a loss or a gain.

The coefficient of friction between the block and the plane is 0.6. The object at A hangs freely but the object at D slides in the groove against a constant frictional force of 15 N. You may assume that air resistance is negligible

The system is released from rest from its initial position.

ii) Show that D hits the floor with a speed of about $2.20 \, \text{m s}^{-1}$.
iii) After the impact of D with the floor, how much further does the block move up the plane?

(MEI)

14

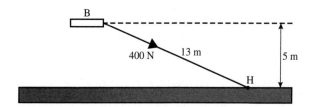

A barge B is pulled along a straight canal by a horse H on the towpath. The barge and the horse move in parallel straight lines 5 m apart. The towrope is 13 m long and it remains taut and horizontal (see diagram). The horse and the barge each move at a constant speed of $0.78 \, \text{m s}^{-1}$ and the towrope has a constant tension of 400 N. Calculate the work done by the horse on the barge in 10 minutes. (OCR)

15 A particle P of mass 0.5 kg moves along a straight line L from the point A (3, −2, 2) to the point B (6, −4, 3), where the unit of length is the metre. During the motion, P is acted upon by a force **F** and a force **R**, where $\mathbf{F} = 3\mathbf{i} + 5\mathbf{j} + 5\mathbf{k}$ newtons and **R** is perpendicular to L.

a) Explain why the force **R** does no work on P during the motion.
b) Show that the work done by the force **F** is 4 J.
c) Given that **F** and **R** are the only forces acting on the particle P, and that it starts from rest at A, find its speed when it reaches B. (NEAB)

16 A cork of length 0.06 m is to be slowly drawn out of the neck of a bottle. Initially the top of the cork is level with the top of the neck of the bottle.

a) Assuming that the force exerted is of constant magnitude 36 N, find the work done in drawing out the cork.
b) A more realistic model is to assume that the magnitude of the force is $kx^2 \, \text{N}$, where x m denotes the length of cork left in the neck of the bottle and k is a constant.

Initially when $x = 0.06$ the force is of magnitude 36 N.

i) Find the value of k.
ii) Find, by integration, the work done in drawing out the cork. (WJEC)

17 a) The diagram on the right shows a particle P free to move along a straight wire and acted on by a force of magnitude 130 N acting at an angle of 50° to the wire. Find the work done by the force when P is moved a distance of 2.4 m to the right along the wire.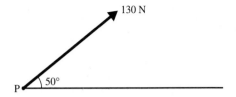

b) A particle Q of mass 0.4 kg moves on the x-axis under the action of a single force acting in the positive x-direction and of variable magnitude $\dfrac{24}{x^4}$ N, where x m is the distance of Q from the origin O.

i) Using the definition of work done as an integral, show that the work done by the force as x increases from 1 to 2 is 7 J.

ii) Given that the speed of Q when $x = 1$ is $2\,\text{m s}^{-1}$, find the speed of Q when $x = 2$.

(WJEC)

18 A jogger of mass 80 kg runs at a steady speed of $4\,\text{m s}^{-1}$ on a horizontal road and works at a constant rate of 140 W. Find the magnitude of the resistive forces acting.

He then comes to a hill inclined at an angle α to the horizontal where $\sin\alpha = \frac{1}{14}$ and runs up this hill at a steady speed such that his rate of working will be increased by 30%. Assuming that the resistive forces are unchanged, find this speed. (WJEC)

19 a) A particle P, of mass m, moves under the action of a single force **F**. The velocity of P at time t is **v**.

i) Show that

$$\frac{\mathrm{d}}{\mathrm{d}t}\left(\tfrac{1}{2}m\mathbf{v}.\mathbf{v}\right) = \mathbf{F}.\mathbf{v}$$

ii) Deduce that the work done by **F** during any given time interval is equal to the gain in the kinetic energy of P during that time interval.

b) A swimming pool has a chute in the form of a spiral down which children slide into the water. A child of mass 30 kg slides down the chute, her descent taking 3π seconds. The velocity, $\mathbf{v}\,\text{m s}^{-1}$, of the child at time t seconds is given by

$$\mathbf{v} = (2\cos t)\,\mathbf{i} - (2\sin t)\,\mathbf{j} + 0.1(2t+1)\,\mathbf{k}$$

where the vector **k** is vertically downwards.

i) Show that the power of the resultant force **F** acting on the child at time t is $0.6(2t+1)$ watts.

ii) Find, by integration, the total work done by **F** during the time interval $0 \leqslant t \leqslant 3\pi$.

iii) Verify directly by calculation that the result in part **a ii** holds for the given time interval.

(NEAB)

20 Assume that the fuel consumption of a car is directly proportional to its power output at any time. A car, of mass 800 kg, experiences a resistance force that is assumed to have a magnitude of $20v$ when the car is travelling at a speed v.

a) When the car is travelling on a horizontal road, find

i) the power output of the car at a constant speed of $20\,\text{m s}^{-1}$

ii) the constant speed at which the power output is 19 000 W.

b) The power output of the car, when travelling at a constant speed of $20\,\text{m s}^{-1}$ up a slope, is 19 000 W. Find the angle between the slope and the horizontal.

c) Describe three situations where the fuel consumption of the car would increase compared with that experienced when the power output is 19 000 W. (AEB 97)

Chapter 13

21 a) List **four** factors that could influence the outcome of a collisions between two snooker balls that are rolling along a straight line towards each other.

b) A red snooker ball that was at rest is hit directly by a white ball moving at $0.8\,\mathrm{m\,s^{-1}}$. After the collision the red ball moves at $0.75\,\mathrm{m\,s^{-1}}$. Find the speed of the white ball after the collision, assuming that momentum is conserved. State clearly any further assumption you have made.

c) Find the energy lost during the collision, defining any variables or constants that you introduce.   (AEB 97)

22 A small smooth sphere A of mass $0.2\,\mathrm{kg}$ and moving with speed $10\,\mathrm{m\,s^{-1}}$ catches up, and collides directly with a second smooth sphere B, of mass $0.4\,\mathrm{kg}$ and moving with speed $2\,\mathrm{m\,s^{-1}}$. The coefficient of restitution between the spheres is $\frac{1}{4}$. Find

a) the speed of A immediately after collision

b) the magnitude of the impulse on A.   (WJEC)

23 A spaceship of mass $80\,000\,\mathrm{kg}$ docks with a space station of mass $400\,000\,\mathrm{kg}$. The space station is travelling at $200\,\mathrm{m\,s^{-1}}$ immediately before docking takes place, and the spaceship is travelling $0.6\,\mathrm{m\,s^{-1}}$ faster. In a model for the docking, two particles, moving in the same direction in the same straight line, collide and coalesce.

i) Show that the speed of the spaceship is reduced by $0.5\,\mathrm{m\,s^{-1}}$ by the docking.

ii) Calculate the total loss in kinetic energy during the docking.   (OCR)

24

A car A of mass $1200\,\mathrm{kg}$ is about to tow another car B of mass $800\,\mathrm{kg}$ by means of a towrope linked between them. Just before the towrope tightens, A is travelling at a speed of $1.5\,\mathrm{m\,s^{-1}}$ and B is at rest. Just after the towrope tightens, both cars have a speed of $v\,\mathrm{m\,s^{-1}}$.

a) Show that $v = 0.9$.

b) Calculate the magnitude of the impulse on A when the towrope tightens.   (NEAB)

25 A truck A of mass 15 tonnes moves along a straight railway line with a speed of $8\,\mathrm{m\,s^{-1}}$. It collides with a truck B of mass 10 tonnes which is moving with a speed of $4\,\mathrm{m\,s^{-1}}$ along the same line and in the same direction. During the collision the impulse acting on each truck has magnitude $30\,000\,\mathrm{N\,s}$.

Calculate

a) the speed of truck A immediately after the collision

b) the speed of truck B immediately after the collision

c) the total loss of kinetic energy due to the collision.   (EDEXCEL)

26 A truck A of mass 6000 kg is moving with a speed of $12\,\mathrm{m\,s^{-1}}$ along a straight horizontal railway line when it collides with another truck B of mass 9000 kg which is stationary. After the collision the two trucks move on together.

  a) Find the speed of the trucks immediately after the collision.
  b) Find the magnitude of the impulse exerted on B when the trucks collide, stating the units in which your answer is given.

After the collision, the motion of the two trucks is opposed by a constant horizontal resistance of magnitude R newtons. The trucks come to rest 20 s after the collision.

  c) Find R.   (EDEXCEL)

27 Abi and Bill are ice skating and they collide. You may assume that both Abi and Bill are simply sliding and not propelling themselves by using their skates, that there is negligible resistance to their motion and that their motion is always in the same straight line with Abi on the same side of Bill.

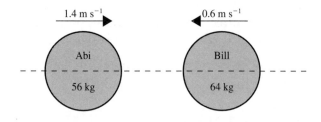

Abi has a mass of 56 kg and Bill a mass of 64 kg. Just before they collide they are sliding directly towards one another with speeds of $1.4\,\mathrm{m\,s^{-1}}$ and $0.6\,\mathrm{m\,s^{-1}}$ respectively, as shown in the figure above. Immediately after the collision, Abi has a speed of $0.28\,\mathrm{m\,s^{-1}}$ in the direction of the original motion.

  i) Find Bill's velocity immediately after the collision.
  ii) Find the coefficient of restitution in the collision.
  iii) What impulse is given to Bill in the collision?

Shortly after the collision, Bill pulls Abi with a horizontal impulse of 4.48 N s towards him and then lets go.

  iv) Calculate Abi's velocity after the pull.

When they collide again they embrace and move on together with a common velocity.

  v) Calculate the total energy lost from before their first collision to after their embrace.

                                                                                              (MEI)

28 Two spacecraft A and B, with masses m and M respectively, are moving along the same straight line during an attempt to link up. There is a direct collision which occurs when the two spacecraft are approaching each other with speed

u, as shown in the figure on the right. The spacecraft fail to link up and after the collision each moves in the opposite direction to its original motion. Spacecraft A now has a speed of $\dfrac{u}{2}$.

  i) Draw diagrams indicating the velocities of A and B before and after the collision.

    Show that after the collision the speed, v, of B is $u\left(\dfrac{3m}{2M} - 1\right)$ and explain why $\dfrac{m}{M} > \dfrac{2}{3}$.

ii) The coefficient of restitution between the spacecraft in the collision is e. Use the fact that $e \leqslant 1$ to show that $v \leqslant \dfrac{3u}{2}$ and hence $\dfrac{m}{M} \leqslant \dfrac{5}{3}$.

The rocket motor of spacecraft B is now fired so that its direction of motion is reversed. It catches up spacecraft A and links up with it. You may assume that there is a negligible loss of mass from spacecraft B when its rocket motor is used.

iii) If the combined spacecraft has a speed of $\dfrac{3u}{4}$, calculate the impulse given to B by its rocket motor in terms of m and M.   (MEI)

29 Particles A and B are moving vertically upwards under gravity, in the same straight line, with B above A. The mass of A is 0.03 kg and the mass of B is 0.05 kg. Immediately before the particles collide, A has a speed of $4.5 \, \text{m s}^{-1}$ and B has a speed of $3.1 \, \text{m s}^{-1}$. The speed of B is increased to $4.0 \, \text{m s}^{-1}$ by the collision. Calculate the speed of A immediately after the collision.

At a certain instant after the collision, B is at a point P below the point of collision. The gravitational potential energy of B at P is 0.8 J less than the gravitational potential energy of B at the point of collision. Find the speed of B at P.   (OCR)

30 a) A boy holds a ball at a height of 1.8 m above a horizontal floor and then releases it from rest. Taking $g = 10 \, \text{m s}^{-2}$, find the speed of the ball just as it is about to hit the floor.
  b) The coefficient of restitution between the ball and the floor is e. Show that the time interval between the first and second impacts of the ball with the floor is $1.2e$ seconds.
  c) Write down the time intervals, in terms of e, between the second and third and between the third and fourth impacts.
  d) The boy uses a stop-watch to measure the time taken from the moment the ball first hits the floor until bouncing ceases. He finds this time to be 3.6 seconds. Find the value of e.
  e) State one modelling assumption used in this question.   (NEAB)

Chapter 14

31 Two uniform rods AB and BC, of weights 80 N and 190 N respectively, are freely jointed to each other at B and to fixed points at A and C, where C is 2.4 m vertically above A. The length AB is 0.8 m. The rods are in a vertical plane, with AB horizontal, and a weight of 240 N is attached to the point X on AB, where AX = 0.6 m (see diagram on the right). Find the vertical and horizontal components of the force acting on BC at B.   (OCR)

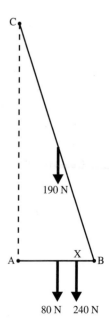

32 The diagram shows a light, smoothly jointed framework, that is at rest and freely pivoted at C. The 100 N force and the force of magnitude F are both perpendicular to AB and act in the same plane as ABC. The length of the rod AB is 2 metres.

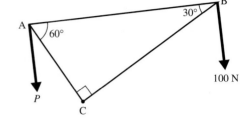

a) Find P.

b) Show that the magnitude of the force in BC is 200 N. Find the force in the rod AB and state whether it is in tension or compression.

(AEB 97)

33 The diagram shows a television aerial QR mounted on a mast OP. The mast is bolted to the side of a building at O and is supported by a light cable attached to it at Q and to the building at S. The angle between the cable and the wall is 60°.

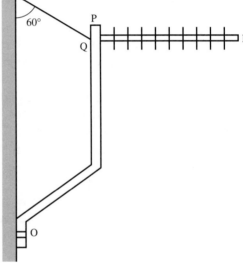

The weights of the mast and the aerial are 40 N and 30 N, respectively, and the tension in the cable is 30 N. The system of forces acting on the mast at O is equivalent to a single force **F** and a couple.

a) i) Find the horizontal and vertical components of **F** and hence show that the magnitude of **F** is approximately 61 N.

ii) Find, to the nearest degree, the angle which **F** makes with the horizontal.

b) The centres of mass of the mast and the aerial are at horizontal distances of 0.75 m and 1.5 m from the wall, respectively. The point Q is at a horizontal distance of 1 m from the wall and at a height of 2 m above the level of O. Show that the magnitude of the couple acting on the mast at O is approximately 8 N m, and indicate the sense in which it acts.   (NEAB)

34 The diagram shows a framework made up of five light rods that are smoothly jointed at A, B, C and D. The framework is smoothly hinged to a fixed point at D and lies in a vertical plane with BD vertical. The lengths of AD and BC are both a.

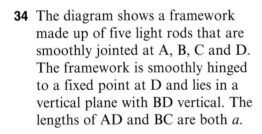

Forces of magnitude P and Q act on the framework as shown, in horizontal and vertical directions respectively.

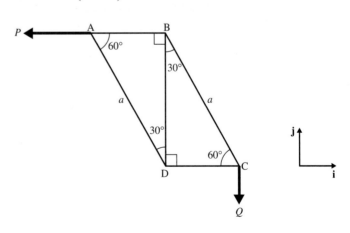

a) Express Q in terms of P.

b) Show that the tension in BC is $2P$ and find the forces in BD and CD, in terms of P.

c) Which rods could be replaced by ropes and which could be removed, without disturbing the equilibrium of the framework?

d) State the reaction force that acts on the framework at D, in terms of **i** and **j**. (AEB 96)

35 A light framework ABCDE consists of seven light pin-jointed rods as shown in the figure below.

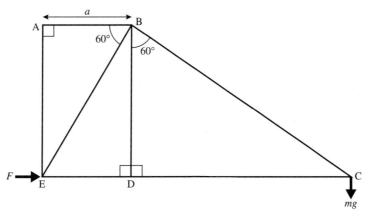

ABDE is a rectangle and BDC is a right-angled triangle. AB has length a and $\widehat{ABE} = \widehat{DBC} = 60°$. A mass m is attached at C. The framework is suspended from A and hangs in a vertical plane with AE being kept vertical by a horizontal force F applied at E.

i) Find the force F.

ii) Find the magnitude and direction of the reaction at the point of suspension.

iii) Find the forces in the rods AB, AE and BC. (NICCEA)

36 A framework used for supporting loads is modelled as five light rods AB, BC, CD, AD and DB freely pin-jointed together. ABCD is a parallelogram with $AB = CD = 4\,\text{m}$ and $AD = BC = 2\,\text{m}$; angle $DAB = 60°$. The framework rests on smooth horizontal supports at A and B which are at the same height. The framework supports vertical loads of $200\,\text{N}$ and $100\,\text{N}$ at the points D and C respectively, as shown in the figure on the right.

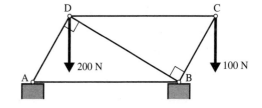

i) Mark in all the external forces acting on the framework.

ii) Show that the supports at A and B exert forces of $125\,\text{N}$ and $175\,\text{N}$ vertically upwards respectively on the framework.

iii) Mark in all the forces acting on each of the pin-joints of the framework, including those due to the internal forces in the rods.

iv) Calculate the magnitudes of the internal forces in the rods and state for each rod whether it is in tension or compression. You may express your answers in terms of surds where appropriate. (MEI)

15 Dimensional analysis

Introduction

In the following expressions, r, b and h represent lengths. Which expressions could represent an area?

a) πr^2 **b)** πrh **c)** $2bh + 4h^2$ **d)** $3b^2h + 5h$ **e)** $\pi r(2b + h^2)$

You may be familiar with questions like this. You will recognise **a** as the formula for the area of a circle. Of the remainder, two are areas, though of less obvious shapes, and two are not.

Calculating an area always involves multiplying a length by a length. For example, the area of a rectangle is length × breadth.

Similarly, the area of a circle is π (a pure number) × radius × radius.

It follows that any expression representing an area must consist only of terms of the form constant × length × length.

The expressions **b** and **c** above both pass this test. Expression **b** is in fact the curved surface area of a trough of length h and having a semicircular cross-section of radius r. Expression **c** could, for example, be the area of a rectangle with length $2h + b$ and breadth $2h$.

The first term in expression **d** involves three lengths multiplied together. This represents a volume. The second term is just a length. The expression therefore corresponds to a volume plus a length and is meaningless.

Similarly in expression **e**, if we expand the bracket, we see that the expression consists of an area plus a volume. Again, this is meaningless.

Dimensional analysis seeks to extend this idea to embrace all physical quantities. Clearly, we cannot restrict ourselves to the consideration of length, since we will need to include, for example, velocity, which is length divided by time.

In fact, all the quantities we meet in mechanics are defined in terms of three basic measures: **mass**, **length** and **time**. They are independent (any one of them cannot be defined in terms of the others). They are the physical **dimensions**, and are denoted in dimensional analysis by M, L and T respectively.

Units and dimensions

You need to be clear about the distinction between units and dimensions.

A unit of length, for example, is an arbitrary, agreed standard length. Other lengths are then expressed as multiples of this unit. However, when calculating an area or a volume, it does not matter whether the lengths are measured in centimetres, metres, feet or inches, as long as only one unit is employed. In the case of area, we will always be multiplying a length by a length, and so the dimensions of area will always be L^2. In the case of volume, we will always be multiplying a length by a length by a length, and the dimensions of volume will always be L^3.

The dimensions of a quantity are, therefore, much more fundamental to its nature than the units used to measure it.

In practice, we use the SI units (Système International d'Unités). The three base units which apply to mechanics are:

Mass   kilogram (kg)
Length   metre (m)
Time   second (s)

Notation

We use square brackets to refer to the dimensions of a quantity, so [force] means 'the dimensions of force'. Accordingly:

if h is the height of a cylinder $[h] \equiv L$

if m is the mass of a ball-bearing $[m] \equiv M$

Compound or derived units

As already stated, all quantities in mechanics are defined in terms of the three basic measures: mass, length and time. Their units often reflect their definition. For example:

- Area is found by multiplying length by length:

$$[\text{Area}] \equiv [\text{Length}] \times [\text{Length}] \equiv L^2$$

The SI unit of area is $1\,\text{m}^2$.

- Velocity is found by dividing length by time:

$$[\text{Velocity}] \equiv [\text{Length}] \div [\text{Time}] \equiv LT^{-1}$$

The SI unit of velocity is $1\,\text{m s}^{-1}$.

- Acceleration is found by dividing (change of) velocity by time:

$$[\text{Acceleration}] \equiv [\text{Velocity}] \div [\text{Time}] \equiv LT^{-2}$$

The SI unit of acceleration is $1\,\text{m s}^{-2}$.

- Force is found by multiplying mass by acceleration:

$$[\text{Force}] \equiv [\text{Mass}] \times [\text{Acceleration}] \equiv \text{MLT}^{-2}$$

We would be justified in calling the SI unit of force $1\,\text{kg}\,\text{m}\,\text{s}^{-2}$, but such an important quantity merits a special unit name – the **newton** (N).

- Work is found by multiplying force by length:

$$[\text{Work}] \equiv [\text{Force}] \times [\text{Length}] \equiv \text{ML}^2\,\text{T}^{-2}$$

Again, rather than calling the unit of work $1\,\text{kg}\,\text{m}^2\,\text{s}^{-2}$, we have the SI unit of work – the **joule** (J).

Dimensions in the calculus

Derivatives

The derivative $\dfrac{\mathrm{d}y}{\mathrm{d}x}$ is the limit of the ratio $\dfrac{\delta y}{\delta x}$ of small changes in the values of the quantities y and x. Dimensionally, therefore, we have

$$\left[\frac{\mathrm{d}y}{\mathrm{d}x}\right] = \frac{[y]}{[x]}$$

For example, if v is velocity and x is displacement, we get

$$v \equiv \frac{\mathrm{d}x}{\mathrm{d}t}$$

which gives

$$[v] \equiv \frac{[x]}{[t]} \equiv \text{LT}^{-1}$$

Care must be taken with second derivatives:

$$\frac{\mathrm{d}^2 y}{\mathrm{d}x^2} \quad \text{is really} \quad \frac{\mathrm{d}\left(\dfrac{\mathrm{d}y}{\mathrm{d}x}\right)}{\mathrm{d}x}$$

and so we have

$$\left[\frac{\mathrm{d}^2 y}{\mathrm{d}x^2}\right] \equiv \frac{\left[\dfrac{\mathrm{d}y}{\mathrm{d}x}\right]}{[x]} \equiv \frac{[y]}{[x]^2}$$

For example, if a is acceleration and x is displacement, we get

$$a \equiv \frac{\mathrm{d}^2 x}{\mathrm{d}t^2} \quad \Rightarrow \quad [a] \equiv \frac{[x]}{[t]^2} \equiv \text{LT}^{-2}$$

Integrals

The integral $\int y\,\mathrm{d}x$ is the limit of the sum $\sum y\,\delta x$. Dimensionally, therefore, we have

$$\left[\int y\,\mathrm{d}x\right] = [y] \times [x]$$

For example, if F is a variable force, the work it does is given by

$$W \equiv \int F\,\mathrm{d}x$$

and so we have

$$\begin{aligned}[W] &\equiv [F] \times [x] \\ &\equiv (\mathrm{MLT}^{-2}) \times \mathrm{L} \\ &\equiv \mathrm{ML}^2\mathrm{T}^{-2}\end{aligned}$$

Dimensionless quantities

Some quantities have no dimensions. They include, for example, the π in the expression πr^2 for the area of a circle, but there are other, less obvious, dimensionless quantities. For example:

Angle In circular measure (radians), we find an angle by dividing the arc length by the radius. It follows that

$$[\text{Angle}] \equiv \frac{[\text{Arc length}]}{[\text{Radius}]} \equiv \mathrm{L} \times \mathrm{L}^{-1} \equiv \mathrm{L}^0$$

so angle has no dimensions.

Expressing the angle in degrees is just a change of units. The underlying dimensionality is unchanged.

Coefficient of friction (μ) This is defined as friction force $\div$ normal reaction force, and so

$$[\mu] \equiv \frac{[\text{Force}]}{[\text{Force}]}$$

which is clearly dimensionless.

Coefficient of restitution (e) This is defined as separation speed $\div$ approach speed, and so

$$[e] \equiv \frac{[\text{Speed}]}{[\text{Speed}]}$$

which is clearly dimensionless.

Dimensions of constants

In the previous section, we saw that μ and e, the coefficients of friction and restitution, which are both constant in a given situation, are dimensionless. We might be tempted to suppose that this extends to other such constants, but this is not so. For example:

Modulus of elasticity (λ) This appears in the formula

$$\text{Tension} \equiv \frac{\lambda \times \text{Extension}}{\text{Natural length}}$$

As [extension] and [natural length] are both L, we have

$$[\text{Tension}] \equiv \frac{[\lambda] \times \text{L}}{\text{L}}$$

It follows that λ has the same dimensions as the tension force. That is,

$$[\lambda] \equiv \text{MLT}^{-2}$$

Gravitational constant (G) Newton's law of gravitation states that the force of attraction (F) between two bodies of mass m_1 and m_2 placed a distance r apart is given by

$$F = \frac{Gm_1m_2}{r^2}$$

Putting in known dimensions, we have

$$\text{MLT}^{-2} \equiv \frac{[\text{G}] \times \text{M}^2}{\text{L}^2}$$
$$\Rightarrow \quad [\text{G}] \equiv \text{M}^{-1}\text{L}^3\text{T}^{-2}$$

Dimensional consistency

We began this chapter by stating that πr^2 could represent an area, but that $3b^2h + 5h$ could not represent anything sensible. This illustrates a general idea:

In any equation or formula relating physical quantities, we cannot equate or add terms which are not dimensionally the same. That is,

If $\quad a \equiv b + c \quad$ then $\quad [a] \equiv [b] \equiv [c]$

Confirming the dimensional consistency of any result we obtain is a useful check on accuracy.

Example 1 When solving a problem to find the maximum range, R, of a projectile fired with initial speed v up a slope inclined at angle θ to the horizontal, we obtain

$$R = \frac{v^2(1 - \sin\theta)}{g\cos\theta}$$

Check this result for dimensional consistency.

SOLUTION

The left-hand side has $\quad [R] \equiv L$

The right-hand side has $\quad [v] \equiv LT^{-1}$

$$[g] \equiv LT^{-2} \quad \text{(because } g \text{ is an acceleration)}$$

$\sin\theta$ and $\cos\theta$ are both dimensionless, since they are each the ratio of two lengths. It follows that

$$\left[\frac{v^2(1 - \sin\theta)}{g\cos\theta}\right] \equiv \frac{(LT^{-1})^2}{LT^{-2}} \equiv L$$

The result is therefore dimensionally consistent. Note that this does not mean that the result is correct, merely that it could be correct.

Finding a formula

Provided we can decide which quantities should appear in a formula, we can use its dimensional consistency to determine what form the expression should take. The formula we obtain will involve an arbitrary, dimensionless constant, and the value of this would need to be found by other, possibly experimental, methods.

Example 2 It is believed that the speed of sound, c, in a gas depends only on the density, ρ, and the pressure, p, of the gas. That is,

$$c \equiv K\rho^\alpha p^\beta$$

where K is a dimensionless constant. Find the values of α and β so that the formula is dimensionally consistent.

SOLUTION

c is a speed, so $\quad [c] \equiv LT^{-1}$

ρ is a density, so $\quad [\rho] \equiv ML^{-3}$

p is a pressure, so $\quad [p] \equiv \dfrac{[\text{Force}]}{[\text{Area}]} \equiv ML^{-1}T^{-2}$

If $c \equiv K\rho^\alpha p^\beta$, then we have

$$[c] \equiv [K] \times [\rho]^\alpha \times [p]^\beta$$

$$\Rightarrow \quad LT^{-1} \equiv (ML^{-3})^\alpha \times (ML^{-1}T^{-2})^\beta$$

$$\Rightarrow \quad LT^{-1} \equiv M^{(\alpha+\beta)} L^{(-3\alpha-\beta)} T^{-2\beta}$$

For dimensional consistency, the power of each dimension on either side of the equation must be the same. Therefore, we have

$$0 = \alpha + \beta \qquad [1]$$

$$1 = -3\alpha - \beta \qquad [2]$$

$$-1 = -2\beta \qquad [3]$$

From [3], we get $\beta = \frac{1}{2}$. Substituting this in [1] gives $\alpha = -\frac{1}{2}$.

Checking these values in [2], we get

$$-3\alpha - \beta = -3 \times \left(-\frac{1}{2}\right) - \frac{1}{2} = 1$$

So equation [2] is satisfied. The required formula is, therefore,

$$c \equiv K\rho^{-\frac{1}{2}} p^{\frac{1}{2}} \quad \text{or} \quad c \equiv K\sqrt{\frac{p}{\rho}}$$

Note

- We do not have sufficient information to decide the value of K.

- As there were three equations and only two unknowns, it may have been that no values of α and β would have satisfied all three equations. Had this been the case, we would have known that our initial belief regarding the nature of the formula was ill-founded. However, the fact that we could satisfy all three equations does not guarantee that that belief was correct, although it is good evidence for it.

Example 3 When a body falls through the air, it experiences a retarding force (air resistance), F, which, it is believed, depends upon the mass of the body, m, its velocity, v, and its cross-sectional area, A. The following formula is proposed:

$$F \equiv K m^\alpha v^\beta A^\gamma$$

where K is a dimensionless constant. By considering the dimensional consistency of the formula, find the values of α, β and γ, and hence write down a possible formula.

SOLUTION

F is a force, so $\quad[F] \equiv MLT^{-2}$

m is a mass, so $\quad[m] \equiv M$

v is a velocity, so $\quad[v] \equiv LT^{-1}$

A is an area, so $\quad[A] \equiv L^2$

If $F \equiv Km^\alpha v^\beta A^\gamma$, then we have

$$[F] \equiv [K][m]^\alpha [v]^\beta [A]^\gamma$$
$$\Rightarrow \quad MLT^{-2} \equiv M^\alpha \times (LT^{-1})^\beta \times (L^2)^\gamma$$
$$\Rightarrow \quad MLT^{-2} \equiv M^\alpha L^{(\beta+2\gamma)} T^{-\beta}$$

Equating the powers of the three dimensions, we obtain

$$1 = \alpha \qquad [1]$$
$$1 = \beta + 2\gamma \qquad [2]$$
$$-2 = -\beta \qquad [3]$$

Solving these equations gives $\alpha = 1$, $\beta = 2$, $\gamma = -\frac{1}{2}$. A possible formula is, therefore,

$$F \equiv Kmv^2 A^{-\frac{1}{2}} \quad \text{or} \quad F \equiv \frac{Kmv^2}{\sqrt{A}}$$

Note

- The formula found in Example 3 is only a possible formula. It ignores some factors, such as the density and viscosity of the air, which may well have a bearing on the air resistance. It also excludes the possibility that the form of the expression may be different, involving perhaps terms like e^A.

  To decide on the usefulness of our formula as a model for air resistance, we would need to obtain experimental results. These would also be needed to find the value of the constant K.

- When we perform dimensional analysis, we have to solve three simultaneous equations. It follows that our proposed formula cannot involve more that three variables, otherwise we will not obtain a unique solution to the problem.

  Even if we have only three variables, the equations we obtain may be inconsistent and give no solution. On the other hand, they may have an infinite number of different solutions. In these cases, our proposed formula would clearly be inappropriate. Dimensional analysis, therefore, can tell us when a possible formula is wrong, but it cannot tell us that a formula is correct.

Exercise 15

1 The formula for kinetic energy is $\frac{1}{2}mv^2$ and the formula for gravitational potential energy is mgh. Check that these formulae are dimensionally equivalent.

2 Momentum is defined as mass × velocity. Impulse is defined as force × time. Show that the relationship

$$\text{Impulse} = \text{Change of momentum}$$

is dimensionally consistent.

3 Show that the formula $s = ut + \frac{1}{2}at^2$ is dimensionally consistent.

4 Power is the rate of doing work. Use this definition to find the dimensions of power, and show that this definition is dimensionally equivalent to the expression

$$\text{Power} = \text{Force} \times \text{Velocity}$$

5 When you meet simple harmonic motion on page 393, you will find that the acceleration of an object at displacement x from the centre of the motion is given by

$$\frac{\mathrm{d}^2x}{\mathrm{d}t^2} = -\omega^2 x$$

Find the dimensions of the constant ω.

6 The viscosity, η, of a fluid is a measure of its 'stickiness' and ability to flow. Newton proposed a formula relating the friction force, F, between two layers of fluid to the viscosity, the area, A, of the layers being considered, and the velocity gradient (the rate of change of velocity, v, with distance, x, across the flow), as follows:

$$F \equiv \eta A \frac{\mathrm{d}v}{\mathrm{d}x}$$

Show that the dimensions of η are $\mathrm{ML^{-1}\,T^{-1}}$.

7 The force, F, acting on a body of mass m, travelling in a circle of radius r with angular speed ω is given by $F \equiv mr\omega^2$. What are the dimensions of ω?

8 The following are results obtained as solutions to problems. Check the results for dimensional consistency.

a) The time taken for a particle to reach the top of a slope of length l and inclination α is

$$t = \sqrt{\frac{2l}{g \sin \alpha}}$$

b) The highest point on a wall reachable by a projectile with initial speed u fired from a distance a from the base of the wall is

$$h = \frac{u^4 - g^2 a^2}{2gu}$$

c) Two particles, masses m_1 and m_2, are connected by an elastic string of length l and modulus λ. One particle is set in motion with velocity u. The maximum extension, x, reached by the string is

$$= \left[\frac{m_1 m_2 \, l u^2}{(m_1 + m_2)\lambda} \right]^{\frac{1}{2}}$$

(The dimensions of λ are those of a force).

9 The time of oscillation, T, of a particle of mass m forming a simple pendulum with a light string of length l is believed to be given by a formula $T = K m^\alpha l^\beta g^\gamma$. Use dimensional analysis to find the values of α, β and γ, and hence write down the possible formula.

10 The velocity, v, of waves on an ocean is believed to be related to the water density, ρ, the wavelength, λ, of the waves, and g. Use dimensional analysis to find a possible formula for v.

11 Liquid flows through a pipe because there is a pressure difference between the ends of the pipe. The rate of flow, $V \, \text{m}^3 \, \text{s}^{-1}$, depends upon this pressure difference, $p \, \text{Nm}^{-2}$, the viscosity of the liquid, $\eta \, \text{kg} \, \text{m}^{-1} \, \text{s}^{-1}$, the length of the pipe, $l \, \text{m}$, and the radius of the pipe, $r \, \text{m}$.

a) Explain why it would be impossible to consider as a formula $V \equiv K \eta^\alpha r^\beta p^\gamma l^\delta$.

b) It is decided that the important thing is the pressure gradient p/l. Use dimensional analysis to obtain a possible formula relating the velocity, V, to the viscosity of the liquid, the radius of the pipe and the pressure gradient.

16 Circular motion

Round and round in circles
Completing the charm.
T.S. ELIOT

Circular motion is a significant aspect of motion in two dimensions. Many objects move in a circular path: for example, a car on a roundabout, a pendulum on a clock or a sock in a spin dryer.

Linear and angular speed

We can describe the motion of a particle travelling in a circle in terms of its displacement, speed and acceleration along the arc of the circle. However, it is often more appropriate to describe it in terms of the angle subtended at the centre of the circle by its path.

For example, consider points A and B on an old vinyl record revolving on a turntable at 33 revolutions per minute (the preferred abbreviation for which is rev min^{-1}, although rpm is more widely used). Suppose that A is 5 cm and B is 10 cm from the centre of the disc.

$$\text{In 1 minute:}\quad \text{A travels } 33 \times 2\pi \times 0.05 = 10.367 \text{ m}$$
$$\text{B travels } 33 \times 2\pi \times 0.10 = 20.735 \text{ m}$$
$$\Rightarrow\quad \text{Speed of A} = 10.367 \div 60 = 0.173 \text{ m s}^{-1}$$
$$\text{Speed of B} = 20.735 \div 60 = 0.346 \text{ m s}^{-1}$$

This means that points on the disc have different linear speeds depending on their distance from the centre. To describe the rate at which the disc is rotating, it is better to consider its angle of rotation.

In 1 rotation the disc turns through an angle of 2π radians.
Hence, in 1 minute it turns through $33 \times 2\pi = 66\pi$ radians.
Therefore, the angular speed of the disc is

$$66\pi \div 60 = 1.1\,\pi \text{ radians per second (written } 1.1\,\pi \text{ rad s}^{-1})$$

Angular displacement, speed and acceleration

The **angular displacement**, θ, of a particle P is the angle in **radians** rotated by the radius OP from some reference position (usually the position when time is zero). We could in theory use degrees, but, because we wish to use calculus methods, in practice we **always** use radians.

The rate of change of the angular displacement gives the **angular speed** (in rad s^{-1}). This is usually denoted by ω. Thus we have

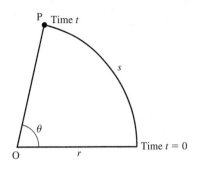

$$\frac{\mathrm{d}\theta}{\mathrm{d}t} = \dot{\theta} = \omega$$

The rate of change of the angular speed gives the **angular acceleration** (in rad s^{-2}). Thus we have

$$\frac{\mathrm{d}^2\theta}{\mathrm{d}t^2} = \ddot{\theta} = \dot{\omega}$$

Note that, as angle is dimensionless, the dimensions of angular speed are T^{-1}, and of angular acceleration are T^{-2}.

Relation between linear and angular measures

If a particle travels along the arc shown in the diagram above right, it defines a sector with angle θ, radius r and arc length s. We know that

$$s = r\theta$$

As r is constant, differentiating this formula gives

$$\frac{\mathrm{d}s}{\mathrm{d}t} = r\frac{\mathrm{d}\theta}{\mathrm{d}t}$$

If the speed of the particle is v, this becomes

$$v = r\dot{\theta} \quad \text{or} \quad v = r\omega$$

Differentiating again gives the relation between the linear and angular accelerations:

$$\frac{\mathrm{d}v}{\mathrm{d}t} = r\ddot{\theta} = r\dot{\omega}$$

Example 1 A particle travels round the circumference of a circle of radius 6 m at a rate of 30 rev min^{-1}. Find **a)** its angular speed in rad s^{-1} and **b)** its linear speed around the circle.

SOLUTION

a) One revolution $= 2\pi$ rad

The particle has angular speed $2\pi \times 30 = 60\pi$ rad min^{-1}

$$= 60\pi \div 60 = \pi \text{ rad s}^{-1}$$

b) Using $v = r\omega$, we have

$$v = 6 \times \pi = 6\pi \text{ m s}^{-1} \quad \text{or} \quad 18.85 \text{ m s}^{-1}$$

Exercise 16A

1 Convert an angular speed of $15\,\text{rev}\,\text{min}^{-1}$ to $\text{rad}\,\text{s}^{-1}$.

2 Convert an angular speed of $25\,\text{rad}\,\text{s}^{-1}$ to $\text{rev}\,\text{min}^{-1}$.

3 Find the speed in $\text{m}\,\text{s}^{-1}$ of a particle moving in a circle of radius $60\,\text{cm}$ at $15\,\text{rad}\,\text{s}^{-1}$.

4 Find the speed in $\text{m}\,\text{s}^{-1}$ of a particle moving in a circle of radius $2.5\,\text{m}$ at $5\pi\,\text{rev}\,\text{min}^{-1}$.

5 Mobusar and Samantha are on a roundabout. Their distances from the centre of the roundabout are $1.3\,\text{m}$ and $1.8\,\text{m}$ respectively. Mobusar is travelling at $3.5\,\text{m}\,\text{s}^{-1}$. Find

   a) the angular speed of the roundabout in $\text{rad}\,\text{s}^{-1}$
   b) Samantha's speed in $\text{m}\,\text{s}^{-1}$.

6 Find the angular speed in $\text{rad}\,\text{s}^{-1}$ of a bicycle wheel of radius $35\,\text{cm}$ when the bicycle is travelling at $18\,\text{km}\,\text{h}^{-1}$.

7 The hands of a large public clock are $70\,\text{cm}$ and $110\,\text{cm}$ long. Find in $\text{m}\,\text{s}^{-1}$ the speed of the tip of each hand.

Modelling circular motion

Circular motion can be quite complicated and, as usual, we will need to make various modelling assumptions in different situations. However, the following considerations will underlie all our models:

- Velocity is a vector quantity, which is changed if either its magnitude or its direction is altered.

- When a body moves along a circular path, its velocity at any instant is directed along the tangent to the circle at that point. This direction is constantly changing, which means that its velocity is changing.

- A change in velocity means that the body is accelerating.

- By Newton's second law, there must be a force acting on the body to cause this acceleration.

Motion with uniform speed

The simplest modelling assumptions we can make are that we are dealing with a particle which is travelling around a circle at uniform speed. We will examine the consequences of these two assumptions.

A particle, P, is moving in a circle, radius r and centre O, with constant angular speed ω rad s^{-1}.

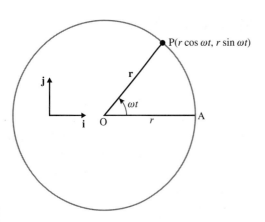

The particle is initially at point A. After a time t s, the particle is at P. Its angular displacement, the angle AOP in the diagram, is then ωt.

Taking O as the origin and with unit vectors as shown, the coordinates of P are $(r\cos\omega t, r\sin\omega t)$, and its position vector is therefore given by

$$\mathbf{r} = r\cos\omega t\,\mathbf{i} + r\sin\omega t\,\mathbf{j} \qquad [1]$$

We find the velocity of the particle by differentiating $\mathbf{r}$ with respect to t. This gives

$$\mathbf{v} = -r\omega\sin\omega t\,\mathbf{i} + r\omega\cos\omega t\,\mathbf{j} \qquad [2]$$

Similarly, we find the acceleration of the particle by a second differentiation. This gives

$$\mathbf{a} = -r\omega^2\cos\omega t\,\mathbf{i} - r\omega^2\sin\omega t\,\mathbf{j} \qquad [3]$$

We can now find the magnitude and direction of the velocity and the acceleration.

Velocity

The magnitude of the velocity, the speed, can be found from [2]:

$$|\mathbf{v}| = \sqrt{(-r\omega\sin\omega t)^2 + (r\omega\cos\omega t)^2}$$

$$= \sqrt{(r\omega)^2(\sin^2\omega t + \cos^2\omega t)}$$

$$\Rightarrow \quad |\mathbf{v}| = r\omega$$

Since ω and r are constants, this confirms that the speed is uniform, and that $v = r\omega$.

To find the direction of the motion, we can use the scalar product to find the angle between the velocity vector, $\mathbf{v}$, and the position vector, $\mathbf{r}$. From [1] and [2], we have

$$\mathbf{r.v} = (r\cos\omega t\,\mathbf{i} + r\sin\omega t\,\mathbf{j}).(-r\omega\sin\omega t\,\mathbf{i} + r\omega\cos\omega t\,\mathbf{j})$$

$$= -r^2\omega\cos\omega t\sin\omega t + r^2\omega\cos\omega t\sin\omega t$$

$$\Rightarrow \quad \mathbf{r.v} = 0$$

This means that the position vector and the velocity vector are perpendicular. The direction of the velocity must therefore be along the tangent to the circle. This is what we expected. The velocity of a body is **always tangential to its path**.

Acceleration

From equations [1] and [3], we can see that

$$\mathbf{a} = -\omega^2 \mathbf{r}$$

The acceleration therefore acts in the opposite direction to $\mathbf{r}$. Since $\mathbf{r}$ is directed away from the centre of the circle, it follows that the **acceleration acts towards the centre of the circle**, whatever the position of P on the circle.

We also have

$$|\mathbf{a}| = |-\omega^2 \mathbf{r}|$$

$$\Rightarrow \quad a = r\omega^2 \qquad [4]$$

Since ω and r are constants, the magnitude of the acceleration is constant, but its direction is constantly changing.

We can obtain an alternative form of the expression for the magnitude of the acceleration as follows:

$$v = r\omega \quad \Rightarrow \quad \omega = \frac{v}{r}$$

Substituting for ω in equation [4], we get

$$a = \frac{v^2}{r}$$

Summary

For a particle moving in a circle of radius r with uniform angular speed ω:

- The velocity is tangential to the circle.
- The speed around the circle, $v = r\omega$.
- The acceleration acts towards the centre of the circle.
- The magnitude of the acceleration is $r\omega^2$ or $\dfrac{v^2}{r}$.

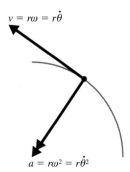

$v = r\omega = r\dot{\theta}$

$a = r\omega^2 = r\dot{\theta}^2$

Exercise 16B

1 Find the magnitude of the acceleration of a particle travelling

   a) around a circle of radius 2.5 m at a uniform speed of $8\,\mathrm{m\,s^{-1}}$
   b) around a circle of radius 4.2 m with uniform angular velocity $3\,\mathrm{rad\,s^{-1}}$
   c) around a circle of diameter 2.6 m at a constant $12\,\mathrm{rev\,min^{-1}}$.

2 The Earth rotates once in every 24 hours (approximately). Its radius is approximately 6400 km.

   a) Find the angular speed of the Earth in rad s^{-1}.

   b) Find the acceleration of a person standing on the Equator.

   c) At what latitude would a person be standing if the person's acceleration were $0.02\,\text{m s}^{-2}$?

   d) What would be the direction of the acceleration of the person in part **c**?

3 The Earth orbits the Sun once every 365 days (approximately). Assuming that its orbit is circular, with a radius of 1.49×10^6 km, find its acceleration.

4 A car is travelling round a roundabout of radius 10 m with a uniform speed of $6\,\text{m s}^{-1}$.

   a) Find its angular speed, ω.

   b) Taking the centre of the roundabout as the origin, the car is initially at the point with position vector 10**i** and is moving in an anticlockwise direction. Find an expression for the position vector, **r**, of the car at time t.

   c) Differentiate the position vector to find expressions for the velocity and acceleration vectors of the car.

5 A boy is swinging a conker so that it moves in a horizontal circle above his head with uniform speed. At time t, the position vector of the conker is given by

$$\mathbf{r} = (0.3 \cos 10t\, \mathbf{i} + 0.3 \sin 10t\, \mathbf{j})\,\text{m}$$

   a) Find the acceleration vector of the conker.

   b) Find the magnitude of the conker's acceleration.

   c) When $t = 10$ s, the boy releases the string. What is the velocity vector of the conker at this time?

Mechanics of circular motion

We have established that when a body is travelling in a circle of radius r with uniform speed v, it has an acceleration of magnitude $\dfrac{v^2}{r}$, or $r\omega^2$, towards the centre of the circle, where ω is the uniform angular speed.

We have also stated that, as a consequence of Newton's second law, there must be a force acting to cause the acceleration. A body on which no force acts will move in a straight line. Indeed, if a body travelling in a circle suddenly has the accelerating force removed (such as when a conker string breaks), the body will move along the tangent at that point.

In solving a practical problem, we need to identify all the forces acting and then write down the equation of motion. Because the acceleration acts towards the centre of the circle, it must be in the plane of the circle. There can be no acceleration perpendicular to this plane.

We will continue to make the modelling assumption that speed is uniform and that the bodies involved are particles. That is, the mass of a body is considered as though it is concentrated at a single point.

Example 2 A particle of mass 3 kg is attached to a light, inextensible string of length 2 m, which is fastened to a point on a smooth table. The particle is set in motion in such a way that it describes circles around the fixed point with a speed of 6 m s^{-1}. Find the tension in the string.

SOLUTION

For the motion described to be possible, the string must be taut. The radius of the circle is therefore 2 m.

There is no acceleration vertically and the tension has no vertical component.

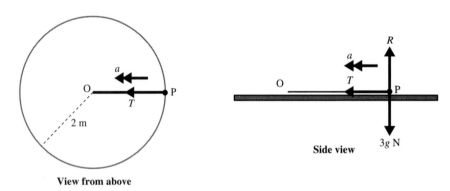

View from above

Side view

The reaction R is $3g$ N. We only need consider the equation of motion for the horizontal direction.

The acceleration towards the centre of the circle is $\dfrac{v^2}{r} = \dfrac{6^2}{2} = 18 \text{ m s}^{-2}$

Resolving in the direction PO and using Force = Mass × Acceleration, we have

$$T = 3 \times 18 = 54 \text{ N}$$

So, the tension in the string is 54 N.

Example 3 Two particles, each of mass m kg, are attached to the ends of a light, inextensible string which passes through a hole in a smooth table. One particle moves in a circular path of radius r m around the hole, so that the string is taut. The other particle hangs freely. What is the speed of the particle on the table?

SOLUTION

Particle P is moving in a circular path and Q is stationary. The tension is the same throughout the string.

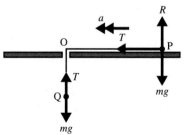

Resolving vertically for Q, we have

$$T - mg = 0$$

$$\Rightarrow \quad T = mg \qquad \text{[1]}$$

P has no acceleration vertically, so we only need consider its equation of motion horizontally.

The acceleration of P is $\dfrac{v^2}{r}$ towards O.

Resolving for P along PO, we have

$$T = m\frac{v^2}{r} \qquad \text{[2]}$$

Substituting for T from [1] into [2], we have

$$mg = \frac{mv^2}{r}$$

$$\Rightarrow \quad v^2 = rg \quad \Rightarrow \quad v = \sqrt{rg}$$

So, the velocity of the particle P is $\sqrt{rg}$ m s^{-1}.

Note Example 3 is a classic problem and the situation could not be achieved in practice. Friction between the particle and the table, and between the string and the edge of the hole, have been ignored. We have also ignored the tendency for the string to wind or unwind as the particle moves round the hole.

Example 4 A particle of mass m is placed on a rough horizontal turntable at a distance of 0.4 m from its centre, O. The coefficient of friction between the particle and the turntable is $\frac{1}{4}$. What is the maximum angular speed at which the turntable can move if the particle is not to slide?

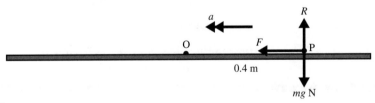

SOLUTION

Resolving vertically, we get

$$R - mg = 0 \quad \Rightarrow \quad R = mg \qquad \text{[1]}$$

If the particle is on the point of sliding, we have

$$F = \tfrac{1}{4}R$$

So, substituting from [1], we get

$$F = \tfrac{1}{4}mg \qquad\qquad [2]$$

The acceleration towards O is $r\omega^2$, where ω is the angular speed of the turntable. Therefore, resolving in the direction PO, we have

$$F = mr\omega^2 \qquad\qquad [3]$$

Substituting from [2], we get

$$\tfrac{1}{4}mg = mr\omega^2$$

$$\Rightarrow \quad \omega^2 = \frac{g}{4r} \quad \Rightarrow \quad \omega = \frac{1}{2}\sqrt{\frac{g}{r}}$$

So, the maximum angular speed is $\dfrac{1}{2}\sqrt{\dfrac{g}{r}}$.

Satellites in circular orbits

We saw on page 51 that the gravitational force between objects of masses m_1 and m_2 separated by a distance d is given by

$$F = \frac{Gm_1 m_2}{d^2}$$

where $G = 6.67 \times 10^{-11}\,\mathrm{N\,m^2\,kg^{-2}}$ is the universal gravitational constant. The mass of the Earth is approximately $5.98 \times 10^{24}\,\mathrm{kg}$. Hence, a satellite of mass $m\,\mathrm{kg}$ at a distance $r\,\mathrm{m}$ from the centre of the Earth is attracted towards the Earth by a force

$$F = \frac{6.67 \times 10^{-11} \times 5.98 \times 10^{24}\,m}{r^2} = \frac{3.99 \times 10^{14}\,m}{r^2}\,\mathrm{N}$$

In general, the orbit of a satellite is elliptical, but many satellites are placed in orbits which are approximately circular. In particular, it is useful for communications and related purposes to have a satellite moving in a circular orbit so that it remains stationary relative to the surface of the Earth. This is called a **parking** or **geostationary orbit**.

Suppose that a satellite of mass $m\,\mathrm{kg}$ moves with angular velocity $\omega\,\mathrm{rad\,s^{-1}}$ in a circular orbit of radius $r\,\mathrm{m}$ (where r is obviously greater than the radius of the Earth). From Newton's second law, we have

$$\frac{3.99 \times 10^{14}\,m}{r^2} = mr\omega^2$$

We are, therefore, able to find the value of ω, and hence the period of the orbit, for a given value of r.

Example 5 Given that the radius of the Earth is 6.37×10^6 m, find the orbital period of a satellite in a circular orbit at a height of 500 km above the suface of the Earth.

SOLUTION

Let the mass of the satellite be m and its angular velocity be ω.

The radius of the orbit is $6.37 \times 10^6 + 500\,000 = 6.87 \times 10^6$ m.

The gravitational force acting on the satellite is

$$\frac{6.67 \times 10^{-11} \times 5.98 \times 10^{24}\, m}{(6.87 \times 10^6)^2} = 8.45m\,\text{N}$$

By Newton's second law, we have

$$8.45m = 6.87 \times 10^6 m\omega^2$$

$$\Rightarrow \quad \omega = 0.0011\,\text{rad s}^{-1}$$

The period of the orbit is, therefore, $\dfrac{2\pi}{0.0011} = 5665\,\text{s}$ or 1 h 34 min to the nearest minute.

Exercise 16C

1 A particle of mass 2 kg is attached by a light, inextensible string of length 1.2 m to a fixed point on the surface of a horizontal, smooth table. The particle travels in a circle on the table at a speed of $2.5\,\text{m s}^{-1}$. Find the tension in the string.

2 A railway engine of mass 80 tonnes is travelling in an arc of a horizontal circle of radius 150 m at a speed of $72\,\text{km h}^{-1}$. What total sideways force is being exerted on the wheels by the rails?

3 A string of length 80 cm can just support a suspended mass of 40 kg without breaking. A mass of 2 kg is attached to the string and the other end of the string is fastened to a point on the surface of a smooth, horizontal table. The mass is made to move in a circle on the table. Find, in rev min^{-1}, the maximum rate at which the mass can revolve without breaking the string.

4 A coin of mass 0.005 kg is placed on a turntable, which is then rotated at $30\,\text{rev min}^{-1}$. If the coin does not slip, what is the minimum coefficient of friction when

a) the coin is placed at a distance of 6 cm from the centre
b) the coin is placed at a distance of 8 cm from the centre
c) a coin of mass 0.01 kg is used in each of the cases **a** and **b**?

5 A railway engine of mass 50 tonnes is rounding a horizontal curve of radius 200 m. The track can withstand a sideways force of 36 kN before it buckles.

a) What is the maximum speed at which the train can travel safely?
b) What would be the maximum speed for an engine of twice the mass?

6 Four particles, each of mass 3 kg, are connected by light, inextensible strings, each of length 0.08 m, so that they form a square with the particles at the corners and the strings forming the sides. The particles are placed in this configuration symmetrically on a smooth turntable, which is made to rotate with angular speed 2 rad s^{-1}. Find the tension in the strings.

7 A rough, horizontal disc rotates at 2 rev s^{-1}. A particle is to be placed on the disc. The coefficient of friction between the particle and the disc is μ. Find, in terms of μ, the maximum distance from the centre at which the particle can be placed without slipping.

8 A particle of mass 3 kg is placed on a rough, horizontal turntable and is connected to its centre by a light, inextensible string of length 0.8 m. The coefficient of friction between the particle and the turntable is 0.4. The turntable is made to rotate at a uniform speed. If the tension in the string is 50 N, find the angular speed of the turntable.

9 A particle, A, of mass m, is travelling in a horizontal circle on the surface of a smooth, horizontal table, on the end of a light, inextensible string, OA, of length $3a$, which is fastened to a point O on the table. The angular speed of the particle is ω. The string then catches on a peg, P, which is at a distance of $2a$ from O. Find the new angular speed of the particle and the new tension in the string.

10 Two particles, each of mass m, are attached to the ends of a light, inextensible string. The string passes through a hole in the centre of a rough, horizontal turntable. One particle is placed on the turntable at a distance a from its centre, and the other hangs freely below the turntable. The coefficient of friction between the particle and the turntable is μ. The contact between the string and the hole is smooth. The turntable is rotating with angular speed ω. Find the maximum and minimum values of ω if the particle on the turntable does not slip.

11 Newton's law of gravitation states that the force of attraction between two bodies of masses M and m is given by

$$F = \frac{GMm}{r^2}$$

where $G = 6.67 \times 10^{-11}$ N m^2 kg^{-2} is the universal gravitational constant, and r is the distance between the centres of the bodies.

a) A communications satellite is in geostationary orbit above the Equator (that is, it moves in a circular orbit so as to always be above the same point on the Earth). What is the angular speed of the satellite?

b) Taking the mass of the Earth to be 5.97×10^{24} kg, find the radius of the satellite's orbit.

c) A second satellite moves in a circular orbit at a height of 1200 km above the surface of the Earth. Find the time taken (to the nearest minute) for it to complete one orbit.

12 A fairground ride consists of a large hollow cylinder placed with its axis vertical. The customers stand against the inside wall of the cylinder, and the whole assembly rotates. When it reaches top speed, the floor is lowered, leaving the riders supported only by the friction between themselves and the wall. The radius of the cylinder is 3 m and the coefficient of friction between a rider and the wall is $\frac{1}{3}$. What is the minimum angular speed at which the ride will operate effectively?

Problems involving non-horizontal forces

In all the situations we have examined so far, the forces causing the acceleration towards the centre of the circle have been horizontal. There are many situations where a body travels in a horizontal circle but where the constraining force has a vertical component.

Conical pendulum

The term 'conical pendulum' refers to the situation in which a body is attached by a string to a fixed point, and travels in a horizontal circle below that point. We usually make the modelling assumptions that the string is light and inextensible, that the body is a particle and that air resistance can be neglected.

Example 6 A pendulum bob, P, of mass 4 kg, hangs at the end of a light, inextensible string of length 5 m. The other end of the string is fixed to a point O. The bob is made to describe a horizontal circle of radius 3 m with uniform speed. Find the tension in the string and the angular speed of the bob.

SOLUTION

By Pythagoras' theorem, we can see that the circle is 4 m below point O. Hence, if θ is the angle between the string and the horizontal, we have

$$\cos\theta = \tfrac{3}{5} \quad \text{and} \quad \sin\theta = \tfrac{4}{5}$$

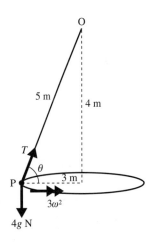

Let the tension in the string be T N and the angular speed be ω rad s^{-1}.

The bob has zero vertical acceleration, and an acceleration of $r\omega^2 = 3\omega^2$ towards the centre of the circle.

Resolving vertically, we have

$$T\sin\theta - 4g = 0$$
$$\Rightarrow \quad \tfrac{4}{5}T = 4g$$
$$\Rightarrow \quad T = 5g\,\text{N} = 49\,\text{N} \qquad [1]$$

So the tension in the string is 49 N.

Resolving towards the centre of the circle, we have

$$T\cos\theta = 4 \times 3\omega^2$$
$$\Rightarrow \quad \omega^2 = \tfrac{1}{20}T$$

Substituting from [1], we have

$$\omega^2 = 2.45$$

$$\Rightarrow \quad \omega = 1.57\,\text{rad s}^{-1}$$

So, the angular speed of the bob is $1.57\,\text{rad s}^{-1}$.

Example 7 A mass of $3\,\text{kg}$ is attached to the mid-point, P, of a light, inextensible string AB of length $4\,\text{m}$. The ends A and B are attached to two fixed points, with A a distance $2\,\text{m}$ vertically above B. The mass travels in a horizontal circle with angular speed ω, so that both parts of the string are taut. Find the minimum value of ω.

SOLUTION

APB is an equilateral triangle, so angle APO is $30°$ and the radius, OP, of the circle is $\sqrt{3}\,\text{m}$.

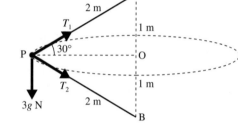

The tensions in the strings AP and PB are T_1 and T_2 respectively.

The mass has zero acceleration vertically. So, resolving vertically, we have

$$T_1 \sin 30° - T_2 \sin 30° - 3g = 0$$

$$\Rightarrow \quad T_1 - T_2 = 6g \qquad [1]$$

The acceleration towards the centre of the circle is $r\omega^2 = \omega^2\sqrt{3}$. So, resolving towards O, we have

$$T_1 \cos 30° + T_2 \cos 30° = 3\omega^2\sqrt{3}$$

$$\Rightarrow \quad T_1 + T_2 = 6\omega^2 \qquad [2]$$

Subtracting [1] from [2], we get

$$2\,T_2 = 6(\omega^2 - g)$$

If the portion PB of the string is to be taut, T_2 must be positive (or zero if it is just going slack). Therefore, we have

$$\omega^2 \geqslant g \quad \Rightarrow \quad \omega \geqslant \sqrt{g}\,\text{rad s}^{-1}$$

So, the minimum value of ω is $\sqrt{g}\,\text{rad s}^{-1}$.

Example 8 A light, inextensible string, of length 0.72 m, is attached to two points A and B, where A is vertically above B and AB = 0.48 m. A smooth ring, P, of mass 0.05 kg, is threaded on the string and is made to move in a horizontal circle about B. Find the angular speed of the ring and the tension in the string. Take $g = 9.8\,\mathrm{m\,s^{-2}}$.

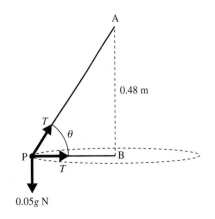

SOLUTION

Because the ring can move freely on the string, the tension in both portions of the string is the same.

We know

$$AP + PB = 0.72 \qquad\qquad [1]$$

By Pythagoras' theorem, we have

$$AP^2 - PB^2 = 0.2304$$

$$\Rightarrow \quad (AP + PB)(AP - PB) = 0.2304$$

Substituting from [1], we get

$$0.72\,(AP - PB) = 0.2304$$

$$\Rightarrow \quad AP - PB = 0.32 \qquad\qquad [2]$$

Adding [1] and [2], we get

$$2AP = 1.04 \quad \Rightarrow \quad AP = 0.52\,\mathrm{m} \quad \text{and} \quad PB = 0.2\,\mathrm{m}$$

If θ is the angle APB, then we have

$$\cos\theta = \frac{0.2}{0.52} = \frac{5}{13} \quad \text{and} \quad \sin\theta = \frac{0.48}{0.52} = \frac{12}{13}$$

Vertically, the acceleration is zero. So, resolving vertically, we have

$$T\sin\theta = 0.05\,g$$

$$\Rightarrow \quad T = 0.05g \div \frac{12}{13} = 0.531\,\mathrm{N}$$

The acceleration towards the centre, B, is $0.2\,\omega^2$. So, resolving towards B, we have

$$T + T\cos\theta = 0.05 \times 0.2\,\omega^2$$

$$\Rightarrow \quad \omega^2 = \frac{T\left(1 + \frac{5}{13}\right)}{0.05 \times 0.2}$$

Substituting for T, we get

$$\omega^2 = 73.5 \quad \Rightarrow \quad \omega = 8.57\,\mathrm{rad\,s^{-1}}$$

So, the tension is 0.531 N and the angular speed is $8.57\,\mathrm{rad\,s^{-1}}$.

Banked curves

Many racing tracks have their curved sections banked. This enables the cars to travel more quickly round the curves and still be safe. The same principle is also applied on some railway tracks and public roads. However, there is a potential danger in that a vehicle travelling at too slow a speed may be unable to stay on a very steeply banked track.

The reason banking is an advantage is that the normal reaction between the vehicle and the track has a horizontal component when the track is banked. This component helps to provide the central force needed to keep the vehicle travelling in a circle.

Example 9 A car is travelling round a circular bend of radius 30 m. The coefficient of friction between the car and the road is 0.3. The car has a mass of 600 kg.

a) What is the maximum safe speed for the car if
   i) the road is unbanked
   ii) the road is banked at 20° to the horizontal?
b) What is the minimum possible speed of the car if the road is banked at 20°?

SOLUTION

a) i) In the diagram, the centre of the bend is towards the left.

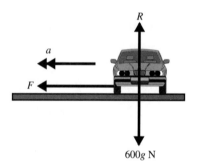

There is no vertical acceleration. So, resolving vertically, we have

$$R - 600g = 0 \quad \Rightarrow \quad R = 600g$$

When the car is on the point of slipping, we have

$$F = 0.3R = 180g \qquad\qquad [1]$$

Let the speed of the car be $v \, \text{m s}^{-1}$. Therefore, the acceleration towards the centre of the circle is $\dfrac{v^2}{30} \, \text{m s}^{-2}$. So, resolving towards the centre, we have

$$F = 600 \times \frac{v^2}{30}$$

Substituting from [1], we get

$$180g = 20v^2$$
$$\Rightarrow \quad v^2 = 9g \quad \Rightarrow \quad v = 9.39 \, \text{m s}^{-1}.$$

So, the maximum safe speed is $9.39 \, \text{m s}^{-1}$, or $33.8 \, \text{km h}^{-1}$.

ii) The forces acting on the car in this situation are shown in the diagram. If the car is travelling at maximum speed, it will tend to slip up the slope, so the friction force must act down the slope.

When the car is about to slip, $F = 0.3R$.

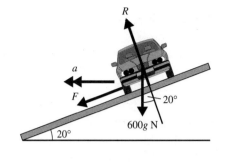

There is no vertical acceleration. So, resolving vertically, we have

$$R\cos 20° - F\sin 20° - 600g = 0$$
$$\Rightarrow \quad R(\cos 20° - 0.3\sin 20°) = 600g \qquad [2]$$

As before, the acceleration towards the centre of the circle is $\dfrac{v^2}{30}$.

Resolving towards the centre of the circle, we have

$$R\sin 20° + F\cos 20° = 600 \times \frac{v^2}{30}$$
$$\Rightarrow \quad R(\sin 20° + 0.3\cos 20°) = 20v^2 \qquad [3]$$

Dividing [3] by [2], we get

$$\frac{v^2}{30g} = \frac{\sin 20° + 0.3\cos 20°}{\cos 20° - 0.3\sin 20°}$$
$$\Rightarrow \quad v^2 = 219.13 \quad \Rightarrow \quad v = 14.8\,\mathrm{m\,s^{-1}}$$

So, the maximum speed in this situation is $14.8\,\mathrm{m\,s^{-1}}$, or $53.3\,\mathrm{km\,h^{-1}}$.

b) If the car is travelling slowly, its tendency is to slip down the slope, so the friction force acts up the slope, as shown.

Resolving vertically, we have

$$R\cos 20° + F\sin 20° - 600g = 0$$
$$\Rightarrow \quad R(\cos 20° + 0.3\sin 20°) = 600g \qquad [4]$$

Resolving towards the centre of the circle, we have

$$R\sin 20° - F\cos 20° = 600 \times \frac{v^2}{30}$$
$$\Rightarrow \quad R(\sin 20° - 0.3\cos 20°) = 20v^2 \qquad [5]$$

Dividing [5] by [4], we get

$$\frac{v^2}{30g} = \frac{\sin 20° - 0.3\cos 20°}{\cos 20° + 0.3\sin 20°}$$
$$\Rightarrow \quad v^2 = 16.96 \quad \Rightarrow \quad v = 4.12\,\mathrm{m\,s^{-1}}$$

So, the minimum safe speed is $4.12\,\mathrm{m\,s^{-1}}$, or $14.8\,\mathrm{km\,h^{-1}}$.

Note When a train goes round a curve, the sideways force needed to keep it on the track is provided by the flanges on the wheels. There is very little friction force to assist in this. Banking is introduced to minimise this sideways force.

Example10 A railway track has a curve with radius 600 m. At what angle should the track be banked so that a train travelling at $30\,\mathrm{m\,s^{-1}}$ has no sideways reaction between the wheel flanges and the track?

SOLUTION

When there is no sideways force between the wheels and the track, there are only two forces acting on the train: its weight, mg, and the normal reaction, R, as shown.

There is no vertical acceleration. So, resolving vertically, we have

$$R\cos\theta - mg = 0$$
$$\Rightarrow \quad R\cos\theta = mg \qquad [1]$$

The acceleration towards the centre of the circle is $\dfrac{30^2}{600} = 1.5\,\mathrm{m\,s^{-2}}$

Resolving towards the centre of the circle, we have

$$R\sin\theta = 1.5m \qquad [2]$$

Dividing [2] by [1], we get

$$\tan\theta = \frac{1.5}{g} = 0.153 \quad \Rightarrow \quad \theta = 8.7°$$

So, the track must be banked at 8.7° to the horizontal.

Exercise 16D

1 A ball of mass 3 kg is fastened to one end of a rope of length 0.5 m. The other end of the rope is fixed and the ball rotates as a conical pendulum at a rate of $5\,\mathrm{rad\,s^{-1}}$.

a) How far below the fixed end of the rope is the centre of the circle traced out by the ball?
b) What assumptions have you made in forming your solution to the problem?

2 A particle moves as a conical pendulum at the end of a light, inextensible string of length 40 cm. If the string makes an angle of 30° with the horizontal, find the angular speed of the particle.

3 A mass of 0.5 kg, suspended by a light, inextensible string of length 1.5 m, revolves as a conical pendulum at $30\,\mathrm{rev\,min^{-1}}$. Find the radius of the circle it travels and the tension in the string.

4 A particle moves as a conical pendulum at the end of a light, inextensible string, which has its fixed end at point A. The angular speed of the particle is ω. The centre of the circle travelled by the particle is O.

a) Show that $AO = g/\omega^2$.
b) Explain why the string cannot be horizontal.

5 A ball of mass 1 kg is fastened to one end of a light, inextensible string of length 1 m. The ball is placed on a smooth, horizontal table. The string is suspended from a point above the table so that the particle moves as a conical pendulum whilst in contact with the table. The radius of the circle travelled by the particle is 0.5 m.

a) If the speed of the particle is $1.5\,\mathrm{m\,s^{-1}}$, what is the normal reaction between the ball and the table?
b) What is the maximum speed the ball could travel without lifting off the table?

6 Two points, A and B, are on a vertical pole, 9 m apart with A above B. A rope of length 27 m is fastened at its ends to A and B. A smooth, heavy metal ring of mass m is threaded onto the rope. The ring is made to move in a horizontal circle about the pole. The upper section, AS, of the rope makes an angle θ with the vertical, as shown. Find the speed of the ring and the tension in the rope when

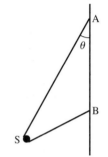

a) $\tan\theta = \frac{8}{15}$　　**b)** $\tan\theta = \frac{4}{3}$

7 A ball, B, of mass 2 kg is attached to one end of a light, inextensible string. The string passes through a smooth, fixed ring, O, and a second ball, A, of mass 4 kg, is attached to the other end. B is made to move as a conical pendulum while A hangs vertically below the ring, as shown. If the speed of B is $7\,\mathrm{m\,s^{-1}}$, how long is the section BO of the string?

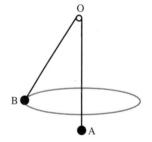

8 A pendulum bob, P, of mass 1.2 kg, hangs at one end of a light, inextensible string which passes through a smooth hole in a table at a point O. The length of OP is 0.7 m. The other end of the string is attached to a particle, Q, of mass 5.2 kg, which is resting on the rough horizontal surface of the table. The coefficient of friction between Q and the table is 0.25. The bob, P, is made to move as a conical pendulum below O. Find the maximum angular speed at which it can move without making Q slip.

9 A car is travelling round a bend of radius 50 m, which is banked at 15° to the horizontal. At what speed should the car travel if there is to be no lateral force on the car?

10 A car is travelling round a curve of radius 150 m banked at 25° to the horizontal. The coefficient of friction between the wheels and the road is 0.4.

a) What is the ideal speed of the car round the curve: that is, when there is no lateral friction force?
b) What is the maximum safe speed of the car?
c) What is the minimum safe speed of the car?

11 The distance between the rails of a track is 1.5 m. The track is banked round a curve of radius 1500 m. The track has been designed for a train to travel at 50 km h^{-1} round the curve. At this speed, there is no lateral force on the rails.

a) How much higher than the inner rail should the outer rail be?

b) The maximum safe lateral force on the rails is 100 kN. What is the maximum safe speed that an engine of mass 60 tonnes can attain round the curve?

c) Is there a minimum safe speed?

12 A smooth wire has a bead of mass 0.005 kg threaded onto it. It is then bent round to form a circular hoop of radius 0.2 m. The hoop is fastened in a horizontal position, whilst the bead travels round it at a constant speed of 1 m s^{-1}. Find the magnitude and direction of the reaction force between the hoop and the bead.

13 The device illustrated is used as a governor for a steam engine. AB is a shaft which is turned by the engine. C is a collar which can slide on the shaft AB and which, in its lowest position, holds down a valve controlling the flow of steam to the piston. Spheres P and Q rotate with the shaft. As the rate of rotation increases, so does the tension in the rods connecting the spheres to the point B and to the collar C. Eventually the speed is great enough for the tensions in the lower rods to lift the collar, reducing the flow of steam. The spheres each have mass 3 kg and the collar has mass 6 kg. PBQC is a rhombus of side length 0.5 m. Angle PBC is 30°. Assuming that the rods are light and inextensible and that the spheres can be modelled as particles:

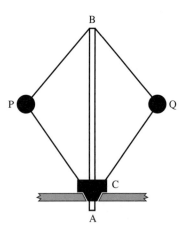

a) Find the tension in the lower rods in terms of ω, the angular speed of the shaft.

b) Find the value of ω if the collar is on the point of moving up the shaft.

14 A hemispherical bowl of radius 13 cm is fixed with its rim horizontal. A ball-bearing of negligible diameter is made to travel in a horizontal circle inside the bowl at a speed of 1.68 m s^{-1}. How far is the centre of the circle described by the ball-bearing above the bottom of the bowl?

15 A car is on a circular bend of radius r in a track banked at an angle θ to the horizontal. If the coefficient of friction between the car and the track is μ, show that, for the car to be secure on the track, its speed, v, must satisfy

$$\frac{rg(\sin\theta - \mu\cos\theta)}{\cos\theta + \mu\sin\theta} \leqslant v^2 \leqslant \frac{rg(\sin\theta + \mu\cos\theta)}{\cos\theta - \mu\sin\theta}$$

16 A hollow cone with semi-vertical angle θ is placed with its axis vertical and its vertex at the bottom. A smooth particle travels in a horizontal circle inside the cone. The angular speed of the particle is ω and the centre of the circle is a distance h above the vertex of the cone. Show that

$$h = \frac{g}{\omega^2 \tan^2\theta}$$

17 A railway track is banked round a circular bend. If a train travels on the track at speed V_1, there is a lateral reaction force, F, between the inner wheels and the rail. If the speed is increased to V_2, there is now a lateral reaction force, F, between the outer wheels and the rail. Find, in terms of V_1 and V_2, the speed at which the train must travel for there to be no lateral force between the wheels and the rails.

18 A car is travelling so that its centre of mass moves in a circle of radius 8 m. Its inner and outer wheels are 1.5 m apart. Its centre of mass is midway between the wheels and is 1 m above the ground. At what speed would the car have to travel to overturn, assuming that no side-slipping occurs?

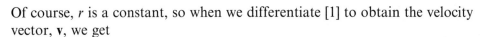

Circular motion with non-uniform speed

So far, we have considered only motion with uniform speed. We now examine the more general situation.

Suppose the particle, P, moves in a circle so that at time t its position vector is

$$\mathbf{r} = r(\cos\theta\,\mathbf{i} + \sin\theta\,\mathbf{j}) \qquad [1]$$

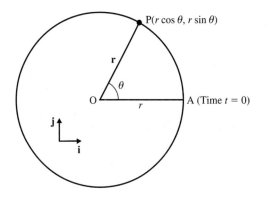

as shown, where θ is a function of t and $\dfrac{\mathrm{d}\theta}{\mathrm{d}t} = \dot{\theta}$ is not, in general, constant.

Notice that, by the chain rule,

$$\frac{\mathrm{d}(\cos\theta)}{\mathrm{d}t} = \frac{\mathrm{d}(\cos\theta)}{\mathrm{d}\theta} \times \frac{\mathrm{d}\theta}{\mathrm{d}t} = -\sin\theta\,\dot{\theta}$$

and

$$\frac{\mathrm{d}(\sin\theta)}{\mathrm{d}t} = \frac{\mathrm{d}(\sin\theta)}{\mathrm{d}\theta} \times \frac{\mathrm{d}\theta}{\mathrm{d}t} = \cos\theta\,\dot{\theta}$$

Of course, r is a constant, so when we differentiate [1] to obtain the velocity vector, $\mathbf{v}$, we get

$$\mathbf{v} = r\left(\frac{\mathrm{d}(\cos\theta)}{\mathrm{d}t}\mathbf{i} + \frac{\mathrm{d}(\sin\theta)}{\mathrm{d}t}\mathbf{j}\right)$$

$$\Rightarrow \quad \mathbf{v} = r\dot{\theta}(-\sin\theta\,\mathbf{i} + \cos\theta\,\mathbf{j}) \qquad [2]$$

From [2] we get two important facts:

- The speed of the particle is $|\mathbf{v}| = r\dot{\theta}$.

  This corresponds to $v = r\omega$, which we used for circular motion with uniform speed, but here $\dot{\theta}$ is not constant.

- $\mathbf{r.v} = r^2\dot{\theta}(-\sin\theta\cos\theta + \sin\theta\cos\theta) = 0$

  This means that the velocity is perpendicular to the radius. In other words, the velocity is **tangential** to the circle.

We now differentiate [2] to obtain the acceleration vector, $\mathbf{a}$. We need to use the product rule:

$$\mathbf{a} = r\frac{d\dot{\theta}}{dt}(-\sin\theta\,\mathbf{i} + \cos\theta\,\mathbf{j}) + r\dot{\theta}\left(-\frac{d(\sin\theta)}{dt}\mathbf{i} + \frac{d(\cos\theta)}{dt}\mathbf{j}\right)$$

$$= r\ddot{\theta}(-\sin\theta\,\mathbf{i} + \cos\theta\,\mathbf{j}) - r\dot{\theta}^2(\cos\theta\,\mathbf{i} + \sin\theta\,\mathbf{j})$$

Now, $(\cos\theta\,\mathbf{i} + \sin\theta\,\mathbf{j})$ is a unit vector in the direction of $\mathbf{r}$. So, we have

$$(\cos\theta\,\mathbf{i} + \sin\theta\,\mathbf{j}) = \hat{\mathbf{r}}$$

Similarly, $(-\sin\theta\,\mathbf{i} + \cos\theta\,\mathbf{j})$ is a unit vector in the direction of $\mathbf{v}$, and so tangential to the circle. Let us write

$$(-\sin\theta\,\mathbf{i} + \cos\theta\,\mathbf{j}) = \hat{\mathbf{t}}$$

We can then write the acceleration vector as

$$\mathbf{a} = r\ddot{\theta}\,\hat{\mathbf{t}} - r\dot{\theta}^2\,\hat{\mathbf{r}} \qquad\qquad [3]$$

From [3] we get two important facts:

- The particle has a **tangential component** of acceleration of $r\ddot{\theta}$.
- The particle has a **radial component** of acceleration of $-r\dot{\theta}^2$. This is directed towards the centre of the circle and corresponds to the $r\omega^2$ which we used for circular motion with uniform speed, but here $\dot{\theta}^2$ is not constant.

Motion in a vertical circle

A common example of circular motion with non-uniform speed is when a body is moving in a vertical circle. This occurs in many circumstances. Typically, a body may be rotating on the end of a string or rod, may be threaded onto a hoop or may be sliding on the inner or outer surface of a circular object. In all cases, the path may be a complete circle, or just a circular arc.

The situations listed can be divided into two categories. In the first, the body cannot leave the circular path. In the second, the body may leave the circular path at some stage of its motion.

The body cannot leave the circle

This includes bodies rotating on the end of rigid rods and beads threaded onto hoops. Such systems can behave in one of two ways:

- If the energy of the system is sufficient, the body rotates in a complete circle.

- If the energy is insufficient, the body cannot reach the highest point of the circle, but oscillates between two symmetrical positions, at each of which its speed is instantaneously zero, as shown.

The body can leave the circle

This includes bodies rotating on the end of strings, or sliding on the surface of circular objects. Except for the case of a body sliding on the outer surface of a circular object, which is bound to leave the circle at some point, such systems can behave in any of three ways.

- If the energy of the system is sufficient, the body rotates in a complete circle.

- If the energy of the system is so low that the body cannot rise beyond the level of the centre of the circle, it oscillates between two symmetrical positions, at each of which its speed is instantaneously zero, as shown.

- If the energy of the system is such that the body can rise above the level of the centre of the circle without being enough to carry it completely round the circle, it will leave the circle at some point and its motion will become that of a projectile, as shown.

The mechanics of motion in a vertical circle

When modelling motion in a vertical circle, we will make the following assumptions:

- The body is a particle.
- There is no air resistance.
- There is no loss of energy through any other resistance.
- If the particle is attached to a string or rod, this is light and inextensible.
- The path is a perfect circle. This assumption is often made when modelling situations such as roller-coaster loop-the-loops, in which the entry and exit paths do not coincide, but where the car travels in something very close to a circle.

In most problems concerning motion in a vertical circle, we make use of the principle of conservation of energy.

Example 11 A pendulum consists of a bob, P, of mass 2 kg, attached to a light rod of length 1 m, the other end of which is freely hinged to a fixed point, O. The bob rests vertically below O, and is then given an impulse so that it starts moving with speed u. Taking $g = 10 \, \text{m s}^{-2}$, find the angle between the rod and the downward vertical when the bob reaches its highest point if

a) $u = 3 \, \text{m s}^{-1}$ **b)** $u = 5 \, \text{m s}^{-1}$

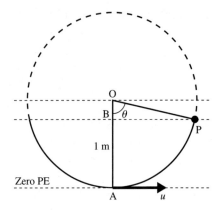

SOLUTION

We take the gravitational potential energy to be zero at the point A, where the motion commences. Let P be the highest point reached by the bob. The speed of the bob at P is therefore zero.

$OP = 1 \, \text{m}$, so $OB = \cos\theta \, \text{m}$. Hence, $AB = (1 - \cos\theta) \, \text{m}$

When the bob is at A, we have

$$\text{Kinetic energy} = \tfrac{1}{2} \times 2 \times u^2 = u^2$$

$$\text{Potential energy} = 0$$

When the bob is its highest point, P, we have

$$\text{Kinetic energy} = 0$$

$$\text{Potential energy} = 2g \times AB = 20(1 - \cos\theta)$$

Energy is conserved, so we have

$$u^2 = 20(1 - \cos\theta)$$

$$\Rightarrow \quad \cos\theta = 1 - \frac{u^2}{20}$$

a) When $u = 3$, we have

$$\cos\theta = 1 - \frac{9}{20} = 0.55 \quad \Rightarrow \quad \theta = 56.6°$$

b) When $u = 5$, we have

$$\cos\theta = 1 - \frac{25}{20} = -0.25 \quad \Rightarrow \quad \theta = 104.5°$$

Note In part **b**, the bob is above the level of O when it comes to rest. This is only possible when the pendulum involves a rod. Had string been used, the bob would have left the circle.

Example 12 A hollow cylinder, of internal radius 2 m, rests with its axis horizontal. O is a point on that axis. A particle, P, of mass 5 kg, rests inside the cylinder vertically below O. It is given an impulse so that it starts to move on the smooth surface of the cylinder in a circle about O. Its initial speed is $8\,\mathrm{m\,s^{-1}}$. Take $g = 10\,\mathrm{m\,s^{-2}}$.

a) What is the normal reaction between the cylinder and the particle when the line OP makes an angle of 60° with the downward vertical?
b) What angle does OP make with the downward vertical when the normal reaction is zero?
c) What is the speed of the particle when the normal reaction is zero?

SOLUTION

First, we use conservation of energy to find the speed of the particle when OP makes an angle θ with the downward vertical.

OP is 2 m, so $OB = 2\cos\theta$. Hence, $AB = 2 - 2\cos\theta = 2(1 - \cos\theta)\,\mathrm{m}$

When the particle is at A, we have

$$\text{Kinetic energy} = \tfrac{1}{2} \times 5 \times 8^2 = 160\,\mathrm{J}$$

$$\text{Potential energy} = 0$$

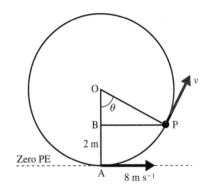

When the particle is at a general point, P, we have

$$\text{Kinetic energy} = \tfrac{1}{2} \times 5 \times v^2 = 2.5v^2$$

$$\text{Potential energy} = 5g \times AB = 100(1 - \cos\theta)$$

Energy is conserved, so we have

$$2.5\,v^2 + 100(1 - \cos\theta) = 160$$

$$\Rightarrow \quad v^2 = 8(3 + 5\cos\theta) \qquad [1]$$

We now need to consider the forces acting on the particle.

Because the surface is smooth, the only forces acting on the particle are its weight, 50 N, and the normal reaction, R.

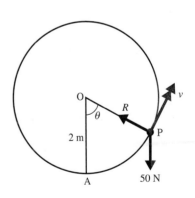

The component of acceleration in the direction PO is $\dfrac{v^2}{2}$.

So, resolving in the direction PO, we have

$$R - 50\cos\theta = 5 \times \frac{v^2}{2}$$

$$\Rightarrow \quad R = 2.5v^2 + 50\cos\theta$$

Substituting from [1], we have

$$R = 20(3 + 5\cos\theta) + 50\cos\theta$$

$$\Rightarrow \quad R = 60 + 150\cos\theta \qquad [2]$$

a) When $\theta = 60°$, we have from [2]

$$R = 60 + 150 \times \tfrac{1}{2} = 135\,\text{N}$$

b) When $R = 0$, we have from [2]

$$60 + 150 \cos\theta = 0$$

$$\Rightarrow \quad \cos\theta = -0.4 \quad \Rightarrow \quad \theta = 113.6°$$

c) Substituting $\cos\theta = -0.4$ into [1], we get

$$v^2 = 8(3 + 5 \times (-0.4)) = 8$$

$$\Rightarrow \quad v = 2\sqrt{2} = 2.83\,\text{m s}^{-1}$$

Notice that the normal reaction becomes zero at the point where the particle leaves the circle. At this point, it is travelling with a speed of $2.83\,\text{m s}^{-1}$ at an angle of $66.4°$ to the horizontal. Its subsequent motion is as a projectile with this initial velocity.

Example 13 A pendulum consists of a bob, P, of mass 3 kg, attached to the end of a light rod of length 2 m, whose other end is freely hinged at the point O. The bob rests at its lowest point. It is then given an impulse and commences to move at speed $u\,\text{m s}^{-1}$. Take $g = 10\,\text{m s}^{-2}$.

a) Find the minimum value of u, given that the pendulum performs a complete circle.

b) If the rod were replaced by a string, find the minimum value of u for the pendulum to perform a complete circle.

SOLUTION

a) Suppose the pendulum reaches the top, B, with speed $v \geqslant 0$. Take the zero level for potential energy as shown.

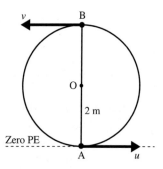

When the bob is at A, we have

$$\text{Kinetic energy} = \tfrac{1}{2} \times 3 \times u^2 = 1.5u^2$$
$$\text{Potential energy} = 0$$

When the bob is at B, we have

$$\text{Kinetic energy} = \tfrac{1}{2} \times 3 \times v^2 = 1.5v^2$$
$$\text{Potential energy} = 3g \times 4 = 120\,\text{J}$$

Energy is conserved, so we have

$$1.5u^2 = 1.5v^2 + 120 \qquad\qquad [1]$$

As $v \geqslant 0$, we have

$$1.5u^2 \geqslant 120 \quad \Rightarrow \quad u \geqslant 4\sqrt{5}\,\text{m s}^{-1} \quad \text{or} \quad u \geqslant 8.944\,\text{m s}^{-1}$$

So, the initial speed of the bob should be at least $8.944\,\text{m s}^{-1}$.

b) If the rod is replaced by a string, the energy equation [1] still applies. However, the speed, v, of the bob when it reaches the top must be great enough for the string to remain taut. In other words, if T is the tension in the string, then $T \geqslant 0$.

At B, the acceleration of the bob towards the centre of the circle is $v^2/2$. So, resolving in the direction BO, we have

$$T + 30 = 3 \times \frac{v^2}{2}$$

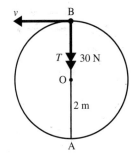

As $T \geqslant 0$, we have

$$\frac{3v^2}{2} \geqslant 30 \quad \Rightarrow \quad v^2 \geqslant 20 \qquad [2]$$

Substituting from [2] into [1], we get

$$1.5u^2 \geqslant 1.5 \times 20 + 120$$

$$u^2 \geqslant 100 \quad \Rightarrow \quad u \geqslant 10\,\mathrm{m\,s}^{-1}$$

So, for a string, the initial speed of the bob should be at least $10\,\mathrm{m\,s}^{-1}$.

Example 14 A particle of mass $0.01\,\mathrm{kg}$ rests at the top of a smooth sphere of radius $0.2\,\mathrm{m}$ which is fixed to a horizontal table. It is displaced slightly so that it slides off the sphere. How far from the point of contact between the sphere and the table does the particle strike the table? Take $g = 9.8\,\mathrm{m\,s}^{-2}$.

SOLUTION

There are two stages to the motion of the particle. In the first stage, the particle slides down the outside of the sphere. At some point it leaves the surface, leading to the second stage in which the particle moves as a projectile.

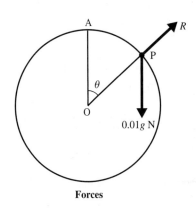

Forces

We need to find the position of the particle and its velocity at the point where it leaves the surface. This will provide the initial conditions for the projectile motion.

The particle will leave the surface at the point where the reaction between the particle and the surface becomes zero.

At point A, we have

Kinetic energy $= 0$

Potential energy $= 0.01 \times g \times 0.2 = 0.002g$

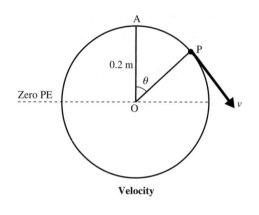

At a general point, P, we have

Kinetic energy $= \frac{1}{2} \times 0.01 \times v^2 = 0.005v^2$

Potential energy $= 0.01 \times g \times 0.2 \cos \theta$

$= 0.002 \, g \cos \theta$

Energy is conserved, so we have

$0.005v^2 + 0.002 \, g \cos \theta = 0.002 \, g$

$\Rightarrow \quad v^2 = 0.4g(1 - \cos \theta)$ [1]

The acceleration towards the centre of the circle is $\dfrac{v^2}{0.2} = 5 \, v^2$.

So, resolving in the direction PO, we have

$0.01 \, g \cos \theta - R = 0.01 \times 5v^2$

$\Rightarrow \quad v^2 = 0.2 \, g \cos \theta - 20 \, R$

When $R = 0$, we have

$v^2 = 0.2 \, g \cos \theta$ [2]

From [1] and [2], we have

$\cos \theta = \frac{2}{3} \quad \Rightarrow \quad \theta = 48.2°$

Substituting into [2], we get

$v^2 = 1.31 \quad \Rightarrow \quad v = 1.143 \, \mathrm{m \, s^{-1}}$

The projectile stage of the motion starts as shown in the diagram.

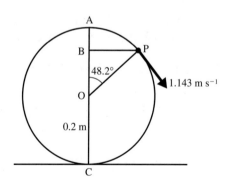

The distance BC $= 0.2 \, (1 + \cos 48.2°) = \frac{1}{3}$ m

The vertical component of the initial velocity is $1.143 \sin 48.2° = 0.852 \, \mathrm{m \, s^{-1}}$.

Using $s = ut + \frac{1}{2}at^2$, the particle reaches the table at time t, where

$\frac{1}{3} = 0.852t + \frac{1}{2}gt^2$

$\Rightarrow \quad 4.9t^2 + 0.852t - \frac{1}{3} = 0$

Solving this quadratic equation, we get $t = 0.188 \, \mathrm{s}$ (or $-0.362 \, \mathrm{s}$).

The horizontal component of the initial velocity is
$1.143 \cos 48.2° = 0.762 \, \text{m s}^{-1}$.

In 0.188 s, the particle travels $0.762 \times 0.188 = 0.143 \, \text{m}$
horizontally from the point where it left the sphere.

As $BP = 0.2 \sin 48.2° = 0.149 \, \text{m}$, the particle lands
$0.149 + 0.143 = 0.292 \, \text{m}$ from C.

Exercise 16E

1 A pendulum consists of a rod of length 1 m with a bob of mass 2 kg attached to one end. The rod is freely pivoted at the other end, O, so that it can rotate in a vertical circle. The pendulum rests with the bob vertically below O. The bob is then given an impulse so that it starts to move with speed $6.5 \, \text{m s}^{-1}$. Assuming that the rod is light and that the bob can be modelled as a particle, what is

a) the speed of the bob
b) the force in the rod

when the pendulum makes an angle of **i)** 30°, **ii)** 90° and **iii)** 150° with the downward vertical?

2 A particle of mass 0.1 kg is attached by a string of length 1.5 m to a fixed point, and is made to travel in a vertical circle about that point.

a) Find the minimum velocity the particle must have at the lowest point of the circle if it is to make complete revolutions.
b) For this velocity, find
 i) the tension in the string when the particle is at point A, a distance of 75 cm above the lowest point
 ii) the tangential component of the particle's acceleration when it is at point A.

3 A particle of mass 0.01 kg is placed on the topmost point, A, of a smooth sphere of centre O and radius 0.5 m. It is slightly displaced. When it reaches point B it is about to leave the surface of the sphere. Calculate the angle AOB.

4 A particle of mass m hangs at rest, suspended from a point, O, by a light, inextensible string of length a. The particle receives an impulse so that it starts moving with speed $\sqrt{3ga}$. Find the angle between the string and the vertical when it goes slack.

5 A stone of mass 0.5 kg performs complete revolutions in a vertical circle on the end of a light, inextensible string of length 1 m. Show that the string must be strong enough to support a tension of at least 29.4 N.

6 A particle of mass m travels in complete vertical circles on the end of a light, inextensible string of length a. If the maximum tension in the string is three times the minimum tension, find the speed of the particle as it passes through the lowest point on the circle.

7 A pendulum of length a has a bob of mass m. The speed of the bob at the lowest point of its path is U. Find the condition which U must satisfy for the bob to make complete revolutions if the pendulum consists of **a)** a rod, **b)** a string.

8 A particle of mass m is projected horizontally with speed v from the topmost point, A, of a sphere of radius a and centre O. It remains in contact with the sphere until leaving the surface at point B. If angle AOB is $30°$, find v.

9 A particle of mass $2\,\text{kg}$ is attached to the end of a light, inextensible string of length $1\,\text{m}$, the other end of which is attached to a fixed point, O. The particle is held with the string taut and horizontal, and is released from rest. When the string reaches the vertical position, it meets a fixed pin, A, a distance x below O. Given that the particle just completes a circle about A, find the value of x.

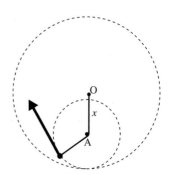

10 A bead of mass m is threaded onto a smooth, circular hoop of radius a, which is fixed in a vertical plane. The bead is displaced from rest at the top of the hoop. Find the **resultant** acceleration of the bead when it has reached a point which is a vertical distance $\frac{3}{4}a$ below its starting point.

11 A body of mass $40\,\text{kg}$ is swinging on the end of a light rope of length $3\,\text{m}$, which is attached to a fixed point $3.6\,\text{m}$ above horizontal ground. The body moves so that at the extreme positions of its motion, the rope makes an angle of $60°$ with the downward vertical through O. At an instant when the pendulum makes an angle of $30°$ with the downward vertical through O and the body is rising, the body breaks free from the rope. Calculate the horizontal displacement of the body from O at the point where it hits the ground.

12 A smooth wire forms a circular hoop of radius $1\,\text{m}$. It is fixed in a vertical plane. Two beads, A and B, of masses m and $2m$ respectively, are threaded onto the wire. The coefficient of restitution between the beads is 0.5. Bead B rests at the bottom of the hoop. Bead A is projected from the topmost point with speed u, and subsequently collides with B.

a) Show that bead A is brought to rest by the collision.
b) Find the value of u, given that the collision imparts just enough speed to bead B for it to make a complete revolution.
c) Investigate whether bead A makes a complete revolution after the second collision.

13 A pendulum bob of mass m is fastened to one end of a string of length r whose other end is fixed at a point O. The bob is at rest in its lowest position when it is set in motion with initial

speed $\sqrt{\dfrac{7gr}{2}}$. As it swings upwards, the string meets a small, fixed peg, P, on the same level as O. The string then wraps round P. What is the closest that P can be to O so that the bob make a complete revolution about P?

14 A and B are two perfectly elastic particles. A is attached to one end of a rod of length *a*. B is attached to one end of a string of length *a*. The other ends of the rod and the string are attached to a fixed point O. B hangs at rest below O. A is displaced from rest vertically above O and swings down to strike B. If the collision imparts just enough speed to B for it to make a complete revolution, find the ratio of the masses of the particles.

15 A ring of mass 5 kg is threaded onto a rope of length 10 m, whose ends are attached to two fixed points 6 m apart and on the same level. The ring hangs at rest. It is then set in motion so that it travels on a circular path whose plane is perpendicular to the line joining the two ends of the rope. Given that the ring can just make complete revolutions, find the maximum tension in the rope. State any modelling assumptions you have made in reaching your answer.

16 Particles A and B, of masses *m* and 2*m* respectively, are connected by a light, inextensible string of length π*a*. The particles are placed symmetrically, and with the string taut, on the smooth outer surface of a cylinder of radius 3*a*, as shown, and the system is released from rest. Find the reactions between the cylinder and the particles at the moment when A reaches the topmost point.

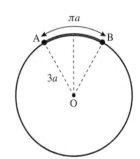

17 The diagram shows a loop-the-loop on a roller-coaster ride. The car approaches the loop on a horizontal track. The maximum speed at which the car can enter the loop is 80 km h^{-1}. What is the greatest radius with which the loop can be constructed if the car is not to leave the track?

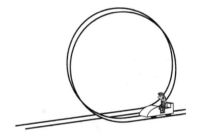

17 Elasticity

There was things which he stretched, but mainly he told the truth.
MARK TWAIN

Elastic strings and springs

In the problems we have so far solved, we have made the modelling assumption that any strings or rods involved are inextensible. That is, any stretching which takes place is negligible compared with the overall length of the string or rod.

Clearly, the above modelling assumption will not always be justified. Many strings, wires and rods can be stretched by significant amounts. We also have to deal with springs, which are designed to stretch and which can also be compressed.

A string or spring which can be stretched, and which regains its original length once the stretching force is removed, is said to be **elastic**. This is to distinguish it from **plastic** items, which can be stretched, but which do not recover when the force is removed. It is common experience that the amount of stretching produced in a string or spring depends on the magnitude of the force exerted. We will investigate suitable models for the relation between the applied force and the extension produced.

Experiments

Experiment 1

You need a spring of reasonable length and 'stretchiness', a mass-hanger and masses, and a metre rule or a tape measure.

Measure the unstretched length of the spring. Fasten the top of the spring to a fixed point and attach the mass-hanger. Hang a mass on the spring, and measure the overall length of the spring. Repeat this for a total of eight different masses.

When the spring is hanging in equilibrium, the tension in it is equal to the downward force supplied by the weight of the attached masses.

(As an alternative, you could use a force-meter or spring balance to measure the tension in the spring. In this case, it may be more convenient to carry out the experiment on a flat surface.)

Tabulate your results as shown and draw a scatter graph of length against tension. If you have access to the spreadsheet ELASTIC, you can enter your data there. The spreadsheet is on the OUP website: http://www.oup.co.uk/mechanics

You should notice that the graph you obtain is approximately linear. The spreadsheet draws a scatter graph and a line of best fit, as shown.

Suspended mass (kg)	Tension (N)	Length (m)
0.000	0.000	0.170
0.020	0.196	0.194
0.040	0.392	0.229
0.060	0.588	0.278
0.080	0.784	0.328
0.100	0.980	0.386
0.120	1.176	0.439
0.140	1.372	0.491
0.160	1.568	0.550

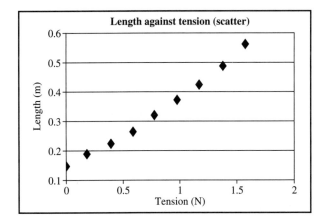

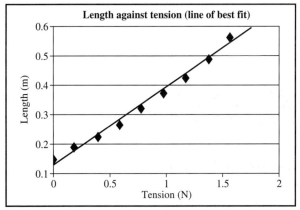

Experiment 2

Repeat Experiment 1, but replace the spring with a piece of elastic, or with a number of elastic bands 'chained' together. The table and graphs shown illustrate the results obtained by the author.

You will probably find that the graph is less obviously linear in this situation. For low tensions, the graph is usually approximately linear, but departs from this as the tension becomes larger. The spreadsheet draws a scattergraph and the line of best fit for the first six points.

Suspended mass (kg)	Tension (N)	Length (m)
0.000	0.000	0.265
0.020	0.196	0.285
0.040	0.392	0.312
0.060	0.588	0.347
0.080	0.784	0.390
0.100	0.980	0.446
0.120	1.176	0.510
0.140	1.372	0.585
0.160	1.568	0.665

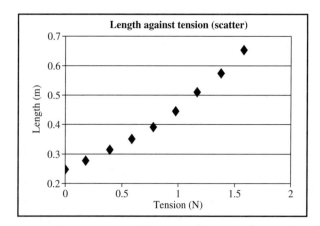

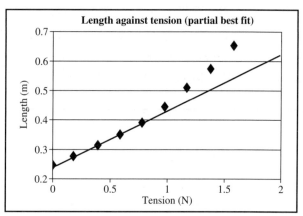

Our experimental design is too crude to make more precise measurements on a range of materials. More sophisticated equipment can be used to show that when, for example, a steel wire is stretched, the graph of its tension against its extension is linear. This is true up to a certain point, called the **elastic limit**. A wire stretched beyond this point loses its elasticity and cannot regain its natural length.

Interpretation

From out experimental results, particularly for Experiment 1, there appears to be a linear relation between the length of the spring, l, and the tension, T. Obviously, if the tension is zero, the length is the unstretched, or **natural length**, l_0. The relationship can therefore be written as

$$l = cT + l_o$$
$$\Rightarrow \quad cT = l - l_o$$
$$\Rightarrow \quad T = k(l - l_o)$$

Notice that $(l - l_o)$ is the **extension** produced in the spring.

This is the usual model adopted for elastic strings and springs. That is, there is a linear relation between the tension in the string or spring and the extension produced. This model is known as **Hooke's law**, after Robert Hooke, who formulated it in 1678. The model is represented by the equation

$$T = kx$$

where T is the **magnitude of the tension**
x is the **extension** from the natural length of the string or spring
k is a constant for a given string or spring, and is called the **stiffness**.

The main assumption embodied in the model is that the spring or string is light. This is clearly reasonable for most strings but a spring probably has a significant mass. In fact, if you suspend a spring without a load, you can probably see that the gaps between the coils are greater near the top of the spring, because there is a greater mass of spring below this point than near the bottom. This would have had an effect on your results if your experimental design involved suspending the spring and hanging a load on the end. If, on the other hand, you used a force-meter with the spring on a horizontal surface, the mass of the spring would not have affected your results in the same way. It is, however, possible that friction between the coils of the spring and the surface would have had an effect.

Compressing a spring

Springs differ from strings in that they can be compressed. In this case, we have a reduction, x, in the length (a negative extension) and a thrust force (a negative tension) in the spring. It can be shown experimentally that Hooke's law continues to be appropriate.

In practice, there is a limit to the amount by which a spring may be compressed, as eventually the coils touch. Our mathematical model, however, would allow the possibility of compressing the spring until its length was zero or even negative. It is important, therefore, that we check that the predictions of the model are realistic.

Units

From the modelling equation $T = kx$, we obtain

$$k = \frac{T}{x}$$

As a result of this, the SI unit for stiffness, k, is the **newton per metre** (N m^{-1}).

Example 1 An object of mass 4 kg is attached to the end of an elastic string whose unstretched length is 3 m. When the string is hung from a beam so that the suspended object is stationary, the string has a length of 3.5 m. What is the stiffness of the string? Take $g = 9.8 \, \mathrm{m \, s^{-2}}$.

SOLUTION

As the object is in equilibrium, the tension in the string must equal the weight of the object. Therefore, we have

$$T = 4 \times 9.8 = 39.2 \, \mathrm{N}$$

The extension in the string is $\quad x = 3.5 - 3 = 0.5 \, \mathrm{m}$

From $T = kx$, we have $\quad 39.2 = 0.5 \, k \quad \Rightarrow \quad k = 78.4 \, \mathrm{N \, m^{-1}}$

So, the stiffness of the string is $78.4 \, \mathrm{N \, m^{-1}}$.

Example 2 A block is fastened to two springs whose natural lengths are 0.5 m and 0.3 m. The springs have stiffness $30 \, \mathrm{N \, m^{-1}}$ and $50 \, \mathrm{N \, m^{-1}}$ respectively. The other ends of the springs are fastened to two points 0.6 m apart on a smooth table, and the block rests on the table so that the springs lie along a straight line. Find the compressed lengths of the springs.

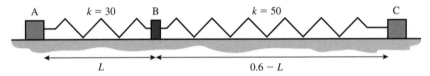

SOLUTION

We assume that the thickness of the block is zero. Let the length AB be L m. The length of BC is then $(0.6 - L)$ m.

Thus, we have

	AB	BC
Length	L	$0.6 - L$
Compression	$0.5 - L$	$0.3 - (0.6 - L) = L - 0.3$
Thrust (using $T = kx$)	$30(0.5 - L)$	$50(L - 0.3)$

As the block is in equilibrium, the two thrusts must be equal. This gives

$$30(0.5 - L) = 50(L - 0.3)$$

$$\Rightarrow \quad 15 - 30L = 50L - 15$$

$$\Rightarrow \quad 80L = 30$$

$$\Rightarrow \quad L = 0.375 \, \mathrm{m}$$

The compressed lengths of the springs are, therefore, AB $= 0.375$ m and BC $= 0.225$ m.

Example 3 A ball of mass 4 kg is fastened to two springs whose natural lengths are 1 m and 0.5 m. The other end of the first spring is fastened to a point, A, and the other end of the second spring to a point B, a distance of 3 m vertically below A. The ball rests in equilibrium. The stiffness of the springs is $30\,\mathrm{N\,m^{-1}}$ and $20\,\mathrm{N\,m^{-1}}$ respectively. Find the lengths of the springs. Take $g = 10\,\mathrm{m\,s^{-2}}$.

SOLUTION

Assume that the spring AM is extended a distance x m, so that its new length is $(1 + x)$ m.

The new length of spring BM is then $3 - (1 + x) = (2 - x)$ m

The extension of spring BM is then $(2 - x) - 0.5 = (1.5 - x)$ m

Let the tensions in the springs be T_1 and T_2, as shown in the diagram. By Hooke's law, we have

$$T_1 = 30x \quad \text{and} \quad T_2 = 20(1.5 - x)$$

The ball is in equilibrium. Resolving vertically, we have

$$T_1 - T_2 - 40 = 0$$
$$\Rightarrow \quad 30x - 20(1.5 - x) - 40 = 0$$
$$\Rightarrow \quad 50x - 70 = 0$$
$$\Rightarrow \quad x = 1.4$$

So, the spring AM is 2.4 m long and the spring BM is 0.6 m long.

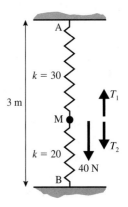

Exercise 17A

1 a) A ball of mass 4 kg is attached to one end of an elastic string whose other end is attached to the ceiling. The string has a natural length of 1 m and a stiffness of $32\,\mathrm{N\,m^{-1}}$ What is the stretched length of the string when the ball hangs at rest?

b) A second, identical, elastic string is fastened between the ball and the first elastic string. What will now be the distance below the ceiling at which the ball will hang at rest?

2 A spring of natural length 0.9 m is compressed by a force of 40 N to a length of 0.7 m.

a) What is the stiffness of the spring?
b) If it were stretched by the same force, how long would it be?

3 A spring is stretched by a force of 36 N to a length of 1.2 m. When it is compressed by a force of 24 N, its length is 0.6 m. What are the natural length and stiffness of the spring?

4 a) A block of mass 2 kg is attached to one end of an elastic string of length 1 m and stiffness $15\,\mathrm{N\,m^{-1}}$. The other end of the string is fastened to the ceiling and the block is lowered until it hangs at rest. What is the length of the string in this position?

b) A second elastic string is attached to the same point on the block and on the ceiling. This second string has a natural length of 0.8 m and a stiffness of $10 \, \text{N m}^{-1}$. The block is again allowed to hang at rest.

  i) How far below the ceiling does the block now hang?
  ii) What are the tensions in the strings?

5 A ball M of mass 5 kg is attached to the ends of two springs whose other ends are fixed to points A and B, with A 4 m vertically above B. Spring AM has natural length 2 m and stiffness $50 \, \text{N m}^{-1}$. Spring BM has natural length 1 m and stiffness $20 \, \text{N m}^{-1}$. The ball rests in its equilibrium position. What are the lengths of the springs?

6 A block of mass 5 kg is attached to one end of an elastic string of natural length 2 m and stiffness $40 \, \text{N m}^{-1}$. The other end of the string is attached to a fixed point. The block is allowed to hang at rest and is then pulled aside by a horizontal force of 30 N. Taking g to be $10 \, \text{m s}^{-2}$, find the stretched length of the string.

7 A block of mass 3 kg rests on a smooth plane which is inclined at $40°$ to the horizontal. The block is attached to a point at the top of the plane by means of an elastic string of natural length 1.6 m and stiffness $30 \, \text{N m}^{-1}$. Find the stretched length of the string.

8 A block of mass 2 kg rests on a rough plane which is inclined at $30°$ to the horizontal. The block is attached to a point at the top of the plane by means of an elastic string of natural length 2 m and stiffness $50 \, \text{N m}^{-1}$. The coefficient of friction between the block and the plane is 0.25. Find the distance between the lowest and highest positions in which the block will rest in equilibrium.

9 Each of the diagrams shows a compound spring made from identical springs each having stiffness k. In each case, find in terms of k the stiffness of the single spring which would be equivalent to the compound spring.

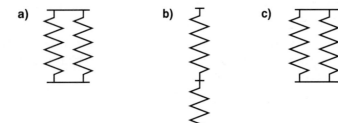

a)   b)   c)

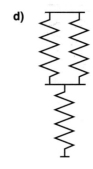

d)

10 Springs AB and BC are connected to a block B of mass m. The springs both have stiffness k. The natural length of spring AB is $3l$ and of BC is l. The ends A and C are fastened to points a distance $4l$ apart on a rough horizontal plane. The coefficient of friction between the block and the plane is μ. The block can just rest in limiting equilibrium at the mid-point of AC.

  a) Find an expression for k in terms of m, g, l and μ.
  b) Find the position of the other point on the line AC at which the block would be in limiting equilibrium.

11 Two identical elastic strings, of stiffness $25\,\mathrm{N\,m^{-1}}$ and natural length $1.5\,\mathrm{m}$, are fastened to a block of mass $2\,\mathrm{kg}$. The other end of one string is fastened to a beam and the other end to the floor, $5\,\mathrm{m}$ below the beam.

    a) Find the lengths of the strings when the block hangs in equilibrium.

    b) If the beam is now gradually lowered, how far above the floor will it be when the lower string becomes slack?

12 A block of mass m is attached by an elastic string of stiffness k to a point, O, on a smooth horizontal plane. The block is travelling in a horizontal circle about O. Find the angular speed of the block in terms of m and k if the radius of the circle is twice the natural length of the string.

Other ways of expressing the linear model

So far, we have expressed Hooke's law by using the equation $T = kx$. The stiffness, k, is a property of the particular spring or piece of string. Springs and strings which are identical except for their lengths have different stiffnesses, as you saw in Question **9b** of Exercise 17A.

There are two ways in which Hooke's law can be expressed so that it is not tied so closely to the particular spring or string. These make use respectively of the **modulus of elasticity** and **Young's modulus**.

Modulus of elasticity

The modulus of elasticity, λ, provides a measure of the elastic strength of a spring or string which does not depend upon its length. It is constant for strings of the same material and cross-section, and for springs of the same construction, regardless of their length. Using λ, the model for tension T is expressed as

$$T = \frac{\lambda x}{l}$$

where x is the extension and l is the natural length of the spring or string.

The modulus of elasticity, λ, and the stiffness, k, are related by $\lambda = kl$.

As the SI unit for k is the newton per metre $(\mathrm{N\,m^{-1}})$, it follows that the SI unit for λ is the **newton** (N). In fact, λ corresponds to the tension in the string when it is stretched to exactly twice its natural length.

Young's modulus

The modulus of elasticity, λ, is different for strings which are made of the same material but which have different cross-sectional areas. A string which has cross-sectional area $2A$ is twice as stiff as one with a cross-sectional area A

(analogous to the springs in Question **9a** of Exercise 17A). To take account of this, we can use Young's modulus, E, which is related to λ by

$$E = \frac{\lambda}{A}$$

Hooke's law is then expressed as

$$T = \frac{EAx}{l}$$

This only applies to strings. Young's modulus is a property of the material from which the string is made. The SI unit for E is the newton per square metre ($N\,m^{-2}$).

The equation for T in this form only represents a straight line provided we can regard A as constant. In fact, as a string stretches, it becomes thinner, and so A is a function of x. This gives us some insight into why our experimental results on p. 372 showed that the linear model became less appropriate for larger extensions of an elastic string.

You will not meet Young's modulus in mathematics examinations, though you may encounter it in physics or engineering.

Problems expressed in terms of the modulus of elasticity, λ, are solved in exactly the same way as those using the stiffness, k. The only difference is in the form of the modelling equation for the tension or compression force.

Example 4 An elastic string of length 2 m and modulus of elasticity 50 N has a block of mass 3 kg attached to one end. The other end is fastened to a hook and the block is lowered into its equilibrium position. If $g = 10\,m\,s^{-2}$, what is the length of the string in this position?

SOLUTION

As the block is in equilibrium, the tension must equal the weight of the block. So, we have

$$T = 3 \times g = 30\,N$$

In the Hooke's law equation $T = \frac{\lambda x}{l}$, we have $\lambda = 50$ and $l = 2$, which give

$$30 = \frac{50x}{2} \quad \Rightarrow \quad x = 1.2\,m$$

So, the total length of the string is $2 + 1.2 = 3.2\,m$.

Example 5 Two springs, each of natural length 0.1 m and modulus of elasticity λ N, are fastened at one end to a block, M, of weight 30 N. The other ends are fastened to two hooks, A and B, fixed to the ceiling and 0.16 m apart. The block is lowered until it rests in equilibrium at a distance 0.15 m below the ceiling. Calculate the value of λ.

SOLUTION

By symmetry, the tensions in the springs are equal.

We can find the stretched length of the springs by using Pythagoras' theorem on the triangle ACM, which gives

$$AM^2 = 0.08^2 + 0.15^2$$

$$\Rightarrow \quad AM = 0.17\,m$$

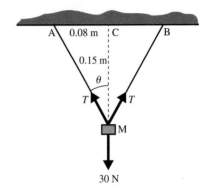

If the springs are inclined at angle θ to the vertical, as shown, we have

$$\cos\theta = \frac{0.15}{0.17} = \frac{15}{17}$$

Resolving vertically, we have

$$2T\cos\theta - 30 = 0$$

$$\Rightarrow \quad 2T \times \frac{15}{17} = 30 \quad \Rightarrow T = 17\,N$$

The natural length of the spring is 0.1 m. Its stretched length is 0.17 m, so the extension is 0.07 m.

The Hooke's law equation $T = \dfrac{\lambda x}{l}$ then gives

$$17 = \frac{\lambda \times 0.07}{0.1}$$

$$\Rightarrow \quad \lambda = \frac{17}{0.7} = 24.29\,N \quad \text{(to 2 dp)}$$

So, the modulus of elasticity of the springs is 24.29 N.

Exercise 17B

1 A block of mass 4 kg is attached to one end of an elastic string whose other end is fixed to the ceiling. The string has a natural length of 2 m and a modulus of elasticity of 90 N. What is the length of the string when the block hangs in its equilibrium position?

2 A spring is designed so that its length doubles when a weight of 50 N is attached to it so that it hangs vertically. When in this position, an extra 20 N is exerted and the extension increases by 10 cm. Find the modulus of elasticity and the natural length of the spring.

3 A spring is used in a weighing machine, as shown in the diagram. A 2 kg bag of sugar is placed on the scale pan and its height above the table top is measured as 15 cm. When the sugar is replaced by a 1.5 kg bag of flour, the scale pan is 18 cm above the table top.

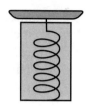

   a) Assuming that the scale pan has negligible weight, what is the height of the scale pan above the table top when there is nothing on the scale pan?

   b) What is the modulus of elasticity of the spring?

4 Two springs, the first of natural length 1 m and modulus of elasticity 50 N, the second of natural length 1.5 m and modulus of elasticity 80 N, are fastened to a hook on the ceiling and to a block of mass 5 kg, which hangs at rest below the hook.

   a) How far below the ceiling does the block hang?

   b) What are the tensions in the springs?

5 A light rod, AB, of length 2 m, is hinged to a vertical wall at A. An elastic string connects B to a point C on the wall, 2 m above A. The string has natural length 1 m and modulus of elasticity 100 N. A block of mass M kg is suspended from B. The system rests in equilibrium. Find the value of M if the angle BAC is

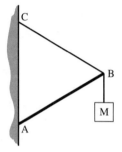

   a) $60°$ **b)** $90°$ **c)** θ

6 A block, A, of mass M kg rests on a smooth plane inclined at an angle θ to the horizontal. It is held in position by an elastic string AB, of natural length l and modulus of elasticity λ, the other end of which is fastened to a point B on the plane. Find, in terms of M, g, l and θ, the stretched length of the string.

7 A ball of mass 4 kg is attached to an elastic string whose other end is fixed to a point P on a smooth horizontal table. The string has a natural length 1.2 m and modulus of elasticity 28 N. The ball is pulled away from P until it is at a distance of 1.65 m. It is then released. What is the initial acceleration of the ball?

8 A block of wood of mass 2 kg is fastened to a spring of natural length 50 cm and modulus of elasticity 20 N. It is placed on a rough table, with the other end of the spring fastened to a vertical support, as shown. The coefficient of friction between the block and the table is 0.4. If the block is pushed towards the support, what is the closest it can be from the support and remain at rest on the table?

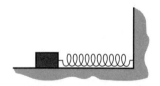

9 Two elastic strings, AB and BC, are joined at B, and the free ends are fixed to points A and C on a smooth horizontal table. The natural length of AB is 85 cm, and of BC is 45 cm. The modulus of elasticity of AB is 45 N and of BC is 65 N. The distance AC is 2.4 m. Find the stretched lengths.

10 An elastic string, of natural length $2l$ and modulus of elasticity λ, is stretched between points A and B, where A is a distance of $4l$ vertically above B. A particle of mass m is then attached to the mid-point of the string and is then lowered at a distance d until it is in equilibrium. Find d in terms of m, g, l and λ.

11 A block, A, of mass m rests on a rough plane inclined at an angle θ to the horizontal. It is on the point of sliding down the slope, and is prevented from doing so by an elastic string, AB, of natural length l and modulus of elasticity λ, which is attached to a point B on the plane above A. The coefficient of friction between the block and the plane is μ. The point B is then gradually moved up the plane. Show that it can be moved a distance d before the block starts to move up the plane, where

$$d = \frac{2\mu mgl\cos\theta}{\lambda}$$

Tension, work and energy

Stretching a spring or elastic string, or compressing a spring, requires the application of a force. The point of application of the force moves in the direction of the force during the stretching or compression, and so work is done by the force.

The tension or compression is a conservative force. The work it does depends only on the initial and final extensions, and is zero if the initial and final extensions are the same. It follows that a stretched string or spring, or a compressed spring, has stored or potential energy. This energy is 'recovered' in the form of kinetic energy when the string or spring is released.

For example, suppose we attach one end of an elastic string to an object on a smooth table and the other end to a fixed point on the table. We then pull the string into a stretched position and let go. The object will start to move as the string returns to its original length. The stored energy in the string is being converted into the kinetic energy of the moving object.

We saw on page 255 that when a variable force F is a function of displacement x, the work done by F is given by

$$W = \int F \, dx$$

In the Hooke's law model that we are using, the variable force is the tension $T = kx$, which gives

$$W = \int kx \, dx = \tfrac{1}{2}kx^2 + c$$

When the extension, x, is zero, no work has been done by the force, so $c = 0$.

So, the work done in stretching a string or spring is given by

$$W = \tfrac{1}{2}kx^2 \quad \text{or} \quad W = \frac{\lambda x^2}{2l}$$

Although we have talked here about stretching a spring, the same equation gives the work done in compressing a spring by an amount x from its natural length.

As mentioned above, when a string or spring is stretched, or a spring compressed, it has stored energy equal to the work done in stretching or compressing it. This is called **elastic potential energy** (EPE), and is the third form of mechanical energy, taking its place in the energy equation alongside the gravitational potential energy (GPE) and the kinetic energy (KE), which we have already met (see pages 257 and 259).

Example 6 A block of mass 4 kg is attached to one end of a spring, whose natural length is 2 m and whose stiffness is 64 N m^{-1}. The other end of the spring is fixed to a point, O, on a smooth table. The block is held 3 m from O and then released. Find the speed of the block when the spring reaches its natural length, and investigate the subsequent motion of the block.

SOLUTION

Assuming we take the table top to be the zero level for GPE, we have GPE = 0 throughout the motion.

At B, we have

$$\text{EPE} = \tfrac{1}{2}kx^2 = \tfrac{1}{2} \times 64 \times 1^2 = 32\,\text{J}$$
$$\text{KE} = 0\,\text{J}$$

So the total energy at B is 32 J.

Let the speed of the block at L be v. Then at L we have

$$\text{EPE} = 0\,\text{J}$$
$$\text{KE} = \tfrac{1}{2}mv^2 = \tfrac{1}{2} \times 4 \times v^2 = 2v^2$$

So, the total energy at L is $2v^2$.

There are no outside forces or sudden changes, so energy is conserved. Therefore, we have

$$2v^2 = 32 \quad \Rightarrow \quad v = 4$$

So, the speed of the block at L is 4 m s^{-1}.

Now let us consider the subsequent motion. The block passes through L and the spring begins to compress. This slows the block, as kinetic energy is converted to elastic potential energy. The block will become stationary when the EPE is again 32 J, which occurs when the spring is compressed by 1 m.

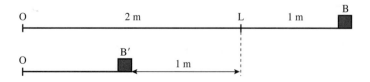

This, of course, assumes that the design of the spring is such as to allow this degree of compression without the coils meeting. Were they to do so, there would be an impulse on the block, and the principle of conservation of energy would cease to apply.

This situation would have been different had the spring been replaced by an elastic string of equivalent length and stiffness. The string would go slack when the block reached L. The block would continue to move through O at $4\,\mathrm{m\,s^{-1}}$.

Eventually, the string would become taut. The block would slow and become stationary when the string was again extended by $1\,\mathrm{m}$.

Example 7 A ball of mass $2\,\mathrm{kg}$ is attached to one end of an elastic string, whose other end is fixed to the ceiling. The string has a natural length of $2\,\mathrm{m}$ and a modulus of elasticity of $100\,\mathrm{N}$. The ball is held so that the string is at its natural length and is then released from rest. Stating any assumptions made, find the distance the ball drops before coming instantaneously to rest.

SOLUTION

Assumptions:
- The ball is a particle.
- The string is light.
- The string does not deform as it stretches, so that we can reasonably assume Hooke's law is a suitable model throughout the motion.
- There is no air resistance.
- $g = 10\,\mathrm{m\,s^{-2}}$.

At L, we have

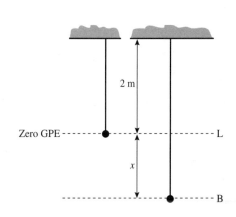

$$\mathrm{KE} = 0 \quad \text{since the ball starts from rest}$$

$$\mathrm{GPE} = 0 \quad \text{chosen zero level}$$

$$\mathrm{EPE} = 0 \quad \text{since the string is unstretched}$$

So, the total energy at L is 0.

At B, the lowest position of the ball, we have

$$\mathrm{KE} = 0 \quad \text{since the ball is again at rest}$$

$$\mathrm{GPE} = -mgx = -20x$$

$$\mathrm{EPE} = \frac{\lambda x^2}{2l} = \frac{100x^2}{2 \times 2} = 25x^2$$

So, the total energy at B is $25x^2 - 20x$.

Energy is conserved. So, we have

$$25x^2 - 20x = 0$$

$$\Rightarrow \quad x = 0 \quad \text{or} \quad x = 0.8\,\text{m}$$

This shows that the ball is at rest at its starting point L (when $x = 0$) and again at B when it has dropped a distance of $0.8\,\text{m}$.

Example 8 With the situation as described in Example 7, find an expression for v, the speed of the ball when it is at a general point P, in terms of the extension, x, of the string. Hence find:

a) the speed when $x = 0.2\,\text{m}$
b) the extension when the speed is $1\,\text{m}\,\text{s}^{-1}$
c) the maximum speed of the ball.

SOLUTION

As before, the total energy at L $= 0$.

At P, we have

$$\text{KE} = \tfrac{1}{2}mv^2 = v^2$$

$$\text{GPE} = -mgx = -20x$$

$$\text{EPE} = \frac{\lambda x^2}{2l} = \frac{100x^2}{2 \times 2} = 25x^2$$

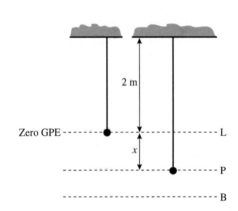

Energy is conserved, So, we have

$$v^2 - 20x + 25x^2 = 0 \qquad\qquad [1]$$

$$\Rightarrow \quad v = \pm\sqrt{20x - 25x^2}$$

a) When $x = 0.2$, we have

$$v = \pm\sqrt{20 \times 0.2 - 25 \times 0.2^2} = \pm 1.732\,\text{m}\,\text{s}^{-1}$$

So, when the extension is $0.2\,\text{m}$, the speed is $1.732\,\text{m}\,\text{s}^{-1}$ and the ball may be travelling either up or down.

b) When the speed is $1\,\text{m}\,\text{s}^{-1}$, we have from [1]

$$1^2 - 20x + 25x^2 = 0$$

$$\Rightarrow \quad 25x^2 - 20x + 1 = 0$$

Solving this quadratic equation, we get

$$x = 0.054\,\text{m} \quad \text{or} \quad x = 0.746\,\text{m}$$

So, there are two positions at which the speed is $1\,\mathrm{m\,s^{-1}}$, and these are symmetrically placed in the motion of the ball.

c) If v is a maximum, so is v^2. Letting $V = v^2$, we have from [1]

$$V = 20x - 25x^2$$

$$\Rightarrow \quad \frac{\mathrm{d}V}{\mathrm{d}x} = 20 - 50x = 0 \quad \text{for a maximum}$$

$$\Rightarrow \quad x = 0.4\,\mathrm{m}$$

$$\Rightarrow \quad V = 20 \times 0.4 - 25 \times 0.4^2 = 4$$

$$\Rightarrow \quad v = \pm 2\,\mathrm{m\,s^{-1}}$$

We can confirm that this is a maximum by differentiating a second time:

$$\frac{\mathrm{d}^2 V}{\mathrm{d}x^2} = -50 < 0$$

So, the point is a maximum.

The maximum speed of the ball is, therefore, $2\,\mathrm{m\,s^{-1}}$.

Example 9 A scale pan of mass 50 grams is hanging in equilibrium on an elastic string of natural length 60 cm and modulus of elasticity 10 N. An object of mass 200 grams is gently placed on the pan. How far does the pan drop before coming instantaneously to rest? Take $g = 10\,\mathrm{m\,s^{-2}}$.

SOLUTION

In the diagram, the pan hangs in equilibrium at A, when its extension is e. Its lowest point after the object is placed on it is B, when it has dropped a distance h.

First, we find e. Resolving vertically when the pan is in equilibrium, we have $T = 0.05g = 0.5\,\mathrm{N}$.

Using $T = \dfrac{\lambda x}{l}$, we have

$$0.5 = \frac{10e}{0.6} \quad \Rightarrow \quad e = 0.03\,\mathrm{m}$$

The object is now placed on the pan, and we use conservation of energy to find how far it falls.

We take the zero level for GPE at A, as shown.

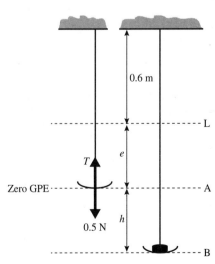

At A, we have

$$KE = 0$$

$$GPE = 0$$

$$EPE = \frac{\lambda x^2}{2l} = \frac{10 \times 0.03^2}{2 \times 0.6} = 0.0075\,J$$

At B, we have

$$KE = 0$$

$$GPE = -mgh = -0.25 \times 10 \times h = -2.5h$$

$$EPE = \frac{10 \times (0.03 + h)^2}{2 \times 0.6} = \frac{25(0.03 + h)^2}{3}$$

Energy is conserved. So, we have

$$\frac{25(0.03 + h)^2}{3} - 2.5h = 0.0075$$

$$\Rightarrow \quad (0.03 + h)^2 - 0.3h = 0.0009$$

$$\Rightarrow \quad h^2 - 0.24h = 0$$

$$\Rightarrow \quad h = 0 \quad \text{or} \quad h = 0.24\,m$$

The solution $h = 0$ corresponds to the starting position, A, so the pan falls by 0.24 m before coming instantaneously to rest at B.

Example 10 One end of an elastic string, of natural length 1.2 m and modulus of elasticity 150 N, is fixed to a point A. A particle of mass 0.75 kg is attached to the other end. The particle is held at A and then released from rest. Find how far the particle falls before coming instantaneously to rest. Take $g = 10\,\text{m}\,\text{s}^{-2}$.

SOLUTION

Take the zero level for GPE to be at A, as shown. The particle comes to rest at B, when the extension of the string is x.

At A, we have KE = 0, GPE = 0 and EPE = 0.

At B, we have

$$KE = 0$$

$$GPE = -mg(1.2 + x) = -7.5(1.2 + x)$$

$$EPE = \frac{\lambda x^2}{2l} = \frac{150x^2}{2 \times 1.2} = 62.5x^2$$

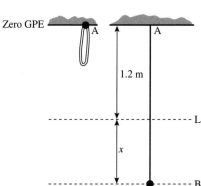

Energy is conserved. So, we have

$$62.5x^2 - 7.5(1.2 + x) = 0$$

$$\Rightarrow \quad 125x^2 - 15x - 18 = 0$$

Solving this quadratic equation, we get

$$x = 0.444\,\text{m} \quad \text{or} \quad x = -0.324\,\text{m}$$

As the required extension is positive, the particle falls a distance of $1.2 + 0.444 = 1.644\,\text{m}$.

Note In Example 10, the second solution, $x = -0.324\,\text{m}$, does not correspond to the starting point, A. To interpret it, we need to remember that our model does not distinguish between a string and a spring. Had we used a spring, the model predicts that, after coming instantaneously to rest at B, the particle would then rise above L, compressing the spring and coming to rest once more when the compression reached $0.324\,\text{m}$. It would be difficult to demonstrate this effect in practice, though if we were to start the motion by pulling the particle down to B it might be possible to achieve it.

As the problem specified a string, after reaching B the particle would rise through L and the string would go slack. Thereafter, it would continue to rise, coming to rest once more at A.

Example 11 Particles of mass 2 kg and 3 kg are attached to either end of a light, elastic string of natural length 2 m and modulus of elasticity 200 N. The particles are placed on a smooth horizontal surface. They are pulled to a distance of 5 m apart and released from rest. Find the speeds of the particles when the string goes slack and find the point at which the particles collide.

SOLUTION

Let the final speeds of the particles be v_1 and v_2, as shown.

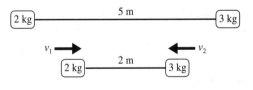

Initially, we have

$$\text{KE} = 0$$

$$\text{EPE} = \frac{\lambda x^2}{2l} = \frac{200 \times 3^2}{2 \times 2} = 450\,\text{J}$$

When string goes slack, we have

$$\text{KE} = \tfrac{1}{2} \times 2 \times v_1^2 + \tfrac{1}{2} \times 3 \times v_2^2 = \frac{2v_1^2 + 3v_2^2}{2}$$

$$\text{EPE} = 0$$

Energy is conserved. So, we have

$$\frac{2v_1^2 + 3v_2^2}{2} = 450$$

$$\Rightarrow \quad 2v_1^2 + 3v_2^2 = 900 \qquad\qquad\qquad [1]$$

To solve the problem, we need a second equation in v_1 and v_2. To obtain this, we notice that all the forces involved in the problem are internal to the system. This means that the total momentum of the system remains constant. Therefore, we have

$$\text{Initially:} \quad \text{momentum} = 0$$

$$\text{When string goes slack:} \quad \text{momentum} = 2v_1 - 3v_2$$

$$\Rightarrow \quad 2v_1 - 3v_2 = 0 \qquad\qquad\qquad [2]$$

From [2], we have $v_1 = \dfrac{3v_2}{2}$. Substituting this into [1], we obtain

$$2 \times \left(\frac{3v_2}{2}\right)^2 + 3v_2^2 = 900$$

$$\Rightarrow \quad v_2^2 = 120 \quad \Rightarrow \quad v_2 = 10.95\,\mathrm{m\,s^{-1}}$$

$$\Rightarrow \quad v_1 = \frac{3v_2}{2} = 16.43\,\mathrm{m\,s^{-1}}$$

So, the final speeds are $16.43\,\mathrm{m\,s^{-1}}$ for the 2 kg particle, and $10.95\,\mathrm{m\,s^{-1}}$ for the 3 kg particle.

As there are no external forces, the centre of mass of the system undergoes no acceleration. It was initially stationary at a point dividing the line joining the particles in the ratio 3 : 2. For it to remain stationary, the particles must collide at that point. That is, at the point 3 m from the initial position of the 2 kg particle.

Example 12 A particle of mass 0.5 kg is attached to one end of an elastic string of natural length 1 m and modulus of elasticity 50 N. The other end of the string is attached to a fixed point O on a rough horizontal plane. The coefficient of friction between the particle and the plane is 0.4. The particle is projected from O along the plane with initial speed $6\,\mathrm{m\,s^{-1}}$. Find

a) the greatest distance from O achieved by the particle
b) the speed of the particle when it returns to O.

SOLUTION

The friction force acting on the particle is

$$F = 0.4 \times 0.5g = 1.96\,\mathrm{N}$$

a) When it leaves O, the energy of the particle is

$$\tfrac{1}{2} \times 0.5 \times 6^2 = 9\,\text{J}$$

When the string reaches it greatest extension, x, the EPE is

$$\frac{50x^2}{2 \times 1} = 25x^2$$

The work done against friction is $1.96(x + 1)$.

By the work–energy principle, we have

$$25x^2 + 1.96(x + 1) = 9$$
$$\Rightarrow \quad 25x^2 + 1.96x - 7.04 = 0$$

Solving this equation gives $x = 0.493$ or -0.571.

Clearly, the negative root is inappropriate, so the greatest distance from O achieved by the particle is $1.49\,\text{m}$.

b) When it returns to O, the particle has travelled $2 \times 1.493 = 2.986\,\text{m}$ against friction. Hence, the total work done against friction is $1.96 \times 2.986 = 5.852\,\text{J}$.

The KE of the particle when it arrives back at O is, therefore, $9 - 5.852 = 3.148\,\text{J}$.

If the speed of the particle is then v, we have

$$\tfrac{1}{2} \times 0.5v^2 = 3.148 \quad \Rightarrow \quad v = 3.548$$

So, the particle returns to O with a speed of $3.55\,\text{m\,s}^{-1}$.

Exercise 17C

In this exercise, take $g = 9.8\,\text{m\,s}^{-2}$ unless a value is given.

1 A ball of mass $500\,\text{g}$ is fastened to one end of a light, elastic rope, whose unstretched length is $3\,\text{m}$ and whose modulus of elasticity is $90\,\text{N}$. The other end of the rope is fastened to a bridge. The ball is held level with the fixed end, and is released from rest.

   a) What will be the speed of the ball when it has fallen to the point where the rope is taut but unstretched?

   b) How far below the bridge is the lowest point reached by the ball?

2 A catapult is made by fastening an elastic string of natural length $10\,\text{cm}$ to points A and B, a distance of $6\,\text{cm}$ apart. The stiffness of the string is $50\,\text{N\,m}^{-1}$. A stone of mass $10\,\text{g}$ is placed at the centre of the string, which is then pulled back until the stone is $25\,\text{cm}$ from the centre of AB. Find the greatest speed reached by the stone when it is released.

3 Find the work done in stretching a spring of natural length 2.5 m and modulus of elasticity 160 N from a length of 3 m to a length of 3.5 m.

4 A particle of mass 2 kg is suspended from a point A on the end of a spring of natural length 1 m and modulus of elasticity 196 N.

a) Find the length of the spring when the particle hangs in equilibrium.

The particle is now pulled down a distance of 0.5 m and released from rest.

b) Find the distance below A at which the particle next comes instantaneously to rest, assuming that the spring can compress to that point without the coils touching.
c) Find the highest position reached by the particle if, instead of the spring, we had used an elastic string of the same natural length and modulus.

5 A block, B, of mass 5 kg is fastened to one end of each of two springs. The other ends are fastened to two points, A and C, 4 m apart on a smooth, horizontal surface, as shown.

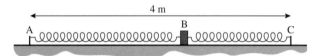

Spring AB has natural length 2 m and modulus of elasticity 30 N. Spring BC has natural length 1 m and modulus of elasticity 40 N.

a) Find the length AB when the block rests in its equilibrium position.
b) Find the tensions in the springs in this position.
c) Find the total elastic potential energy of the system in this position.

The block is moved 0.5 m towards C from the equilibrium position, and is held there.

d) What are the tensions or compressions in the springs in this position?
e) What is the total elastic potential energy of the system in this position?

The block is now released.

f) What is the speed of the block as it passes its equilibrium position?
g) How far beyond its equilibrium position does it travel before coming instantaneously to rest?

6 An elastic string has natural length a and modulus of elasticity mg. Particles of masses m and $2m$ are attached to its ends. The particles are held at rest a distance $3a$ apart on a smooth, horizontal surface and are then released. Find the speeds of the particles at the moment when the string goes slack, and find the point where they collide.

7 A particle of mass 1.5 kg is attached to one end of a light, elastic string of natural length 0.5 m and modulus of elasticity 160 N. The other end of the string is attached to a point at the base of a vertical wall. The particle is placed on the smooth, horizontal floor at a distance of 1 m from the wall, so that the string is perpendicular to the wall. The particle is released from rest. If the coefficient of restitution between the particle and the wall is 0.6, find the greatest distance from the wall achieved by the particle after its first impact with the wall.

8 A ring of mass m is threaded onto a smooth hoop of radius a, which is fixed in a vertical plane. The ring is attached to the lowest point of the hoop by means of an elastic string of natural length a and modulus mg. The ring is initially at rest at the highest point of the hoop, and is then slightly displaced. Find the speed of the ring and the reaction between the ring and the hoop at the point where the ring is level with the centre of the hoop.

9 A particle of mass 2 kg is attached to one end of an elastic string of natural length 1.2 m and modulus of elasticity 240 N. The other end of the string is fixed to a point A on a rough, horizontal plane. The particle is held at rest on the plane with the string stretched and is then released. The particle just reaches A before coming to rest. The coefficient of friction between the particle and the plane is 0.5. Take $g = 10\,\mathrm{m\,s^{-2}}$

a) Find the initial extension of the string.

b) Find the speed of the particle at the moment when the string went slack.

10 A particle of mass 2 kg is attached to one end of an elastic string of natural length 2 m and modulus of elasticity 40 N. The other end of the string is attached to a fixed point O on a rough horizontal plane. The coefficient of friction between the particle and the plane is 0.5. The particle is projected from O along the plane with initial velocity $v\,\mathrm{m\,s^{-1}}$. The particle returns and comes to rest exactly at O. Find

a) its furthest distance from O

b) the value of v.

11 A particle of mass m is attached by means of a light elastic string of natural length l and modulus of elasticity λ to a fixed point O on a rough plane inclined at an angle α to the horizontal. The coefficient of friction between the plane and the particle is μ. Initially, the particle is held at point A directly down the slope from O, such that $OA = l$, and is then released from rest. Show that the distance moved by the particle before coming to rest again is

$$\frac{2mgl(\sin\alpha - u\cos\alpha)}{\lambda}$$

18 Oscillatory motion

God order'd motion, but ordain'd no rest.
HENRY VAUGHAN

When a mass, attached to the free end of a vertical elastic string, is pulled down and released, it performs oscillations. This is one example of oscillatory motion. In this chapter, we extend the study of elastic strings and springs to model such oscillations, and look at other examples of oscillatory motion. It turns out that all such examples have the same mathematical solution.

Modelling vertical oscillations

A ball, of mass M kg, is attached to one end of an elastic string, whose other end is fixed to the ceiling. The string has natural length l and stiffness k. The ball is pulled down from its equilibrium position and released. Our task is to describe the subsequent motion of the ball.

Assumptions

- There is no air resistance.
- The string is of negligible mass.
- The string stretches without deforming, so that Hooke's law is an appropriate model throughout the motion
- The ball is a particle

Setting up the model

All variables in this model are measured positive downwards.

First, we consider the equilibrium position. In the diagram, L corresponds to the unstretched position of the string. E corresponds to the equilibrium position, where the extension is e.

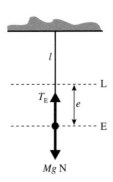

The resultant force is zero. So, we have

$$Mg - T_E = 0$$
$$\Rightarrow \quad T_E = Mg$$

From Hooke's law, $T_E = ke$. Hence, we have

$$ke = Mg \qquad\qquad [1]$$

Suppose the ball is now pulled down to B, a distance a below E, and released. We need to consider forces acting on the ball in a general position, P, a distance x below E. In this position, the ball has acceleration $\ddot{x}$.

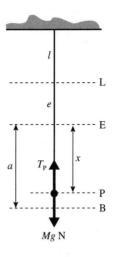

By Newton's second law, we have

$$Mg - T_P = M\ddot{x}$$

From Hooke's law, we have

$$T_P = k(e + x)$$

Hence, we have

$$Mg - k(e + x) = M\ddot{x} \qquad [2]$$

Substituting from [1] into [2], we obtain

$$ke - k(e + x) = M\ddot{x}$$
$$\Rightarrow \quad -kx = M\ddot{x}$$
$$\Rightarrow \quad \ddot{x} = -\left(\frac{k}{M}\right)x \qquad [3]$$

This differential equation relates the acceleration of the ball to its position, and so provides a model for the motion.

Simple harmonic motion

In the example above, we have arrived at a model in which the acceleration of the ball is directed towards the equilibrium position and is proportional to the displacement of the ball from it. An oscillatory motion for which this is true is called **simple harmonic motion** (SHM). We have the following definition:

A particle undergoes simple harmonic motion if its acceleration is **directed towards a fixed point** and is **proportional to the displacement** of the particle from that point.

In general, we express the model for simple harmonic motion as the second-order differential equation

$$\ddot{x} = -\omega^2 x$$

We choose to use ω^2 as the constant of proportionality because it is always positive. This means that the sign of the acceleration, $\ddot{x}$, is always opposite to that of the displacement, x, and so the acceleration is always directed towards the origin.

Note Not all oscillatory motions are simple harmonic. It does, however, provide a good model in a wide variety of situations.

Exercise 18A

1 A particle of mass $5\,\text{kg}$ is suspended from a fixed point on the end of a light spring of natural length $2\,\text{m}$ and modulus of elasticity $90\,\text{N}$. Take $g = 10\,\text{m}\,\text{s}^{-2}$.

a) Find the length of the spring when the particle hangs in equilibrium.

The particle oscillates in a vertical direction. At time t, it is a distance $x\,\text{m}$ vertically below the equilibrium position.

b) Draw a diagram of the system, including all the forces acting on the particle, and the relevant lengths. Choose a suitable direction to be positive.

c) Write down the equation of motion of the particle. Show that the acceleration of the particle is given by $\ddot{x} = -9x$ and hence that the particle moves with SHM. State the value of ω^2 for this motion.

2 An object of mass $0.5\,\text{kg}$ is fastened to the mid-point of a spring of natural length $2\,\text{m}$ and modulus of elasticity $8\,\text{N}$. The ends of the spring are fixed to two points, A and B, on a smooth table, such that AB is $2\,\text{m}$, as shown.

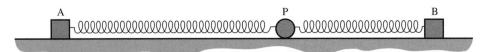

The object is moved away from its equilibrium position and released. Some time later, it is at the point P, a distance $x\,\text{m}$ to the right of the equilibrium point.

a) Draw a diagram of the system, showing the forces acting on the object. Indicate, for each part of the spring, whether it is extended or compressed.

b) Write down the equation of motion of the object and use it to show that the object moves with SHM. State the value of ω^2 for this motion.

c) The ends of the spring, A and B, are now fixed to points $3\,\text{m}$ apart, and the experiment is repeated. By finding the equation of motion of the object in this new situation, show that it still moves with SHM. What is now the value of ω^2?

d) Is the equation of motion in part **c** affected by whether the value of x is such that both springs are stretched, or one stretched and one compressed? Explain your answer.

3 A ball of mass m is suspended from a fixed point by means of a spring of natural length l and modulus of elasticity λ.

a) By how much is the spring extended when the ball is in its equilibrium position?

The ball is now pulled down and released. The displacement of the ball **below** its equilibrium position is x.

b) Write down the equation of motion of the ball and from it find its acceleration. Describe the motion of the ball.

4 A ball of mass M is suspended from a fixed point by means of a spring of natural length l and stiffness k. The displacement of the ball **below the fixed end** of the spring is x.

a) Write down the equation of motion of the ball.

b) Find an expression for the acceleration of the ball.

c) A body is in equilibrium when its acceleration is zero. Find the equilibrium position of the ball.

5 In the spring–mass system shown below, the modulus of elasticity and the natural length of each spring are as shown. $AB = 5\,m$ and the mass is $0.5\,kg$.

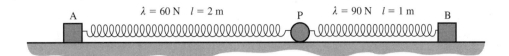

A $\lambda = 60\,N$ $l = 2\,m$ P $\lambda = 90\,N$ $l = 1\,m$ B

a) Draw a diagram of the forces acting on the mass.
b) i) If x is the displacement of the mass **from A**, write down the force in each spring.
   ii) Find the equation of motion of the mass.
   iii) Use the equation of motion to find the equilibrium position of the mass.
c) i) If x is the displacement of the mass **from the equilibrium position**, write down the forces in each spring.
   ii) Find the equation of motion of the mass using this new origin.

Choice of origin

In Exercise 18A, Question **4**, we took the fixed end of the spring as the origin. This resulted in the following equation for the acceleration of the body:

$$\ddot{x} = -\frac{k}{M}x + kl + g$$

In our original model for the motion, we took the origin to be the equilibrium position of the object. This resulted in the following equation for the acceleration of the body:

$$\ddot{x} = -\frac{k}{M}x$$

We can see from this that taking the equilibrium position as the origin results in a simpler equation. We will therefore choose to take the **origin at the equilibrium position** from now on.

Solving the model

To find expressions for the velocity and displacement of the body at a time t, we need to solve the differential equation $\ddot{x} = -\omega^2 x$.

The formal methods for solving this type of differential equation are given on pages 433–9. Another way is to propose and test possible solutions. The fact that the motion is an oscillation, and therefore periodic, suggests that the most familiar periodic functions, the sine and cosine functions, might be the basis of the solution. You can explore this in the following exercise.

Exercise 18B

1 a) By differentiating twice with respect to t, show that $x = a \cos \omega t$ is a solution of the differential equation $\ddot{x} = -\omega^2 x$.

b) By differentiating twice with respect to t, show that $x = a \sin \omega t$ is a solution of the differential equation $\ddot{x} = -\omega^2 x$.

c) By differentiating twice with respect to t, show that $x = a \cos(\omega t + \alpha)$ is a solution of the differential equation $\ddot{x} = -\omega^2 x$.

d) Expand $a \cos(\omega t + \alpha)$ using a compound-angle formula. Hence find expressions for A and B which allow x to be written in the form

$$x = A \cos \omega t + B \sin \omega t$$

Show that the solutions indicated in parts **a** and **b** are special cases of this more general solution. In each case, find the value of α needed to turn this general formula into the special case.

2 Differentiate each of these formulae once with respect to t:

a) $x = a \cos \omega t$ **b)** $x = a \sin \omega t$ **c)** $x = a \cos(\omega t + a)$

By writing $\dot{x} = v$, confirm that in each case the velocity and displacement of the body satisfy the equation

$$v^2 = \omega^2(a^2 - x^2)$$

Interpreting the solutions of the equation

In Exercise 18B, you found three possible solutions of the SHM equation. All were sinusoidal. (We will establish on pages 435–6 that the general solution to such equations is sinusoidal.)

We now investigate each of these in turn.

$x = a \cos \omega t$

The graph of $x = a \cos \omega t$ is shown below.

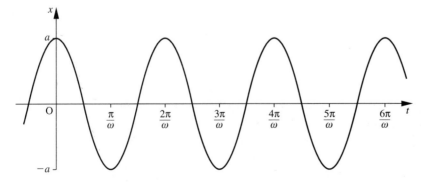

The graph describes an oscillation between the values $x = a$ and $x = -a$. The value a is called the **amplitude** of the motion. When $t = 0$, $x = a$.

This solution represents the case where a body is initially at the positive extreme of its motion. For example, if we pull a mass attached to a spring down a distance a from its equilibrium position and start timing as we release it, $x = a \cos \omega t$ describes the subsequent motion.

$x = a \sin \omega t$

The graph of $x = a \sin \omega t$ is shown below

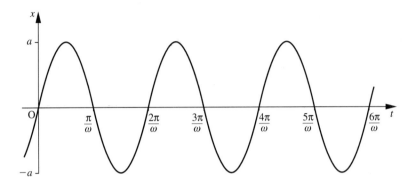

This graph also describes an oscillation between the values $x = a$ and $x = -a$. In this case, however, when $t = 0$, $x = 0$. This solution therefore represents the case where the body is initially at the origin and travelling in the positive direction.

$x = a \cos (\omega t + \alpha)$

The graph of $x = a \cos (\omega t + \alpha)$ is shown below

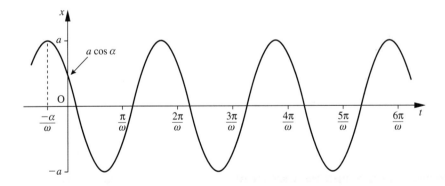

Again, the graph describes an oscillation between the values $x = a$ and $x = -a$. In this case, however, when $t = 0$, $x = a \cos \alpha$. This solution represents the case where the body is initially at some intermediate point.

Note The three forms of the solution differ only in the initial conditions. That is, in the position and direction of motion of the body at the moment when timing commences.

Period of the motion

Oscillations are periodic. That is, a body at any point P will, after a fixed period of time, again be at P and travelling in the same direction as before.

Sine and cosine are periodic functions. For example,

$$\cos \omega t = \cos(\omega t + 2\pi) = \cos\left(\omega\left(t + \frac{2\pi}{\omega}\right)\right)$$

This indicates that, if t is increased by an amount $\dfrac{2\pi}{\omega}$, the body will be at the same stage of its motion. This can also be seen from the graph below, where it is clear that the time between successive maximum displacements is $\dfrac{2\pi}{\omega}$.

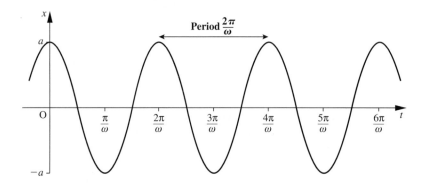

This time needed for one complete oscillation is called the **period** of the oscillation (also called the **periodic time**), and is denoted by T, where

$$T = \frac{2\pi}{\omega}$$

We also refer to the **frequency** of the oscillation. This is the number of oscillations, or **cycles**, per second. The frequency is given by

$$\frac{1}{T} = \frac{\omega}{2\pi}$$

Extreme values of speed, velocity and acceleration

The acceleration of a body moving with SHM has its greatest magnitude at the extreme positions. That is, when $x = a$, the acceleration is $-\omega^2 a$, and when $x = -a$, the acceleration is $\omega^2 a$.

The acceleration is zero as the body passes through the centre of the oscillation.

The speed of the body is greatest as it passes through the centre of the oscillation. Using $v^2 = \omega^2(a^2 - x^2)$, we can see that when $x = 0$, the velocity is either ωa or $-\omega a$, depending on the direction of travel.

The speed of the body is zero at the extreme positions $x = a$ and $x = -a$.

Summary

For a body moving with SHM, the following equations apply, and can usually be quoted without proof in the solution of a problem.

Acceleration: $\ddot{x} = -\omega^2 x$ (This defines SHM)

Displacement: This equation has solutions of the form

$$x = a \cos \omega t$$

or $x = a \sin \omega t$

or $x = a \cos(\omega t + \alpha)$

or $x = A \cos \omega t + B \sin \omega t$

where a is the amplitude of the motion, and the choice of solution depends on the initial conditions.

Velocity: The velocity at time t is found by differentiating the chosen solution above.

The velocity at a given position x is found from

$$v^2 = \omega^2(a^2 - x^2)$$

Period: $T = \dfrac{2\pi}{\omega}$

Frequency: $f = \dfrac{1}{T} = \dfrac{\omega}{2\pi}$

Example 1 A block of mass 3 kg, is suspended from a fixed point by means of a spring of natural length 2 m and stiffness 27 N m^{-1}. Take $g = 10 \, \text{m s}^{-2}$.

a) Find the length of the spring when the block rests in equilibrium.

The block is then pulled down so that the spring has length 4 m, and is released. At time t, the block has displacement x below the equilibrium position.

b) Find an expression for the acceleration of the block, and hence show that it moves with SHM.

c) What is the period of the oscillations?

d) Find an expression for the position of the block at time t.

SOLUTION

a) The forces on the block when it is in equilibrium are shown in the diagram. The extension in this position is e.

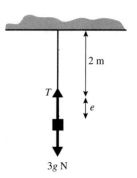

Resolving downwards, we have

$$3g - T = 0$$
$$\Rightarrow \quad T = 30\,\text{N}$$

By Hooke's law, we have

$$T = ke = 27e$$
$$\Rightarrow \quad 27e = 30 \quad \Rightarrow \quad e = 1\tfrac{1}{9}\,\text{m}$$

So, the length of the spring in the equilibrium position is $3\tfrac{1}{9}\,\text{m}$.

b) When the block is x below its equilibrium position, its extension is $(e + x)$. Hence, by Hooke's law, its tension is $27(e + x)$.

Resolving downwards, we find that the equation of motion is

$$3g - 27(e + x) = 3\ddot{x}$$
$$\Rightarrow \quad \ddot{x} = -9x \quad \text{(since } 3g - 27e = 0\text{)}$$

This is SHM with $\omega^2 = 9$, giving $\omega = 3$.

c) The period of the motion is $T = \dfrac{2\pi}{\omega} = \dfrac{2\pi}{3}\,\text{s}$.

(Note that it is usual to leave the period in terms of π.)

d) The amplitude of the motion is $4 - 3\tfrac{1}{9} = \tfrac{8}{9}\,\text{m}$. Timing commences when the displacement is at its maximum, so the appropriate form of the displacement equation is $x = a \cos \omega t$. Therefore, we have $x = \tfrac{8}{9} \cos 3t$.

Example 2 A block of mass $2\,\text{kg}$ is suspended from a fixed point on the end of a spring of stiffness $50\,\text{N}\,\text{m}^{-1}$. When the block is at rest in its equilibrium position, it is hit from below so that it starts to move with a speed of $2.5\,\text{m}\,\text{s}^{-1}$.

a) Find the amplitude of the resulting oscillations.
b) Find the maximum acceleration during the subsequent motion.
c) Find the period of the oscillations.

SOLUTION

a) Let e be the extension of the spring when the block is in equilibrium. The tension in the spring is then $50e$. Resolving downwards, we have

$$2g - 50e = 0$$

When the block has displacement x below the equilibrium position, the tension is $50(e + x)$. Resolving downwards, we find that the equation of motion is

$$2g - 50(e + x) = 2\ddot{x}$$
$$\Rightarrow \quad \ddot{x} = -25x \quad (\text{since } 2g - 50e = 0)$$

This is SHM with $\omega^2 = 25$, giving $\omega = 5$.

Putting $x = 0$ and $v = -2.5$ into the equation $v^2 = \omega^2(a^2 - x^2)$, we obtain

$$6.25 = 25(a^2 - 0) \quad \Rightarrow \quad a = 0.5$$

So, the amplitude of the motion is $0.5\,\text{m}$.

b) The maximum acceleration occurs when $x = -0.5$. Hence, we have

maximum acceleration $= -25 \times -0.5 = 12.5\,\text{m s}^{-2}$.

c) The period of the motion is $T = \dfrac{2\pi}{\omega} = \dfrac{2\pi}{5}\,\text{s}$.

Exercise 18C

1 Each of the following gives the equation of motion and the initial conditions for a body moving with SHM. In each case, find

i) the period of the oscillation

ii) the maximum velocity or the amplitude of the motion, as appropriate

iii) an expression for x, the displacement of the body from the centre of the oscillation, in terms of t.

a) $\ddot{x} = -9x$, when $t = 0\,\text{s}$, $x = 0\,\text{m}$ and $v = 3\,\text{m s}^{-1}$.

b) $\ddot{x} = -25x$, when $t = 0\,\text{s}$, $x = 2\,\text{m}$ and $v = 0\,\text{m s}^{-1}$.

c) $\ddot{x} + 16x = 0$, when $t = 0\,\text{s}$, $x = 3\,\text{m}$ and $v = 0\,\text{m s}^{-1}$.

d) $\ddot{x} + 9x = 0$, when $t = 0\,\text{s}$, $x = 0\,\text{m}$ and $v = 6\,\text{m s}^{-1}$.

2 A block of mass M is suspended from a fixed point on the end of a spring of natural length l. It is made to perform vertical oscillations. In each of the following cases, find

i) the period of the oscillation

ii) the maximum speed of the block.

a) $M = 2\,\text{kg}$, $a = 0.5\,\text{m}$, $k = 18\,\text{N}\,\text{m}^{-1}$.

b) $M = 3\,\text{kg}$, $a = 0.4\,\text{m}$, $l = 2\,\text{m}$, $\lambda = 96\,\text{N}$.

c) $M = 0.05\,\text{kg}$, $a = 0.05\,\text{m}$, $k = 180\,\text{N}\,\text{m}^{-1}$.

d) $M = 0.4\,\text{kg}$, $a = 0.5\,\text{m}$, $l = 2.5\,\text{m}$, $\lambda = 36\,\text{N}$.

3

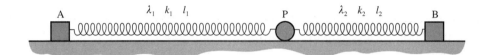

The diagram shows a ball P, of mass M, on a smooth horizontal surface and moving with SHM attached to two springs, AP and BP, in a straight line. The natural lengths are l_1 and l_2, the moduli of elasticity λ_1 and λ_2, and the stiffnesses k_1 and k_2. The amplitude of the motion is a. In each of the following cases, find

i) the period of the oscillation

ii) the maximum speed of the ball.

a) $M = 4\,\text{kg}$, $k_1 = 20\,\text{N}\,\text{m}^{-1}$, $k_2 = 16\,\text{N}\,\text{m}^{-1}$, $a = 0.5\,\text{m}$.

b) $M = 2\,\text{kg}$, $\lambda_1 = 20\,\text{N}$, $\lambda_2 = 40\,\text{N}$, $l_1 = 2\,\text{m}$, $l_2 = 1\,\text{m}$, $a = 1.4\,\text{m}$.

c) $M = 0.04\,\text{kg}$, $k_1 = 60\,\text{N}\,\text{m}^{-1}$, $k_2 = 40\,\text{N}\,\text{m}^{-1}$, $a = 0.1\,\text{m}$.

d) $M = 0.6\,\text{kg}$, $\lambda_1 = 30\,\text{N}$, $\lambda_2 = 27\,\text{N}$, $l_1 = 2.5\,\text{m}$, $l_2 = 1.5\,\text{m}$, $a = 0.4\,\text{m}$.

4 A sphere is suspended from a fixed point by means of a spring. The sphere moves with SHM. The period of the motion is $0.5\,\text{s}$ and the distance between the extremes of the motion is $0.4\,\text{m}$.

a) Find the frequency of the oscillations.

b) Find the maximum speed attained by the sphere.

c) Find the maximum acceleration of the sphere.

5 A particle of mass $100\,\text{g}$ is attached to the mid-point of a spring of natural length $80\,\text{cm}$ and modulus of elasticity $15\,\text{N}$. The ends of the spring are attached to points A and B, a distance of $1\,\text{m}$ apart, on a smooth horizontal plane. The particle rests in equilibrium.

a) Find the elastic potential energy stored in the spring.

The particle is now pulled a distance of $15\,\text{cm}$ towards B.

b) Find the elastic potential energy stored in the spring in this position.

The particle is now released and performs SHM.

c) Use conservation of energy to find the speed at which it is travelling when it passes through the mid-point of AB.

d) Use your answer to part **c** to find the period of the motion

e) How far is the particle from the mid-point of AB when it is travelling at half its maximum speed?

6 Points A, B, C, D and E lie in a straight line. $AB = BC = 15\,\text{cm}$, $CD = 10\,\text{cm}$ and $DE = 20\,\text{cm}$. A particle is moving with SHM so that A and E are the extreme positions of its motion. The period of the motion is $0.2\,\text{s}$.

a) Find the maximum speed attained by the particle.

b) Find the speed of the particle as it passes through B and through D.

c) Find the time the particle takes to get from B to D
  i) if it is travelling towards D as it passes through B
  ii) if it is travelling away from D as it passes through B.

7 A coaster needs a 3 m depth of water before it can enter or leave a harbour. The depth of water in the region of the harbour changes by 8 m from high tide to low tide. At low tide, the harbour bottom is 2 m above the sea level, so no vessels can move. There is a high tide at 0600 and the next is at 1830. Assuming that the sea level rises and falls with SHM, find

a) the latest time that the coaster can leave the harbour after 0600

b) the earliest time that the coaster can return to harbour after low tide.

8 A mass, M kg, is attached to the bottom of each spring system shown below. The component springs are identical and of stiffness k. The mass is set moving vertically with SHM. Find the period of each system.

a)
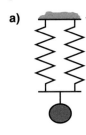

b)
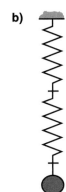

c)
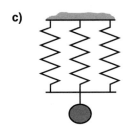

d)

9 A canoe travelling along a canal forms ripples in the water. A piece of wood is floating on the surface of the water when the ripples pass. The height of the piece of wood above the bottom of the canal is modelled by the equation

$$x = 2 + 0.04 \sin\left(\frac{t}{2\pi}\right)$$

where t is the time after the piece of wood first starts to move.

a) Sketch the graph of x against t, showing the important features of the motion.

b) Find the vertical speed of the piece of wood as it passes through the original level of the water.

Associated circular motion

Simple harmonic motion and circular motion are closely related. Suppose that a particle, P, moves in a circle of radius a with constant angular speed ω, as shown in the diagram at the top of the next page.

When $t = 0$, the particle is at B. At time t, the angle POB $= \omega t$.

The point Q is the projection of P onto the diameter AB. As P moves round the circle, Q moves along AB, 'keeping pace' with P.

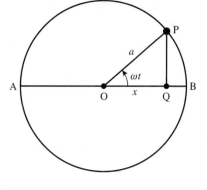

If the displacement of Q from O is x, we have

$$x = a\cos\omega t$$

This is a solution of the SHM equation.

As P moves round the circle, Q performs SHM along the diameter.
The motion has centre O, amplitude a and period $\dfrac{2\pi}{\omega}$.

You may be able to show this in practice. You need an object moving in a circle with constant angular speed – an object placed on a record turntable would do nicely. Viewed from the level of the turntable, the object appears to be moving from side to side with SHM. The effect can be enhanced by using a lamp at the same level as the turntable to project the shadow of the object onto a screen. The shadow will move in a straight line with SHM.

The associated circular motion is sometimes used to solve problems in SHM, although the same results are always obtainable using the standard SHM equations.

Example 3 A particle is performing SHM along a line from A to B, where AB $= 8$ m. O is the mid-point of AB. Points P and Q lie on OB such that OP $= 2$ m and OQ $= 1$ m. The period of the motion is 12 s. How long does it take the particle to travel from P to Q?

SOLUTION

Since we have SHM, the imaginary particle undergoing the associated circular motion moves with angular speed ω in a circle with radius 4 m, as shown. This gives

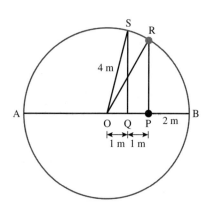

$$T = \frac{2\pi}{\omega} = 12 \quad \Rightarrow \quad \omega = \frac{\pi}{6}$$

In the diagram, we have

$$\cos \hat{ROP} = 0.5 \quad \Rightarrow \quad \hat{ROP} = \frac{\pi}{3} \text{ rad}$$

$$\cos \hat{SOQ} = 0.25 \quad \Rightarrow \quad \hat{SOQ} = 1.318 \text{ rad}$$

Hence, we have

$$\hat{SOR} = 1.318 - \frac{\pi}{3} = 0.271 \text{ rad}$$

The time taken for the particle moving with SHM to travel from P to Q is the same as that taken by the associated particle to travel from R to S, which is

$$\frac{\hat{SOR}}{\omega} = 0.271 \times \frac{6}{\pi} = 0.52\,\text{s}$$

Simple pendulum

A simple pendulum consists of a mass (called a bob) attached to the end of a rod or string.

Imagine the bob hanging in equilibrium with the string vertical. It is then pulled to one side, keeping the string taut, and released. The pendulum swings from side to side through the equilibrium position.

With a suitable set of modelling assumptions, we can show that the motion of the pendulum is approximately SHM.

Modelling assumptions

- The string or rod is inextensible.
- The string or rod has zero mass.
- The bob is a particle.
- There is no air resistance acting on the string or rod and the bob.
- There are no frictional forces acting on the fixed end of the string or rod.
- The angle of swing is small.

Setting up the model

Let l be the length of the string or rod, m the mass of the bob and θ the angle between the string or rod and the vertical at time t, as shown.

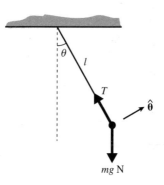

We will consider the equation of motion in the tangential direction, as indicated by the unit vector $\hat{\boldsymbol{\theta}}$ in the diagram.

The tangential acceleration is $l\ddot{\theta}$.

The resultant force in the direction of $\hat{\boldsymbol{\theta}}$ is $-mg\sin\theta$.

By Newton's second law, we have

$$-mg\sin\theta = m\,l\ddot{\theta}$$

$$\Rightarrow \quad \ddot{\theta} = -\frac{g}{l}\sin\theta$$

Our assumption that θ is small means that $\sin\theta \approx \theta$, which gives

$$\ddot{\theta} = -\frac{g}{l}\theta$$

This is the equation for SHM with $\omega^2 = \dfrac{g}{l}$.

If the maximum angle of swing is Θ, and $\theta = \Theta$ when $t = 0$, the solution of this equation is

$$\theta = \Theta\cos\left(\frac{gt}{l}\right)$$

The period of the pendulum's motion is $T = \dfrac{2\pi}{\omega}$, which gives

$$T = 2\pi\sqrt{\frac{l}{g}}$$

Note

- The pendulum's motion is independent of the mass of the bob, as this does not appear in the solution of the equation.

- The assumption that the angle is small is generally taken to mean that Θ is no more than $0.3\,\text{rad}$ ($17.2°$) for two decimal-place accuracy, and approximately half this for three decimal-place accuracy. (You can test this with your calculator.)

Experiment

You may wish to test the validity of the above model with an experiment. Set up a simple pendulum using a length of thin cord and a small mass. For ten different lengths of the pendulum, time how long it takes to do 20 complete swings. (One twentieth of this will give a more accurate result for the period than trying to time one swing.)

From the model, we have the equation

$$T^2 = \frac{4\pi^2}{g}l$$

This suggests that plotting a graph of T^2 against l should result in a straight line through the origin with gradient $\dfrac{4\pi^2}{g}$. You should do this by hand or on a spreadsheet. (The spreadsheet PENDULUM can be downloaded from the Oxford University Press website: http://www.oup.co.uk/mechanics.) From the gradient of your graph, you can obtain an estimate of g.

The results obtained by the author are shown below. The line **almost** passes through the origin: the intercept is -0.0265. The gradient of the line is $\dfrac{4\pi^2}{g} = 3.87$, which gives an estimate for g of $10.2\,\mathrm{m\,s}^{-2}$.

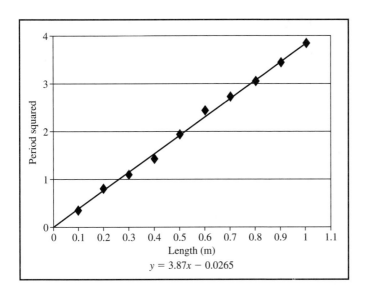

$y = 3.87x - 0.0265$

Time for 20 swings (s)	Period	Length (m)	Period squared
11	0.55	0.1	0.3025
18	0.9	0.2	0.81
21	1.05	0.3	1.1025
24	1.2	0.4	1.44
28	1.4	0.5	1.96
31	1.55	0.6	2.4025
33	1.65	0.7	2.7225
35	1.75	0.8	3.0625
37	1.85	0.9	3.4225
39	1.95	1.0	3.8025

Example 4 A pendulum is used to keep time for a group of musicians. They require 100 beats per minute. How long should the pendulum be? Take $g = 9.8\,\text{m}\,\text{s}^{-2}$.

SOLUTION

To beat 100 times per minute, the pendulum must make 50 complete oscillations in 1 minute.

Using $T = 2\pi\sqrt{\dfrac{l}{g}}$, we have

$$\frac{60}{50} = 2\pi\sqrt{\frac{l}{g}}$$

$$\Rightarrow \quad 1.44 = 4\pi^2\frac{l}{9.8}$$

$$\Rightarrow \quad l = \frac{1.44 \times 9.8}{4\pi^2} = 0.357\,\text{m} \quad \text{(to 3 dp)}$$

So, the length of the pendulum must be 35.7 cm.

Example 5 A seconds pendulum is so called because it beats seconds. That is, it has a period of 2 s. If a seconds pendulum is designed to be accurate in a place where $g = 9.8\,\text{m}\,\text{s}^{-2}$, how much time will it gain in an hour in a place where $g = 9.81\,\text{m}\,\text{s}^{-2}$?

SOLUTION

Using $T = 2\pi\sqrt{\dfrac{l}{g}}$, with $T = 2$ and $g = 9.8$, we have

$$2 = \frac{2\pi\sqrt{l}}{\sqrt{9.8}} \qquad\qquad [1]$$

If the period is T when $g = 9.81$, we have

$$T = \frac{2\pi\sqrt{l}}{\sqrt{9.81}} \qquad\qquad [2]$$

Dividing [2] by [1], we get

$$\frac{T}{2} = \frac{\sqrt{9.8}}{\sqrt{9.81}} \quad \Rightarrow \quad T = 1.998\,980\,372\,\text{s}$$

Therefore, in 1 hour the pendulum will register

$$\frac{3600}{1.998\,980\,372} = 1800.918\,133 \text{ oscillations}$$

This corresponds to 3601.836 266 s. The pendulum will therefore gain 1.836 s in 1 hour.

Incomplete oscillations

When a mass on the end of a spring is oscillating vertically, it performs SHM throughout its travel provided that the coils of the spring do not collide as it shortens. However, if the spring is replaced by an elastic string, the motion is simple harmonic only while the string is stretched.

If the amplitude of the oscillations is less than the extension of the string in the equilibrium position, the whole motion is SHM. For larger amplitudes, the string is slack for part of the motion, and during this time the mass moves as a projectile. As the mass falls back, the string again becomes taut and the SHM is resumed. If no energy is lost, this cycle of SHM and projectile motion will continue indefinitely.

In order to solve problems such as this, we must split the motion into two parts: the SHM phase and the projectile phase.

Example 6 A particle of mass 3 kg is suspended from a fixed support by means of an elastic string, of natural length 2 m and stiffness 90 N m^{-1}. The mass is originally hanging in equilibrium. It is then pulled down until the length of the string is 3 m, and released. Taking $g = 10\,\text{m}\,\text{s}^{-2}$, find

a) the position of equilibrium of the particle
b) the equation of motion and the acceleration of the particle whilst in a general position
c) the time taken for the mass to rise to the position where the string is unstretched
d) the velocity of the particle in this position
e) the height to which the particle rises before coming to rest
f) the total time for one complete oscillation of the particle.

SOLUTION

a) In the equilibrium position, we have

$$3g - 9e = 0$$
$$\Rightarrow \quad e = \tfrac{1}{3}$$

Therefore, the equilibrium position is $2\tfrac{1}{3}$ m below the support, A.

b) In the general position, P, the equation of motion is

$$3g - 90(e + x) = 3\ddot{x}$$
$$\Rightarrow \quad \ddot{x} = -30\,x$$

This is SHM with $\omega^2 = 30$.

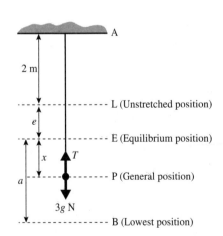

c) As the particle was at maximum displacement when $t = 0$, the appropriate solution to the SHM equation is

$$x = a \cos \sqrt{30}\, t$$

where a is the amplitude of the motion.

In this case, $a = 3 - 2\frac{1}{3} = \frac{2}{3}$ m. The string reaches the unstretched position, L, when $x = -\frac{1}{3}$ m. So, we have

$$-\frac{1}{3} = \frac{2}{3} \cos \sqrt{30}\, t$$

$$\Rightarrow \quad \cos \sqrt{30}\, t = -\frac{1}{2}$$

$$\Rightarrow \quad \sqrt{30}\, t = \frac{2\pi}{3}$$

$$\Rightarrow \quad t = \frac{2\pi}{3\sqrt{30}} = 0.382\, \text{s}$$

d) The velocity of the particle is given by $v^2 = \omega^2(a^2 - x^2)$.

At the unstretched position, $a = \frac{2}{3}$ m and $x = -\frac{1}{3}$ m. So, we have

$$v^2 = 30\left(\frac{4}{9} - \frac{1}{9}\right) = 10$$

$$\Rightarrow \quad v = \sqrt{10} = 3.16\, \text{m s}^{-1}$$

e) The particle now moves as a projectile. The extra height, s, it gains is given by $v^2 = u^2 + 2as$, where $v = 0$ and $u = \sqrt{10}\, \text{m s}^{-1}$. So, we have

$$0 = 10 - 2 \times 10 \times s$$

$$\Rightarrow \quad s = 0.5\, \text{m}$$

f) The time taken to travel from L to the highest point of the motion is given by

$$v = u + at$$

$$\Rightarrow \quad 0 = \sqrt{10} - 10t$$

$$\Rightarrow \quad t = 0.316\, \text{s}$$

The total time for one oscillation is, therefore, $2(0.382 + 0.316) = 1.40\, \text{s}$

In Example 6, notice that the highest point in the motion is 1.5 m above the lowest point, B. Had the particle been moving on the end of an equivalent spring, so that the motion was simple harmonic throughout, it would only have risen to a height of $2a = 1\frac{1}{3}$ m above B.

Damped oscillations

The most important assumption made in our model of oscillations is that no energy is lost during the motion. In practice, energy will be lost due to air resistance and to friction, both internal and external.

In some cases, we want the oscillations to die down, and steps are taken to ensure that this happens quickly. For example, when a car goes over a hump, it would clearly be undesirable for it to continue bouncing up and down for a significant time afterwards. The suspension system in a car therefore consists of two parts.

- For each wheel there is a spring designed to absorb the initial shock of the bump from the road. This will react in exactly the same way as a spring with a mass on its end.
- For each spring there is a shock absorber whose purpose is to prevent the spring from vibrating for too long. This is called **damping**, and shock absorbers are also called suspension dampers.

We can represent a spring and a damper system diagramatically, as shown on the right.

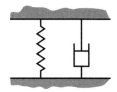

In order to model damping, the simplest assumption we make is that the resistive force acting on an oscillating particle is proportional to the velocity of the particle. This is **linear damping**.

It follows that the resistive force, **R**, is given by

$$\mathbf{R} = -r\mathbf{v}$$

where r is the constant of proportionality – the **damping constant** – and **v** is the velocity of the oscillating particle.

Alternatively, we can write the resistive force as

$$\mathbf{R} = -r\frac{\mathrm{d}x}{\mathrm{d}t} \quad \text{(for one-dimensional motion)}$$

Let us consider the example of a particle of mass m, suspended from a fixed point O by a perfect spring of stiffness k and natural length l_o. A damper provides linear damping whose damping constant is r.

The forces on the particle are the tension, T, in the spring, the resistance, R, of the damper and the weight, mg, of the particle.

First consider the mass hanging at rest in the equilibrium position. Taking downwards as positive, and using Hooke's law, we have

$$T = -ke$$

As the particle is stationary, we have $R = 0$. Therefore, resolving vertically, we obtain

$$mg - ke = 0 \qquad [4]$$
$$\Rightarrow \quad e = \frac{mg}{k}$$

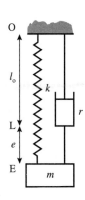

Now consider the particle at P, with displacement x below E. Again taking downwards as positive, we have

$$T = -k(e + x) \quad \text{and} \quad R = -r\frac{dx}{dt}$$

The equation of motion is

$$mg - k(e + x) - r\frac{dx}{dt} = m\frac{d^2x}{dt^2}$$

Substituting from [4] and rearranging, we obtain

$$m\frac{d^2x}{dt^2} + r\frac{dx}{dt} + kx = 0 \qquad [5]$$

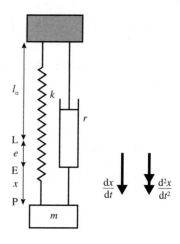

Solving the mathematical model

In order to solve equation [5]

$$m\frac{d^2x}{dt^2} + r\frac{dx}{dt} + kx = 0$$

we solve the auxiliary equation $m\lambda^2 + r\lambda + k = 0$ (see pages 433–6 for more details). The roots of this are

$$\lambda = \frac{-r \pm \sqrt{r^2 - 4mk}}{2m}$$

We will call the two roots λ_1 and λ_2. Where it helps to simplify the expressions, we will write

$$\rho = -\frac{r}{2m} \quad \text{and} \quad \Omega = \frac{\sqrt{4mk - r^2}}{2m}$$

We have three cases to consider:

- $r^2 - 4mk > 0$: The general solution has the form $x = Ae^{\lambda_1 t} + Be^{\lambda_2 t}$.
- $r^2 - 4mk = 0$: The general solution has the form $x = e^{\rho t}(A + Bt)$.
- $r^2 - 4mk < 0$: The general solution has the form
$$x = e^{\rho t}(A \cos \Omega t + B \sin \Omega t) \quad \text{or} \quad x = Re^{\rho t} \cos(\Omega t + \phi)$$

A, B, R and ϕ are arbitrary constants.

Interpreting the solution

There is a spreadsheet DAMPEDHM available on the Oxford University Press website (http://www.oup.co.uk/mechanics) with which to investigate the solutions to this equation.

You may like to try the following suggestions.

- With $m = 2$, $k = 2$, $A = 1$ and $B = 0$, investigate the effects on the solution graph of changing the parameter r. In particular, what happens when $r = 4$?
- With $m = 2$, $k = 2$, $A = 4$ and $B = -2$, investigate the effects on the solution graph of changing the parameter r. In particular, what happens when $r = 4$?
- Try other combinations of the parameters.
- Try $r = 0$, which corresponds to undamped motion. The oscillations should continue for ever.

We can draw the following conclusions about the different forms of the general solution.

$r^2 - 4mk > 0$

The general solution has the form $x = Ae^{\lambda_1 t} + Be^{\lambda_2 t}$. This is the condition for **strong damping**.

When $x = Ae^{\lambda_1 t} + Be^{\lambda_2 t}$, we have

$$\frac{\mathrm{d}x}{\mathrm{d}t} = \lambda_1 Ae^{\lambda_1 t} + \lambda_2 Be^{\lambda_2 t}$$

λ_1 and λ_2 are both negative, so we have a combination of diminishing exponential terms. Hence, as $t \to \infty$, $x \to 0$ and $\dfrac{\mathrm{d}x}{\mathrm{d}t} \to 0$.

In strong damping, the system does not perform a complete oscillation, but approaches the equilibrium position asymptotically, having passed through it at most once. Two examples of strong damping are shown in the figure below.

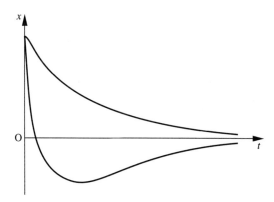

$r^2 - 4mk = 0$

The general solution has the form $x = e^{\rho t}(A + Bt)$, giving

$$\frac{\mathrm{d}x}{\mathrm{d}t} = e^{\rho t}(\rho A + B + \rho Bt)$$

This is the condition for **critical damping**. Since ρ is negative, we have a diminishing exponential term. Hence, as $t \to \infty$, $x \to 0$ and $\dfrac{\mathrm{d}x}{\mathrm{d}t} \to 0$.

This means that in critical damping, as in strong damping, the system does not perform a complete oscillation, but approaches the equilibrium position asymptotically, having passed through it at most once. The importance of critical damping is that the system approaches the equilibrium position at the fastest possible rate. Three examples of critical damping are shown in the figure below.

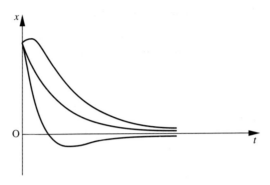

$r^2 - 4mk < 0$

The general solution has the form $x = Re^{\rho t} \cos(\Omega t + \phi)$, giving

$$\frac{\mathrm{d}x}{\mathrm{d}t} = Re^{\rho t}[\rho \cos(\Omega t + \phi) - \Omega \sin(\Omega t + \phi)]$$

This is the condition for **weak damping**. The motion will therefore be oscillatory, but as $e^{\rho t}$ is a diminishing exponential, the amplitude of the oscillation will diminish with time. Successive maxima occur when $(\Omega t + \phi)$ has increased by 2π. This corresponds to an increase in t of $\dfrac{2\pi}{\Omega}$. The motion therefore has a constant period of $\dfrac{2\pi}{\Omega}$.

An example of weak damping, with the bounding curves, is shown below.

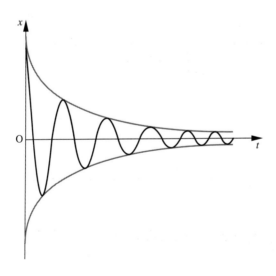

The arbitrary constants

The values of A and B, or R and ϕ, depend upon the particular situation we are dealing with. We need initial conditions to find the **particular solution** (see pages 438–9).

Example 7 A particle, P, of mass 1 kg, rests on a smooth horizontal surface. It is attached to two points, A and B, 3 m apart, by two springs AP and BP, as shown. The natural length of each spring is 1 m and the stiffnesses are $2\,\mathrm{N\,m^{-1}}$ and $3\,\mathrm{N\,m^{-1}}$ respectively. The particle is subject to linear damping with a damping constant $r = 4\,\mathrm{N\,m^{-1}\,s}$.

a) Find the position of equilibrium of the particle.

The particle is then displaced a distance 0.5 m from the equilibrium position and released.

b) Examine the subsequent motion of the particle.

SOLUTION

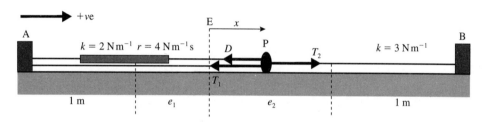

a) The forces acting on the particle are the tensions, T_1 and T_2, and the resistance, D. When the particle is in the equilibrium position, E, we have

$$T_1 = 2e_1 \qquad T_2 = 3e_2 \qquad D = 0$$

where e_1 and e_2 are the extensions of the springs, as shown in the diagram.

Taking positive to the right, when the particle is in equilibrium, the equation of motion is

$$T_2 - T_1 = 0$$
$$\Rightarrow \quad 3e_2 - 2e_1 = 0 \qquad [1]$$

Also we have $AB = 3\,\mathrm{m}$, giving

$$e_1 + e_2 = 1 \qquad [2]$$

Solving [1] and [2], we get $e_1 = \frac{3}{5}$, $e_2 = \frac{2}{5}$. So, the equilibrium position is $AE = 1.6\,\mathrm{m}$.

b) Suppose the particle is in a general position, P, x m to the right of the equilibrium position, E, as shown.

The forces acting on the particle are

$$T_1 = 2(e_1 + x) \qquad T_2 = 3(e_2 - x) \qquad D = 4\frac{dx}{dt}$$

The equation of motion is

$$T_2 - T_1 - D = \frac{d^2x}{dt^2}$$

$$\Rightarrow \quad 3(e_2 - x) - 2(e_1 + x) - 4\frac{dx}{dt} = \frac{d^2x}{dt^2}$$

Substituting the values of e_1 and e_2 obtained in part **a**, we have

$$\frac{d^2x}{dt^2} + 4\frac{dx}{dt} + 5x = 0$$

which is the equation of motion of the particle.

The auxiliary equation $\lambda^2 + 4\lambda + 5 = 0$ gives $\lambda = -2 \pm i$. So, the general solution is

$$x = e^{-2t}(A\cos t + B\sin t)$$

Differentiating, we get

$$\frac{dx}{dt} = e^{-2t}[(B - 2A)\cos t - (A + 2B)\sin t]$$

The initial conditions are $t = 0$, $x = 0.5$, $\frac{dx}{dt} = 0$. Hence, we have

$$t = 0, x = 0.5 \quad \Rightarrow \quad 0.5 = A$$
$$t = 0, \frac{dx}{dt} = 0 \quad \Rightarrow \quad 0 = B - 2A \quad \Rightarrow \quad B = 1$$

So, the particular solution is

$$x = e^{-2t}(\tfrac{1}{2}\cos t + \sin t)$$

The motion is therefore weakly damped and the particle will perform oscillations of diminishing amplitude. The periodic time of the oscillations is $\frac{2\pi}{1} = 2\pi$ s.

Exercise 18D

1 A pendulum has a length of 2 m. What is the period of the oscillations if the pendulum is in a place, where $g = 9.80\,\text{m s}^{-2}$?

2 A seconds pendulum is one which has a period of 2 s. How long should such a pendulum be if it is in a place, where $g = 9.82\,\text{m s}^{-2}$?

3 A clock is controlled by a pendulum. The clock gains 10 s in every hour. What percentage change in the length of the pendulum is needed for the clock to keep perfect time?

4 A clock keeps accurate time in a place where $g = 9.81\,\mathrm{m\,s^{-2}}$. It is then moved to a place where $g = 9.80\,\mathrm{m\,s^{-2}}$. By how much will it be wrong after one day?

5 A pendulum clock, which is accurate on the Earth, is taken to the Moon, where the acceleration due to gravity is one sixth that of the Earth. What time interval will the clock register during one hour?

6 A pendulum is of length l m. The pendulum is hanging at rest when the bob is given an impulse, causing it to start moving with speed $u\,\mathrm{m\,s^{-1}}$. Taking $g = 9.8\,\mathrm{m\,s^{-2}}$, use conservation of energy to find the maximum angular displacement achieved by the pendulum in each of the following cases, and hence decide if the SHM model is appropriate.

i) $l = 2$, $u = 5$ **ii)** $l = 6$, $u = 2$ **iii)** $l = 0.5$, $u = 0.5$

If it is, find

a) the period of the motion
b) the time taken to reach half the maximum angular displacement.

7 A ball of mass 2 kg is attached to one end of a light, elastic rope, whose other end is fixed to the ceiling. The unstretched length of the rope is 2 m and its modulus of elasticity is 100 N. The ball hangs in equilibrium. It is then pulled down a further 0.6 m and released. Take $g = 10\,\mathrm{m\,s^{-2}}$.

a) Find the equation of motion of the ball in a general position with the rope stretched.
b) Find the speed of the ball at the moment when the rope becomes slack.
c) Find the time taken for the ball to reach this point.
d) Find the distance travelled by the ball from its lowest point to its highest.
e) Find the time for one complete cycle of the motion.

8 A bungee jumper of mass 50 kg is attached to a bungee rope whose stiffness is $75\,\mathrm{N\,m^{-1}}$ and whose unstretched length is 10 m. The other end of the rope is attached to the jumping platform. The jumper leaps from the platform. Take $g = 10\,\mathrm{m\,s^{-2}}$.

a) State the assumptions needed to model the motion.
b) How far does the jumper fall before coming to rest?
c) How long does it take for the jumper to reach the lowest point?

9 A ball of mass 3 kg is attached to one end of an elastic rope whose unstretched length is 2 m and whose modulus of elasticity is 90 N. The other end of the rope is attached to a horizontal beam. The ball is pulled down to a point 4 m below the beam and released. Take $g = 10\,\mathrm{m\,s^{-2}}$.

a) Show that the ball performs SHM while the rope is taut.
b) What is the position of the ball when it first comes to rest?
c) How long after release does the ball first come to rest?
d) How far below the beam should the ball be pulled initially if it is to just reach the beam?

10 A scale pan of mass m is suspended from a fixed point on the end of a spring of natural length l and modulus of elasticity $6mg$. A particle of mass $3m$ is placed on the pan and the system hangs in equilibrium. The pan is then pulled down a further distance l and released.

a) Find how far the pan rises before the particle loses contact with it.
b) Find the greatest height to which the particle rises, assuming that it can do so without colliding with the apparatus.

11 The weight of an object below ground level is proportional to its distance from the centre of the Earth. That is, its weight is kr, where r is its distance from the centre of the Earth and k is a constant. A tunnel, in the form of a smooth, straight tube, is constructed joining two points on the surface of the Earth and an object is released from rest at one end of the tunnel.

a) Show that the object moves with SHM about the mid-point of the tunnel.
b) Find the time taken for the object to reach the other end of the tunnel.

Take the radius of the Earth to be 6.4×10^6 m and g to be $9.8\,\mathrm{m\,s^{-2}}$.

12 A ball of mass 1 kg is fastened to the mid-point of an elastic string of length 2 m and modulus of elasticity 40 N, whose ends are fixed to two points, P and Q, 3 m apart on a smooth table. The ball is held at point P and released.

a) Show that, for the first 1 m of its motion, the ball performs SHM. Find the period if it were able to complete oscillations with this SHM.
b) Find the ball's speed when it is 1 m from P and the time taken to reach this point.
c) Show that the ball also performs SHM for the next 1 m of its motion, but with a different period.
d) Find the speed of the ball when it is midway between P and Q.
e) Find the time taken for a complete cycle from P back to P.

13 A ball of mass m is attached to the centre of an elastic string of length $2l$ and stiffness k, whose ends are attached to two points, P and Q, a distance $4l$ apart on a smooth table. The ball is initially at rest at the mid-point of PQ. It is then pulled at right angles to PQ and released, so that it oscillates along the perpendicular bisector of PQ.

a) If, at a general point in the motion, the distance of the ball from PQ is x, show that the equation of motion can be written as

$$m\ddot{x} = -4k\left(1 - \frac{l}{\sqrt{4l^2 + x^2}}\right)x$$

b) Use the binomial theorem to expand $(4l^2 + x^2)^{-\frac{1}{2}}$ as far as the term in x^2.
c) If x is small enough so that terms in x^2 and greater powers can be ignored, show that the motion of the ball is SHM and find its period.

14 A particle moves along a straight line Ox such that its displacement, x, from the equilibrium point, O, is satisfied by the equation of motion

a) $\dfrac{d^2x}{dt^2} + 3\dfrac{dx}{dt} + 5x = 0$ b) $2\dfrac{d^2x}{dt^2} + 5\dfrac{dx}{dt} + 2x = 0$ c) $3\dfrac{d^2x}{dt^2} + 6\dfrac{dx}{dt} + 3x = 0$

d) $2\dfrac{d^2x}{dt^2} + 5\dfrac{dx}{dt} + 3x = 0$ e) $2\dfrac{d^2x}{dt^2} + 4\dfrac{dx}{dt} + 4x = 0$ f) $8\dfrac{d^2x}{dt^2} + 8\dfrac{dx}{dt} + 2x = 0$

In each case, determine whether the motion will be strongly damped, critically damped or oscillatory. If it is oscillatory, write down the period of the motion.

15 A block of mass 2 kg is attached to a spring of stiffness $10\,\mathrm{N\,m^{-1}}$. The block is suspended in a resistive medium producing linear damping with a damping constant of $4\,\mathrm{N\,m^{-1}\,s}$. The block is pulled 3 m down from the equilibrium position and released with an initial velocity of $3\,\mathrm{m\,s^{-1}}$ upwards.

  a) Determine the equation of motion of the block.

  b) By finding the particular solution of the equation of motion, show that the motion is weakly damped harmonic and find the period.

16 A particle of mass 2 kg is attached to the end of a light elastic string of natural length 1 m and modulus of elasticity 40 N. The string is hung from a fixed point A and the particle is held at rest a distance of 1 m below A. It is then released. Its motion is subject to a resistance force of magnitude rv, where v is the speed of the particle and r is the damping constants.

  a) Find the equation of motion of the particle.

  b) Find the value of r which corresponds to critical damping. Solve the equation of motion for this value of r.

  c) Find the position of the particle after 2 seconds.

19 Differential equations

Wander in my words and dream about the pictures that I play of changes.
PHIL OCHS

Definitions and classification

A differential equation is any equation containing a derivative. Here are four examples:

i) $\dfrac{dy}{dx} + 2y = \sin x$ ii) $3\dfrac{d^2x}{dt^2} - 3\dfrac{dx}{dt} + 6x = 4\cos 3t$

iii) $\left(\dfrac{dy}{dt}\right)^2 - y\dfrac{d^2y}{dt^2} = \dfrac{dy}{dt}$ iv) $\dfrac{dy}{dx} - x^2y = x^3 - 3x + 4$

When we speak of 'solving' a differential equation, we mean obtaining a relationship connecting the two variables which does not involve a derivative.

A derivative corresponds to a rate of change. Many real-life problems involve quantities which are changing continuously, and the associated mathematical models will be expressed in the form of differential equations. The solution of such equations is therefore an important branch of mathematics. In this chapter, we will encounter applications not only in mechanics but in a range of other fields.

You have already met simple differential equations such as $\dfrac{dy}{dx} = 2x$ and can solve these by integration. Some other types of differential equation can be solved using analytical techniques, but many differential equations can only be solved by numerical methods.

Differential equations are categorised by their style and complexity.

- The **order** of a differential equation is the **order of the highest derivative** appearing in the equation. Of the examples above, **i** and **iv** are first-order equations, while **ii** and **iii** are second-order equations.
- The **degree** of a differential equation in y is the **highest power of y or its derivatives** appearing in the equation. Of the examples above, **i**, **ii** and **iv** are first-degree equations, while **iii** is a second-degree equation because it contains $\left(\dfrac{dy}{dt}\right)^2$. An equation of the first degree is called **linear**.

In this chapter, we will concentrate on linear equations.

Forming differential equations

Before we consider how to solve the various types of differential equation, we should look at some situations in which they arise.

Any situation involving a continuous rate of change can be modelled using differential equations. Often, the rate of change is with respect to time. For example, we might consider the rate at which the temperature, T, of an object is increasing as it is heated. This appears in the equation as $\dfrac{dT}{dt}$.

When time is the independent variable, it is not usual to say so explicitly – we just refer to the **rate of change**. In other cases, we have to be explicit. For example, we might consider the rate at which temperature changes as we bore down into the surface of the Earth. This would appear as $\dfrac{dT}{dx}$, say, where x is the depth below the surface. We would refer to this as the **rate of change of temperature with respect to depth**.

Example 1 The rate at which the population of mice on an island is increasing is modelled as being proportional to the size of the population. It is estimated that when the population reaches 3000, it is increasing at a rate of 900 per week. Express this model in the form of a differential equation.

SOLUTION

Let t be the time in weeks and p be the population. The rate of increase of the population is then expressed as $\dfrac{dp}{dt}$. As this is proportional to p, we have

$$\frac{dp}{dt} = kp$$

where k is a positive constant.

We know that $\dfrac{dp}{dt} = 900$ when $p = 3000$. Substituting these, we have

$$900 = 3000k \quad \Rightarrow \quad k = 0.3$$

So, the required differential equation is

$$\frac{dp}{dt} = 0.3p$$

Example 2 A spherical air freshener is evaporating at a rate proportional to its surface area. Find an expression for the rate of change of its radius.

SOLUTION

Let the volume be V, the surface area be A and the radius be r at time t. The information given corresponds to the differential equation

$$\frac{dV}{dt} = -kA$$

where k is a positive constant. (The negative sign indicates that the volume is decreasing.)

We know that $A = 4\pi r^2$, giving

$$\frac{dV}{dt} = -4\pi k r^2$$

We also know that $V = \frac{4}{3}\pi r^3$, giving

$$\frac{dV}{dr} = 4\pi r^2$$

We require an expression for $\dfrac{dr}{dt}$. These three rates of change are related by the chain rule

$$\frac{dV}{dt} = \frac{dV}{dr} \times \frac{dr}{dt}$$

$$\Rightarrow \quad -4\pi k r^2 = 4\pi r^2 \times \frac{dr}{dt}$$

which gives

$$\frac{dr}{dt} = -k$$

The radius is therefore decreasing at a constant rate.

Exercise 19A

1 An object of mass 5 kg is dropped from a hot air balloon. As it falls, it is subjected to air resistance R N proportional to its speed v m s^{-1}. By considering Newton's second law, express this as a differential equation connecting v, the time t s and a constant k. As it falls, the object's acceleration decreases and it approaches terminal velocity. If the terminal velocity of the object is 60 m s^{-1}, find the value of k.

2 If a capacitor is subjected to a potential difference it gains charge up to a maximum, Q_{MAX}. The rate at which it gains charge is proportional to the difference between its charge and the maximum charge. Express this as a differential equation connecting the time, t, the charge, Q, on the capacitor at that time, and a constant k.

3 The radius of a sphere is increasing at a constant rate, k. Find a differential equation connecting the time, t, and the volume, V, of the sphere.

4 Salt is being poured at the rate of $2\,\text{cm}^3\,\text{s}^{-1}$ onto a conical pile. Assuming that the cone always has equal height and base diameter, find a differential equation connecting the time, $t\,$s, and the height, $h\,$cm, of the cone.

5 In a population of size N, the number of people, n, who have heard a particular joke increases at a rate proportional to the number of people who have heard it, but also to the number of people who have yet to hear it. Express this as a differential equation.

6 A moorland fire, burning in windless conditions, advances outwards from the point at which it started at a constant rate, $k\,\text{m}\,\text{s}^{-1}$. Find a differential equation connecting the time, $t\,$s, and the area $A\,\text{m}^2$, which has been burned.

7 A particle of mass m moves along a straight line. At time $t\,$s, its displacement from a fixed point O is $x\,$m. It is repelled from O by a force which is inversely proportional to x.

a) Express the situation as a second-order differential equation connecting x and t.
b) Modify your equation to include a resistive force proportional to the velocity of the particle.

First-order equations

Separation of variables

We now look at the method of solution for a particular group of first-order differential equations.

The term **separation of variables** is used to cover equations which are given or can be rewritten in the form

$$g(y)\frac{dy}{dx} = f(x)$$

We then integrate both sides with respect to x, which gives

$$\int g(y)\frac{dy}{dx}\,dx = \int f(x)\,dx$$

The left-hand side of this can be simplified to give

$$\int g(y)\,dy = \int f(x)\,dx$$

and in practice we move directly to this without the intervening statement.

Once the equation is written in this form, we can obtain the general solution provided that we can integrate the functions f and g. The resulting relationship between y and x contains an arbitrary constant of integration, and is called the **general solution** of the equation. It defines a whole family of solutions which differ only in the value of the constant.

Example 3 Find the general solution of the equation $\dfrac{dy}{dx} = xy$.

SOLUTION

First, we separate the variables to give

$$\frac{1}{y}\frac{dy}{dx} = x$$

Then we integrate, getting

$$\int \frac{1}{y}\,dy = \int x\,dx$$

$$\Rightarrow \quad \ln y = \tfrac{1}{2}x^2 + C$$

Note We need only one arbitrary constant of integration. Had we introduced one for each integral, we could have subtracted one of them from both sides and simplified to give the above result.

We have

$$\ln y = \tfrac{1}{2}x^2 + C \quad \Rightarrow \quad y = e^{\frac{1}{2}x^2 + C}$$

which is commonly written as

$$y = e^{\frac{1}{2}x^2}e^C$$

giving the general solution as

$$y = A\,e^{\frac{1}{2}x^2}$$

where $A = e^C$

The particular solution

The general solution of a first-order differential equation contains an arbitrary constant. If we have further information about the problem, we can find the value of the constant. We then have the **particular solution** to the problem.

We find the value of the arbitrary constant which is consistent with a known pair of values of the variables. Often the information is given in the form 'the value of x when $t = 0$ is ...' or 'the value of y when $x = 0$ is ...'.

These are usually called **initial conditions** if the stated value of the independent variable is zero, and more generally are called **boundary conditions**.

Initial conditions such as '$x = 2$ when $t = 0$' are sometimes given in the form $x(0) = 2$.

Example 4 Find the general solution of the equation

$$\frac{dy}{dx} = \frac{\cos x}{y}$$

Hence find the particular solution if the boundary conditions are $y = 2$ when $x = \frac{\pi}{6}$.

SOLUTION

First, we separate the variables to give

$$y\frac{dy}{dx} = \cos x$$

Then we integrate, getting

$$\int y\,dy = \int \cos x\,dx$$

$$\Rightarrow \quad \tfrac{1}{2}y^2 = \sin x + C$$
$$\Rightarrow \quad y^2 = 2\sin x + K \quad \text{where } K = 2C$$
$$\Rightarrow \quad y = \sqrt{2\sin x + K}$$

This is the general solution.

For the particular solution, we substitute $y = 2$ and $x = \frac{\pi}{6}$, which gives

$$2 = \sqrt{1 + K}$$
$$\Rightarrow \quad K = 3$$

The particular solution is, therefore, $y = \sqrt{2\sin x + 3}$.

Exercise 19B

1 Find the general solution of the differential equation $\frac{dy}{dx} = f(x, y)$ by separating the variables, where $f(x, y)$ is

a) $\dfrac{x}{y}$

b) $\dfrac{x}{y^2 - 2y}$

c) $(x - 2)(y - 3)$

d) $x(y^2 - 4)$

2 Find the particular solution of the equation $\dfrac{dy}{dx} = \dfrac{\sin x}{\cos y}$ given that $y = 0$ when $x = 0$.

3 Given that $\dfrac{dy}{dx} = e^{x+y}$ and that $y = 0$ when $x = 0$, show that $y = \ln\left(\dfrac{1}{2 - e^x}\right)$.

4 Find the general solution of the following differential equations by separating the variables.

a) $x\dfrac{dy}{dx} + y^2 = 1$

b) $\dfrac{dy}{dx} - 2y = xy$

c) $x\tan y\dfrac{dy}{dx} = 1$

d) $(1 - \cos x)\dfrac{dy}{dx} = \sin x \cot y$

5 Use the substitution $z = y - x$ to rewrite $\dfrac{dy}{dx} = (y - x)^2$ as a differential equation in z and x.

Hence

a) find the general solution of the original equation in the form $y = f(x)$

b) find the particular solution, given that $y = 0$ when $x = 0$

c) show that, in this case, $y = \dfrac{2}{1 + e^2}$ when $x = 1$.

6 An object of mass 1 kg falls from rest. Air resistance is of magnitude $0.1v$ N, where v is the speed of the object. Write a differential equation connecting v and t, and solve it to find v in terms of t. Hence show that v cannot exceed $98\,\text{m s}^{-1}$.

7 The motion of a go-cart with a laden weight of 300 kg is modelled by assuming that its engine exerts a constant power of 4 kW, and that when travelling at a speed of $v\,\text{m s}^{-1}$ it is subject to a resistance force of $10v$ N.

a) Use this model to write a differential equation connecting v and t.

b) Find the particular solution if the go-cart starts from rest.

c) How long does the go-cart take to reach half its maximum speed?

First-order linear equations

A first-order linear equation can be written in the form

$$\frac{dy}{dx} + P(x)\,y = Q(x) \qquad\qquad [1]$$

Such equations can be solved by using an **integrating factor**.

Consider the expression $e^{\int P(x)\,dx}y$. Differentiating this with respect to x, we have

$$\frac{d}{dx}\left(e^{\int P(x)\,dx}y\right) = e^{\int P(x)\,dx}\frac{dy}{dx} + P(x)\,e^{\int P(x)\,dx}y \qquad\qquad [2]$$

Multiplying through equation [1] by $e^{\int P(x)\,dx}$, we get

$$e^{\int P(x)\,dx}\frac{dy}{dx} + P(x)\,e^{\int P(x)\,dx}y = e^{\int P(x)\,dx}Q(x)$$

which from [2] gives

$$\frac{d}{dx}\left(e^{\int P(x)\,dx}y\right) = e^{\int P(x)\,dx}Q(x)$$

Integrating, we have

$$e^{\int P(x)\,dx}y = \int e^{\int P(x)\,dx}Q(x)\,dx$$

and, provided we can integrate the right-hand side, the equation is solved.

The expression $e^{\int P(x)\,dx}$ is called the **integrating factor**.

Example 5 Find the general solution of the differential equation

$$\frac{dy}{dx} + 2xy = x$$

SOLUTION

Comparing this with equation [1], we have $P(x) = 2x$, which gives

$$\int P(x)\,dx = \int 2x\,dx = x^2$$

The integrating factor is, therefore, $e^{\int P(x)\,dx} = e^{x^2}$. Multiplying through by this, we get

$$e^{x^2}\frac{dy}{dx} + 2x\,e^{x^2}y = x\,e^{x^2}$$

$$\Rightarrow \quad \frac{d}{dx}(e^{x^2}y) = x\,e^{x^2}$$

$$\Rightarrow \quad e^{x^2}y = \int x\,e^{x^2}\,dx$$

$$\Rightarrow \quad e^{x^2}y = \tfrac{1}{2}e^{x^2} + C$$

which gives the general solution as

$$y = \tfrac{1}{2} + C\,e^{-x^2}$$

Example 6 Find the general solution of the differential equation

$$(x^2 + 1)\frac{dy}{dx} + xy = 3x$$

and hence find the particular solution for which $y = 2$ when $x = 0$.

SOLUTION

The equation is not yet in the correct format. To correct this, we divide by $(x^2 + 1)$ to get

$$\frac{dy}{dx} + \frac{xy}{x^2 + 1} = \frac{3x}{x^2 + 1}$$

Comparing this with equation [1], we have $P(x) = \dfrac{x}{x^2 + 1}$, which gives

$$\int P(x)\,dx = \int \frac{x}{x^2 + 1}\,dx = \tfrac{1}{2}\ln(x^2 + 1) = \ln\sqrt{x^2 + 1}$$

The integrating factor is, therefore,

$$e^{\int P(x)\,dx} = e^{\ln\sqrt{x^2+1}} = \sqrt{x^2 + 1}$$

Multiplying through by this, we get

$$\sqrt{x^2+1}\,\frac{dy}{dx}+\frac{xy}{\sqrt{x^2+1}}=\frac{3x}{\sqrt{x^2+1}}$$

$$\Rightarrow \quad \frac{d}{dx}\left(y\sqrt{x^2+1}\right)=\frac{3x}{\sqrt{x^2+1}}$$

$$\Rightarrow \quad y\sqrt{x^2+1}=\int\frac{3x}{\sqrt{x^2+1}}\,dx=3\sqrt{x^2+1}+C$$

$$\Rightarrow \quad y=3+\frac{C}{\sqrt{x^2+1}}$$

which is the general solution.

To find the particular solution, we substitute $x=0$ and $y=2$, giving

$$2=3+C \quad \Rightarrow \quad C=-1$$

So, the particular solution is

$$y=3-\frac{1}{\sqrt{x^2+1}}$$

Exercise 19C

1 Find the general solution of the following first-order linear differential equations by using an integrating factor.

a) $\dfrac{dy}{dx}+2y=4$

b) $\dfrac{dy}{dx}+\dfrac{y}{x}=x+2$

c) $\dfrac{dy}{dx}+y=x$

d) $\dfrac{dy}{dx}+3y=e^{-x}$

e) $\dfrac{dy}{dx}+2xy=4x$

f) $\dfrac{dy}{dx}-y\sin x=\sin x$

g) $\dfrac{dy}{dx}+3x^2y=18x^2$

h) $x\dfrac{dy}{dx}+y=x^2$

i) $x^2\dfrac{dy}{dx}-2xy=x^2\ln x$

j) $\cos x\dfrac{dy}{dx}+y\sin x=\cos^2x$

2 A particle has a mass of 1 kg and starts moving from rest. For the first 10 s of its motion it is subject at time t s to a force of magnitude $(10-t)$ N in a constant direction, and to a resistance force of magnitude $0.2v$ N, where v is the velocity of the particle. Express this information as a differential equation connecting v and t, and, by solving it, find the speed at which the particle is moving at the end of the 10 s period.

Linear differential equations with constant coefficients

The integrating factor technique is applicable to any first-order linear differential equation, but if the equation has constant coefficients there is an alternative technique.

A constant-coefficient linear differential equation has the form

$$a_0 y + a_1 \frac{dy}{dx} + a_2 \frac{d^2y}{dx^2} + a^3 \frac{d^3y}{dx^3} + \ldots = f(x)$$

where $a_0, a_1, a_2, a_3, \ldots$ are constants.

If the function $f(x) = 0$, the equation is called a **homogeneous linear equation**.

Initially, we will consider a homogeneous first-order linear equation with constant coefficients, such as

$$\frac{dy}{dx} + 5y = 0$$

We already know how to solve this using the integrating factor technique.

The integrating factor is e^{5x}. Multiplying through by this, we get

$$e^{5x} \frac{dy}{dx} + 5e^{5x}y = 0$$

$$\Rightarrow \quad \frac{d}{dx}(e^{5x}y) = 0$$

$$\Rightarrow \quad e^{5x}y = C$$

$$\Rightarrow \quad y = C e^{-5x}$$

However, it is easy to see that **all** such equations will yield an exponential function, and knowing this, we can solve such equations as follows.

Consider again the equation

$$\frac{dy}{dx} + 5y = 0$$

We assume that the solution to this equation is of the form $y = A e^{mx}$.

Differentiating this, we get

$$\frac{dy}{dx} = Am e^{mx} = my$$

Substituting in the differential equation, we have

$$my + 5y = 0$$

Now, y cannot be zero (or we have the trivial particular solution $y = 0$) and so

$$m + 5 = 0 \qquad [3]$$

$$\Rightarrow \quad m = -5$$

The general solution is, therefore, $y = A\,\mathrm{e}^{-5x}$.

Equation [3] is called the **auxiliary equation**.

In general, for the equation $\dfrac{\mathrm{d}y}{\mathrm{d}x} + ky = 0$, the auxiliary equation is $m + k = 0$.

Example 7 Find the general solution of the homogeneous differential equation $\dfrac{\mathrm{d}y}{\mathrm{d}x} + 4y = 0$.

SOLUTION

The auxiliary equation is $m + 4 = 0$, from which $m = -4$.

Thus the general solution is $y = A\,\mathrm{e}^{-4x}$.

Exercise 19D

1 Find the general solution of the differential equation $\dfrac{\mathrm{d}y}{\mathrm{d}x} - 4y = 0$

    a) by separating the variables
    b) by using an integrating factor
    c) by using the auxiliary equation.

2 Find the general solution of the following first-order homogeneous linear differential equations by using the auxiliary equation.

  a) $\dfrac{\mathrm{d}y}{\mathrm{d}x} + y = 0$ **b)** $\dfrac{\mathrm{d}x}{\mathrm{d}t} - 5x = 0$ **c)** $\dfrac{\mathrm{d}p}{\mathrm{d}t} + 2p = 0$

  d) $\dfrac{\mathrm{d}x}{\mathrm{d}t} - 0.5x = 0$ **e)** $\dfrac{\mathrm{d}y}{\mathrm{d}x} + 3y = 0$ **f)** $\dfrac{\mathrm{d}z}{\mathrm{d}y} = 3z$

The complementary function and the particular integral

We now have to find the solution to the **non-homogeneous linear equation**

$$\frac{\mathrm{d}y}{\mathrm{d}x} + py = \mathrm{f}(x)$$

To see what form our solution should take, let us use the integrating factor method to solve a simple example in which $\mathrm{f}(x)$ is a constant:

$$\frac{\mathrm{d}y}{\mathrm{d}x} + 5y = 3$$

The integrating factor is e^{5x}. Multiplying through by this, we get

$$e^{5x}\frac{dy}{dx} + 5e^{5x}y = 3e^{5x}$$

$$\Rightarrow \quad \frac{d}{dx}(e^{5x}y) = 3e^{5x}$$

$$\Rightarrow \quad e^{5x}y = \frac{3e^{5x}}{5} + C$$

$$\Rightarrow \quad y = Ce^{-5x} + \tfrac{3}{5}$$

which is the general solution.

The solution we have found consists of two parts:

- $y = Ce^{-5x}$ is the solution to the **associated homogeneous equation** $\frac{dy}{dx} + 5y = 0$.

  Ce^{-5x} is called the **complementary function**.

- $y = \tfrac{3}{5}$ is one possible solution to the original equation $\frac{dy}{dx} + 5y = 3$.

  (This can be checked by differentiating it.)

  $\tfrac{3}{5}$ is called a **particular integral**.

If we can find these two parts for a given equation, our task is complete. Our approach is, therefore, as follows.

- First, solve the associated homogeneous equation to find the complementary function (CF).

- Then find a particular integral (PI) which satisfies the original equation. The general solution is then

$$y = CF + PI$$

Note You need to be clear as to the difference between the terms **particular integral**, as used above, and **particular solution**, used to indicate the solution consistent with a given set of boundary conditions.

We find the particular integral by deciding on its likely form and substituting our trial solution into the equation.

Example 8 Find the general solution to the homogeneous differential equation

$$\frac{dy}{dx} - 4y = 2x - 8$$

SOLUTION

The complementary function is found from the associated homogenous equation

$$\frac{dy}{dx} - 4y = 0$$

The auxiliary equation $m - 4 = 0$ gives $m = 4$, and so the CF is $y = A\,e^{4x}$.

We now need a particular integral which will generate the $(2x - 8)$ term. Because this is a linear expression, we try the general linear expression

$$y = Px + Q$$

Differentiating this, we get

$$\frac{dy}{dx} = P$$

Substituting in the differential equation, we have

$$P - 4(Px + Q) = 2x - 8$$

This is an identity and so we can compare coefficients:

$$x^1: \qquad -4P = 2 \qquad \text{giving} \quad P = -\tfrac{1}{2}$$
$$x^0: \qquad P - 4Q = -8 \quad \text{giving} \quad Q = 1\tfrac{7}{8}$$

Hence, the PI is $y = 1\tfrac{7}{8} - \tfrac{1}{2}x$.

We find the general solution by combining the CF and the PI to give

$$y = A\,e^{4x} + 1\tfrac{7}{8} - \tfrac{1}{2}x$$

Finding a particular integral

When using the above technique, our trial solution should be the most general form of the right-hand side of the original equation. In this book, we only consider equations in which the right-hand side is either a constant or a linear expression. (Other situations are dealt with in *Further Mechanics*.) Hence, we have

Right-hand side is a constant	Trial solution $y = P$
Right-hand side is a linear expression	Trial solution $y = Px + Q$

Exercise 19E

1 In each of the following, use the auxiliary equation method to find the general solution.

a) $\dfrac{dy}{dx} + 3y = 2x - 4$ **b)** $\dfrac{dx}{dt} - 2x = 6$ **c)** $\dfrac{dy}{dx} + y = 1 - x$ **d)** $\dfrac{dp}{dt} - 5p = t + 5$

e) $\dfrac{dz}{dx} + 2z = 3x$ **f)** $\dfrac{dx}{dt} - 4x = 3 - 4t$ **g)** $\dfrac{dy}{dx} + 4y = 4x - 2$ **h)** $\dfrac{dx}{dt} = x - 5t$

2 Obtain the general solution of each of the following differential equations by finding the complementary function and the particular integral. Hence find in each case the particular solution consistent with the given initial conditions.

a) $\dfrac{dy}{dx} - 4y = 8$, given that $y = 4$ when $x = 0$

b) $\dfrac{dx}{dt} + 5x = 10 + 2t$, given that $x = 2$ when $t = 0$

c) $\dfrac{dy}{dx} = 2y - 6$, given that $y = 2$ when $x = 0$

d) $\dfrac{dx}{dt} = 3t - 3x + 2$, with initial condition $x(0) = 6$

e) $\dfrac{dy}{dx} - 2y = 4 - 8x$, with initial condition $y(0) = 8$

f) $\dfrac{dx}{dt} + 3x = 6t$, with initial condition $x(0) = 9$

Second-order linear equations with constant coefficients

With first-order linear differential equations with constant coefficients, it is a matter of individual choice whether to use the auxiliary equation technique or the integrating factor technique. However, when we progress to second-order equations, we cannot use the integrating factor technique but the auxiliary equation technique still holds good.

The complementary function

Consider the non-homogeneous second-order differential equation with constant coefficients

$$a\frac{d^2y}{dx^2} + b\frac{dy}{dx} + cy = f(x)$$

The associated homogeneous equation is

$$a\frac{d^2y}{dx^2} + b\frac{dy}{dx} + cy = 0 \qquad [4]$$

Let us assume that this has a solution of the form $y = A\,e^{mx}$. Differentiating this, we have

$$\frac{dy}{dx} = Am\,e^{mx} = my \quad \text{and} \quad \frac{d^2y}{dx^2} = Am^2e^{mx} = m^2y$$

Substituting into equation [4], we obtain

$$am^2 y + bmy + cy = 0$$

Since $y \neq 0$ (this would be a trivial particular solution), we have

$$am^2 + bm + c = 0$$

This is the quadratic auxiliary equation derived from the second-order homogeneous differential equation.

If we assume that its solutions are m_1 and m_2, then we have two solutions to the differential equation:

$$y_1 = A\,e^{m_1 x} \quad \text{and} \quad y_2 = B\,e^{m_2 x}$$

If y_1 and y_2 are two solutions to a differential equation, then $y_1 + y_2$ is also a solution.

Hence, the **most general form** of the complementary function is

$$y = A\,e^{m_1 x} + B\,e^{m_2 x}$$

Note that this equation contains two arbitrary constants. Solving a second-order differential equation is equivalent to integrating twice, generating an arbitrary constant each time we integrate. (In general, the solution to an nth order differential equation has n arbitrary constants.)

However, the quadratic auxiliary equation $am^2 + bm + c = 0$ has three types of solution:

- Two real, distinct roots.
- A real, repeated root.
- Two complex roots.

We will deal with each case separately.

Two real, distinct roots

Essentially, this is the case dealt with above. If the two roots are m_1 and m_2 then the complementary function takes the form

$$y = A\,e^{m_1 x} + B\,e^{m_2 x}$$

A real, repeated root

The problem here is that if the repeated root is m, say, then the complementary function would be

$$
\begin{aligned}
& y = A\,e^{mx} + B\,e^{mx} \\
\Rightarrow\quad & y = (A + B)\,e^{mx} \\
\Rightarrow\quad & y = C\,e^{mx}
\end{aligned}
$$

This cannot be the complete complementary function, since it has only one arbitrary constant.

If the auxiliary equation has a repeated root m, the associated equation is

$$\frac{d^2y}{dx^2} - 2m\frac{dy}{dx} + m^2y = 0$$

Consider $y = Bx\,e^{mx}$. Differentiating this, we obtain

$$\frac{dy}{dx} = B(mxe^{mx} + e^{mx}) \quad \text{and} \quad \frac{d^2y}{dx^2} = B(m^2xe^{mx} + 2m\,e^{mx})$$

We can establish by substitution that these satisfy the equation. (You should check that this is so.) Hence, the most general form of solution is

$$y = A\,e^{mx} + Bx\,e^{mx} \quad \text{or} \quad y = (A + Bx)\,e^{mx}$$

Two complex roots

Suppose the roots of the auxiliary equation are

$$m_1 = \lambda + \mu i \quad \text{and} \quad m_2 = \lambda - \mu i$$

where i represents $\sqrt{-1}$. (Some examination boards use j to represent $\sqrt{-1}$.)

These correspond to an auxiliary equation

$$[m - (\lambda + \mu i)]\,[m - (\lambda - \mu i)] = 0$$

which expands to give

$$m^2 - 2\lambda m + (\lambda^2 + \mu^2) = 0$$

This would arise from the differential equation

$$\frac{d^2y}{dx^2} - 2\lambda\frac{dy}{dx} + (\lambda^2 + \mu^2)y = 0 \qquad [5]$$

Now consider the function

$$y = e^{\lambda x}(A\cos\mu x + B\sin\mu x)$$

Differentiating, we obtain

$$\frac{dy}{dx} = \lambda y + \mu e^{\lambda x}(B\cos\mu x - A\sin\mu x) \qquad [6]$$

Differentiating again, we obtain

$$\frac{d^2y}{dx^2} = \lambda\frac{dy}{dx} - \mu^2 y + \lambda\mu\,e^{\lambda x}(B\cos\mu x - A\sin\mu x) \qquad [7]$$

From equation [6], we have

$$\mu e^{\lambda x}(B\cos\mu x - A\sin\mu x) = \frac{dy}{dx} - \lambda y$$

Substituting into equation [7], we obtain

$$\frac{d^2y}{dx^2} = \lambda\frac{dy}{dx} - \mu^2 y + \lambda\frac{dy}{dx} - \lambda^2 y$$

$$\Rightarrow \quad \frac{d^2y}{dx^2} - 2\lambda\frac{dy}{dx} + (\lambda^2 + \mu^2)\,y = 0$$

which corresponds to equation [5].

Hence, the complete complementary function is given by

$$y = e^{\lambda x}(A \cos \mu x + B \sin \mu x)$$

where λ and μ are the real and imaginary parts of the roots of the auxiliary equation. This is sometimes written in the alternative form

$$y = R e^{\lambda x} \cos(\mu x + \phi)$$

where the arbitrary constants are now R and ϕ.

Example 9 Find the general solution to the homogeneous equation

$$\frac{d^2 y}{dx^2} - 5\frac{dy}{dx} + 6y = 0$$

SOLUTION

The auxiliary equation is

$$m^2 - 5m + 6 = 0$$
$$\Rightarrow \quad (m-3)(m-2) = 0$$
$$\Rightarrow \quad m = 3 \quad \text{or} \quad m = 2$$

Thus the general solution is $y = A e^{3x} + B e^{2x}$.

Example 10 Find the general solution to the homogeneous equation

$$\frac{d^2 y}{dx^2} - 6\frac{dy}{dx} + 9y = 0$$

SOLUTION

The auxiliary equation is

$$m^2 - 6m + 9 = 0$$
$$\Rightarrow \quad (m-3)^2 = 0$$
$$\Rightarrow \quad m = 3 \quad \text{(repeated root)}$$

Thus, the general solution is $y = e^{3x}(Ax + B)$.

Example 11 Find the general solution to the homogeneous equation

$$\frac{d^2 y}{dx^2} - 6\frac{dy}{dx} + 13y = 0$$

SOLUTION

The auxiliary equation is

$$m^2 - 6m + 13 = 0$$
$$\Rightarrow \quad m = \frac{6 \pm \sqrt{36 - 52}}{2} = 3 \pm 2i$$

Thus, the general solution is $y = e^{3x}(A \cos 2x + B \sin 2x)$.

The particular integral

The approach to the particular integral is the same for second-order equations as for first. The trial solution is a general form of the right-hand side of the equation.

Example 12 Find the general solution of the differential equation

$$\frac{d^2x}{dt^2} + 6\frac{dx}{dt} + 13x = 13t + 32$$

SOLUTION

The auxiliary equation is

$$m^2 + 6m + 13 = 0$$

which gives $m = -3 \pm 2i$.

The complementary function is

$$x = e^{-3t}(A\cos 2t + B\sin 2t)$$

The trial solution is

$$x = Pt + Q$$
$$\Rightarrow \quad \frac{dx}{dt} = P \quad \text{and} \quad \frac{d^2x}{dt^2} = 0$$

Substituting these into the differential equation, we obtain

$$6P + 13(Pt + Q) = 13t + 32$$

Comparing coefficients, we have

$$t^1: \qquad 13P = 13 \quad \text{giving} \quad P = 1$$
$$t^0: \quad 6P + 13Q = 32 \quad \text{giving} \quad Q = 2$$

So, the general solution is

$$x = e^{-3t}(A\cos 2t + B\sin 2t) + t + 2$$

A problem case

If the coefficient of y in the differential equation is zero, a clash arises between the complementary function and the particular integral, and we need to modify the form of our trial solution.

Example 13 Find the general solution of the differential equation

$$\frac{d^2y}{dx^2} + 4\frac{dy}{dx} = 4x - 7$$

SOLUTION

The auxiliary equation is

$$m^2 + 4m = 0$$

$$\Rightarrow \quad m = 0 \quad \text{or} \quad m = -4$$

Thus, the complementary function is $y = A + B\mathrm{e}^{-4x}$.

Note that this contains a constant term, which conflicts with our first-choice trial solution of

$$y = Px + Q$$

We overcome this by using the trial solution

$$y = Px^2 + Qx$$
$$\Rightarrow \quad \frac{\mathrm{d}y}{\mathrm{d}x} = 2Px + Q \quad \text{and} \quad \frac{\mathrm{d}^2y}{\mathrm{d}x^2} = 2P$$

Substituting into the differential equation, we obtain

$$2P + 4(2Px + Q) = 4x - 7$$

Comparing coefficients, we have

$$x^1 : \qquad 8P = 4 \qquad \text{giving} \quad P = \tfrac{1}{2}$$
$$x^0 : \quad 2P + 4Q = -7 \quad \text{giving} \quad Q = -2$$

Hence, the particular integral is $y = \tfrac{1}{2}x^2 - 2x$.
The general solution is CF + PI, which gives

$$y = A + B\mathrm{e}^{-4x} + \tfrac{1}{2}x^2 - 2x$$

The particular solution

The general solution of a second-order differential equation contains two arbitrary constants, and so we need two sets of conditions to find their values. We can do this in two ways:

- Give the values of y and $\dfrac{\mathrm{d}y}{\mathrm{d}x}$ when x takes a particular value (usually $x = 0$). These are called **initial conditions**. This is the most common situation.

- Give the value of y for each of two different values of x (usually the end points of the interval being used). These are called **boundary conditions**.

Note For a given set of conditions, there will usually be a unique solution, but the simultaneous equations to find the constants may be inconsistent (no solution) or not independent (an infinite number of solutions).

Example 14 Find the particular solution of the differential equation

$$\frac{\mathrm{d}^2y}{\mathrm{d}x^2} + 4\frac{\mathrm{d}y}{\mathrm{d}x} = 4x - 7$$

given that $y = 0$ and $\dfrac{\mathrm{d}y}{\mathrm{d}x} = 6$ when $x = 0$.

From Example 13, the general solution is

$$y = A + Be^{-4x} + \tfrac{1}{2}x^2 - 2x$$

$$\Rightarrow \quad \frac{dy}{dx} = -4Be^{-4x} + x - 2$$

When $x = 0$, $y = 0$, so we have

$$0 = A + B \qquad\qquad [1]$$

When $x = 0$, $\dfrac{dy}{dx} = 6$, so we have

$$6 = -4B - 2 \qquad\qquad [2]$$

Solving [1] and [2], we obtain $A = 2$, $B = -2$.

So, the particular solution is

$$y = 2 - 2e^{-4x} + \tfrac{1}{2}x^2 - 2x$$

Example 15 Find the particular solution of the differential equation

$$\frac{d^2x}{dt^2} + 6\frac{dx}{dt} + 13x = 13t + 32$$

given that $x = 1$ when $t = 0$, and $x = 2$ when $t = \dfrac{\pi}{4}$.

From Example 12, the general solution is

$$x = e^{-3t}(A\cos 2t + B\sin 2t) + t + 2$$

When $t = 0$, $x = 1$, so we have

$$1 = A + 2 \quad\Rightarrow\quad A = -1$$

When $t = \dfrac{\pi}{4}$, $x = 2$, so we have

$$2 = Be^{-\frac{3\pi}{4}} + \frac{\pi}{4} + 2 \quad\Rightarrow\quad B = -\frac{\pi e^{\frac{3\pi}{4}}}{4}$$

So, the particular solution is

$$x = t + 2 - e^{-3t}\left(\cos 2t + \frac{\pi e^{\frac{3\pi}{4}}}{4}\sin 2t\right)$$

Exercise 19F

1 Find the general solution to the following homogeneous differential equations.

a) $\dfrac{d^2y}{dx^2} + 2\dfrac{dy}{dx} - 3y = 0$

b) $\dfrac{d^2y}{dx^2} - 4\dfrac{dy}{dx} + 4y = 0$

c) $\dfrac{d^2x}{dt^2} + 4\dfrac{dx}{dt} + 8x = 0$

d) $\dfrac{d^2y}{dx^2} + 6\dfrac{dy}{dx} + 9y = 0$

e) $\dfrac{d^2x}{dt^2} - 2\dfrac{dx}{dt} + 5x = 0$

f) $\dfrac{d^2y}{dx^2} + 6\dfrac{dy}{dx} + 8y = 0$

g) $\dfrac{d^2x}{dt^2} + 2\dfrac{dx}{dt} + 2x = 0$

h) $\dfrac{d^2y}{dx^2} - 3\dfrac{dy}{dx} - 4y = 0$

2 Find the general solution to the following non-homogeneous differential equations. Hence find the particular solution corresponding to the stated conditions.

a) $\dfrac{d^2y}{dx^2} + 2\dfrac{dy}{dx} - 8y = 4x + 5$

$y = 0$ and $\dfrac{dy}{dx} = 0$ when $x = 0$

b) $\dfrac{d^2y}{dx^2} - 8\dfrac{dy}{dx} + 16y = 8$

$y = 1.5$ and $\dfrac{dy}{dx} = 5$ when $x = 0$

c) $\dfrac{d^2y}{dx^2} + 2\dfrac{dy}{dx} + y = 2x$

$y = -2$ and $\dfrac{dy}{dx} = 4$ when $x = 0$

d) $\dfrac{d^2y}{dx^2} + 4\dfrac{dy}{dx} = 3 - x$

$y = 2$ and $\dfrac{dy}{dx} = \dfrac{11}{16}$ when $x = 0$

e) $\dfrac{d^2y}{dx^2} - 9y = 3x - 18$

$y = 2$ and $\dfrac{dy}{dx} = 8$ when $x = 0$

f) $\dfrac{d^2x}{dt^2} + 4x = 8t - 4$

$x = 3$ when $t = 0$ and

$x = \dfrac{\pi}{2}$ when $t = \dfrac{\pi}{4}$

g) $\dfrac{d^2y}{dx^2} - 3\dfrac{dy}{dx} = 9$

$y = 0$ when $x = 0$ and
$y = e^3 - 4$ when $x = 1$

h) $\dfrac{d^2x}{dt^2} + 4\dfrac{dx}{dt} + 13x = 18 - 26t$

$x = 2$ when $t = 0$ and
$x = \sin 3$ when $t = 1$

3 A particle of mass 10 kg moves along a straight line through a point O under the influence of a constant force of 120 N. There is also an attraction towards O proportional to the displacement from O. The particle starts from rest at O. When it reaches a point 1 m from O, its acceleration is 8 m s^{-2}. Write a second-order differential equation for the motion of the particle, and hence show that at time t s the particle has displacement $x = 3(1 - \cos 2t)$ m.

Examination questions

Chapters 15 to 19

Chapter 15

1 Show that the constant acceleration equation $v^2 = u^2 + 2as$ is dimensionally correct.

(NICCEA)

2 The formula $\mu = \dfrac{l - x}{x}$ is used to calculate the coefficient of friction for a uniform chain of length l metres about to slip over the edge of a rough, horizontal table. A length x metres rests on the table while the remainder hangs vertically over the edge. Show that the formula is dimensionally correct.   (NICCEA)

3 When a pendulum is suspended and allowed to swing, its motion can be described by the equation

$$\tfrac{1}{2}I\omega^2 - mgh\cos\theta = \text{constant}$$

where m is its mass, h is the distance from its centre of mass to the point from which it is suspended, ω is the angular velocity, θ is the angle the axis of symmetry of the pendulum makes with the vertical and I is a quantity called the 'moment of inertia' of the pendulum.

i) Write down the dimensions of ω. Show that $\cos\theta$ is dimensionless.

ii) Hence show that, for the equation of motion to be dimensionally consistent, I must have dimensions ML^2.

For small oscillations, the period, T, of the pendulum is believed to depend on its moment of inertia, its weight and the distance of its centre of mass from the point of suspension. The formula $T = kI^\alpha(mg)^\beta h^\gamma$ is proposed, where k is a dimensionless constant.

iii) Use dimensional analysis to determine α, β and γ.   (MEI)

Chapter 16

4 A roundabout in a playground moves in a horizontal circle with centre O. The roundabout completes one revolution every 5 seconds. A child sits on the roundabout at a horizontal distance of 2 metres from O. Calculate, giving your answers in terms of π,

a) the speed of the child

b) the magnitude of the acceleration of the child.   (NEAB)

5 A railway engine travels at a constant speed of $v\,\mathrm{m\,s^{-1}}$ on a curved track. The curve is an arc of a horizontal circle of radius $550\,\mathrm{m}$. The magnitude of the acceleration of the engine is $0.22\,\mathrm{m\,s^{-2}}$. Making a suitable modelling assumption, which should be stated, calculate v.

The mass of the engine is $45\,000\,\mathrm{kg}$. Calculate the magnitude of the resultant horizontal force on the engine.   (OCR)

6 In some amusement parks there is a ride which is effectively a hollow cylinder which can rotate about its vertical axis. The riders stand on the horizontal base of the cylinder and in contact with the curved surface of the cylinder. When the angular speed reaches a certain value the floor is dropped but the riders remain in contact with the curved surface of the cylinder. The radius of the cylinder is 2.5 m and the speed of rotation is 30 revolutions per minute. Find the smallest possible coefficient of friction between the rider and the cylinder surface so that the ride works effectively.   (WJEC)

7 A conical pendulum consists of a particle attached to the end of a light inextensible string of length 0.8 m. The particle moves in a horizontal circle, and the system rotates at a constant angular speed of 4 rad s^{-1} (see diagram on the right). Find the angle that the string makes with the vertical.   (OCR)

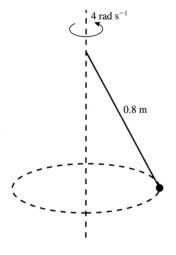

8 OP is a light, inelastic string of length 0.5 m. A particle of mass 5 kg is fastened at the end P and the end O is held at a fixed point A on a smooth, horizontal surface. The particle is made to rotate in horizontal circles on the surface with angular speed 2 rad s^{-1} about A.

i) Find the tension in the string.

The end O of the string is now raised until it is 0.3 m vertically above A. The particle continues to rotate on this surface in horizontal circles about A. The figure below shows all the forces acting on P while it is in motion.

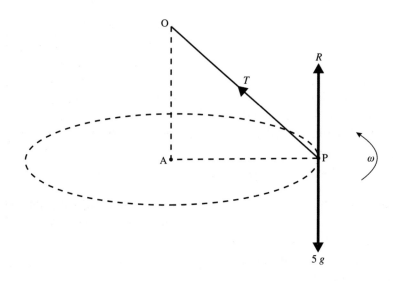

The angular speed of the particle about A is ω rad s^{-1}.

ii) Find, in terms of ω, the tension in the string.

iii) Find, in terms of ω, the reaction between the particle and the horizontal surface.

iv) Find the maximum value of ω, correct to two decimal places, for which the particle will remain in contact with the surface.   (NICCEA)

9 A particle, P, is attached to two light, inelastic strings AP and BP. A and B are two fixed points on a vertical wire with A 2.5 m above B. The particle is made to move uniformly in a horizontal circle about O, where O is the mid-point of AB. This arrangement is shown in the figure on the right.

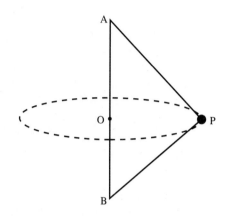

i) Draw a diagram which shows the forces acting on P while it is in motion.

ii) If the angular speed of P about O is 5.6 radians per second and the tension in the string AP is 50 N, find the tension in the string BP.   (NICCEA)

10

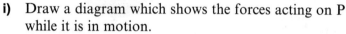
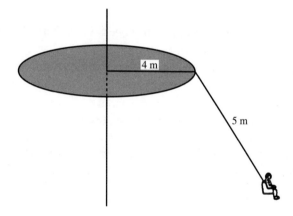

In a fun ride at a theme park, children sit in seats attached by chains of length 5 m to points on the rim of a horizontal circular disc of radius 4 m which rotates at a constant angular speed round a vertical axis through its centre, as shown in the figure above. Safety rules specify that, when the disc rotates during the ride, the chains must not make an angle greater than α to the vertical, where $\tan \alpha = \frac{3}{4}$.

The chains are modelled as light, inextensible strings which lie in a vertical plane containing the axis. A child and a seat are modelled together as a single particle, and any effect of air resistance is ignored.

Find the maximum angular speed with which the disc can rotate in order to comply with the safety rules, giving your answer in rad s^{-1} to three significant figures.   (EDEXCEL)

11

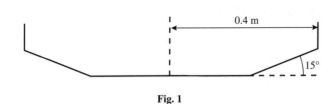

Fig. 1

Fig. 1 shows the cross-section of a hollow container. The base of the container is circular, and is horizontal. The sloping part of the side makes an angle of 15° with the horizontal, and the vertical part of the side forms a circular cylinder of radius 0.4 m. A small steel ball of mass 0.1 kg moves in a horizontal circle inside the container, in contact with the vertical and sloping parts of the side at A and B respectively, as shown in Fig. 2.

It is assumed that all contacts are smooth and that the radius of the ball is negligible compared to 0.4 m.

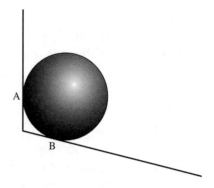

Fig. 2

i) Given that the ball is moving with constant speed $3\,\mathrm{m\,s^{-1}}$, find the magnitudes of the contact forces acting on the ball at A and at B.

ii) Calculate the least speed that the ball can have while remaining in contact with the vertical part of the side of the container.   (OCR)

12 A bend in a road is in the shape of a circular arc of radius r and is banked at 20° to the horizontal. A lorry, travelling around the bend at $20\,\mathrm{m\,s^{-1}}$, is on the verge of slipping down this slope. The coefficient of friction between the lorry's tyres and the road is 0.2.

i) Model the lorry as a particle and draw a diagram to show all the forces acting on it.

ii) Find the value of r.   (NICCEA)

13 A car is travelling round a circular bend on a road which is banked at an angle of 10° to the horizontal, as shown in the figure on the right. The car is modelled as a particle moving in a horizontal circle of radius 80 m. When the car is moving at a constant speed of $v\,\mathrm{m\,s^{-1}}$ there is no sideways frictional force acting on the car.

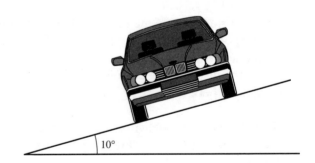

Find, to three significant figures, the value of v.   (EDEXCEL)

14 A racing car is travelling around a racing track. The coefficient of friction between the car's tyres and the track is 0.8.

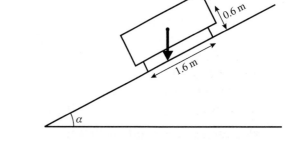

i) Find the maximum speed at which the racing car can negotiate an unbanked bend of radius 100 m without sliding. (Assume toppling does not occur.)

The racing car now moves onto a banked bend of this curved track. The angle of banking is α to the horizontal and the radius is again 100 m. The wheels of the racing car are 1.6 m apart and the centre of gravity of the car is 0.6 m above the track as shown in the figure on the right.

ii) Show that when the car is about to topple its speed will be

$$\sqrt{\frac{(3\tan\alpha + 4)100g}{3 - 4\tan\alpha}}\ \text{m s}^{-1}$$

(You may assume in this case that friction has not reached its limiting value.)

iii) Use your answer to part **ii** to show that the assumption about toppling in part **i** was correct.   (NICCEA)

15

The diagram above shows a railway engine of mass m kg travelling round a curved track. The centre of mass of the engine is moving along the arc of a horizontal circle of radius 500 m. The track is banked at an angle θ to the horizontal.

i) Calculate the acceleration of the engine when it travels at a constant speed of 12.5 m s^{-1}.

At this speed, the engine experiences no sideways force from the rails up or down the slope.

ii) Write down the radial equation of motion and the vertical equilibrium equation. Hence show that $\theta \approx 1.8°$.

When the engine is stationary, the magnitude of the sideways force that the rails exert on the engine up the slope is F N.

iii) Find F in terms of m.

There is a speed limit, V m s^{-1}, for this section of the track. It is set such that, at the speed limit, the magnitude of the sideways force exerted by the rails **down** the bank at the speed limit should also be F N.

iv) Calculate V.   (MEI)

16 In this question assume $g = 10\,\text{m}\,\text{s}^{-2}$.

At an adventure park a child, of mass 46 kg, swings on the end of a rope of length 6 m. All the motion takes place in a vertical plane. Initially the rope is at an angle of 40° to the vertical and the child has a velocity of $2\,\text{m}\,\text{s}^{-1}$ at right angles to the rope. Ignore the effects of air resistance on the child.

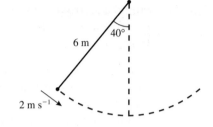

a) State **two** assumptions that it would be appropriate to make about the rope in this problem.
b) Show that the maximum speed of the child is approximately $5.7\,\text{m}\,\text{s}^{-1}$.
c) Find the maximum tension in the rope.   (AEB 96)

17 The pilot of a stunt aircraft flies it in a vertical circle of radius 200 m. Near the top of the circle, the aircraft is flying upside down at a constant speed of $40\,\text{m}\,\text{s}^{-1}$. The mass of the pilot is 61 kg. Show that, at the top of the circle, the magnitude of the reaction between the pilot, modelled as a particle, and her aircraft is approximately 110 N.

State whether, at the top of the circle, the pilot feels that she is being pressed into her seat, or is hanging suspended in the safety harness.   (OCR)

18 In this question take the value of g, the acceleration due to gravity, as $10\,\text{m}\,\text{s}^{-2}$.

A wall at the end of a horizontal track can be modelled as half of a smooth cylindrical shell, centre O and radius $5\sqrt{2}\,\text{m}$, as shown in Fig. 1 below.

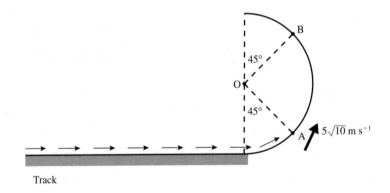

Fig. 1

A stuntman on a skateboard skates along the track and skates up the wall. A is a point on the wall such that OA makes an angle of 45° with the downward vertical. The stuntman has a speed of $5\sqrt{10}\,\text{m}\,\text{s}^{-1}$ at A.

i) Find the speed of the stuntman when he is at point B on the wall where OB makes an angle of 45° with the upward vertical.
ii) Prove that he is just about to lose contact with the wall when he reaches B.

A horizontal platform is mounted at the same level as A and ends at a point vertically below O, as shown in Fig. 2 below.

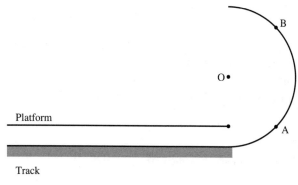

Fig. 2

iii) How far from the end of the platform does the stuntman land?   (NICCEA)

Chapter 17

19 A student hangs a 0.2 kg mass on a spring and notes that the length of the spring increases by 4 cm from its natural length.

a) Use this information to write down a possible relationship between the tension in the spring and its extension.

The 0.2 kg mass is removed from the spring and a 0.5 kg mass is hung on the spring. Its length increases by 12 cm from its natural length.

b) Show that this does not agree with the previous model. Sketch a graph to show how tension varies with extension and suggest an alternative linear model that both observations satisfy.

(AEB 97)

20

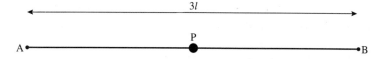

The ends of a light elastic string, of natural length $2l$ and modulus of elasticity λ, are attached to fixed points A and B, where AB is horizontal and $AB = 3l$. A particle P of mass m, is attached to the mid-point of the string, as shown in the figure above. The particle is held at rest with the string horizontal and is then released. In the subsequent motion air resistance may be neglected,

The particle falls a distance $2l$ before first coming instantaneously to rest.

a) Show that $\lambda = mg$.
b) Find the magnitude of the acceleration of P when it first comes to instantaneous rest.

(EDEXCEL)

21

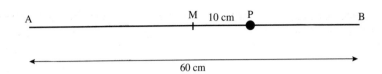

A particle P, of mass 1.2 kg, is attached to fixed points A and B by two light elastic strings, each of natural length 10 cm and modulus of elasticity 1.5 N. The points A and B are 60 cm apart on a smooth horizontal table, and the mid-point of AB is M. The particle P is released from rest at a point on the line AB which is 10 cm from M (see diagram above). Neglecting air resistance, calculate

i) the initial acceleration of P
ii) the speed with which P passes through M.   (OCR)

22 A light elastic spring of modulus of elasticity 80 N and natural length 20 cm is held horizontally in the jaws of a vice, as shown in the diagram. Given that the distance between the jaws is 16 cm, calculate the magnitude of the force exerted by each of the jaws on the spring.   (OCR)

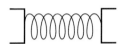

23

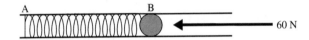

a) Assuming Hooke's law show, by integration, that the work in extending a spring of modulus λ and natural length l a distance x beyond its natural length is $\dfrac{\lambda x^2}{2l}$.

b) The diagram above shows a spring of natural length 0.15 m in a smooth horizontal tube with its end A fixed and a small bead B of mass 0.2 kg held in equilibrium by a force of magnitude 60 N pressing it against the free end of the spring. The compression of the spring in this position is 0.03 m.

i) Find the modulus of elasticity of the spring.
ii) The bead is released. Find, using the conservation of energy, the speed of the bead just as the spring attains its natural length.   (WJEC)

24 A particle P, of mass 0.2 kg, hangs in equilibrium suspended by two light strings attached to fixed points A and B at the same horizontal level and 0.5 m apart. The string AP is inextensible and has length 0.5 m. The string BP is elastic, with unstretched length 0.5 m and modulus of elasticity λ newtons. In the equilibrium position, angle PAB is 80° and angle APB = angle ABP = 50° (see diagram on the right).

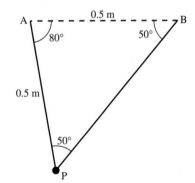

i) By considering the equilibrium of P, find the tension in BP
ii) Hence find the value of λ.   (OCR)

25 A spring of natural length 0.6 m and modulus of elasticity 60 N obeys Hooke's Law in extension and compression. It hangs vertically with one end attached to a fixed point O. A particle of mass 0.5 kg is attached to the other end. The particle is pulled vertically down and released from rest.

Take the gravitational potential energy of the particle to be zero at the horizontal through O.

When the spring is stretched x metres vertically downwards beyond its natural length, the particle has speed v m s^{-1}.

i) Show that the energy stored in the spring is $50x^2$ J.

ii) Find, in terms of x and v, an expression for the total mechanical energy, in joules, of the system.

iii) If the particle is released from rest when the total length of the spring is 0.7 m, use the principle of conservation of mechanical energy to show that the length of the spring is 0.598 m when the particle next comes to rest.   (NICCEA)

26

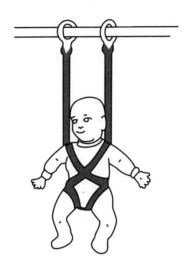

A manufacturer of nursery equipment is using a large doll to test the safety of a 'baby bouncer'. The 'baby bouncer' consists of a light harness, which is attached to one end of each of two identical elastic cords. The other end of each cord is fastened to a fixed horizontal rail, as shown in the diagram. Each cord may be modelled by a light elastic string of natural length 1.2 metres and modulus λ newtons. The mass of the doll is 8 kg.

a) When the doll is suspended in equilibrium, the extension of each of the cords is 20 cm. Taking $g = 10$ m s^{-2}, show that $\lambda = 240$.

b) The doll is now held at a distance of 1.2 metres vertically below the rail, so that both cords are vertical and unstretched, and released from rest. Find the vertical distance through which the doll falls before coming momentarily to rest.

c) The doll is pulled vertically downwards, so that the extension of each of the cords is 60 cm, and released from rest.

i) Show that, when the extension of the cords is x metres, where $x > 0$, the kinetic energy of the doll is

$$(24 + 80x - 200x^2) \text{ J}$$

ii) Find, in terms of x, the acceleration of the doll.

iii) Calculate the greatest speed of the doll during its upward motion.   (NEAB)

27

Fig. 1

Carol plays a bat and ball game in her garden, as shown in Fig. 1.

The ball of mass 0.05 kg is attached to one end of an elastic string of natural length 1 m and modulus of elasticity 4.9 N and hangs vertically in equilibrium. The other end of the string is attached to a fixed point A. Carol strikes the ball and gives it a speed of 15 m s^{-1}.

i) Taking the gravitational potential energy of the ball to be zero at the equilibrium position, show that the total energy of the ball and string is 5.6495 J.

At the greatest extension of the string, the ball is stationary and at a vertical height of 0.1 m above the equilibrium position, as shown in Fig. 2.

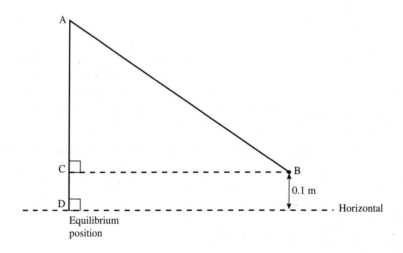

Fig. 2

ii) Using the principle of conservation of mechanical energy, find the horizontal distance, BC, in metres, correct to two decimal places, of the ball from the equilibrium position.

(NICCEA)

Chapter 18

28 A body moves with simple harmonic motion in a straight line. The amplitude of the motion is 3.4 m, and the equation of motion is $\dfrac{d^2x}{dt^2} = -9.9x$, where x metres is the displacement of the body from the centre of motion at time t seconds.

   i) Find the period of the motion of the body.
   ii) Find the maximum speed of the body.   (OCR)

29 The diagram on the right shows a particle P of mass 0.2 kg on a horizontal platform. The platform oscillates vertically in such a manner that the motion of P is simple harmonic with centre A and period $\dfrac{2\pi}{3}$ s. The amplitude of the oscillations is denoted by a m.

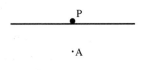

   a) Find, in terms of a, the speed of the platform when at a height of $\dfrac{3a}{5}$ m above the level of A.

   b) Find the greatest value of a such that the particle does not leave the platform.   (WJEC)

30 A sailing ship is rolling in heavy seas. A sailor is at the top of a tall mast which swings from side to side. While the sailor is in contact with the mast, his motion is modelled as that of a particle moving horizontally with simple harmonic motion of period 10 s and amplitude 7 m. The mass of the sailor is 100 kg and he loses his grip when the horizontal force acting on him is 245 N.

   a) Find, to two significant figures, his distance from the centre of oscillation at the moment when he loses his grip.

   b) Find, to two significant figures, his speed when he loses his grip.   (EDEXCEL)

31 The blade of an electric saw moves up and down describing simple harmonic motion. The tip of the blade moves between two points a distance d apart. The amplitude, speed and period of the motion can all be varied.

   a) For one particular job the maximum speed of the blade is set to 5 m s^{-1} and d is set to 5 cm. Find the period of the motion.

   b) The maximum speed of the blade can be set to any value. Show that the speed of the blade always drops to half of its maximum value when it is at a distance of $\dfrac{d\sqrt{3}}{4}$ from the mid-point of its motion.   (AEB 98)

32 In this question you may quote, without proof, any SHM formulae.

The diagram on the right shows a cylindrical buoy of height 2 m and mass 440 kg floating vertically in equilibrium in a calm sea, the point marked A on the cylinder being at sea level. The upward buoyancy force due to the sea, when the length of the buoy beneath sea level is d m, is 2750d N. Find the height of A above the base of the buoy.

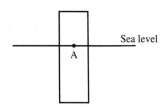

a) The top of the buoy is then moved downwards a distance of 0.2 m and released from rest at time $t = 0$ s. During the subsequent motion the downward displacement of A from sea level at time t s is denoted by x m. Assuming that

- the motion of the buoy can be modelled by the motion of a particle of mass 440 kg at the centre of gravity under the action of gravity and the buoyancy force
- the motion of the buoy does not affect the sea level

show that

$$\frac{d^2x}{dt^2} = -6.25x$$

   i) Write down an expression for x in terms of t.
   ii) Find the time taken before A first returns to sea level and the maximum speed of the buoy.
   iii) Find the time taken until A is at a depth of 0.1 m below sea level.

b) A passing ship disturbs the sea level so that at time t s the displacement of the sea level below the original level of the calm sea is $0.4 \sin 2t$. The downward displacement of A below the original level of the calm sea is again denoted by x m. Obtain, but do not attempt to solve, the differential equation satisfied by x.   (WJEC)

33 An engineer is observing a machine component performing simple harmonic motion about a fixed point O. She uses x to denote the displacement (in cm) of the component from O, and t to denote the time (in seconds) from the first observation.

She first observes the component when $x = 3$ and it is moving away from O. Two seconds later she observes it pass through O for the first time. She observes that the period of the motion is 6 seconds.

i) Sketch a graph of x against t for $0 \leqslant t \leqslant 6$, showing the values of x and t where the graph crosses the axes.

An expression for x in terms of t of the form $A \cos \omega t + B \sin \omega t$ is to be found.

ii) Calculate ω and A and show that $B = \sqrt{3}$.
iii) Calculate the amplitude of the motion and hence, or otherwise, calculate the speed of the component when it was first observed.
iv) Find the greatest acceleration of the component.   (MEI)

34 A plumber, working on a plumbing system, causes the water in a U-bend to oscillate to and fro in simple harmonic motion between the two positions shown in the figure below.

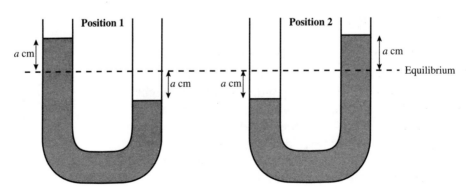

At a particular instant the water level in the left hand side of the U-bend is 0.5 cm above the equilibrium position and rising at $\dfrac{\sqrt{15}}{4}$ cm s^{-1}. A short time later it is 1 cm above the equilibrium position and rising at $\dfrac{\sqrt{3}}{2}$ cm s^{-1}.

i) Find the amplitude, a centimetres, of the motion.
ii) Find the time which elapses from the instant the water level is rising through the point 1 cm above the equilibrium position until it next reaches the highest point. (NICCEA)

35 The graph shows how the displacement, s, of the tip of a needle in a sewing machine varies with time, t. The displacement, s, is measured from the level of the cloth in the machine.

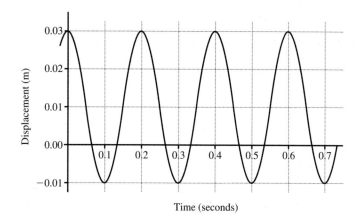

Assume that the motion of the tip of the needle is simple harmonic and that its displacement can be modelled by $s = A\cos(\omega t) + d$.

a) Find the values of A, ω and d.
b) Find the time when the needle first pierces the cloth.
c) Find the maximum speed of the needle and the speed when it pierces the cloth. (AEB 98)

36 A simple pendulum consists of a mass on the end of a string of length $2a$.

a) Draw a diagram to show the forces acting on the mass when the string is inclined at an angle θ to the vertical.
b) By considering the component of the resultant force perpendicular to the string, show that at time t

$$\frac{d^2\theta}{dt^2} \approx -\frac{g\theta}{2a}$$

stating clearly any assumptions you have made.
c) Find the period of the simple pendulum in terms of a, g, and π.

d) The pendulum swings through a small angle from A, where it was at rest, to C where it comes to rest again. It describes an arc of radius $2a$ between A and B and an arc of radius a between B and C.

Find the time that it takes for the pendulum to get from A to C.　　(AEB 97)

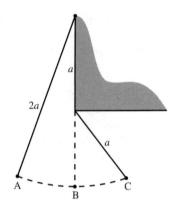

Chapter 19

37 Show that the differential equation

$$\frac{dy}{dx} + \frac{2}{1+x}y = 7(1+x)^4$$

has an integrating factor $(1+x)^n$, where n is an integer whose value is to be found. Given that $y = 2$ when $x = 0$, find the solution of the equation.　　(WJEC)

38 Find an integrating factor for the equation

$$\frac{dy}{dx} + 2xy = e^{-x^2}$$

and hence obtain the general solution of the equation.　　(WJEC)

39 Show that an appropriate integrating factor for the differential equation

$$\frac{dy}{dx} + \frac{5y}{x} = x^2$$

is x^5. Hence, or otherwise, find the solution of the differential equation for which $y = 1$ when $x = 1$.　　(WJEC)

40 a) Find the general solution of the differential equation

$$\frac{d^2x}{dt^2} + 4\frac{dx}{dt} + 8x = 0$$

b) In economic modelling of investment it is assumed, with a particular choice of units, that

　i)　the rate of increase of excess capital k is equal to the investment I
　ii)　the rate of decrease of the investment I is equal to the sum of $8k$ and $4I$.

Express $\dfrac{dk}{dt}$ in terms of I, and $\dfrac{dI}{dt}$ in terms of k and I. Hence show that k satisfies the differential equation in part **a**. Comment on the behaviour of k for large values of t.　　(WJEC)

41

A   O   P   B

In the diagram, A and B are two fixed points at a distance of 1 m apart on a smooth horizontal wire. A bead of mass 0.25 kg is threaded on the wire. When the bead is at any point P on the wire it is attracted towards A by a force of magnitude 3PA newtons and towards B by a force of magnitude 2PB newtons (in these expressions PA and PB denote the distances between P and A and P and B respectively, measured in metres). The point O where the bead is in equilibrium is at a distance of d metres from A. The displacement of P from O in the direction OB is denoted by x metres.

a) Show that $d = 0.4$.

b) Show that the resultant of the above two forces on the bead when it is at P is $-5x$ in the direction OB.

c) At time $t = 0$ the bead is displaced a distance 0.1 m from O towards B and is released from rest. Subsequently when it is moving with speed v m s^{-1} it experiences a resistance to its motion of magnitude $2.25v$ newtons. Show that

$$\frac{\mathrm{d}^2 x}{\mathrm{d}t^2} + 9\frac{\mathrm{d}x}{\mathrm{d}t} + 20x = 0$$

d) Find the general solution of the differential equations in part **c** and hence find the distance of the bead from O at time t.

42 Find the general solution of the differential equation

$$\frac{\mathrm{d}^2 y}{\mathrm{d}x^2} - 9\frac{\mathrm{d}y}{\mathrm{d}x} + 20y = 60x + 13 \qquad \text{(WJEC)}$$

Answers

Exercise 1

1 a) Yes **b)** Probably not **c)** No **d)** Yes **e)** Probably **f)** Yes **g)** Yes **h)** No **2 a)** Yes **b)** No **c)** No **d)** Yes

2 e) No, in the sense that friction is needed for there to be a forward force at the wheels driving the car forward

2 f) No, because in addition to **e** there must be a friction force towards the centre of curvature causing the car to change direction

Exercise 2A

1 a) i) $p + \frac{1}{2}q$ ii) $\frac{1}{2}q - p$ iii) $q - p$ **2 a)** $2q$ **b)** $p + q$ **c)** $q - 2p$ **d)** $2q - 2p$ **e)** $p - 2q$ **3 a)** $2q$ **b)** $p + q$ **c)** $-q$

3 d) $2q - p$ **e)** $\frac{1}{2}p + 1\frac{1}{2}q$ **f)** $\frac{1}{2}p - 1\frac{1}{2}q$ **4 a)** $c - a$ **b)** $b - a$ **c)** $\frac{1}{2}(b - a)$ **d)** $c - b$ **e)** $\frac{2}{3}(c - b)$ **f)** $\frac{1}{3}b + \frac{2}{3}c$

4 g) $\frac{2}{3}c - \frac{1}{2}a - \frac{1}{6}b$ **6 a)** i) $-\frac{1}{2}q$ ii) $-p - \frac{1}{2}q$ iii) $\frac{1}{3}(q - p)$ iv) $\frac{1}{6}(q - p)$, showing that B, G and E collinear

9 a) 7.02 km, 098.8° **b)** 23.4 km, 217.3° **c)** 23.5 km h^{-1}, 290.6° **d)** 529.1 N, 347.5° **10** 221.8° or 288.2°

11 a) $\overrightarrow{AB} = \overrightarrow{OB} - \overrightarrow{OA}$ **b)** i) 20.6 km, 140.9° ii) 61.8 km, 140.9° iii) 20.6t km, 140.9° **c)** 5.82 h (5 h 49 min)

11 d) 10.49 h (10 h 29 min) **12 a)** 5.385 m s^{-1}, 80 m downstream from B **b)** Steer upstream at 23.6° to AB, 4.583 m s^{-1}

12 c) Steer upstream at 55.5° to AB, 70.7 s

Exercise 2B

1 a) $2j$ **b)** $-2i - 2j$ **c)** $-6i + 7j$ **d)** $16i + j$ **e)** $\sqrt{5}$ **f)** $2\sqrt{5}$ **2 a)** $u = -10, v = 7$ **b)** $u = -4.5$ **3 a)** $-12i + 16j$

3 b) $-0.6i + 0.8j$ **4 a)** $5.803i + 3.914j$ **b)** $1.996i + 9.39j$ **c)** $10.064i + 6.536j$ **d)** $4j$ **e)** $-3.638i + 7.46j$ **f)** $-10.143i + 21.751j$

4 g) $-3.556i - 5.079j$ **h)** $12.474i - 6.536j$ **i)** $-2.057i - 2.832j$ **5 a)** $5.385, 21.8°$ **b)** $11.402, 52.12°$ **c)** $5, -90°$

5 d) $3.606, 123.69°$ **e)** $5.831, -59.04°$ **f)** $7.81, -140.19°$ **g)** $2, 180°$ **6 a)** $\overrightarrow{OA} = 19.021i + 6.18j$, $\overrightarrow{AB} = 11.389i + 25.579j$

6 b) $\overrightarrow{OB} = 30.41i + 31.76j$ **c)** 43.97 km, 043.8° **7 a)** $\overrightarrow{OA} = -34.468i + 6.078j$, $\overrightarrow{OB} = 25i + 43.301j$

7 b) $\overrightarrow{AB} = \overrightarrow{OB} - \overrightarrow{OA} = 59.468i + 37.224j$ **c)** 70.157 km, 058° **8 a)** $\sqrt{5}$ m s^{-1}

8 b) A at 63.435°, B at 18.435° to x-direction, angle between 45° **c)** $\overrightarrow{OA} = ti + 2tj$, $\overrightarrow{OB} = 3ti + tj$, $\overrightarrow{AB} = 2ti - tj$ **d)** 40.25 s

Exercise 2C

1 a) i) $-3i + 2j + 4k$ ii) $11i + 3j - 11k$ iii) 7 iv) $\sqrt{33}$ v) $\frac{6}{7}i - \frac{3}{7}j - \frac{2}{7}k$ vi) $31°, 115.4°, 106.6°$ **b)** $24i - 12j - 8k$

1 c) $p = 2, q = -3, r = 13$ **2** $7.128i + 0.697j + 3.507k$ **3** $27.1°$ or $152.9°$, $p = \pm 10.684i + 4.495j + 3.106k$

4 a) $a = 5.516, b = 2.509, c = 3.5$ **b)** $a = -2.536, b = 3.936, c = -3.751$ **c)** $\gamma = 41.6°$ or $138.4°$, $a = 5.29, b = 2.78, c = \pm 6.73$

4 d) $r = 14, \alpha = 73.4°, \beta = 64.6°, \gamma = 31°$ **e)** $r = \sqrt{50}, \alpha = 115.1°, \beta = 55.6°, \gamma = 45°$

4 f) $a = 60°, \beta = 104.5°, \gamma = 33.99°$ or $146.01°, c = \pm 3.32$ **g)** $r = 5.962, \gamma = 79.8°$ or $100.2°, b = 3.07, c = \pm 1.06$

4 h) $r = 12.97, \gamma = 128.1°, a = 3.36, b = 9.64$ **5 a)** $44.7°, 54.4°, 66.7°$ **b)** 3.095 m

Exercise 2D

1 a) 10 **b)** 0 **c)** 2 **d)** 3 **e)** 0 **f)** -12 Vectors in **b** and **e** are perpendicular to each other **2 a)** 19.02 **b)** -10.14 **c)** 3.44

2 d) 0 **e)** -94.56 **f)** -44 **3 a)** $109.7°$ **b)** $92.7°$ **c)** $98.1°$ **d)** $73.98°$ **e)** $101.95°$ **f)** $148.05°$ **g)** $97.42°$ **h)** $90°$ **4** $34.51°$

5 $53.08°$ **7 a)** $81.04°$ **b)** 0.268 s and 3.732 s

Exercise 3A

1 3 km h^{-1} **2 a)** 9 km h^{-1} **b)** $2\frac{1}{4}$ h **c)** 8.47 km h^{-1} **3** 9.49 m s^{-1}, 9.37 m s^{-1}, 8.22 m s^{-1}, 6.89 m s^{-1}, 6.41 m s^{-1}, 5.92 m s^{-1}

4 90 km h^{-1} **5** $58\frac{1}{3}$ km h^{-1}

Exercise 3B

1 a)

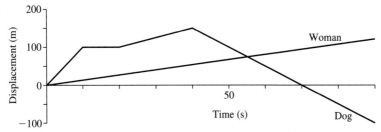

1 b) Pass after 54.7 s, 76.6 m from A   **c)** Average speed $4\frac{4}{9}$ m s^{-1}, average velocity $-1\frac{1}{9}$ m s^{-1}

2 a), b)

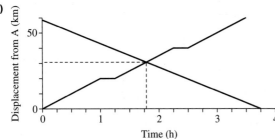

2 c) Pass after 1 h $48\frac{1}{3}$ min, $31\frac{1}{9}$ km from A

3 a)

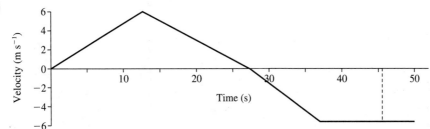

3 b) 81 m   **c)** 45.5 s   **4 a)** Constant speed of 5 km h^{-1} for 2 h, stationary for 1 h, return to start at $3\frac{1}{3}$ km h^{-1}

4 b)

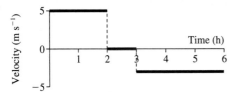

5 a) 5 min 30 s

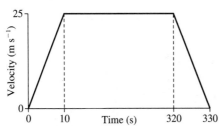

5 b) Maximum 7 min 33.6 s, when limit starts more than 125 m from Chulchit's home and ends more than 230 m from his work; minimum 7 min 28.8 s, when limit starts at Chulchit's home or ends at his work

Exercise 3C

1 $s = ut + \frac{1}{2}at^2$ **2** $s = vt - \frac{1}{2}at^2$ **3** $v^2 = u^2 + 2as$

Exercise 3D

1 a) 1350 m   **b)** 90 m s^{-1} **2 a)** 10 m s^{-2} **b)** 125 m   **3 a)** 3 m s^{-2} **b)** 30 m s^{-1} **4** 7392 m   **5** 0.139 m s^{-2}, -0.069 m s^{-2} '
6 a) $23\frac{5}{6}$ m   **b)** $3\frac{2}{3}$ s   **7** 15 m   **8 a)** 5 s   **b)** 25 m

ANSWERS

Exercise 3E

1 a) 3.19 s **b)** 31.3 m s^{-1} **2 a)** 11.5 m **b)** 1.53 s **c)** 3.06 s **3 a)** 19.8 m s^{-1} **b)** 4.04 s (total time in air)

3 c) 19.8 m s^{-1} downwards **4 a)** 1.28 m above top of cliff **b)** −34.7 m s^{-1} **c)** 4.05 s **5 a)** 3.88 m **b)** 1.52 s **6** 45 m

7 Time 4 s, height 8g m **8** $\frac{5}{16}$ m above top of window **9** $V = h − 5.5g$ **10** $\dfrac{u}{g} − \dfrac{t}{2}$

Exercise 4A

1 In parts **d** and **e**, air resistance has been included

1 a) 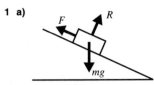 **b)** 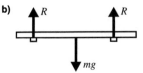 **c)**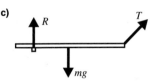

1 d) 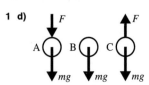 **e)** 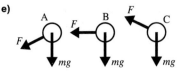 **f)**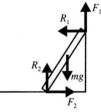

2 a) 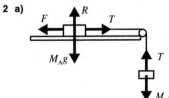 **b)** $T = M_B g, F = M_B g$ **3 a)** **b)** **4**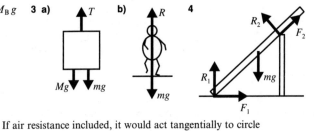

5 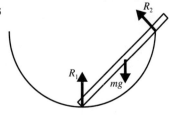 **6** 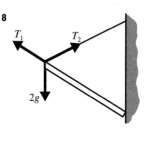 If air resistance included, it would act tangentially to circle

7 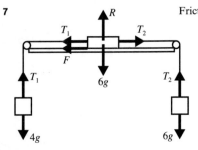 Friction $F = 3g$ N **8**

Exercise 4B

1 a) Resultant: $(4.312\mathbf{i} + 4.095\mathbf{j})$ N or 5.947 N at 43.5° to x-direction, equilibrant: $(−4.312\mathbf{i} − 4.095\mathbf{j})$ N or 5.947 N at −136.5° to x-direction

1 b) Resultant: $(−1.186\mathbf{i} + 2.981\mathbf{j})$ N or 3.208 N at 111.7° to x-direction, equilibrant: $(1.186\mathbf{i} − 2.981\mathbf{j})$ N or 3.208 N at −68.3° to x-direction

1 c) Resultant: $(−1.793P\mathbf{i} + 0.109P\mathbf{j})$ N or 1.796P N at 176.5° to x-direction, equilibrant: $(1.793P\mathbf{i} − 0.109P\mathbf{j})$ N or 1.796P N at −3.5° to x-direction

2 a) $P = 9.178\,\text{N}$, $Q = 9.664\,\text{N}$ **b)** $P = 7.660\,\text{N}$, $Q = 6.732\,\text{N}$ **c)** $P = -3.230\,\text{N}$, $Q = -12.102\,\text{N}$ **3** $10\,\text{N}$ at $36.9°$ to direction BA

4 a) $P = 12.856\,\text{N}$, $Q = 15.321\,\text{N}$ **b)** $P = 9.434\,\text{N}$, $\theta = 148°$ **c)** $P = 17.174\,\text{N}$, $Q = 15.827\,\text{N}$ **5** As Question **4**

6 a) $P = 2.536\,\text{N}$, $Q = 5.438\,\text{N}$ **b)** $P = 17.341\,\text{N}$, $Q = 24.508\,\text{N}$ **c)** $P = 17.264\,\text{N}$, $Q = 24.045\,\text{N}$ **7 a)** $P = 16\frac{1}{3}\,\text{N}$, $T = 42.467\,\text{N}$

7 b) $15.077\,\text{N}$, $T = 36.185\,\text{N}$ **8** $60.058\,\text{N}$, $50.395\,\text{N}$ **9 a)** $17.925\,\text{N}$, $25.313\,\text{N}$ **b)** $P = 5.597\,\text{N}$, $T = 22.403\,\text{N}$

10 a)

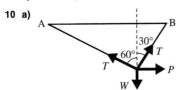 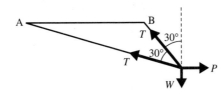

10 c) Case 1: $P = W(2 - \sqrt{3})$, Case 2: $P = W$ **11** $P = \dfrac{Wb}{a}$ **12** $\dfrac{Wb}{a}$

Exercise 5A

1 $2.25\,\text{m s}^{-2}$ **2** $52.5\,\text{N}$ **3** $6\frac{2}{3}\,\text{kg}$ **4 a)** $0.8\,\text{m s}^{-2}$ **b)** $400\,\text{kg}$ **c)** $380\,\text{N}$ **d)** $1290\,\text{N}$ **e)** $-0.4\,\text{m s}^{-2}$ **5 a)** $12\,\text{m s}^{-1}$ **b)** $99\,\text{m}$

6 $(0.75\mathbf{i} + 1.25\mathbf{j})\,\text{m s}^{-2}$ **7** $P = (-\mathbf{i} - \mathbf{j})\,\text{N}$ **8** $5\sqrt{13}\,\text{N}$ **9** $702.36\,\text{N}$ **10 a)** $9.497\,\text{m s}^{-2}$ at $51.2°$ to x-direction

10 b) $11.77\,\text{m s}^{-2}$ at $100.7°$ to x-direction **11 a)** $538.9\,\text{N}$ **b)** $15.6\,\text{m s}^{-2}$ **12** $0.725\,\text{m s}^{-2}$ on bearing $073.3°$

Exercise 5B

1 a) i) $2.7\,\text{m s}^{-2}$ upwards **ii)** $2.3\,\text{m s}^{-2}$ downwards **b) i)** $196\,\text{N}$ **ii)** $196\,\text{N}$ **iii)** $236\,\text{N}$ **iv)** $182\frac{2}{3}\,\text{N}$ **v)** $241\,\text{N}$

2 a) $2640\,\text{N}$, $440\,\text{N}$ **b)** $4200\,\text{N}$ **3 a)** $550\,\text{N}$ **b)** $490\,\text{N}$ **c)** $452.5\,\text{N}$ **d)** $390\,\text{N}$ **4 a)** $13.06\,\text{kg}$ **b)** $92.24\,\text{kg}$ **c)** $73.47\,\text{kg}$

4 d) $80\,\text{kg}$ **e)** $72.5\,\text{kg}$ **5** $1.96\,\text{m s}^{-2}$ upwards **6 a)** $39.2\,\text{N}$, $68.6\,\text{N}$ **b)** $39.2\,\text{N}$, $68.6\,\text{N}$ **c)** $51.2\,\text{N}$, $89.6\,\text{N}$ **7 a)** $90.4\,\text{N}$, $56.5\,\text{N}$

7 b) $5.2\,\text{m s}^{-2}$, top string breaks **8 a)** $3.07\,\text{m s}^{-2}$ **b)** $59.16\,\text{N}$ **9 a)** 7.8 **b)** $4.6\,\text{m s}^{-2}$ **c)** $3.77\,\text{m s}^{-1}$ **10** $8\,\text{s}$

11 a) $1\,\text{m s}^{-2}$, $1500\,\text{N}$ **b)** $1200\,\text{N}$ **c)** $136.36\,\text{N}$ **12 a)** $2.45\,\text{m s}^{-2}$ **13** $\frac{1}{4}(5R - P)$

Exercise 5C

1 a) $1.633\,\text{N}$ **b)** $57.17\,\text{N}$ **c)** $114.3\,\text{N}$ **2** $1.96\,\text{m s}^{-2}$, $3.96\,\text{m s}^{-1}$ **3** $3.267\,\text{m s}^{-2}$, $13.067m\,\text{N}$ **4** $3.92\,\text{m s}^{-2}$, $11.76\,\text{N}$

5 $3.222\,\text{m s}^{-2}$, $32.889\,\text{N}$ **6** $5.88\,\text{m s}^{-2}$, $0.714\,\text{s}$ **7** $2.285\,\text{m s}^{-2}$, $22.55\,\text{N}$ **8** $\frac{1}{4}g$, $\frac{3}{4}mg$ **9** $4.356\,\text{m s}^{-2}$, $21.778\,\text{N}$, $13.067\,\text{N}$

10 $1.8375\,\text{m s}^{-2}$, $46.55\,\text{N}$, $55.738\,\text{N}$ **11** $3.267\,\text{m s}^{-2}$, $13.067m\,\text{N}$, $19.6m\,\text{N}$ **12** $a = \dfrac{g(r - h)}{2r}$ **13 a)** $4.2\,\text{m s}^{-2}$ **b)** $5.02\,\text{m s}^{-1}$

13 c) $1.286\,\text{m}$ **14 a)** $3.834\,\text{m s}^{-1}$ **b)** $3\,\text{m}$ **16 a)** $0.891\,\text{m s}^{-2}$ **b)** $17.818\,\text{N}$

Exercise 5D

1 a) $19.174\,\text{N}$; $5.965\,\text{m s}^{-2}$ down ($5\,\text{kg}$), $2.983\,\text{m s}^{-2}$ up ($3\,\text{kg}$)

1 b) $59.33\,\text{N}$, $29.665\,\text{N}$; $1.324\,\text{m s}^{-2}$ down ($7\,\text{kg}$), $5.032\,\text{m s}^{-2}$ up ($2\,\text{kg}$), $2.384\,\text{m s}^{-2}$ down ($4\,\text{kg}$)

1 c) $40.091\,\text{N}$; $1.782\,\text{m s}^{-2}$ down ($5\,\text{kg}$), $3.564\,\text{m s}^{-2}$ up ($3\,\text{kg}$), $0.891\,\text{m s}^{-2}$ down ($9\,\text{kg}$)

1 d) $12.166\,\text{N}$, $24.331\,\text{N}$; $5.745\,\text{m s}^{-2}$ down ($3\,\text{kg}$), $3.717\,\text{m s}^{-2}$ down ($2\,\text{kg}$), $2.366\,\text{m s}^{-2}$ up ($4\,\text{kg}$)

1 e) $9.046\,\text{N}$; $1.508\,\text{m s}^{-2}$ ($6\,\text{kg}$), $0.754\,\text{m s}^{-2}$ ($2\,\text{kg}$)

1 f) $7.84\,\text{N}$, $15.68\,\text{N}$; $3.92\,\text{m s}^{-2}$ ($4\,\text{kg}$), $5.88\,\text{m s}^{-2}$ down ($2\,\text{kg}$), $1.96\,\text{m s}^{-2}$ down ($1\,\text{kg}$) **4** $0.7\,\text{m s}^{-2}$, $173.25\,\text{N}$

Exercise 6A

1 a) $30t - 3t^2$, $15t^2 - t^3$ **b)** $75\,\text{m s}^{-1}$, $250\,\text{m}$ **c)** $500\,\text{m}$ **d)** $15\,\text{s}$ **2 a)** $-2\,\text{m}$ **b)** $6\,\text{m}$ **3 a)** $2t^2 + 6t - 8$ **b)** $-4\,\text{s}$ or $1\,\text{s}$

3 c) i) $1\,\text{s}$ **ii)** $4\,\text{s}$ before timing started **4 a)** $2.5\sqrt{3}\,\text{m}$, $2.5\sqrt{3}\,\text{m}$, $0\,\text{m}$, $-2.5\sqrt{3}\,\text{m}$, $-2.5\sqrt{3}\,\text{m}$, $0\,\text{m}$

4 b) $\dfrac{5\pi}{6}\,\text{m s}^{-1}$, $-\dfrac{5\pi}{6}\,\text{m s}^{-1}$, $-\dfrac{5\pi}{3}\,\text{m s}^{-1}$, $-\dfrac{5\pi}{6}\,\text{m s}^{-1}$, $\dfrac{5\pi}{6}\,\text{m s}^{-1}$, $\dfrac{5\pi}{3}\,\text{m s}^{-1}$

4 c) $-\dfrac{2.5\sqrt{3}\pi^2}{9}\,\text{m s}^{-2}$, $-\dfrac{2.5\sqrt{3}\pi^2}{9}\,\text{m s}^{-2}$, $0\,\text{m s}^{-2}$, $\dfrac{2.5\sqrt{3}\pi^2}{9}\,\text{m s}^{-2}$, $\dfrac{2.5\sqrt{3}\pi^2}{9}\,\text{m s}^{-2}$, $0\,\text{m s}^{-2}$ **d)** Cycle repeats

5 a) $0\,\text{s}$, $4\,\text{s}$, $-4\,\text{s}$ (Particle is at O when timing starts, and $4\,\text{s}$ before and after this)

5 b) $-\dfrac{4}{\sqrt{3}}\,\text{s}$ and $\dfrac{4}{\sqrt{3}}\,\text{s}$ (before and after timing starts) **c)** $\pm\dfrac{128}{3\sqrt{3}}\,\text{m}$ **d** $30\,\text{m s}^{-2}$ **6 a)** $t^2 - 5t + 6$ **b)** $2\,\text{s}$ and $3\,\text{s}$

6 c) $7\frac{1}{3}$ m and $7\frac{1}{6}$ m **7 a)** 30 s **b)** 225 m

7 c) Speed of bird at start and end of flight $30\,\mathrm{m\,s^{-1}}$, so acceleration and deceleration must be instantaneous **8 a)** $-1\,\mathrm{s}$, $1\,\mathrm{s}$, $2\,\mathrm{s}$

8 b) $-2\,\mathrm{m\,s^{-1}}$, $6\,\mathrm{m\,s^{-1}}$, $3\,\mathrm{m\,s^{-1}}$; $-10\,\mathrm{m\,s^{-2}}$, $2\,\mathrm{m\,s^{-2}}$, $8\,\mathrm{m\,s^{-2}}$ **9 a)** 2 s **b)** $v = 8 - 10t$ **c)** $-12\,\mathrm{m\,s^{-1}}$ **d)** 7.2 m

9 e) Gives displacement (height above ground) not distance travelled **10 a** 12 m, 8.21 m, 6.81 m, 6.30 m, 6.11 m, 6.04 m

10 b) $-6\,\mathrm{m\,s^{-1}}$, $-2.21\,\mathrm{m\,s^{-1}}$, $-0.812\,\mathrm{m\,s^{-1}}$, $-0.299\,\mathrm{m\,s^{-1}}$, $-0.110\,\mathrm{m\,s^{-1}}$, $-0.040\,\mathrm{m\,s^{-1}}$

10 c) $6\,\mathrm{m\,s^{-2}}$, $2.21\,\mathrm{m\,s^{-2}}$, $0.812\,\mathrm{m\,s^{-2}}$, $0.299\,\mathrm{m\,s^{-2}}$, $0.110\,\mathrm{m\,s^{-2}}$, $0.040\,\mathrm{m\,s^{-2}}$ **d)** Model implies body never reaches complete rest

11 a) $120t^{-1} - 60$ **b)** $120\ln t - 60t + 60$ **c)** Approaches constant velocity $-60\,\mathrm{m\,s^{-1}}$

Exercise 6B

2 a) $50 - s$ **b)** Approaches limit of 50 m **3** $v^2 + 4s^2 = 9$ **4 a)** $v = \sqrt{100 - 4t}$ **b)** $v = \sqrt[3]{1000 - 6s}$ **5 a)** $v = 2\mathrm{e}^{-2s}$

5 b) $v = \dfrac{2}{4t+1}$ **6 a)** $\dfrac{v^2}{2} = \dfrac{gR^2}{s} + c$ **d)** $\sqrt{2gR}$ **7 a) ii)** $v = 5(1 - \mathrm{e}^{-2t})$ **iii)** $s = 5t + \frac{5}{2}(\mathrm{e}^{-2t} - 1)$ **8 a)** $v = -10\sin 2t$

8 b) $s = 5\cos 2t$ **d)** $v^2 = 100 - 4s^2$ **9** $4v\dfrac{\mathrm{d}v}{\mathrm{d}x} = gx$, $\frac{3}{4}\sqrt{7g}\,\mathrm{m\,s^{-1}}$ **10** $30\sqrt{2}\,\mathrm{m\,s^{-1}}$

Exercise 7A

1 a) $\mathbf{v} = 4\mathbf{i} + 4t\mathbf{j}$ **b)** $\mathbf{v} = 2(t - 2)\mathbf{i} + (3t^2 - 4t)\mathbf{j}$ **c)** $\mathbf{a} = 3t^2\mathbf{i} + 6t\mathbf{j}$ **d)** $\mathbf{a} = (1 - 2t)\mathbf{i} + 3\mathbf{j}$ **e)** $\mathbf{v} = 5(3 - 2t)\mathbf{j}$ **f)** $\mathbf{a} = -10\mathbf{j}$

1 g) $\mathbf{a} = 6\cos 2t\,\mathbf{i} - 6\sin 2t\,\mathbf{j} - 5\mathbf{k}$ **h)** $\mathbf{v} = -4\sin t\cos t\,\mathbf{i} + 4\sin t\cos t\,\mathbf{j} + \dfrac{1}{2\sqrt{t}}\mathbf{k}$ **i)** $\mathbf{a} = 3\mathbf{i} - 2\sin t\,\mathbf{j} - 2\cos\frac{1}{2}t\,\mathbf{k}$

2 a) $\mathbf{v} = 3t^2\mathbf{i} + 2t(1 - t)\mathbf{j} + \mathbf{c}$ **b)** $\mathbf{v} = -10t\mathbf{j} + \mathbf{c}$ **c)** $\mathbf{r} = 3t\mathbf{i} + \frac{1}{2}t(8 - 5t)\mathbf{j} + \mathbf{c}$ **d)** $\mathbf{r} = t^2(1 - 2t)\mathbf{i} + t^3(1 - t)\mathbf{j} + \mathbf{c}$

2 e) $\mathbf{v} = 2t\mathbf{i} + 2t^2\mathbf{j} + \mathbf{c}$ **f)** $\mathbf{r} = 3t(1 - t)\mathbf{i} + t^2(1 - 3t)\mathbf{j} + \mathbf{c}$ **g)** $\mathbf{r} = 2t^2\mathbf{i} + 4\sin t\,\mathbf{j} - 2\cos t\,\mathbf{k} + \mathbf{c}$

2 h) $\mathbf{v} = -5\sin t\,\mathbf{i} + 5\cos t\,\mathbf{j} + 3t^2\mathbf{k} + \mathbf{c}$ **i)** $\mathbf{r} = 8\cos t\,\mathbf{i} + 8\sin t\,\mathbf{j} + 2\mathrm{e}^{2t}\mathbf{k} + \mathbf{c}$ **3 a)** $\mathbf{r} = 2(t^2 + 1)\mathbf{i} + (2t^4 - 1)\mathbf{j}$

3 b) $\mathbf{r} = (7 - 4\cos t)\mathbf{i} + 2(t + 1)\mathbf{j}$ **c)** $\mathbf{r} = 5(\cos t - 1)\mathbf{i} + (1 - 5\sin t)\mathbf{j} + \frac{1}{2}(3\sin 2t - 10)\mathbf{k}$

4 a) $\mathbf{v} = (3t - t^2 + 3)\mathbf{i} + \frac{1}{2}t^2(2 - 3t^2)\mathbf{j}$, $\mathbf{r} = \frac{1}{6}(9t^2 - 2t^3 + 18t + 6)\mathbf{i} + \frac{1}{30}(10t^3 - 9t^5 - 60)\mathbf{j}$

4 b) $\mathbf{v} = 2(1 + \sin 2t)\mathbf{i} + (5 - 4\cos 2t)\mathbf{j}$, $\mathbf{r} = (2t - \cos 2t + 1)\mathbf{i} + (5t - 2\sin 2t + 4)\mathbf{j}$

4 c) $\mathbf{v} = (2t^2 - 3t + 2)\mathbf{i} - 3\mathbf{j} + t(3t - 2)\mathbf{k}$, $\mathbf{r} = \frac{1}{6}(4t^3 - 9t^2 + 12t + 6)\mathbf{i} + (1 - 3t)\mathbf{j} + (t^3 - t^2 + 2)\mathbf{k}$

Exercise 7B

1 a) $\mathbf{v} = 3\mathbf{i} + 4(1 - t)\mathbf{j}$ **b)** $\mathbf{a} = -4\mathbf{j}$ **c)**

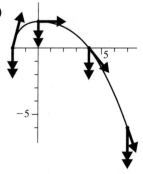

2 a) $\mathbf{a} = 4\mathbf{i} + 6t\mathbf{j}$ **b)** $\mathbf{r} = 2(t^2 - t + 1)\mathbf{i} + (t^3 + 3)\mathbf{j}$ **d)** $4\mathbf{i} + 9\mathbf{j}$ **3 a)** $\mathbf{v} = (3 + 16t)\mathbf{i} + (1 + 24t)\mathbf{j}$ **b)** $\mathbf{a} = 16\mathbf{i} + 24\mathbf{j}$ **c)** 2 s or $-\frac{2}{7}$ s

3 d) $40\mathbf{i} + 56\mathbf{j}$ away, $1.80\mathbf{i} + 6.69\mathbf{j}$ towards **4 a)** $\mathbf{v} = 480(\mathbf{i} + \sqrt{3}\,\mathbf{j})\,\mathrm{km\,h^{-1}}$ **b)** $\mathbf{r} = (480t\mathbf{i} + 480\sqrt{3}t\mathbf{j} + 0.8\mathbf{k})\,\mathrm{km}$

5 a) $5\mathbf{i}$, $2.5\mathbf{i} + 4.33\mathbf{j}$, $-2.5\mathbf{i} + 4.33\mathbf{j}$, $-5\mathbf{i}$, $-2.5\mathbf{i} - 4.33\mathbf{j}$, $2.5\mathbf{i} - 4.33\mathbf{j}$, $5\mathbf{i}$

5 b) $\mathbf{v} = -\dfrac{5\pi}{3}\sin\left(\dfrac{\pi t}{3}\right)\mathbf{i} + \dfrac{5\pi}{3}\cos\left(\dfrac{\pi t}{3}\right)\mathbf{j}$

5 c) $5.24\mathbf{j}$, $-4.53\mathbf{i} + 2.62\mathbf{j}$, $-4.53\mathbf{i} - 2.62\mathbf{j}$, $-5.24\mathbf{j}$, $4.53\mathbf{i} - 2.62\mathbf{j}$, $4.53\mathbf{i} + 2.62\mathbf{j}$, $5.24\mathbf{j}$

5 d) $\dfrac{5\pi}{3}$ throughout **e)** $\mathbf{a} = -\dfrac{5\pi^2}{9}\cos\left(\dfrac{\pi t}{3}\right)\mathbf{i} - \dfrac{5\pi^2}{9}\sin\left(\dfrac{\pi t}{3}\right)\mathbf{j}$

5 f) $-5.48\mathbf{i}$, $-2.74\mathbf{i} - 4.75\mathbf{j}$, $2.74\mathbf{i} - 4.75\mathbf{j}$, $5.48\mathbf{i}$, $2.74\mathbf{i} + 4.75\mathbf{j}$, $-2.74\mathbf{i} + 4.75\mathbf{j}$, $-5.48\mathbf{i}$

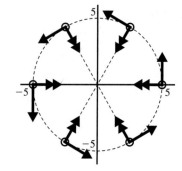

6 a) $\mathbf{v} = 2\mathbf{i} + 2(2 - t)\mathbf{j}$, $\mathbf{r} = 2(t + 1)\mathbf{i} + (4t - t^2 - 3)\mathbf{j}$ **b)** 2 s **c)** 1 s or 3 s **7 a)** 1 s **b)** $3\mathbf{i} + 2.5\mathbf{j}$, $3\mathbf{i} - 4\mathbf{j}$

8 a) $5\mathbf{i}$, $3.54\mathbf{i} + 3.54\mathbf{j} + 2\mathbf{k}$, $5\mathbf{j} + 4\mathbf{k}$, $-3.54\mathbf{i} + 3.54\mathbf{j} + 6\mathbf{k}$, $-5\mathbf{i} + 8\mathbf{k}$, $-3.54\mathbf{i} - 3.54\mathbf{j} + 10\mathbf{k}$, $-5\mathbf{j} + 12\mathbf{k}$, $3.54\mathbf{i} - 3.54\mathbf{j} + 14\mathbf{k}$, $5\mathbf{i} + 16\mathbf{k}$

8 a) $3.54\mathbf{i} + 3.54\mathbf{j} + 18\mathbf{k}$, $5\mathbf{j} + 20\mathbf{k}$ **b)**

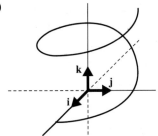

c) A spiral based on circle of radius 5 units and rising 16 units in each complete revolution

8 d) $\mathbf{v} = -\dfrac{5\pi}{4}\sin\left(\dfrac{\pi t}{4}\right)\mathbf{i} + \dfrac{5\pi}{4}\cos\left(\dfrac{\pi t}{4}\right)\mathbf{j} + 2\mathbf{k}$, $\mathbf{a} = -\dfrac{5\pi^2}{16}\cos\left(\dfrac{\pi t}{4}\right)\mathbf{i} - \dfrac{5\pi^2}{16}\sin\left(\dfrac{\pi t}{4}\right)\mathbf{j}$

Exercise 7C

1 a) 1.76 s **b)** 21.6 m **c)** 0.878 s **d)** 3.78 m **2 a)** 1.77 s **b)** 8.84 m **c)** 0.884 s **d)** 3.83 m **3 a)** 17.2 m **b)** 15.4 m s^{-1}

3 c) 2.5° below horizontal **4** 28 m s^{-1}, 4.04 s **5** $\dfrac{U^2}{\sqrt{U^2 - gh}}$ **6 a)** Gradient $\dfrac{\sin\theta - \sin\alpha}{\cos\theta - \cos\alpha}$

6 b) Rate of increase $u\sqrt{2(1 - \cos(\theta - \alpha))}$ **7** 4b **8 a)** $30\mathbf{i} + 35\mathbf{j}$, $60\mathbf{i} + 60\mathbf{j}$ **b)** 8.16 s **c)** 244.9 m **d)** $180y = 240x - x^2$

10 b) $\mathbf{a} = -10\mathbf{j}$, $\mathbf{v} = 2\sqrt{5}\mathbf{i} + 2(2\sqrt{5} - 5t)\mathbf{j}$, $\mathbf{r} = 2\sqrt{5}t\mathbf{i} + t(4\sqrt{5} - 5t)\mathbf{j}$ **c)** $4y = 8x - x^2$ **d)** 3.75 m

Exercise 8A

1 a) 24 m **b)** 53.1°, 5 s **c)** No **d)** 26.7 m **2** 16.1 km h^{-1}, 053° **3** 10 m s^{-1} **4** 096.8°, 320.7°, 212.5°, 15 min 19.3 s

5 a) $\mathbf{v_A} = 33.2\mathbf{i} + 39.6\mathbf{j} + 18.8\mathbf{k}$, $\mathbf{v_B} = 59.6\mathbf{i} - 21.7\mathbf{j} - 29.6\mathbf{k}$, $_A\mathbf{v_B} = -26.4\mathbf{i} + 61.3\mathbf{j} + 48.4\mathbf{k}$ **b)** 82.4 km h^{-1}, 35.9°

Exercise 8B

1 28.3 km h^{-1} on bearing 122° **2** 20 km h^{-1} from bearing 053.1° **3** 30.6 km h^{-1} from bearing 229° **4 b)** $6\mathbf{i} + 9\mathbf{j}$ **5** 112.6°, 1 h

6 Miss by 0.458 km **7** 28 km on bearing 045.5° **8** 41.4° to horizontal, 6.8 s **9** 42.5 km h^{-1} on bearing 105.1° **10 c)** 1 s, 3 m

11 1.5 s, 15.4 m

Exercise 9A

1 a) 0.408 **b)** 0.321 **c)** 0.294 **d)** 0.641 **e)** 0.619 **f)** 0.379 **2 a)** 23.5 N **b)** 35.1 N **c)** 32.0 N **d)** 29.2 N **e)** 52.1 N

2 f) 54.5 N **3** 22.2 N, 0.51 **5 a)** 18.4 N **b)** 4.77 N **6** 49.3 N **7** 34.9° **8** 0.36 **9** $P = mg\sqrt{k^2m^2 - 1}$ **15** $\dfrac{1}{k\sqrt{3}}$

Exercise 9B

1 4.04 m s^{-2} **2** 0.306 **3** 14.6 m **4** Stage 1: 18.7 N, stage 2: 14.7 N, stage 3: -9.3 N **5** 0.092 **6** 10.844 m s^{-1} **7** 67.7 N

8 a) 17.836 N **b)** 21.704 N **c)** 17.693 N **10** 4.592 m, particle remains at rest

Exercise 9C

1 6.087 N **2** 3.91° **3** 0.414 **4 a) i)** 11.113 N **ii)** 28.087 N **iii)** 13.01 N **iv)** 27.316 N **b)** 10.781 N **c)** 27.248 N

5 77.764 N **6** 8.385 N **7** $\alpha = 58.28°$, $\lambda = 13.28°$

Exercise 10A

1 a) $+56$ N m **b)** -87.5 N m **c)** -100.8 N m **d)** -31.5 N m **e)** $+451.2$ N m **f)** $+720$ N m **2 a) i)** $+19$ N m **ii)** -35.6 N m

2 b) i) -18 N m **ii)** -24.4 N m **c) i)** -9.39 N m **ii)** -12.81 N m **3 a)** $+14.4$ N m **b)** -9.6 N m **c)** $+2.4$ N m **d)** -4.8 N m

4 a) 1.1 m **b)** 0.6 m **c)** 0.8 m

Exercise 10B

1 a) $+23.84\,\text{N m}$ **b)** $-26.05\,\text{N m}$ **c)** $-81.35\,\text{N m}$ **d)** $-151.9\,\text{N m}$ **e)** $+169.4\,\text{N m}$ **f)** $+324.5\,\text{N m}$ **2 a) i)** $+7.89\,\text{N m}$

2 a) ii) $-60.37\,\text{N m}$ **b) i)** $-10.19\,\text{N m}$ **ii)** $-17.04\,\text{N m}$ **c) i)** $+9.43\,\text{N m}$ **ii)** $-9.99\,\text{N m}$ **3** $-23.96\,\text{N m}$ **4** $-12\,\text{N m}$

5 $-17\,\text{N m}$ **6** $-8\,\text{N m}$

Exercise 10C

1 a) Resultant $8\,\text{N}$, $1.5\,\text{m}$ from A, $2.5\,\text{m}$ from B **b)** Couple $+12\,\text{N m}$ **c)** Resultant $3\,\text{N}$, $1\frac{1}{3}\,\text{m}$ from A, $2\frac{2}{3}\,\text{m}$ from B

1 d) Resultant $1\,\text{N}$ at mid-point of AB **e)** Equilibrium **f)** Couple $-26\,\text{N m}$ **2 a)** $R = 8\,\text{N}$, $x = 1\,\text{m}$ **b)** $R = 8\,\text{N}$, $x = 1.5\,\text{m}$

2 c) $R = 6\,\text{N}$, $x = 0.5\,\text{m}$ **3 a)** $5\frac{1}{3}\,\text{N}$ **b)** $4\,\text{N}$ **4** Resultant zero so system forms couple, moment $-27\,\text{N m}$

5 $a = -1$, $b = -3$; $-10\,\text{N m}$ **6 a)** $50\,\text{N}$, $150\,\text{N}$ **b)** $0.2\,\text{m}$ **c)** $-40\,\text{N m}$ (if A is to left of B on your diagram) **7** $46g\,\text{N}$, $64g\,\text{N}$

9 $a < x < 10a$ **10** $\dfrac{W(a + c - 2b)}{2c}$, $\dfrac{W(2b + c - a)}{2c}$ **11 a)** $2\frac{2}{3}\,\text{m}$ **b)** $12.5\,\text{kg}$

12 $26.6°$. Swapping $3W$ and $2W$ gives equilibrium with AC at $33.7°$

Exercise 10D

1 Resultant zero so system forms couple, moment $-22\,\text{N m}$

2 Resultant zero so system forms couple, moment $120\,\text{N m}$ in sense ABC

3 Resultant zero so system forms couple, moment $k \times$ area ABC **4** Resultant zero so system forms couple, moment $10.5a\sqrt{3}\,\text{N m}$

5 a) $63.65\,\text{N}$ **b)** $73.5\,\text{N}$ **c)** $127.31\,\text{N}$ **6 b)** $P = \dfrac{2W\sqrt{3}}{3}$, $\text{AD} = \dfrac{a}{3}$ **7** $0.258a$ **8** $P = 21.2\,\text{N}$; $61.2\,\text{N}$ at A, $42.4\,\text{N}$ at B

9 a) $34.6\,\text{N}$ **b)** $2.33\,\text{m}$ **10** $63.4°$ **11** $61.9°$ **12** $P = 29.7\,\text{N}$, $\mu = 0.152$ **14** Horizontal $0.2W$, vertical $0.6W$

Exercise 10E

1 a) $\mathbf{P} = 2\,\text{N}$ in BA direction, $\mathbf{Q} = 2\,\text{N}$ in CB direction **b)** $\mathbf{P} = 2\,\text{N}$ in AB direction, $\mathbf{Q} = 2\,\text{N}$ in BC direction **2 b)** $+66\,\text{N m}$

3 $Q = \sqrt{2}\,\text{N}$; $\sqrt{13}\,\text{N}$, $33.7°$ **4** $X = 6P$, $Y = 5P$, $c = 4a$ **5** $\sqrt{2}\,\text{N}$ at $45°$ to EC, $x = 0.464a$; $-2.536a\,\text{N m}$

6 a) $P = -4W$, $Q = 3W\sqrt{2}$ **b)** $P = 4W$, $Q = 3W\sqrt{2}$ **7 b)** $P = 24$, $Q = 15$ **c)** $P = 16$, resultant $-(Q + 5)$ **8 c)** $17.5°$

Exercise 11A

1 a) $\left(3\frac{4}{15}, 5\frac{8}{15}\right)$ **b)** $(1.3, 3.15)$ **c)** $\left(\frac{4}{11}, -6\frac{7}{11}\right)$ **d)** $\left(2\frac{7}{11}, 0, -1\frac{8}{11}\right)$ **2** $4\frac{7}{16}\mathbf{i} - 2\frac{3}{8}\mathbf{j}$ **3** $-\frac{5}{18}\mathbf{i} - 3\frac{8}{9}\mathbf{j} + \frac{2}{9}\mathbf{k}$ **4** $(-13.6, -8.4)$

5 $m = 4\,\text{kg}$, CE $= 1\,\text{m}$ **6** $\left(2\frac{18}{19}, 2\frac{1}{19}, \frac{18}{19}\right)\,\text{m}$ **7** $42°$ **8** $45.15°$ **9** $2.75\,\text{kg}$

Exercise 11B

1 a) $\left(2\frac{5}{16}, 1\frac{11}{16}\right)\,\text{m}$ **b)** $(1.54, 0.641)\,\text{m}$ **c)** $(0.923, 0.8)\,\text{m}$ **d)** $(-0.049, 0)\,\text{m}$ **e)** $(3.20, 1.81)\,\text{m}$ **f)** $(59.7, 67.8)\,\text{cm}$

2 $(0.578, 0.356)\,\text{m}$ **3** $(41.1, 17.8)\,\text{cm}$ **4** $(0.629, 0.793, 0.214)\,\text{m}$ **5 a)** $\left(\frac{2}{5}, \frac{1}{3}\right)\,\text{m}$ **b)** $13.5°$ **6** $1.83\,\text{m}$ **7** $49.34\,\text{cm}$

8 $54.0 < h < 58.7\,\text{cm}$ **9 a) i)** $13.9\,\text{cm}$ **b)** 20.84

Exercise 11C

1 $(3.36, 0)$ **4** $\left(\dfrac{\pi}{2}, \dfrac{\pi}{8}\right)$ **5 a)** $\left(\dfrac{e^2 + 1}{e^2 - 1}, \frac{1}{4}(e^2 + 1)\right)$ **b)** $\left(\dfrac{(3e^4 + 1)}{2(e^4 - 1)}, 0\right)$ **6 a)** $(0.75, 4)$ **b)** $(0.75, 0)$ **7** $1.5a$

8 $\left(\dfrac{4a}{3\pi}, \dfrac{4b}{3\pi}\right)$, $\left(\dfrac{3a}{8}, 0\right)$ **9 a)** $\left(\frac{4}{5}, \frac{32}{7}\right)$ **b)** $(0, 5)$ **11** $\dfrac{a\sin\alpha}{\alpha}$ from centre of circle **12** $a(1 + \cos\alpha)$ from centre of sphere

14 $(1.66, 0)$

Exercise 11D

1 Slides if $P > 25g\,\text{N}$, topples if $P > 20\frac{5}{6}g\,\text{N}$, so equilibrium broken by toppling **2** $h < 50\,\text{cm}$ **3** $\mu < 0.8$

4 a) $(29.01, 25, 60.82)\,\text{cm}$ **b)** $0.092 < \mu < 0.242$ **6** $\frac{1}{2}(1 - \tan a)$, cannot slide if $a > 45°$ **8** $10\,\text{cm}$

Exercise 12A

1 13 720 J   **2 a)** 1764 J   **b)** 6762 J   **3 a)** 220.5 J   **b)** $0.25gn(n-1)$ J   **4** 1058.4 J   **5** 33 N, 528 J   **6** 2880 J   **7 a)** 2250 J

7 b) 4350 J   **8** 2784.6 J   **9 a)** 1020 m, 102 kJ   **b)** 1252.05 m, 108.43 kJ   **c)** 557.37 m, 52.38 kJ   **10** 36 J   **11** 23 J   **12** 34 J

13 0.5 J   **14** 980 J   **15** 4.25×10^9 J

Exercise 12B

1 15.26 m s^{-1}, mass not needed   **2 a)** 8.85 m s^{-1} **b)** 530 J, $88\frac{1}{3}$ N   **3 a)** 18.4 m s^{-1} **b)** 50.7 m   **c)** 19.0 m s^{-1} **4** 4.70 m s^{-1}

5 a) 3.43 m s^{-1} **b)** Jerk as string went taut   **6 a)** $\sqrt{2ga\cos\theta}$ **b)** $\sqrt{2ga}$ **7** $\sqrt{\dfrac{8ga}{5}}$ **8** $\sqrt{2}:1$

Exercise 12C

1 a) 39.2 W   **b)** $6\frac{8}{15}$ s   **2** 2254.4 W   **3** 6.28 W   **4** 25.03 W   **5 a)** $1998a$ J   **b)** 0.25 m s^{-1} **6** $5\frac{1}{6}$ W   **7** 3.83 m   **8** 500 N

10 1.70×10^6 J   **11 a)** 487.5 J   **b)** 4.42 m s^{-1}

Exercise 12D

1 2 MW   **2** 14 500 W   **3** 20.58 m s^{-1} **4** 30 m s^{-1} **5** -0.030 m s^{-2}, 9.05 m s^{-1}, 0.255 m s^{-2}

6 $216\frac{2}{3}$ kW, $\dfrac{6500}{3(40+5n+5g+ng)}$ m s^{-1} **7 a)** 11 200 W   **b)** 0.773 m s^{-2}, 283.6 N   **8** 0.221 m s^{-2} **9 a)** 2520 N   **b)** 22.3 m s^{-1}

11 837 m   **12** 5.2 m s^{-1} **13** $8P$ W

Exercise 13A

1 7 m s^{-1} **2 a)** 42 N s   **b)** 120 N   **3** 7.2 N s   **4** 50 000 N s, $16 666\frac{2}{3}$ N   **5** $(-12\mathbf{i}+16\mathbf{j})$ N s   **6** 5.795 N s at 15° to horizontal

7 $mu\sqrt{3}$ **8** 6.40 N s   **9** 300 N   **10** 4908.7 N   **11 a)** 1.18 m s^{-1} **b)** 11.76 m s^{-2} **12 a)** $68\mathbf{i}-48\mathbf{j}$ **b)** $10.5\mathbf{i}-3\mathbf{j}$

Exercise 13B

1 3.974 m s^{-1} **2 a)** 3.2 m s^{-1} **b)** 0.8 m s^{-1} **3** 0.45 kg   **4 a)** $\dfrac{10v}{7}$ **b)** $\dfrac{2v}{7}$ **5 a)** 15 m s^{-1} **b)** 30 N s, 6000 N   **6** 717 m s^{-1}

7 $\dfrac{m^2 v^2}{2\mu g(m+M)^2}$ **8 a)** 1.91 m   **b)** 1.25 m   **9** 3.6 m s^{-1} towards smaller mass   **10** 2 m s^{-1} **11** 3600 N   **12** $3.2\mathbf{i}+3.2\mathbf{j}$ m s^{-1}

13 $-8\mathbf{j}$ m s^{-1} **14** $3\frac{1}{3}$ m s^{-1} **15** 2 m s^{-1} **16** 1.48 m s^{-1}

17 1.534 m s^{-1}, 0.6 m. Loss of energy in the jerk so, although A will return to the ground, B will not get that far. The system will repeat a sort of oscillatory motion with the masses travelling less distance each time

Exercise 13C

1 a) 2.51 m s^{-1}, 7.31 m s^{-1} **b)** 2.64 m s^{-1}, 4.64 m s^{-1} **c)** 0.91 m s^{-1}, 5.71 m s^{-1} **d)** -2.71 m s^{-1}, -0.91 m s^{-1} **e)** 1 m s^{-1}, 2 m s^{-1}

1 f) 5.6 m s^{-1}, 0.2   **g)** $\frac{6}{19}$ kg, 21 m s^{-1} **h)** -7.38 m s^{-1}, 4.62 m s^{-1} **2 a)** 2.24 m s^{-1}, 0.32 m s^{-1}, 53.9 J

2 b) 0.932 m s^{-1}, 0.018 m s^{-1}, 126 J   **3** 6 m s^{-1}, 60 J   **5** $1:2$ **6** $\dfrac{13u}{64}, \dfrac{15u}{64}, \dfrac{9u}{16}$ **7** $\dfrac{20}{g}$ s, 2.25 m s^{-1}, 4.25 m s^{-1} **8** $\dfrac{25u}{64}, \dfrac{9u}{16}, \dfrac{81u}{128}$

Exercise 13D

1 11.4 m s^{-1} at 18.1°   **2.** 2.7 J   **3** 5.35 m s^{-1}, 24.2° above horizontal   **4 a)** 0.155   **b)** 4.96 m s^{-1} at 53.8° to line of centres

5 a) $-1.6\mathbf{i}+3\mathbf{j}$, $2.4\mathbf{i}-4\mathbf{j}$ **b)** 5.4 J   **6** $\frac{1}{3}$ **7** $\frac{1}{2}$ **8** 3.28 m s^{-1} at 112° to their original direction

Exercise 14A

The answers to this exercise show only the **magnitude** of the reaction components.

1 $\begin{pmatrix} 56.6 \\ 0 \end{pmatrix}$ N   **2** $\begin{pmatrix} 87.6 \\ 24.5 \end{pmatrix}$ N   **3** A and B: $\begin{pmatrix} \dfrac{W}{2\sqrt{3}} \\ W \end{pmatrix}$, C: $\begin{pmatrix} \dfrac{W}{2\sqrt{3}} \\ W \end{pmatrix}$ **4** Tension 107.4 N, reaction $\begin{pmatrix} 107.4 \\ 163\frac{1}{3} \end{pmatrix}$ N

5 Tension $2W$, reaction $\begin{pmatrix} \dfrac{W}{2\sqrt{3}} \\ 0 \end{pmatrix}$ **6** A: $\begin{pmatrix} 0.6W \\ 0.7W \end{pmatrix}$, B: $\begin{pmatrix} 0.6W \\ 1.7W \end{pmatrix}$, C: $\begin{pmatrix} 0.6W \\ 1.3W \end{pmatrix}$ **7** Reaction horizontal with magnitude $\dfrac{W}{2\sqrt{3}}$

ANSWERS

8 Suspended at 0.1*a* from mid-point of rod. Reactions $\begin{pmatrix} 1.68W \\ 3.26W \end{pmatrix}, \begin{pmatrix} 1.68W \\ 0.74W \end{pmatrix}, \begin{pmatrix} 1.68W \\ 3.74W \end{pmatrix}$ **9** $\begin{pmatrix} 33.4 \\ 0 \end{pmatrix}$ N

10 $\begin{pmatrix} \dfrac{W\tan\beta}{\tan\alpha\tan\beta - 1} \\ \dfrac{W(\tan\alpha\tan\beta + 1)}{2(\tan\alpha\tan\beta - 1)} \end{pmatrix}$ **11** 18.4°, $\begin{pmatrix} 0.3W \\ 0.4W \end{pmatrix}$

Exercise 14B

1 a) AE 400$\sqrt{2}$ N tension, AB 400 N thrust, CD 600$\sqrt{2}$ N tension, BC 600 N thrust, BE 100$\sqrt{2}$ N thrust, DE 500 N tension

1 a) BD 100$\sqrt{2}$ N tension **b)** AB 100$\sqrt{3}$ N thrust, AC 600 N thrust, BC 200$\sqrt{3}$ N tension, CD 400$\sqrt{3}$ N tension, AD 0 N

1 c) AB $\dfrac{500\sqrt{2}}{3}$ N tension, AF $\dfrac{500}{3}$ N thrust, BC $\dfrac{1000}{3}$ N tension, BE $\dfrac{500\sqrt{2}}{3}$ N thrust, BF 0 N, CD $\dfrac{1000\sqrt{2}}{3}$ N tension,

1 c) CE $\dfrac{500}{3}$ N tension, DE $\dfrac{1000}{3}$ N thrust, EF $\dfrac{500}{3}$ N thrust **d)** AB $\dfrac{115\sqrt{3}}{3}$ N tension, AE $\dfrac{230}{3}$ N thrust, BC $\dfrac{95\sqrt{3}}{3}$ N tension

1 d) BE $\dfrac{40\sqrt{3}}{3}$ N tension, CD $\dfrac{95\sqrt{3}}{3}$ N tension, CE 0 N, DE $\dfrac{190}{3}$ N thrust **2** BE 85.07 N thrust, CE 0 N **3** 98.97 N **4 a)** 500 N

4 b) CE $\dfrac{250}{\sqrt{3}}$ N thrust, CF $\dfrac{250}{\sqrt{3}}$ N tension **5** AB 80$\sqrt{3}$ N thrust, AD 80$\sqrt{3}$ N tension, AE 80$\sqrt{3}$ N thrust, BC 160$\sqrt{3}$ N thrust

5 BD 0 N, CD 160 N tension, DE 80 N tension **6 a)** 306.6 N **b)** 377.4 N at 35.7° to AB

6 c) AD 220$\sqrt{2}$ N thrust, BD 50 N tension, BC 100 N thrust

Exercise 14C

1 a) AB 47.6 N tension, AC 15 N thrust, AD 69.3 N thrust, BC 95.3 N thrust, CD 34.6 N thrust

1 b) AC, BC 305.2 N tension, AB 175.1 N thrust, AD, BD, CD 0 N

1 c) AB 120 N thrust, AE 169.7 N tension, BC 70 N tension, BD 28.3 N thrust, BE 240.4 N tension, CD 99 N thrust

1 c) DE 50 N thrust

1 d) AB 223.2 N thrust, AF 446.4 N tension, BC 315.7 N thrust, BF 263.4 N thrust, CD 200 N thrust, CE 173.2 N thrust,

1 d) CF 70.7 N tension, DE 173.2 N tension, EF 244.9 N tension

1 e) AB 70.7 N tension, AE 50 N thrust, BC 290 N thrust, BD 152.1 N tension, BE 70 N thrust, CD 410.1 N thrust, DE 50 N thrust

1 f) AB 519.6 N thrust, AH 259.8 N tension, BC 866 N thrust, BH 173.2 N tension, CD 1039.2 N thrust, CG 173.2 N tension

1 f) CH 173.2 N thrust, DE 692.8 N thrust, DF 346.4 N thrust, DG 346.4 N tension, EF 346.4 N tension, FG 346.4 N tension

1 f) GH 433 N tension

2 a) $W\sqrt{1.5}$ **b)** AB zero, AD 1.23W tension, BC W thrust, BD 1.37W thrust, CD W tension **c)** 366 N **3 a)** $\dfrac{800}{\sqrt{3}}$ N

3 b) In both cases: AB 346.4 N thrust, AE 346.4 N thrust, BC 692.8 N thrust, BD 600 N tension, CD 461.9 N thrust, DE 200 N thrust

3 b) If P is applied at A, AD 115.5 N thrust. If P is applied at D, AD 346.4 N tension. Applying P at A gives some advantage

4 a) 900 N **b)** Reaction at A 948.7 N at 18.4°; internal forces: AB 600 N thrust, AF 424.3 N thrust, BC 848.5 N thrust

4 b) BF 600 N tension, CD 600 N thrust, CE 300 N thrust, CF 80.4 N tension, DE 519.2 N tension, EF 300 N thrust

5 AB, CD, BF and CF 1.41P; AG and DE P; BG and CE 2.41P; BC, EF and FG 2.61P. Maximum P 30.6 N

6 AB 138.6 N thrust, AD 554.3 N thrust, BC 277.1 N thrust, BD 240 N tension, CD 138.6 N tension, DE 480 N tension

Exercise 15

1 Both have dimensions ML^{2}T^{-2} **2** Both have dimensions MLT^{-1} **3** Each component has dimensions L

4 Both have dimensions ML^{2}T^{-3} **5** T^{-1} **7** T^{-1} **8** Parts **a** and **c** consistent **9** $T = K\sqrt{\dfrac{l}{g}}$ **10** $v = k\sqrt{\lambda g}$

11 a) Only three variables possible **b)** $V = \dfrac{Kr^4 p}{\eta l}$

Exercise 16A

1 $\dfrac{\pi}{2}$ rad s^{-1} **2** $\dfrac{750}{\pi}$ rev min^{-1} **3** 9 m s^{-1} **4** $\dfrac{5\pi^2}{12}$ m s^{-1} **5 a)** 2.69 rad s^{-1} **b)** 4.85 m s^{-1} **6** 14.3 rad s^{-1}

7 1.02×10^{-4} m s^{-1}, 1.92×10^{-3} m s^{-1}

Exercise 16B

1 a) $25.6\,\mathrm{m\,s^{-2}}$ **b)** $37.8\,\mathrm{m\,s^{-2}}$ **c)** $4.11\,\mathrm{m\,s^{-2}}$ **2 a)** $7.27\times10^{-5}\,\mathrm{rad\,s^{-1}}$ **b)** $0.034\,\mathrm{m\,s^{-2}}$ **c)** $53.8°$

2 d) Perpendicular to Earth's axis of rotation **3** $5.91\times10^{-5}\,\mathrm{m\,s^{-2}}$ **4 a)** $0.6\,\mathrm{rad\,s^{-1}}$ **b)** $\mathbf{r}=(10\cos0.6t\,\mathbf{i}+10\sin0.6t\,\mathbf{j})\,\mathrm{m}$

4 c) $\mathbf{v}=(-6\sin0.6t\,\mathbf{i}+6\cos0.6t\,\mathbf{j})\,\mathrm{m\,s^{-1}}$, $\mathbf{a}=(-3.6\cos0.6t\,\mathbf{i}-3.6\sin0.6t\,\mathbf{j})\,\mathrm{m\,s^{-2}}$ **5 a)** $\mathbf{a}=(-30\cos10t\,\mathbf{i}-30\sin10t\,\mathbf{j})\,\mathrm{m\,s^{-2}}$

5 b) $|\mathbf{a}|=30\,\mathrm{m\,s^{-2}}$ **c)** $(1.52\,\mathbf{i}+2.59\,\mathbf{j})\,\mathrm{m\,s^{-1}}$

Exercise 16C

1 $10.4\,\mathrm{N}$ **2** $213\,\mathrm{kN}$ **3** $149\,\mathrm{rev\,min^{-1}}$ **4 a)** 0.06 **b)** 0.081 **c)** Same in each case **5 a)** $12\,\mathrm{m\,s^{-1}}$ **b)** $8.49\,\mathrm{m\,s^{-1}}$ **6** $0.48\,\mathrm{N}$

7 $\dfrac{\mu g}{16\pi^2}$ **8** $5.07\,\mathrm{rad\,s^{-1}}$ **9** $3\omega,\,9ma\omega^2$ **10** $\sqrt{\dfrac{g(1-\mu)}{a}}\leqslant\omega\leqslant\sqrt{\dfrac{g(1+\mu)}{a}}$ **11 a)** $7.27\times10^{-5}\,\mathrm{rad\,s^{-1}}$ **b)** $42\,200\,\mathrm{km}$

11 c) $1\,\mathrm{h}\,49\,\mathrm{min}$ **12** $\sqrt{g}\,\mathrm{rad\,s^{-1}}$

Exercise 16D

1 a) $0.392\,\mathrm{m}$ **b)** Ball is a particle, string light and inextensible, no air resistance **2** $7\,\mathrm{rad\,s^{-1}}$ **3** $1.12\,\mathrm{m},\,7.40\,\mathrm{N}$

4 b) $AO>0$ for all finite values of ω. If string horizontal, no vertical component of tension to oppose particle weight **5 a)** $2.01\,\mathrm{N}$

5 b) $1.68\,\mathrm{m\,s^{-1}}$ **6 a)** $8.20\,\mathrm{m\,s^{-1}},\,6\tfrac{11}{18}m\,\mathrm{N}$ **b)** $18.8\,\mathrm{m\,s^{-1}},\,16\tfrac{1}{3}m\,\mathrm{N}$ **7** $3\tfrac{1}{3}\,\mathrm{m}$ **8** $3.89\,\mathrm{rad\,s^{-1}}$ **9** $11.5\,\mathrm{m\,s^{-1}}$ **10 a)** $26.2\,\mathrm{m\,s^{-1}}$

10 b) $39.6\,\mathrm{m\,s^{-1}}$ **c)** $9.06\,\mathrm{m\,s^{-1}}$ **11 a)** $1.97\,\mathrm{cm}$ **b)** $186.8\,\mathrm{km\,h^{-1}}$ **c)** No, rails can support train at rest

12 $0.055\,\mathrm{N}$ at $27°$ to vertical **13 a)** $\tfrac{3}{4}\omega^2-g\sqrt{3}\,\mathrm{N}$ **b)** $8.24\,\mathrm{rad\,s^{-1}}$ **14** $8\,\mathrm{cm}$ **17** $\sqrt{\tfrac{1}{2}(V_1^2+V_2^2)}$ **18** $7.67\,\mathrm{m\,s^{-1}}$

Exercise 16E

1 a) i) $6.29\,\mathrm{m\,s^{-1}}$ **ii)** $4.76\,\mathrm{m\,s^{-1}}$ **iii)** $2.38\,\mathrm{m\,s^{-1}}$ **b) i)** $96.2\,\mathrm{N}$ tension **ii)** $45.3\,\mathrm{N}$ tension **iii)** $5.62\,\mathrm{N}$ thrust **2 a)** $8.57\,\mathrm{m\,s^{-1}}$

2 b) i) $4.41\,\mathrm{N}$ **ii)** $8.49\,\mathrm{m\,s^{-2}}$ **3** $48.2°$ **4** $70.5°$ **6** $\sqrt{8ag}$ **7 a)** $U\geqslant2\sqrt{ag}$ **b)** $U\geqslant\sqrt{5ag}$ **8** $\sqrt{\tfrac{1}{2}ag(3\sqrt{3}-4)}$ **9** $0.6\,\mathrm{m}$

10 $\tfrac{1}{4}g\sqrt{51}\,\mathrm{N}$ at $42.7°$ to vertical **11** $4.5\,\mathrm{m}$ **12 b)** $\sqrt{12g}$ **c)** Bead A has just enough speed to return to top **13** $\tfrac{1}{2}r$

14 $4-\sqrt{5}:\sqrt{5}$ **15** $183.75\,\mathrm{N}$. Assumes mass is a particle, string light and inextensible, no air resistance

16 $R_A=\tfrac{1}{3}mg(7-3\sqrt{3})$, $R_B=\tfrac{1}{3}mg(11-6\sqrt{3})$ **17** $10.1\,\mathrm{m}$

Exercise 17A

1 a) $2.225\,\mathrm{m}$ **b)** $4.45\,\mathrm{m}$ **2 a)** $200\,\mathrm{N\,m^{-1}}$ **b)** $1.1\,\mathrm{m}$ **3** $0.84\,\mathrm{m},\,100\,\mathrm{N\,m^{-1}}$ **4 a)** $2.31\,\mathrm{m}$ **b) i)** $1.704\,\mathrm{m}$ **ii)** $10.6\,\mathrm{N},\,9.04\,\mathrm{N}$

5 AM $2.986\,\mathrm{m}$, BM $1.014\,\mathrm{m}$ **6** $3.46\,\mathrm{m}$ **7** $2.23\,\mathrm{m}$ **8** $0.170\,\mathrm{m}$ **9 a)** $2k$ **b)** $\tfrac{1}{2}k$ **c)** $3k$ **d)** $\tfrac{2}{3}k$ **10 a)** $\dfrac{mg\mu}{2l}$

10 b) At C (unrealistic for real springs) **11 a)** $2.89\,\mathrm{m},\,2.11\,\mathrm{m}$ **b)** $3.78\,\mathrm{m}$ **12** $\sqrt{\dfrac{k}{2m}}$

Exercise 17B

1 $2.87\,\mathrm{m}$ **2** $50\,\mathrm{N},\,0.25\,\mathrm{m}$ **3 a)** $0.27\,\mathrm{m}$ **b)** $44.1\,\mathrm{N}$ **4 a)** $1.732\,\mathrm{m}$ **b)** $36.6\,\mathrm{N},\,12.4\,\mathrm{N}$ **5 a)** $10.2\,\mathrm{kg}$ **b)** $13.2\,\mathrm{kg}$

5 c) $\dfrac{50(4\sin\tfrac{1}{2}\theta-1)}{g\sin\tfrac{1}{2}\theta}$ **6** $\dfrac{l(\lambda+Mg\sin\theta)}{\lambda}$ **7** $2.63\,\mathrm{m\,s^{-2}}$ **8** $30.4\,\mathrm{cm}$ **9** $1.655\,\mathrm{m},\,0.745\,\mathrm{m}$ **10** $d=\dfrac{mgl}{2\lambda}$

Exercise 17C

1 a) $7.67\,\mathrm{m\,s^{-1}}$ **b)** $4\tfrac{1}{6}\,\mathrm{m}$ **2** $28.5\,\mathrm{m\,s^{-1}}$ **3** $24\,\mathrm{J}$ **4 a)** $1.1\,\mathrm{m}$ **b)** $0.6\,\mathrm{m}$ **c)** $0.2\,\mathrm{m}$ above A **5 a)** $2\tfrac{8}{11}\,\mathrm{m}$ **b)** Both $10.9\,\mathrm{N}$

5 c) $5.45\,\mathrm{J}$ **d)** AB $18.4\,\mathrm{N}$ tension, BC $9.09\,\mathrm{N}$ compression **e)** $12.3\,\mathrm{J}$ **f)** $1.66\,\mathrm{m\,s^{-1}}$ **g)** $0.5\,\mathrm{m}$

6 $2\sqrt{\dfrac{2ag}{3}}$ and $\sqrt{\dfrac{2ag}{3}}$. Collide $2a$ from start position of lighter particle **7** $0.8\,\mathrm{m}$ **8** $\sqrt{2ag\sqrt{2}},\,\dfrac{mg(5\sqrt{2}-2)}{2}$ **9 a)** $0.4\,\mathrm{m}$

9 b) $3.46\,\mathrm{m\,s^{-1}}$ **10 a)** $3.97\,\mathrm{m}$ **b)** $8.82\,\mathrm{m\,s^{-1}}$

Exercise 18A

1 a) $3\frac{1}{9}$ m **c)** $\omega^2 = 9$ **2 b)** $\ddot{x} = -32x$, $\omega^2 = 32$ **c)** $\omega^2 = 21\frac{1}{3}$ **d)** No **3 a)** $\dfrac{mgl}{\lambda}$ **b)** $\ddot{x} = -\dfrac{\lambda}{mg}x$, which is SHM with $\omega^2 = \dfrac{\lambda}{mg}$

4 a) $Mg - k(x - l) = M\ddot{x}$ **b)** $\ddot{x} = g + \dfrac{kl}{M} - \dfrac{kx}{M}$ **c)** $x = \dfrac{Mg}{k} + l$ **5 b) i)** $30(x - 2)$, $90(4 - x)$ **ii)** $\ddot{x} = 840 - 240x$ **iii)** $x = 3.5$ m

5 c) i) $45 + 30x$, $45 - 90x$ **ii)** $\ddot{x} = -240x$

Exercise 18B

1 d) $A = a \cos \alpha$, $B = -a \sin \alpha$. Solutions **a** and **b** correspond to $\alpha = 0$, $-\dfrac{\pi}{2}$ respectively **2 a)** $v = -a\omega \sin \omega t$

2 b) $v = a\omega \cos \omega t$ **c)** $v = -a\omega \sin(\omega t + \alpha)$

Exercise 18C

1 a) i) $\dfrac{2\pi}{3}$ s **ii)** $a = 1$ m **iii)** $x = \sin 3t$ **b) i)** $\dfrac{2\pi}{5}$ s **ii)** $v = 10$ m s^{-1} **iii)** $x = 2 \cos 5t$ **c) i)** $\dfrac{\pi}{2}$ s **ii)** $v = 12$ m s^{-1}

1 c) iii) $x = 3 \cos 4t$ **d) i)** $\dfrac{2\pi}{3}$ s **ii)** $a = 2$ m **iii)** $x = 2 \sin 3t$ **2 a) i)** $\dfrac{2\pi}{3}$ s **ii)** 1.5 m s^{-1} **b) i)** $\dfrac{\pi}{2}$ s **ii)** 1.6 m s^{-1} **c) i)** $\dfrac{\pi}{30}$ s

2 c) ii) 3 m s^{-1} **d) i)** $\dfrac{\pi}{3}$ s **ii)** 3 m s^{-1} **3 a) i)** $\dfrac{2\pi}{3}$ s **ii)** 1.5 m s^{-1} **b) i)** $\dfrac{2\pi}{5}$ s **ii)** 7 m s^{-1} **c) i)** 0.04π **ii)** 5 m s^{-1}

3 d) i) $\dfrac{\pi\sqrt{2}}{5}$ s **ii)** $2\sqrt{2}$ m s^{-1} **4 a)** 2 cycles s^{-1} **b)** $0.8\,\pi$ m s^{-1} **c)** $3.2\pi^2$ m s^{-1} **5 a)** 0.38 J **b)** 1.22 J **c)** 4.11 m s^{-1} **d)** 0.23 s

5 e) 13 cm **6 a)** 9.42 m s^{-1} **b)** 8.16 m s^{-1} at B, 8.89 m s^{-1} at D **c) i)** 0.0275 s **ii)** 0.0942 s **7 a)** 4.81 h (4 h 49 min) **b)** 1341

8 a) $2\pi\sqrt{\dfrac{M}{2k}}$ **b)** $2\pi\sqrt{\dfrac{2M}{k}}$ **c)** $2\pi\sqrt{\dfrac{M}{3k}}$ **d)** $2\pi\sqrt{\dfrac{2M}{3k}}$ **9 a)** 2.04 ⌐ **b)** 0.006 m s^{-1}

2 - - - - - - - - - - $4\pi^2$

1.96 ⌐

Exercise 18D

1 2.84 s **2** 0.995 m **3** 0.556% **4** 44 s slow **5** 24.5 min **6 i)** Angle $68.8°$, so model inappropriate

6 ii) Angle $15°$, so **a)** 4.92 s **b)** 0.410 s **iii)** Angle $13°$, so **a)** 1.42 s **b)** 0.118 s **7 a)** $\ddot{x} = -25x$ **b)** $\sqrt{5}$ m s^{-1} **c)** 0.460 s

7 d) 1.25 m **e)** 1.37 s **8 a)** Jumper is a particle falling from rest, rope is light and obeys Hooke's law, no air resistance **b)** 30 m

8 c) 3.12 s **9 b)** 1 m below beam **c)** 0.988 s **d)** 4.43 m below beam **10 a)** $\dfrac{5l}{3}$ **b)** $\dfrac{25l}{12}$ above start **11 b)** 42.3 min

12 a) 0.993 s **b)** 11.0 m s^{-1}, 0.166 s **c)** 0.702 s **d)** 11.8 m s^{-1} **e)** 0.836 s **13 b)** $\dfrac{1}{2l} - \dfrac{x^2}{16l^3}$ **c)** $\pi\sqrt{\dfrac{2m}{k}}$

14 a) Weak, period 3.79 s **b)** Strong **c)** Critical **d)** Strong **e)** Weak, period 2π s **f)** Critical

15 a) $\dfrac{d^2x}{dt^2} + 2\dfrac{dx}{dt} + 5x = 0$ **b)** $x = 3e^{-t} \cos 2t$, period π s **16 a)** $2\dfrac{d^2x}{dt^2} + r\dfrac{dx}{dt} + 40x = 2g$ **b)** $8\sqrt{5}$, $x = (At + B)e^{-2\sqrt{5}t} + 0.05g$

16 c) 0.4894 m

Exercise 19A

1 $5\dfrac{dv}{dt} = 5g - kv$, $k = \dfrac{49}{60}$ **2** $\dfrac{dQ}{dt} = k(Q_{MAX} - Q)$ **3** $\dfrac{dV}{dt} = k(36\pi V^2)^{\frac{1}{3}}$ **4** $\dfrac{dh}{dt} = \dfrac{8}{\pi h^2}$ **5** $\dfrac{dN}{dt} = kn(N - n)$ **6** $\dfrac{dA}{dt} = 2k\sqrt{\pi A}$

7 a) $m\dfrac{d^2x}{dt^2} = \dfrac{k}{x}$ **b)** $m\dfrac{d^2x}{dt^2} = \dfrac{k}{x} - K\dfrac{dx}{dt}$

Exercise 19B

1 a) $y^2 = x^2 + k$ **b)** $2y^3 - 6y^2 = 3x^2 + k$ **c)** $y = Ae^{\frac{1}{2}(x-2)^2} + 3$ or $y = Ae^{\frac{1}{2}x^2 - 2x} + 3$ **d)** $y = \dfrac{2(1 + Ae^{2x^2})}{(1 - Ae^{2x^2})}$

2 $\sin y = 1 - \cos x$ **4 a)** $y = \dfrac{Ax^2 - 1}{Ax^2 + 1}$ **b)** $y = Ae^{\frac{1}{2}(x+2)^2}$ **c)** $\sec y = Ax$ **d)** $\sec y = A(1 - \cos x)$ **5 a)** $y = x + \dfrac{1 + Ae^{2x}}{1 - Ae^{2x}}$

5 b) $y = x + \dfrac{1 - e^{2x}}{1 + e^{2x}}$ **6** $v = 10g(1 - e^{-0.1t})$ **7 a)** $\dfrac{400}{v} - v = 30\dfrac{dV}{dt}$ **b)** $v = 20\sqrt{1 - e^{-\frac{t}{15}}}$ **c)** 4.32 s

Exercise 19C

1 a) $y = Ae^{-2x} + 2$ **b)** $y = \dfrac{x^2}{3} + x + \dfrac{A}{x}$ **c)** $y = Ae^{-x} + x - 1$ **d)** $y = \dfrac{e^{-x}}{2} + Ae^{-3x}$ **e)** $y = Ae^{-x^2} + 2$ **f)** $y = Ae^{-\cos x} - 1$

1 g) $y = Ae^{-x^3} + 6$ **h)** $y = \dfrac{x^2}{3} + \dfrac{A}{x}$ **i)** $y = Ax^2 - x(1 + \ln x)$ **j)** $y = (x + A)\cos x$ **2** $\dfrac{dV}{dt} + 0.2v = 10 - t$, $14.8\,\text{m s}^{-1}$

Exercise 19D

1 $y = Ae^{4x}$ **2 a)** $y = Ae^{-x}$ **b)** $x = Ae^{5t}$ **c)** $p = Ae^{-2t}$ **d)** $x = Ae^{0.5t}$ **e)** $y = Ae^{-3x}$ **f)** $z = Ae^{3y}$

Exercise 19E

1 a) $y = Ae^{-3x} + \frac{2}{9}(3x - 7)$ **b)** $x = Ae^{2t} - 3$ **c)** $y = Ae^{-x} + 2 - x$ **d)** $p = Ae^{5t} - \frac{1}{25}(5t + 26)$ **e)** $z = Ae^{-2x} + \frac{3}{4}(2x - 1)$

1 f) $x = Ae^{4t} + \frac{1}{2}(2t - 1)$ **g)** $y = Ae^{-4x} + \frac{1}{4}(4x - 3)$ **h)** $x = Ae^t + 5(t + 1)$ **2 a)** $y = 2(3e^{4x} - 1)$ **b)** $x = \frac{2}{25}(e^{-5t} + 5t + 24)$

2 c) $y = 3 - e^{2x}$ **d)** $x = \frac{1}{3}(17e^{-3t} + 3t + 1)$ **e)** $y = 4(2e^{2x} + x)$ **f)** $x = \frac{1}{3}(29e^{-3t} + 6t - 2)$

Exercise 19F

1 a) $y = Ae^{-3x} + Be^x$ **b)** $y = (Ax + B)e^{2x}$ **c)** $x = e^{-2t}(A\cos 2t + B\sin 2t)$ **d)** $y = (Ax + B)e^{-3x}$ **e)** $x = e^{2t}(A\cos 2t + B\sin 2t)$

1 f) $y = Ae^{-4x} + Be^{-2x}$ **g)** $x = e^{-t}(A\cos t + B\sin t)$ **h)** $y = Ae^{4x} + Be^{-x}$

2 a) $y = Ae^{-4x} + Be^{2x} - \frac{1}{4}(2x + 3)$, $y = \frac{1}{6}e^{-4x} + \frac{7}{12}e^{2x} - \frac{1}{4}(2x + 3)$ **b)** $y = e^{4x}(Ax + B) + \frac{1}{2}$, $y = e^{4x}(x + 1) + \frac{1}{2}$

2 c) $y = e^{-x}(Ax + B) + 2x - 4$, $y = 2[e^{-x}(2x + 1) + x - 2]$ **d)** $y = Ae^{-4x} + B - \frac{1}{8}x^2 + \frac{13}{16}x$, $y = \frac{1}{32}(63 + 26x - 4x^2 + e^{-4x})$

2 e) $y = Ae^{3x} + Be^{-3x} - \frac{1}{3}x + 2$, $y = \frac{1}{18}(25e^{3x} - 25e^{-3x} - 6x + 36)$ **f)** $x = A\cos 2t + B\sin 2t + 2t - 1$, $x = 4\cos 2t + \sin 2t + 2t - 1$

2 g) $y = Ae^{3x} + B - 3x$, $y = e^{3x} - 3x - 1$ **h)** $x = e^{-2t}(A\cos 3t + B\sin 3t) - 2t + 2$, $x = e^{2(1-t)}\sin 3t - 2t + 2$

3 $\dfrac{d^2 x}{dt^2} + 4x = 12$

Examination questions

Chapters 2 to 4

1 i) $\mathbf{i} - 4\mathbf{j} - 8\mathbf{k}$ **ii)** $\dfrac{1}{3\sqrt{10}}(4\mathbf{i} - 7\mathbf{j} - 5\mathbf{k})$ **2 i)** $p = 4$, $q = -1$, $r = 3$ **ii)** $\dfrac{1}{\sqrt{3}}(\mathbf{i} + \mathbf{j} - \mathbf{k})$ **3 i)** $|\mathbf{a}| = |\mathbf{b}| = 3\sqrt{2}$, $|\mathbf{c}| = 6$

3 ii) $|a|^2 + |b|^2 = |c|^2$ **iii)** $\dfrac{1}{\sqrt{5}}(\mathbf{i} + 2\mathbf{k})$ **4 a)** 5 or -3 **b) i)** $\frac{1}{13}(3\mathbf{i} - 12\mathbf{j} + 4\mathbf{k})$ **ii)** $24\mathbf{i} - 96\mathbf{j} + 32\mathbf{k}$ **5 a)** $63°$ **c)** $2\sqrt{5}\,\text{N}$

6 a) $p = 2$, $q = -6$ **b)** $2\sqrt{10}\,\text{N}$ **c)** $18°$ **7 a)** $36\,\text{m}$ **b)** $40\,\text{m}$, assuming car length $4\,\text{m}$ and humps of zero width

8 a) $12\,\text{m s}^{-2}$ **b)** $80\,\text{m s}^{-1}$ **9 a) i)** $4\,\text{s}$ **ii)** $35.6\,\text{m}$ **b) i)** Sam **ii)** $1.37\,\text{m}$

9 c) Posture of athletes can make a difference, e.g. 'dipping' for tape **10 i)** $4\,\text{s}$ **ii)** $14\,\text{m s}^{-1}$, $20\,\text{m s}^{-1}$

11 a) $10\,\text{m s}^{-1}$, $2.5\,\text{m s}^{-2}$ **b)** $1\frac{7}{9}\,\text{s}$, $5\frac{1}{16}\,\text{m s}^{-2}$

11 c) B builds up lead in first $3.6\,\text{s}$. A reduces this over rest of race to catch B on finishing line

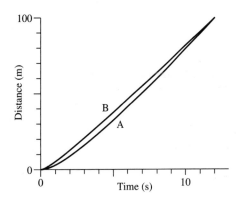

ANSWERS

12 i) $0.3\,\mathrm{m\,s^{-2}}$ **ii)** 75 s **iii)**

(graph: x vs t, reaching 560 m at 40 s, concave increasing curve from O)

13 i)

(graph: v vs t, two lines reaching $30\,\mathrm{m\,s^{-1}}$, marks at 5 s and 20 s)

ii) 40 s

14 a)

(graph: v vs t; line rising to $45\,\mathrm{m\,s^{-1}}$ (PC) at T; horizontal line at $30\,\mathrm{m\,s^{-1}}$ (MC); marks at 5 s, 20 s, T)

b) $45T - 562.5$ **c)** $37.5\,\mathrm{s}$

15 a)

(graph: v vs t, trapezium rising to $4\,\mathrm{m\,s^{-1}}$; marks at 1.5, $t+1.5$, $t+2.5$)

b) 5 s **c)** $3\,\mathrm{m\,s^{-1}}$

16 b) i) $5d = 2u + 8$ **c) i)** $14\,\mathrm{m\,s^{-1}}$ **ii)** 0.24 s   **17 a)** 176.4 m **b)** 146.5 m **c)** $d = 4.9\left(6 - \dfrac{d}{332}\right)^2$

18 i) Uniform acceleration of $10\,\mathrm{m\,s^{-2}}$ for 4 s, parachute opens, uniform acceleration of $-5\,\mathrm{m\,s^{-2}}$ for next 6 s, constant speed of $10\,\mathrm{m\,s^{-1}}$ for last 15 s **ii)** 380 m, 1350 N

Freefall acceleration would decrease due to air resistance. When parachute opens, deceleration takes over quickly but its magnitude decreases as speed falls. Speed decreases asymptotically to $10\,\mathrm{m\,s^{-1}}$

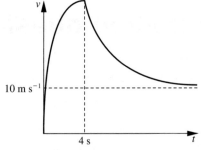

19 a) 19.9 N **b)** 3.46 N   **20 i)**

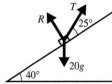

ii) 91.4 N **iii)**

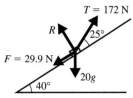

20 iv) Component of T perpendicular to plane. When T changes, normal reaction also changes so equilibrium preserved
21 15.07 N, 15.63 N   **22** 1793 N. 'Light' implies cables straight and tensions constant throughout
23 i)

(diagram: point with T_1 at 45° and T_2 at 30° above horizontal, weight $12g$ downward)

ii) $T_1 = 105.4\,\mathrm{N}, T_2 = 86.09\,\mathrm{N}$ **iii)** None

23 iv) Both tensions increase. Tensions cannot be horizontal as there must be vertical component to support weight

24 17.4 N, 16.7°   **25 i)** 19.5°   **ii)** 9.12 N   **26 a)** 16°   **b)** 1270 N

26 c) Slow speed, so air resistance low compared with other forces   **26 d)** Ropes at angle to ground   **27 i)** 202 N   **ii)** 129.8°

28 150°   **29** 200 N, 193 N   **30 i)** 265.7 N   **ii)** 126.6°   **31 a)** $-90\mathbf{i} + 120\mathbf{j} + 140\mathbf{k}$, 205 N   **b)** 69.7°   **32 i)** $3000\mathbf{i}$

32 ii) $2624\mathbf{i} - 136.8\mathbf{j}$ **iii)** 260 N, 157.4°   **iv)** $-480\mathbf{i} - 310\mathbf{j}$

32 v) a) Required, because otherwise rope sags and forces no longer horizontal

32 b) Not required, because elastic rope could supply required 3000 N tension   **33 i)** 4.42 N   **ii)** 10°

Chapters 5 to 7

1 a) $k = gR^2$ **c)** $4R$ above surface   **2 a)** 1040 N   **b)** 1040 N downwards   **3 ii)** 2800 N   **iii) b)** 2.43 m   **c)** 8.51 m

4 i)

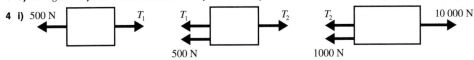

4 ii) $T_1 = 1500$ N, $T_2 = 4000$ N   **iv)** 10 800 N   **5 a) i)** *Light* string   **ii)** *Smooth* peg   **b)** $\frac{3}{5}g$

6 i)

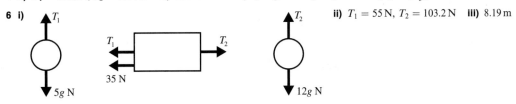

ii) $T_1 = 55$ N, $T_2 = 103.2$ N   **iii)** 8.19 m

7 i) 0.32 m s^{-2}, 0.8 m s^{-1} **ii)** 4.1 N, 4.74 N   **8 a)** String inextensible   **b)** $T - mg = 2mf$, $3mg - T = 3mf$ **d)** 1.6h

9 i) Only one horizontal force   **ii)**

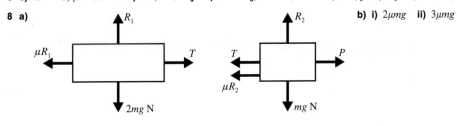

iii) 94.0 N, 750 N   **iv)** 128 N   **v)** 0.144 m s^{-2}

10 a) 514 N   **b)** 296 N   **11 a)** $8\mathbf{i} + 5\mathbf{j}$ **b)** 58°   **12 a)** $\mathbf{v} = 2\mathbf{i} - 12 \sin 3t\,\mathbf{j} + 12 \cos 3t\,\mathbf{k}$ **b)** 0.5 s   **13 a)** $20t$, $15t - \frac{1}{2}gt^2$

13 b) Ball is a particle, air resistance negligible   **c)** 1.5 s, 60 m   **14 i) c)** 86.2°   **ii) a)** 4.0°, 87.9°   **b)** 4.0°   **15 a)** 2 s, 10 m

15 b) 3.92   **16 i)** 30°   **ii)** 61.3°   **iii)** Negligible air resistance   **17 a) i)** 0.75 s   **ii)** Height 1.04 m < 1.4 m   **b) i)** 8.56 m s^{-1}

18 ii) 4.02 m s^{-1} **iii)** 6° above horizontal; U must be > 40 **19 a) i)** $(t^2 + 2)\mathbf{i} + 2t\mathbf{j}$ **ii)** $\sqrt{2}$ s   **b)** $15\sqrt{2}$ m   **20 a)** $\dfrac{dv}{dt} = -kv^3$

20 c) $6\frac{2}{3}$ s   **21 a)** $75f$ **c)** 780 m   **22 i) a)** 1 m   **b)** 1.79 m   **ii)** $\mathbf{v} = 0.25t\mathbf{i} + 2.25t^2\mathbf{j}$, $\mathbf{a} = 0.25\mathbf{i} + 4.5t\mathbf{j}$ **iii)** 85.8° to horizontal

Chapters 8 to 10

1 i) 8 m s^{-1}, 240°   **ii)** 4.33 m   **2 i)** 076.4°   **ii)** 2.21 m s^{-1} **iii)** 384 s   **3 i)** 31.9 m s^{-1}, 086.2°   **ii)** 0712 s   **4 i)** 12.8 m s^{-1}, 248.7°

4 ii) 246.4°   **iii)** 1.02 pm   **5 b) ii)** $\mathbf{r} = -(\mathbf{i} + 30\mathbf{j}) + t(5.25\mathbf{i} + 7\mathbf{j})$ **c)** $t = 4$ **6 a)** Colin: $(3t + 10)\mathbf{i} + 4(2 - t)\mathbf{j}$, David: $3t\mathbf{i} + 5t\mathbf{j}$

6 c) $\lambda = -2$, $\mu = 10$ **7 a)** $10\sqrt{3}\mathbf{i} + 10\mathbf{j}$ **b)** $\mathbf{r} = 30t\mathbf{j}$, $\mathbf{s} = 10\sqrt{3}t\mathbf{i} + 10(t + 7)\mathbf{j}$ **c)** 2 pm (4 h later)

8 a)

b) i) $2\mu mg$ **ii)** $3\mu mg$

9 a) 18.3°   **b)** 15.9 N   **c)** Iron moves very slowly

10 a) i) Rail smooth, so no friction component, only normal component

ANSWERS

10 a) ii) Three force in equilibrium are concurrent   **iii)** See diagram

10 c) Gymnast's hand would hook round rail, so friction not negligible at rail.

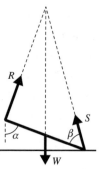

11 b) i) $\dfrac{3g}{25}\,\mathrm{m\,s^{-2}}$ **ii)** $\dfrac{36mg}{25}\,\mathrm{N}$ **c) i)** Same acceleration each side of pulley   **ii)** Same tension each side of pulley   **12 c)** $3.92\,\mathrm{m\,s^{-1}}$

13 a) $44\,\mathrm{N}$ **b)** $45.2\,\mathrm{N}$ **14** $63°$ **15** $W(1 - \tfrac{2}{3}\cos^2\theta), \tfrac{2}{3}W\sin\theta\cos\theta$ **16** $28\,\mathrm{N}, 11\,\mathrm{N}$ **17** $94\,\mathrm{N}$

18 i) $75g\,\mathrm{N}$ **ii)** Man is a particle. Beam is a rigid rod   **iii)** $P = 90g\,\mathrm{N}, Q = 15g\,\mathrm{N}$

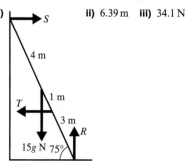

19 a) Plank is rigid rod with weight acting at mid-point, rollers small, roller A right at end of rod   **b)** $2352\,\mathrm{N}$ **c)** $2058\,\mathrm{N}$

20 b) $\dfrac{mg\sqrt{5}}{2}$ **c)** $AB = 2a$ (no bending of pole)   **21 a)** $\tfrac{5}{6}\mathrm{m}$ **b)** $AB = 3\,\mathrm{m}$ (no bending of bar), weight of bar acts at mid-point

22 i) $112\,\mathrm{N}$ **ii)** $39.2\,\mathrm{N\,m}$ **iv)** $49\sqrt{3}\,\mathrm{N}, 12.6\,\mathrm{N}$ **23 i)**   **ii)** $6.39\,\mathrm{m}$ **iii)** $34.1\,\mathrm{N}$

24 $18.4°$ **25 i)** $\dfrac{2b\sqrt{13}}{3}\,\mathrm{N}, 73.9°$ below DB   **ii)** $\dfrac{3bd\sqrt{3}}{200}\,\mathrm{N\,m}$ clockwise   **26 ii)** $4.48\,\mathrm{N}$ **iii)** $27.7\,\mathrm{N\,m}$ anticlockwise

27 i) $R = 10W\cos\alpha$

Chapters 11 to 14

1 $11.8\,\mathrm{cm}$ **2** $(1.4, 2.3)\,\mathrm{cm}$ **3 a)** $5\,\mathrm{cm}$ **b)** $1.5\,\mathrm{cm}$ **c)** $55°$ **4 a)** $\dfrac{2\sqrt{3}}{9}\,\mathrm{cm}$ **b)** $44°$ **5 ii)** $40.6°$ **6 b) ii)** 0.97 **7** $0.9r$

8 ii) $7.64\,\mathrm{cm}$ **9 i)** $(0.3, 0.2)\,\mathrm{m}$ **ii)** $26.6°$ to horizontal   **iii)** $(0.29, 0.21)\,\mathrm{m}$ **10 a)** $21\,\mathrm{kJ}$ **b)** Increase   **11** $5.42\,\mathrm{m\,s^{-1}}$

12 a) $21.6\,\mathrm{J}$ **b)** $21.6\,\mathrm{J}$ **c)** $1.35\,\mathrm{m}$ **13 i)** $114.4\,\mathrm{J}$ loss   **iii)** $0.234\,\mathrm{m}$ **14** $172.8\,\mathrm{kJ}$ **15 a)** No displacement in direction of R

15 c) $4\,\mathrm{m\,s^{-1}}$ **16 a)** $2.16\,\mathrm{J}$ **b) i)** $k = 10\,000$ **ii)** $0.72\,\mathrm{J}$ **17 a)** $200.5\,\mathrm{J}$ **b) ii)** $6.24\,\mathrm{m\,s^{-1}}$ **18** $35\,\mathrm{N}, 2\,\mathrm{m\,s^{-1}}$

19 b) ii) $1.8\pi(3\pi + 1)\,\mathrm{J}$ **20 a) i)** $8\,\mathrm{kW}$ **ii)** $30.8\,\mathrm{m\,s^{-1}}$ **b)** $4°$

20 c) For example: headwind, more passengers, steeper slope, higher speed

21 a) For example: speeds, spin, coefficient of restitution, friction, oblique impact   **b)** $0.05\,\mathrm{m\,s^{-1}}$, assuming balls have same mass

21 c) $0.0375m\,\mathrm{J}$, where m is mass of a ball   **22 a)** $3\tfrac{1}{3}\,\mathrm{m\,s^{-1}}$ **b)** $1\tfrac{1}{3}\,\mathrm{N\,s}$ **23 ii)** $12\,\mathrm{kJ}$ **24 b)** $720\,\mathrm{N\,s}$ **25 a)** $6\,\mathrm{m\,s^{-1}}$ **b)** $7\,\mathrm{m\,s^{-1}}$

25 c) $45\,\mathrm{kJ}$ **26 a)** $4.8\,\mathrm{m\,s^{-1}}$ **b)** $43\,200\,\mathrm{N\,s}$ **c)** $3600\,\mathrm{N}$ **27 i)** $0.38\,\mathrm{m\,s^{-1}}$ to right   **ii)** 0.05 **iii)** $62.72\,\mathrm{N\,s}$ **iv)** $0.36\,\mathrm{m\,s^{-1}}$

27 v) $59.7\,\mathrm{J}$ **28 iii)** $\dfrac{(7m - M)u}{4}$ **29** $3\,\mathrm{m\,s^{-1}}, 6.93\,\mathrm{m\,s^{-1}}$ **30 a)** $6\,\mathrm{m\,s^{-1}}$ **c)** $1.2e^2, 1.2e^3$ **d)** 0.75 **e)** No air resistance

31 $220\,\mathrm{N}, 105\,\mathrm{N}$ **32 a)** $300\,\mathrm{N}$ **b)** $100\sqrt{3}\,\mathrm{N}$ tension   **33 a) i)** $15\sqrt{3}\,\mathrm{N}, 55\,\mathrm{N}$ **ii)** $65°$ **b)** Clockwise   **34 a)** $Q = P\sqrt{3}$

34 b) Both $P\sqrt{3}$ **34 c)** AB, AD, BC   **d)** $Pi + \sqrt{3}Pj$ **35 i)** $\dfrac{4mg}{\sqrt{3}}$ **ii)** $2.52mg$ at $23.4°$ to horizontal

35 iii) AB $\dfrac{4mg\sqrt{3}}{3}$ tension, AE mg tension, BC $2mg$ tension

36 iv) AB $\dfrac{125\sqrt{3}}{3}\,\mathrm{N}$ tension, BC $\dfrac{200\sqrt{3}}{3}\,\mathrm{N}$ compression, CD $\dfrac{100\sqrt{3}}{3}\,\mathrm{N}$ tension, AD $\dfrac{250\sqrt{3}}{3}\,\mathrm{N}$ compression, BD $150\,\mathrm{N}$ compression

Chapters 15 to 19

1 Both sides have dimensions L^2T^{-2} **2** Both sides dimensionless   **3 i)** T^{-1} **iii)** $\alpha = 0.5, \beta = \gamma = -0.5$ **4 a)** $0.8\pi\,\mathrm{m\,s^{-1}}$

4 b) $0.32\pi^2\,\mathrm{m\,s^{-2}}$ **5** $11\,\mathrm{m\,s^{-1}}$ (assuming engine is a particle), $9900\,\mathrm{N}$ **6** 0.397 **7** $40°$ **8 i)** $10\,\mathrm{N}$ **ii)** $2.5\omega^2$ **iii)** $5g - 1.5\omega^2$

8 iv) $5.72\,\mathrm{rad\,s^{-1}}$ **9 i)**   **ii** $38.9\,\mathrm{N}$

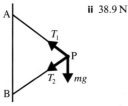

10 $1.02\,\mathrm{rad\,s^{-1}}$ **11 i)** $1.99\,\mathrm{N}, 1.01\,\mathrm{N}$ **ii)** $1.02\,\mathrm{m\,s^{-1}}$ **12 i)**   **ii)** $267\,\mathrm{m}$

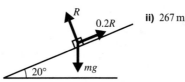

13 $11.8\,\mathrm{m\,s^{-1}}$ **14 i)** $28\,\mathrm{m\,s^{-1}}$ **15 i)** $0.3125\,\mathrm{m\,s^{-2}}$ **iii)** $0.312m$ **iv)** $17.7\,\mathrm{m\,s^{-1}}$ **16 a)** Light, inextensible   **c)** $706\,\mathrm{N}$

17 Hanging   **18 i)** $5\sqrt{2}\,\mathrm{m\,s^{-1}}$ **iii)** $5\,\mathrm{m}$ **19 a)** $T = 50x$ **b)** $0.5g$

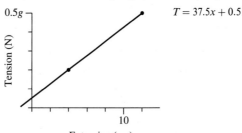

$T = 37.5x + 0.5$

20 b) $1.4g$ **21 i)** $2.5\,\mathrm{m\,s^{-2}}$ **ii)** $0.5\,\mathrm{m\,s^{-1}}$ **22** $16\,\mathrm{N}$ **23 b) i)** $300\,\mathrm{N}$ **ii)** $3\,\mathrm{m\,s^{-1}}$ **24 i)** $0.444\,\mathrm{N}$ **ii)** $1.56\,\mathrm{N}$

25 ii) $50x^2 + \frac{1}{4}v^2 - \frac{1}{2}g(x + 0.6)$ **26 b)** $0.4\,\mathrm{m}$ **c) ii)** $10(1 - 5x)\,\mathrm{m\,s^{-2}}$ **iii)** $2.83\,\mathrm{m\,s^{-1}}$ **27 ii)** $2.30\,\mathrm{m}$ **28 i)** $2.00\,\mathrm{s}$ **ii)** $10.7\,\mathrm{m\,s^{-1}}$

29 a) $2.4a$ **b)** $a < \dfrac{g}{9}$ **30 a)** $6.21\,\mathrm{m}$ **b)** $2.03\,\mathrm{m\,s^{-1}}$ **31 a)** $\dfrac{\pi}{100}\,\mathrm{s}$ **32** $1.57\,\mathrm{m}$ **a) i)** $x = 0.2\cos 2.5t$ **ii)** $0.628\,\mathrm{s}, 0.5\,\mathrm{m\,s^{-1}}$

32 a) iii) $0.419\,\mathrm{s}$ **b)** $\dfrac{d^2x}{dt^2} + 6.25x = 2.5\sin 2t$ **33 i)**

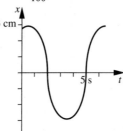

ii) $\omega = \dfrac{\pi}{3}, A = 3$

iii) $2\sqrt{3}\,\mathrm{cm}, \dfrac{\pi}{\sqrt{3}}\,\mathrm{cm\,s^{-1}}$

iv) $\dfrac{2\pi^2}{3\sqrt{3}}\,\mathrm{cm\,s^{-2}}$

34 i) $2\,\mathrm{cm}$ **ii)** $\dfrac{2\pi}{3}\,\mathrm{s}$ **35 a)** $A = 0.02, \omega = 10\pi, d = 0.01$ **b)** $\frac{1}{15}\,\mathrm{s}$ **c)** $0.628\,\mathrm{m\,s^{-1}}, 0.544\,\mathrm{m\,s^{-1}}$

36 a)   **c)** $T = 2\pi\sqrt{\dfrac{2a}{g}}$ **d)** $\dfrac{\pi(1 + \sqrt{2})}{2}\sqrt{\dfrac{a}{g}}$ **37** $n = 2, y = (1 + x)^5 + (1 + x)^{-2}$

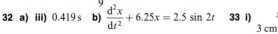

38 Integrating factor e^{x^2}, $y = (x + A)e^{-x^2}$ **39** $y = \frac{1}{8}(x^3 + 7x^{-5})$ **40 a)** $x = e^{-2t}(A\cos 2t + B\sin 2t)$

40 b) $\dfrac{dk}{dt} = I, \dfrac{dI}{dt} = -(8k + 4I)$. Value of k oscillates about zero, but its magnitude approaches zero as t increases

41 d) General solution: $x = Ae^{-4t} + Be^{-5t}$; particular solution: $x = 0.5e^{-4t} - 0.4e^{-5t}$ **42** $y = Ae^{4x} + Be^{5x} + 3x + 2$

Index

acceleration   31
  angular   340
  average   33
  as derivative   97, 104
  as function of displacement
    105
  as function of time   97, 104
  as function of velocity   105
  due to gravity   51, 80
  negative   44
  radial component of   360
  systems with related   92
  tangential component of   360
  uniform   34, 41, 121
  unit of   33
  from velocity–time graph   36
Aristotelian mechanics   75

banked curved tracks   354
boundary conditions   116,
  424, 438
Bow's notation   313

calculus in kinematics   97
  in dimensional analysis   332
centre of gravity   222
centre of mass   221
  of arcs   240
  of circular lamina   228
  of composite bodies   230
  of composite lamina   230
  of cone   230
  of cylinder   229
  of hemisphere   230
  of prism   229
  of rectangular lamina   228
  of rigid body   227
  of rod   227
  of semicircular lamina   229
  of shells   242
  of solid of revolution   238
  of sphere   229
  of system of particles   224
  of tetrahedron   229
  of triangular lamina   228
  of uniform lamina   236
centroid   223
circular motion   340
  banking of tracks   354
  central force   345
  conical pendulum   351
  in horizontal circle   346, 351
  linear/angular speed   340
  modelling   342
  with nonuniform speed   359
  radians   340
  with uniform speed   342
  in vertical circle   360
closest approach, and
  collision   156
coefficient of friction   168
coefficient of restitution   239

commutativity
  of scalar product   28
  of vector addition   10
complementary function   430
composite bodies, centre of
  mass   230
compression
  in rods   307
  of springs   373
connected particles   85
conservation
  of linear momentum   281
  of mechanical energy   260
constants, dimensions of   334
contact forces   53
couple
  definition   191
  moment of   191
  nonparallel forces   195
  parallel forces   191

deceleration   44
differential equations   115, 420
  auxiliary equation   430, 434
  boundary conditions   424,
    438
  classification   420
  complementary function
    431, 434
    with constant coefficients
    429, 433
  first-order equations   423
  forming   421
  homogeneous linear   429,
    433
  initial conditions   424, 433
  integrating factor   426
  particular integral   431, 437
  second-order equations   433
  separation of variables   423
differentiation, of vectors   111
dimensional analysis   330
  in calculus   332
  finding formulae   335
  notation   331
dimensional consistency   334
dimensionless quantities   333
displacement   31, 34
  angular   340
  relative   143
  from velocity–time graph   36
displacement–time graph   35
distance   31, 34
distributivity, of scalar product
  28, 29
drawing diagrams   54
  to find resultants   57

Einstein's relativity theory   75
elastic impact   289
  coefficient of restitution   279
  spreadsheet   289

elastic limit   372
elastic strings   370
  extension   373
  natural length   372
  oscillation of   392
  potential energy   382
  see also springs
elasticity, modulus of   377
energy
  conservation of   260
  and work   262
  see also kinetic energy;
    potential energy
equilibrant   58
equilibrium   58
  conditions for   196, 197
  limiting   168
  of nonparallel forces   58, 196
  of parallel forces   192
  of three forces   58, 199
equivalence of work and energy
  262
equivalent systems of forces
  205
extension, of elastic strings and
  springs   373

force
  central   345
  conservative   258
  dissipative   258
  exerted by water jet   279
  of gravity   50
  line of action   199, 205
  tractive   270
  unit of   50
  variable   106, 271, 278, 381
forces
  components of   59
  contact   53
  equilibrium of   58
  equivalent systems of   205
  explosive   285
  internal   55
  like   190
  modelling by vectors   57
  moments of   184
  parallel   190, 192
  polygon of   58
  resolving   59
  resultant of   50
    by drawing   57
  three in equilibrium   58, 199
  triangle of   58
  turning effect of   183
  types of   51
  unlike   190
frameworks   307
  graphical solution   312
  Bow's notation   313
  method of sections   309
free fall   45

friction   53, 164
  angle of   179
  coefficient of   168
  dynamic   170
  laws of   168
  limiting   165, 168
  sliding and toppling   245
  spreadsheets   165, 167
  static   169

graphs
  displacement–time   35
  force–distance   184
  friction–reaction   166, 168
  length–tension   371, 372
  velocity–time   36
gravitational constant   51
gravity   50, 80
  free fall   45
  motion under   45

hinge, reaction in   198, 301
Hooke's law   373

impact
  direct   289
  oblique   295
impulse   275
  unit of   275
  of variable force   278
impulsive tension   284
initial conditions   116, 424,
  438
integrating factor   426
integration, of vectors   111
interception   156
inverse square law   51, 106

joint, reaction in   198, 301
jointed rods   301

kinematics
  boundary conditions in   116
  calculus in   31
  in two and three dimensions
    111
kinetic energy   259, 382

Lami's theorem   59
laminas, centre of mass   228
  230
line of action of force   199, 205

mass   50
  see also centre of mass
medians   228, 249
  concurrence of   249
modelling   1
  conventional terms in   6
  flowchart   3
  projectiles   123

moment
    and components of a frame
        187
    of a couple   191
    of a force   184
    sense of   185
moments
    principle of   190, 195, 209
    spreadsheets   183, 202
    sum of   185
momentum
    change of   275
    conservation of   281
    definition   275
    and impulse   275
    spreadsheet   289
    unit of   275
motion
    equation of   76
    under gravity   45
    in two and three dimensions
        111, 116
    variable resistance to   271
    with uniform acceleration
        41, 121
    see also circular motion;
        oscillatory motion; simple
        harmonic motion

Newton's law of restitution
    289
Newton's laws of motion   75
    first   75
    second   76
    third   81

orbits
    circular   348
    geostationary   348
oscillatory motion   375
    damped oscillations   411
        spreadsheet   412
    incomplete oscillations   409

parallel forces   190
    and couples   190
    in equilibrium   190
    resultant of   190
particles   6
    connected   85
particle system, centre of mass
    224
particular integral   430
pendulum
    conical   351
    simple   405
    spreadsheet   406

periodic time   398, 406
polygon of forces   58
potential energy
    elastic   382
    gravitational   257
power   266
    spreadsheet   272
    and velocity   269
    and work done   266
    unit of   266
    with variable resistance   271
projectiles   123
    equation of path   130
    maximum height   127
    maximum range   125
    spreadsheet   123, 133
pulley systems   92

radians   340
reaction
    at hinge or joint   198, 301
    normal   53
recoil of guns   285
relative displacement   145
relative motion   144
    interception, collision and
        closest approach   156
relative velocity   145, 147
    notation   146
resistance to motion, variable
    271
resultant
    of forces   57
        by drawing   57
    of like forces   190
    of unlike forces   191
    of vectors   8, 19
retardation   44
rigid body   6
    centre of mass   227
    jointed   301
    light   6
    thrust in   52, 307
rotation, sense of   185

satellite, in circular orbit   348
scalar   8
    vector multiplied by   10, 18
scalar product   25
    commutativity of   28
    distributivity over addition
        28
    properties of   27
simple harmonic motion   393
    amplitude   396
    associated circular motion
        403

definition   393
    equation of   393, 396
    frequency   398
    incomplete oscillations   409
    maximum acceleration   398
    maximum speed   399
    period   398
sliding and toppling see
    friction
smooth surface   6, 53
solids of revolution   238
speed   31
    air   150
    angular   340
    average   31
    ground   150
    uniform   34
    units of   32
    see also velocity
spreadsheets
    DAMPEDHM   412
    ELASTIC   371
    FRIC1   165
    FRIC2   167
    MOM1   183
    MOM2   202
    MOMENTUM   289
    PENDULUM   406
    POWER   272
    SHOT1   123
    SHOT2   133
springs   370
    compression of   373
    extension   373
    natural length   372
    oscillation of   392
    potential energy   382
stiffness   373
strut   307

tension   51, 307
    impulsive   284
thrust   52, 307
tie   307
toppling see friction
torque   185
    see also couple; moments
tractive force   270
triangle of forces   58

units, and dimensions   331

vectors   8
    addition   8, 18
    angle between   25, 26
    associativity of addition   10
    column   16

commutativity
    of addition   10
    of scalar product   28
components of   15, 16
definition   8
differentiation of   111
equality of   9, 19
integration of   111
magnitude   9
multiplication by scalar   10,
    18
negative   10
notation   8, 9
properties of   9
resolving   16
resultant   8, 19
scalar product of   25
    distributivity   28
subtraction   11
    in three dimensions   21
unit vectors   9, 18
zero vector   9
velocity   31
    as derivative   97
    from displacement–time
        graph   35
    as function of displacement
        105
    as function of time   97
    at an instant   97
    and power   269
    relative   145
    uniform   34
    units of   32
velocity–time graphs   36
    from displacement–time
        graph   35

water jet, force exerted
    by   279
wedge, motion of   94
weight   50, 80
work   250
    and energy   262
    definition   251
    and power   266
    rate of doing   266
    unit of   251
work done
    by force   251
    against friction   251
    against gravity   252
    by variable force   255, 381
    in vector terms   254

Young's modulus   377